Planning Support Systems
Best Practice and New Methods

The GeoJournal Library

Volume 95

For other titles published in this series, go to www.springer.com/series/6007

Stan Geertman • John Stillwell
Editors

Planning Support Systems Best Practice and New Methods

Springer

Editors
Dr. Stan Geertman
Utrecht University
Fac. Geosciences
3508 TC Utrecht
The Netherlands
s.geertman@geo.uu.nl

Professor John Stillwell
School of Geography
University of Leeds
Woodhouse Lane
Leeds
LS2 9JT
United Kingdom
j.c.h.stillwell@leeds.ac.uk

ISBN: 978-1-4020-8951-0 e-ISBN: 978-1-4020-8952-7

Library of Congress Control Number: 2008940138

Printed on acid-free paper

9 8 7 6 5 4 3 2 1

springer.com

Preface

Planning Support Systems: Retrospect and Prospect

It has been nearly twenty years since the term 'planning support systems' (PSS) first appeared in an article by Britton Harris (Harris 1989) and more than ten years since the concept was more broadly introduced in the academic literature (Harris and Batty 1993; Batty 1995; Klosterman 1997). As a result, the publication of a new book on PSS provides an excellent opportunity to assess past progress in the field and speculate on future developments.

PSS have clearly become very popular in the academic world. This is the fourth edited book devoted to the topic following Brail and Klosterman (2001), Geertman and Stillwell (2003), and a third by Brail (2008). Papers devoted to PSS have been published in the leading planning journals and the topic has become a regular theme at academic conferences around the world; it has even spawned intellectual offspring such as spatial planning and decision support systems (SPDSS) and public participation planning support systems (PP-PSS).

However, as Geertman and Stillwell point out in their introductory chapter, the experience with PSS in the world of professional practice has been disappointing. A substantial number of PSS have been developed but most of them are academic prototypes or 'one off' professional applications that have not been adopted elsewhere. A handful of commercial off-the-shelf PSS have been created but these systems have been adopted only by a small number of agencies and are rarely used on a routine basis. Geertman and Stillwell offer some important observations on why PSS adoption has been so disappointing but I fear that the obstacles to PSS adoption are inherent in the concepts that comprise planning support systems.

The term 'planning' is notoriously wide-ranging and difficult to define and I am not going to attempt to do that here. Nevertheless, most definitions include the concepts of rational analysis, foresight, evaluation and the attempt to achieve desired outcomes. It is this particular goal of supporting informed forward-looking action that differentiates PSS from geographic information systems (GIS) that organize information to support routine management tasks and decision support systems (DSS) that facilitate executive decision making within an organization. It is important to recognize in this regard that PSS were first proposed as an alternative to GIS that did not adequately meet the needs of planning.

Given this general conception of planning, the adoption of PSS is hindered by the somewhat paradoxical fact that today's planners often do very little planning in the traditional sense, at least in the United States. They are often too busy with routine administrative tasks such as permit processing, code enforcement and infrastructure management to devote much time to analyzing alternatives and planning for the future. Thus, the development of PSS can be seen as part of a larger effort to return the planning profession to its traditional concern with using information and analysis to more effectively engage the future (see, for example, Hopkins and Zapata 2007). As a result, it is perhaps not surprising that PSS are not widely used by planning organizations whose routine land management responsibilities are largely satisfied by today's GIS.

The term 'support' is equally problematic. The computer models of planning's first computer revolution were an important component of the emerging conception of planning as value-neutral and scientific 'planning for the public'. Thus, it was assumed that computers would allow planners, as technical experts on the city, to better understand the urban development process and more accurately forecast the future, with little need to involve the public. Today's PSS are generally designed to support newer modes of citizen-based 'planning for the public' that encourage stakeholders and private citizens to participate more directly in the actions of government by reviewing past and present conditions, identifying problems and opportunities, evaluating alternatives and attempting to make decisions they can all support.

It is important to recognize that new ideals of citizen-based planning are much more challenging to implement in practice than the models of professional expertise and limited citizen participation they replace. Achieving the sustained and meaningful involvement of public officials, stakeholders and other private citizens requires a great deal of time and effort, innovative citizen involvement strategies and carefully designed decision-making procedures. As a result, it is not surprising that planning practitioners are reluctant to abandon the familiar and less demanding – if ultimately less effective – modes of professional practice for the new approaches that may be required to use PSS effectively.

Finally, we come to the concept of 'systems'. Experience with GIS has demonstrated that the effective adoption and use of computer-based systems requires much more than purchasing a computer program and installing it on a computer. It also requires a trained and committed staff, an appropriate organizational setting and support from the top of the organization. Similarly, the 'system' required to support the successful implementation of a PSS that supports community-based planning includes more than planning-related theory, information, methods and models. It must also include planning agencies and public officials who are aware of – and support – the technology, and, most importantly, stakeholders and private citizens who are willing to accept the technology and use it to help guide their collective decision making. Achieving this essential component of a planning support system requires the long-term commitment of organizational champions and community members and a proven record of local success.

It is thus not at all surprising that PSS are so popular in the academic realm and so poorly adopted in the world of professional practice because they embody a new conception of technologically-informed community-based planning that is attractive in theory but difficult to implement in practice. The collection of papers in this book will be helpful in overcoming the obstacles that constrain the effective use of PSS. The book begins with an insightful review of the factors that are stimulating the development of PSS – and limiting their adoption in practice. It describes relatively well-established PSS that have proven to be effective tools for planners and policy makers and new PSS that have been developed for a range of planning applications around the world. It concludes by considering the increasingly important issues of user engagement, stakeholder participation and techniques for integrating analytic and participatory techniques.

Experience has demonstrated that progress in the field of PSS will not be quick or easy. Nevertheless, we live in an era of exciting technical possibilities, an increased appreciation for spatially-referenced information, widespread concern with issues such as global warming and sprawl, and a growing demand for more meaningful public participation. Together these trends will increasingly encourage academic institutions, innovative government agencies and entrepreneurial organizations and individuals around the world to adopt and improve today's planning support systems. As a result, I am confident that the long-term prospects for PSS are bright indeed.

Richard E. Klosterman
President, What if?, Inc.,
78 Hickory Lane, Hudson,
OH 44236, USA
E-mail: kloster@uakron.edu

References

Batty, M. (1995) Planning support systems and the new logic of computation, *Regional Development Dialogue,* 16: 1–17.

Brail, R. and Klosterman, R. (eds.) (2001) *Planning Support Systems: Integrating Geographic Information Systems, Models and Visualization Tools,* ESRI Press, Redlands, California.

Brail, R. (ed.) (2008) *Planning Support Systems for Cities and Regions,* Lincoln Institute of Land Policy, Cambridge, MA.

Geertman, S. and Stillwell, J. (eds.) (2003) *Planning Support Systems in Practice,* Springer, Heidelberg.

Harris, B. (1989) Beyond geographic information systems: computers and the planning professional, *Journal of the American Planning Association,* 55: 85–90.

Harris, B. and Batty, M. (1993) Locational models, geographic information, and planning support systems, *Journal of Planning Education and Research,* 12: 184–189.

Hopkins. L. D. and Zapata, M. A. (eds.) (2007) *Engaging the Future: Forecasts, Scenarios, Plans and Projects,* Lincoln Institute for Land Policy, Cambridge, MA.

Klosterman, R. (1997) Planning support systems: a new perspective on computer-aided planning, *Journal of Planning Education and Research,* 17: 45–54.

Contents

1 **Planning Support Systems: Content, Issues and Trends** 1
Stan Geertman and John Stillwell

Part I: Application and Assessment of Existing PSS

2 **A Use-Driven Approach to Large-Scale Urban Modelling and Planning Support** 29
Brian Deal and Varkki Pallathucheril

3 **Future Land-use in The Netherlands: Evaluation of the National Spatial Strategy** 53
Ton C. M. de Nijs

4 **A Planning Support System Toolkit Approach for Formulating and Evaluating Land-use Change Scenarios** 69
Christopher J. Pettit and Ray Wyatt

Part II: Design, Development and Implementation of Recent PSS

5 **A Motor Vehicle Safety Planning Support System: The Houston Experience** 93
Ned Levine

6 ***Elbe DSS*: A Planning Support System for Strategic River Basin Planning** 113
Bernhard M. Hahn, Sebastian Kofalk, Jean-luc De Kok, Jürgen Berlekamp and Mariele Evers

7 **The *Emission Inventory Water*: A Planning Support System for Reducing Pollution Emissions in the Surface Waters of Flanders** 137
Leen Van Esch, Greta Vos, Liliane Janssen and Guy Engelen

8 An Integrated Discussion Support System for New Dutch Flood Risk Management Strategies **159**
Noor van der Hoeven, Jeroen Aerts, Hanneke van der Klis and Eric Koomen

9 A Spatial Planning Support System for Managing Bangalore's Urban Sprawl **175**
H. S. Sudhira and T. V. Ramachandra

10 *GRAS*: A Spatial Decision Support System for Green Space Planning **191**
Claudia Pelizaro, Theo Arentze and Harry Timmermans

Part III: New Methods for Planning Support

11 Agent-Based Modelling: A Dynamic Scenario Planning Approach to Tourism PSS **211**
Peter A. Johnson and Renee Sieber

12 *Google Maps* Mashups for Local Public Health Service Planning **227**
Maurizio Gibin, Pablo Mateos, Jakob Petersen and Phil Atkinson

13 ArcGIS Processing Models for Regional Environmental Planning in Bavaria **243**
Jörg Schaller, Thomas Gehrke and Nathan Strout

14 Geo-Information Technologies in Support of Environmental and Landscape Planning **265**
Mariolina Besio and Nadia Quadrelli

Part IV: Participation and Collaboration in PSS

15 Enhancing Comprehensive Planning with Public Engagement and Planning Support Integration **295**
Scott N. Lieske, Steve Mullen and Jeffrey D. Hamerlinck

16 Integrating Analytical and Participatory Techniques for Planning the Sustainable Use of Land Resources and Landscapes **317**
David Miller, Nathan Vogt, Maria Nijnik, Eduardo Brondizio and Stefano Fiorini

17 New Ways of Supporting Decision Making: Linking Qualitative Storylines with Quantitative Modelling **347**
Hedwig van Delden and Alex Hagen-Zanker

18 Land Management with the *SMURF* Planning Support System ... **369**
Marc Soutter and Alexandre Repetti

19 SoftGIS as a Bridge-Builder in Collaborative Urban Planning **389**
Maarit Kahila and Marketta Kyttä

20 The *Mainport Planning Suite*: Planning Support Software for Studio-Based Planning **413**
R. Chin

21 Developing Computer-Based Participatory Approaches to Mapping Landscape Values for Landscape and Resource Management **431**
Steve Carver, Alan Watson, Tim Waters, Roian Matt, Kari Gunderson and Brett Davis

Index **449**

Colour Plate Section **459**

Contributors

Jeroen Aerts
Institute for Environmental Studies, Vrije Universiteit Amsterdam, De Boelelaan 1085, 1081 HV Amsterdam, The Netherlands

Theo Arentze
Department of Architecture, Building and Planning, Eindhoven University of Technology, Den Dolech 2, PO Box 513 - VRT 8.08, 5600 MB Eindhoven, The Netherlands

Phil Atkinson
Southwark Primary Care Trust, Woodmill Building, Neckinger, London SE16 3QN, UK

Jürgen Berlekamp
Institute of Environmental Systems Research, Department of Mathematics and Computer Science, University of Osnabrück, 49069 Osnabrück, Germany

Mariolina Besio
Dipartimento DIPARCH, Università di Genova, Genova, Italy

Eduardo Brondizio
Indiana University, Bloomington, Indiana, IN, USA

Steve Carver
School of Geography, University of Leeds, Leeds, UK

Roy Chin
Delft University of Technology, Faculty of Technology, Policy and Management, Jaffalaan 5, 2628BX Delft, The Netherlands

Brett Davis
Aldo Leopold Wilderness Research Institute, USDA Forest Service Rocky Mountain Research Station, Missoula, Montana, MT, USA

Brian Deal
University of Illinois at Urbana-Champaign, Department of Urban and Regional Planning, 611 Taft Drive, Champaign, Illinois 61820, USA

Hedwig van Delden
Research Institute for Knowledge Systems (RIKS), P.O. Box 463, 6200 AL Maastricht, The Netherlands

Guy Engelen
VITO, Vlaamse Instelling voor Technologisch Onderzoek (Flemish Institute for Technological Research), Boeretang 200, 2400 Mol, Belgium

Mariele Evers
Institute for Environmental Strategies, Leuphana University of Lüneburg, 21335 Lüneburg, Germany

Stefano Fiorini
The Macaulay Land Use Research Institute (MLURI), Aberdeen, Scotland, UK; Indiana University, Bloomington, Indiana, IN, USA

Stan Geertman
Faculty of Geosciences, URU, Utrecht University, Heidelberglaan 2, 3508 TC Utrecht, The Netherlands

Thomas Gehrke
Technische Universität München (TUM), Am Hochanger 6, 85354 Freising, Germany

Maurizio Gibin
Southwark Primary Care Trust, Woodmill Building, Neckinger, London SE16 3QN, UK; Department of Geography and Centre for Advanced Spatial Analysis, University College London, Pearson Building, Gower Street, London WC1E 6B, UK

Kari Gunderson
University of Montana, College of Forestry and Conservation, Department of Society and Conservation, Montana, MT, USA

Alex Hagen-Zanker
Research Institute for Knowledge Systems (RIKS), P.O. Box 463, 6200 AL Maastricht, The Netherlands

Bernhard M. Hahn
Research Institute of Knowledge Systems (RIKS), PO Box 463, 6200 AL, Maastricht, The Netherlands

Jeffrey D. Hamerlinck
Wyoming Geographic Information Science Center, Department 4008, 1000 E. University Avenue, University of Wyoming, Laramie, Wyoming 82071, USA

Noor van der Hoeven
Department of Spatial Economics/SPINlab, Vrije Universiteit Amsterdam, De Boelelaan 1105, 1081 HV Amsterdam, The Netherlands

Liliane Janssen
VITO, Vlaamse Instelling voor Technologisch Onderzoek (Flemish Institute for Technological Research), Boeretang 200, 2400 Mol, Belgium

Peter A. Johnson
Department of Geography, McGill University, 805 Sherbrooke Street West, Montreal, Quebec, Canada H3A 2K6

Maarit Kahila
Helsinki University of Technology, Centre for Urban and Regional Studies, PO Box 9300, 02015 TKK, Finland

Hanneke van der Klis
WL/ Delft Hydraulics, Rotterdamseweg 185, 2629 HD Delft, The Netherlands

Sebastian Kofalk
German Federal Institute of Hydrology, PO Box 200253, 56002 Koblenz, Germany

Jean-Luc de Kok
Water Engineering & Management Group, University of Twente, PO Box 217, 7500 AE Enschede, The Netherlands

Eric Koomen
Department of Spatial Economics/SPINlab, Vrije Universiteit Amsterdam, De Boelelaan 1105, 1081 HV Amsterdam, The Netherlands

Marketta Kyttä
Helsinki University of Technology, Centre for Urban and Regional Studies, PO Box 9300, 02015 TKK, Finland

Ned Levine
Ned Levine & Associates, Houston, Texas, USA

Scott N. Lieske
Wyoming Geographic Information Science Center, Department 4008, 1000 E. University Avenue, University of Wyoming, Laramie, Wyoming 82071, USA

Pablo Mateos
Department of Geography and Centre for Advanced Spatial Analysis, University College London, Pearson Building, Gower Street, London WC1E 6B, UK

Roian Matt
Confederated Salish and Kootenai Tribes Forestry Department, Ronan, Montana, MT, USA

David Miller
The Macaulay Land Use Research Institute (MLURI), Aberdeen, Scotland, UK

Steve Mullen
Foresee Consulting, Inc., PO Box 305 Lyons, Colorado 80540, USA

Maria Nijnik
The Macaulay Land Use Research Institute (MLURI), Aberdeen, Scotland, UK

Ton C.M. de Nijs
National Institute of Public Health & Environment, P.O. Box 1, 3720 BA Bilthoven, The Netherlands

Varkki Pallathucheril
American University of Sharjah, School of Architecture and Design, American University of Sharjah, P.O. Box 26666, Sharjah, United Arab Emirates.

Claudia Pelizaro
Landscape Systems, Future Farming Systems Research Division, Department of Primary Industries, 32 Lincoln Square Nth, Carlton, VIC 3052, Australia

Jakob Petersen
Southwark Primary Care Trust, Woodmill Building, Neckinger, London SE16 3QN, UK; Department of Geography and Centre for Advanced Spatial Analysis, University College London, Pearson Building, Gower Street, London WC1E 6B, UK

Christopher J. Pettit
Landscape Systems – Spatial Sciences, Department of Primary Industries, Victoria, Australia

Nadia Quadrelli
Provincia di Varese, SIT, Settore Territorio e Urbanistica, Italy

T.V. Ramachandra
Centre for Ecological Sciences and Centre for Sustainable Technologies, Indian Institute of Science, Bangalore 560 012, India

Alexandre Repetti
Cooperation@epfl, Ecole Polytechnique Fédérale de Lausanne (EPFL), 1015 Lausanne, Switzerland

Jörg Schaller
ESRI Geoinformatik GmbH, Ringstrasse 7, 85402 Kranzberg, Germany

Renee Sieber
Department of Geography, McGill University, 805 Sherbrooke Street West, Montreal, Quebec, Canada H3A 2K6

Marc Soutter
Environmental Science and Technology Institute (ISTE-HYDRAM), Ecole Polytechnique Fédérale de Lausanne (EPFL), 1015 Lausanne, Switzerland

John Stillwell
School of Geography, University of Leeds, Leeds, LS2 9JT, UK

Nathan Strout
Redlands University of California, 1220 East Colton Av., Redlands CA 92373, USA

H.S. Sudhira
Department of Management Studies and Centre for Sustainable Technologies, Indian Institute of Science, Bangalore 560 012, India

Harry Timmermans
Department of Architecture, Building and Planning, Eindhoven University of Technology, Den Dolech 2, PO Box 513 - VRT 8.08, 5600 MB Eindhoven, The Netherlands

Leen Van Esch
VITO, Vlaamse Instelling voor Technologisch Onderzoek (Flemish Institute for Technological Research), Boeretang 200, 2400 Mol, Belgium

Nathan Vogt
Indiana University, Bloomington, Indiana, IN, USA; Nucleo de Altos Estudos de Amazônia (NAEA), Universidade Federal de Parà (UFPA), Belém, Parà, Brazil

Greta Vos
VMM, Vlaamse MilieuMaatschappij (Flemish Environmental Agency), A. Van de Maelestraat 96, 9320 Erembodegem, Belgium

Tim Waters
Independent consultant, Leeds, UK

Alan Watson
Aldo Leopold Wilderness Research Institute, USDA Forest Service Rocky Mountain Research Station, Missoula, Montana, MT, USA

Ray Wyatt
School of Resource Management, University of Melbourne, Victoria, Australia

Abbreviations

AAS	Amsterdam Airport Schiphol
ABM	Agent Based Modelling
ABS	Australian Bureau of Statistics
ACT	Anthropological Center for Training and Research on Global Environmental Change
AGIV	Agency for Geographical Information Flanders
AI	Amazon Initiative
Al	Aluminium
ANZSIC	Australian New Zealand Standard Industry Classification
API	Application Programming Interface
BAU	Business as Usual
BfG	German Federal Institute of Hydrology
BUGIS	Bottom-Up Geographic Information System
CAC	Citizens Advisory Committee
CAD	Computer Aided Design
CASA	Centre for Advanced Spatial Analysis
CBA	Cost-Benefit Analysis
CBD	Central Business District
CBE	Collaborative Business Engineering
CBR	Case-based reasoning
CBS	Statistics Netherlands
CCA	Constrained Cellular Automata model
CEL	Cultural Ethnic and Linguistic
C-GIS	Critical Geographical Information Systems
CHAID	CHi-square-Automatic-Interaction-Detection
CIGAR	Consultative Group on International Agricultural Research
CIPEC	Center for the study of Institutions, Population, and Environmental Change
CLD	Causal Loop Diagrams
COTS	Commercial of The Shelf
CPB	Bureau for Economic Policy Analysis
CPS	Collaborative Planning Systems

CPTED	Crime Prevention Through Environmental Design
Cr	Chromium
CSM	Conceptual System Model
CTS	Canadian Travel Survey
CU	Copper
DBMS	DataBase Management System(s)
DGMS	Dialogue Generation and Management System(s)
DSE	Department of Sustainability and Environment
DSS	Decision Support System(s)
ECW	Enhanced Compressed Wavelet
EEA	European Environment Agency
EEV	Emission Explanatory Variable
EF	Emission Factor
EIW	Emission Inventory Water
EMS	Emergency Medical Services
EpiSims	Epidemiological Simulation System
EU	European Union
EWGateway	East-West Gateway Coordinating Council
ES	Expert System(s)
FAR	Floor Area Ratio
FFH	Flora Fauna Habitat areas
FPA	Facility Planning Area
GAS	Geographic Automata System(s)
GDP	Gross Domestic Product
GeoVUE	Geographic Virtual Urban Environments
GIS	Geographic Information System(s)
GML	Geography Markup Language
GP	General Practitioner
GUI	Graphical User Interface
GV	Geovisualisation
GWR	Geographically Weighted Regression
HALT	Hospital Admissions for Long-Term diseases
HEFCE	Higher Education Funding Council for England
H-GAC	Houston-Galveston Area Council
HPC	High Performance Computing
HSIP	Highway Safety Improvement Program
HTML	HyperText Markup Language
ICT	Information and Communication Technology
INSPIRE	Infrastructure for Spatial Information in Europe
IPCC	Intergovernmental Panel on Climate Change
IPF	Iterative Proportional Fitting
IRBM	Integrated River Basin Management
ISDSS	Integrated Spatial Decision Support System
ISM	Integrated System Model
ISTAT	Italian Statistics Institute

ITS	International Travel Survey
iWWT	individual Waste Water Treatment
KBS	Knowledge Based System(s)
KML	Keyhole Markup Language
KMZ	Zipped Keyhole Markup Language
KNMI	Royal Netherlands Meteorological Institute
LDA	Land Degradation Assessment
LDC	Landscape Development Concept
LEAM	Land-use Evolution and impact Assessment Model
LEI	Agricultural Economics Research Institute
LOAC	London Output Area Classification
LSG	Landscape Protection Areas
LUC	Land-Use Change model
LUCC	Land-Use and Cover Change
MBMS	Model Based Management System(s)
MCA	Multi-Criteria Analysis
MCE	Multi-Criteria Evaluation
MNP	Netherlands Environmental Assessment Agency
MO	Missouri
MoSeS	Modelling and Simulation for e-Social Science
MPO	Metropolitan Planning Organization
MPS	Mainport Planning Suite
NASA	National Aeronautics and Space Administration
NCeSS	National Centre for e-Social Science
NGO	Non-Governmental Organisation
Ni	Nickel
NIMBY	Not In My Back Yard
NIPC	North eastern Illinois Planning Commission
NIROV	Netherlands Institute of Housing and Planning
NSF	National Science Foundation
NSG	Nature Protection Area
NUTS	Nomenclature des Unités Territoriales Statistiques
OAC	Output Area Classification
OGC	Open Geospatial Consortium
ONS	Office for National Statistics
OPUS	Open Platform for Urban Simulation
OWL	Web Ontology Language
PAH	Polycyclic Aromatic Hydrocarbon
Pb	Lead
P-GIS / PGIS	Participatory Geographic Information System(s)
POLAR	Participation of Local Areas
PoR	Port of Rotterdam
PoSS	Policy Support System(s)
PPGIS	Public Participatory Geographic Information System(s)
PPSS	Participatory Planning Support System(s)

PPP	Public Private Partnership
PPPSS	Public Participatory Planning Support System(s)
PSI	Planning Support Instrument(s)
PSS	Planning Support System(s)
PSSD	Planning System for Sustainable Development
RDF	Resource Description Framework
RS	Remote Sensing
RSC	Regional Safety Council
SAS	Story And Simulation approach
SD	System Dynamics
SDSS	Spatial Decision Support System(s)
SGML	Standard Generalized Markup Language
SMURF	System for Monitoring Urban Functionalities
SOSIG	Social Science Information Gateway
SPA	Special Protected Bird areas
SPDSS	Spatial Planning and Decision Support System(s)
SPSS	Spatial Planning Support System(s)
SVG	Scalable Vector Graphics
TAC	Technical Advisory Committee
TAM	Technology Acceptance Model
TAZ	Traffic Analysis Zone
TBT	Tributyltin
TTRA	Travel and Tourism Research Association
TU	Technical University
TUM	University of Technology Munich
UCL	University College London
UK	United Kingdom
UWWTP	Urban Waste Water Treatment Plant
USA	United States of America
USAID	United States Agency for International Development
VUE	Virtual Urban Environments
VMM	Flemish Environmental Agency
VMT	Vehicle Miles Travelled
VR	Virtual Reality
W3C	World Wide Web Consortium
WLC	Weighted Linear Combination
WTA	Water Treatment Areas
WWW	World Wide Web
XML	Extensible Markup Language
Zn	Zinc

Chapter 1
Planning Support Systems: Content, Issues and Trends

Stan Geertman and John Stillwell

1.1 Introduction

Since we edited *Planning Support Systems in Practice* (Geertman and Stillwell 2003), we have become increasingly aware that successful applications of geo-technology by planning practitioners to support their activities are far from commonplace. It is disturbing, in fact, to observe the extent to which new computer-based support systems are developed by researchers to the point of adoption but are never implemented in planning practice or policy making. Similarly, there is evidence to indicate that systems which are made operational are not extensively used, after the initial novelty has passed, by those planning organizations for which they have been developed in the first instance. In terms of application, it is possible to point to more failures than successes, i.e. to more cases where systems have not been implemented than examples where they are used routinely. Moreover, many state-of-the-art systems appear to take a long time to reach the 'market' and this is often a process requiring considerable financial resources. There are a number of reasons for this state of affairs that we will explore in Section 1.2 of this chapter after providing some initial clarification of what we mean by Planning Support Systems (PSS) and how they are distinguished for Geographical Information Systems (GIS) and Spatial Decision Support Systems (SDSS).

In these rather disappointing circumstances, it becomes very important to identify examples of 'best practice' where systems have been developed that are doing the job that they were designed for and have become effective tools for policy makers or practitioners to utilize on a regular basis. This identification process is one of the primary aims of this book and several of the chapters report on systems that fall in this category, that have been applied successfully and that are used as a matter of

S. Geertman (✉)
Faculty of Geosciences
URU, Utrecht University
Heidelberglaan 2, 3508 TC
Utrecht, The Netherlands
E-mail: s.geertman@geo.uu.nl

S. Geertman, J. Stillwell (eds.), *Planning Support Systems Best Practice and New Methods*,
DOI 10.1007/978-1-4020-8952-7_1, © Springer Science+Business Media B.V. 2009

routine to support some planning process or assist the policy maker in reaching a decision on which option to adopt.

We are also conscious of the pace of change in recent years and the way in which technology is driving forward developments at a rapid rate in the face of massive volumes of data and information of relevance to planning becoming more available and more accessible. This is occurring at a time when government pre-occupations with fashionable concepts like sustainability, participation, performance and evaluation are making new demands on human resources, in terms of both skill requirements and technological tools. Thus, in Section 1.3 of the chapter, we provide a discussion of some of the disparate trends that are taking place in the context in which PSS are developed and used. Implicit in this is that new methodologies are being developed continuously and, consequently, it is the second major aim of the book to illustrate the type of innovative technological developments and new processes of using geo-technology in different planning situations that are taking place in different parts of the world. In Section 1.4, we offer a rationale for the remainder of the book and present some introductory comments on each of the chapters, before drawing a short conclusion in the final section.

1.2 What's up with PSS?

In this section, we consider initially the concept of PSS and how this genre of systems is distinguishable from similar geo-technological instruments like geographical information systems (GIS) and spatial decision support systems (SDSS). Thereafter, we focus our attention on their hitherto rather restricted applications in planning practice, the reasons that underlie this disappointing and unsatisfactory situation and the lessons that can be learned from experience to date and applied to improve this state of affairs.

1.2.1 The PSS Concept

PSS are a relatively recent phenomenon, emerging onto the planning scene in the mid-1990s as geo-technological instruments fully dedicated to support and improve the performance of those involved in undertaking specific planning tasks (Batty 1995; Klosterman 1997). In a sense they are related to GIS, but while the latter are general purpose tools for capturing, storing, manipulating, analysing and displaying spatially referenced data, applicable for many different spatially-related problems, PSS distinguish themselves through being focused on supporting specific planning tasks. On many occasions a PSS will contain a GIS, particularly if the task in hand requires geographical/spatial data. PSS are also related to SDSS, although the former generally pay particular attention to long-range problems and strategic issues, while SDSS are generally designed to support shorter-term policy making by isolated individuals or by business organizations (Clarke 1990). Moreover, the prime dedication of SDSS is towards supporting operational decision making rather than strategic planning activities which is the prime focus of PSS. Consequently, PSS usually consist of a combination of planning-related theory, data, information, knowledge, methods and instruments that take the form of an integrated framework with a shared graphical user interface (Geertman and Stillwell 2003). Many regard PSS as valuable support tools that will enable planners to better handle the complexity of planning processes, leading to plans of better quality and saving a lot of time and resources. In this respect, it seems that a new, more positive attitude concerning PSS has emerged since the turn of the century. At present, much more attention is focused on planning support and its technological instruments than has been the case in the past, a trend that appears to be evident from the volume of studies that are being undertaken, the dedicated conferences that are taking place, and the diversity of articles and books that take PSS as their prime focus.

It is important to acknowledge how the definition of what is considered here as PSS has evolved. Harris and Batty (1993) were really the first to associate the concept of PSS with combining a range of computer-based methods and models into an integrated system used to support a particular planning function. More precisely, in their opinion, a single PSS forms the framework in which three sets of components are combined: the specification of the planning tasks and problems at hand, including the assembly of data; the system models and methods that inform the planning process through analysis, prediction and prescription; and the transformation of basic data into information which, in turn, provides the driving force for modelling and design (through a cyclical process). In a similar vein, Klosterman (1997, 1999) and Brail and Klosterman (2001) have more recently described PSS as information technologies that are used specifically by planners to undertake their unique professional responsibilities. They suggest that PSS have matured into frameworks of integrated systems of information and software that synthesize the three components of traditional DSS – information, models, and visualization – and deliver them into the public realm. Earlier, Batty (1995) had suggested PSS to be a subset of geo-information technologies, dedicated to supporting those involved in planning to explore, represent, analyse, visualize, predict, prescribe, design, implement, monitor, and discuss issues associated with the need to plan. Geertman and Stillwell (2003) considered PSS to be geo-information technology-based instruments that incorporate a suite of components (theories, data, information, knowledge, methods, tools, meta-information) which collectively support some specific parts of a unique professional planning task. Brail (cited in Batty 2005), on the other hand, has drawn attention to the fact that many PSS are developed and used to provide projections forward to some point in the future or may involve some estimation of the impacts that result from some form of development. In summary, this kaleidoscopic review indicates that whilst there may be no strict uniform definition of PSS at the moment – a conclusion reached also by Klosterman and Pettit (2005) in their 'Update on PSS' – all definitions tend to coincide by including or touching upon the same kind of required functionalities within this category of support instruments. Many commentators regard PSS as being capable of improving the handling of knowledge and information in planning processes, a function that provides huge assistance to those involved in handling the ever-increasing complexity of planning tasks.

1.2.2 Bottlenecks

In spite of the apparent benefits that various PSS would appear to provide, they have not yet become widely used in planning practice (Brail and Klosterman 2001; Geertman and Stillwell 2003). This neglect or even antipathy is remarkable for at least two reasons. First, planning support instruments are increasingly finding their way from the scientific laboratories into the marketplace. An inventory by Geertman and Stillwell (2003) has concluded that some PSS can be considered 'off-the-shelf' products that can be purchased on the market at a reasonable price. Secondly, planning tasks characterize themselves increasingly by their growing complexity: an increasing number of fields of policy must be taken into account and integrated in order to produce balanced planning proposals. More and more people with divergent interests and agendas are closely involved in participatory planning processes; the involvement of participants is taking place at a much earlier stage in the planning process. In sum, there are plenty of reasons for professional planners to actively embrace all kinds of assistance, including planning support instruments, to enlighten and improve their procedures as the process of practice changes. However, there appear to be a number of important obstacles that hamper the use of PSS in practice and several studies have been undertaken recently to identify these so-called 'bottlenecks'. A brief synopsis of three studies will be summarized here, with more detailed explanation being available in Vonk (2006) and Vonk *et al.* (2005; 2007a,b).

First of all, the views of users of PSS have been gathered through a series of interviews held with 43 employees of 12 highly comparable Dutch regional spatial planning organizations. In particular, interviews were held with three archetypal users that currently fulfill an important role in (potentially) using and evaluating PSS: geo-information specialists, planners and managers. Secondly, given that the views of PSS developers concerning potentials and bottlenecks are well recorded in the scientific literature, a set of 58 PSS and their developers were examined on the basis of a literature review; these constitute a relatively accurate overview of system developers' perspectives. Thirdly, the views of experts in PSS have been gathered by means of conducting two additional World Wide Web (WWW) surveys based on 800 individuals with interests in PSS obtained from various listserv e-mail networks. The first survey had 96 respondents of which 86 were considered experts, since they were familiar with at least two PSS. The second survey had 40 respondents, of which 30 were experts, the majority of which appeared to be university researchers and employees in public planning research and/or advisory bureaux dealing with planning support. The findings of the literature survey, the interviews and the web surveys were combined in order to learn lessons on the usage of PSS, their potential and the bottlenecks leading to non-implementation or usage.

In order to interpret the results of these studies in a structured way, three different viewpoints and their associated conceptual underpinnings were applied: the so-called 'instrument approach', the 'user approach' and the 'transfer approach' (Fig. 1.1). The instrument approach regards usage mainly in terms of the instrumental quality of the PSS, focusing particularly on fitness for use and user friendliness (Dishaw and Strong 1999; Dishaw *et al.* 2002; Goodhue 1995; Goodhue and Thompson 1995). The user approach considers usage in relation to the extent of user acceptance of the PSS, identifying the broader set of factors related to the accepting environment. Finally, the transfer approach explains usage in terms of the extent of diffusion, paying particular attention to the flow of information concerning PSS from the system developer to and among the user and the user community. In fact, the three approaches overlap in that they all look at the same issue of usage, but each emphasizes slightly different aspects.

The instrument approach which indicates instrumental quality is defined here as consisting of a judgment of: (i) how well the instruments are capable of carrying out the tasks that they were developed for; and (ii) how well they fit to

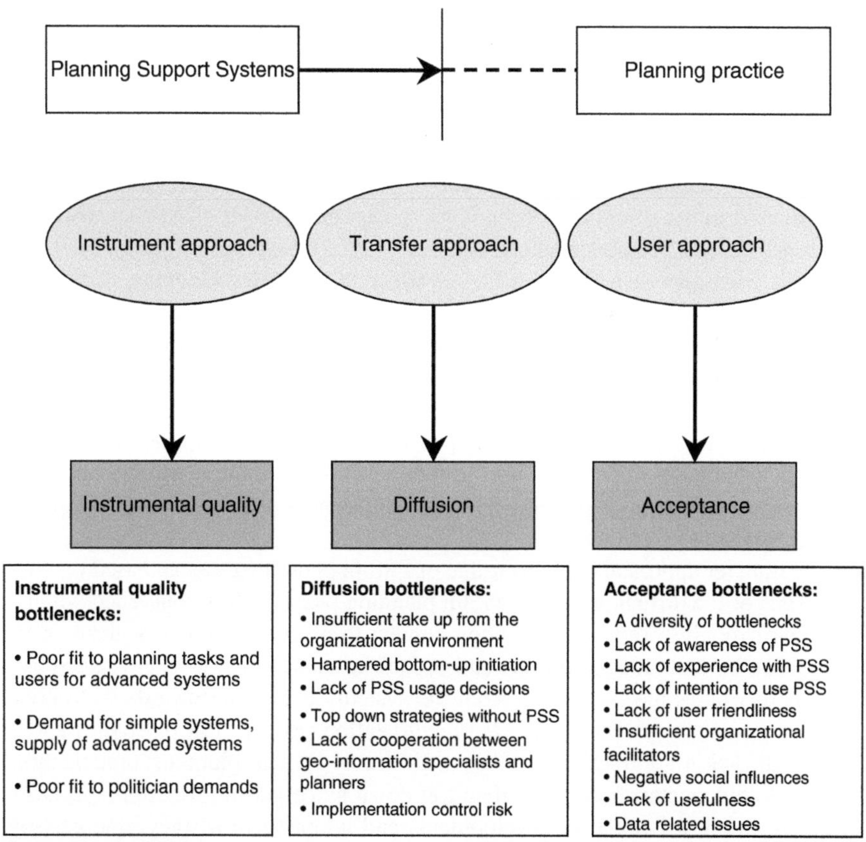

Source: Vonk (2006)

Fig. 1.1 Conceptual framework explaining under-usage of PSS from three different approaches

the capabilities and demands of intended users. Goodhue and Thompson (1995) showed the importance of these characteristics as determinants of usage of information technologies in their model of 'Task-Technology-User fit' (Dishaw and Strong 1999; Dishaw *et al.* 2002; Goodhue 1995; Goodhue and Thompson 1995). The outcomes of the research suggest that PSS technology is still at an early stage of development, with a large variety of systems, very few standards, and characterized by large differences in instrumental quality. Moreover, it indicates that there exists a fundamental dichotomy between those PSS that are demanded for use in practice by potential users and those PSS supplied by system developers according to their perceptions of what is required. In short, while practice generally demands rather simple and straightforward PSS for exploratory tasks such as making an inventory of conditions or recording planning applications, the majority of PSS that have been developed focus on more analytical tasks, including modelling and projection. These more sophisticated systems are seen as a poor match with the demands of planning practice in reality. The instrumental quality of simple instruments is deemed to be acceptable whilst that of advanced instruments is generally considered to be poor. Results suggest that simple instruments have a relative advantage over manual operations; for many existing advanced instruments, the advantage is doubtful at least.

The user approach focuses in particular on characteristics of users that determine the acceptance of PSS. This approach emphasizes how users should change in order to enhance usage of PSS. Factors that influence acceptance include the characteristics of users, instruments, organizations, the social environment, the external environment and facilitating conditions. These factors that influence acceptance have been framed in the so-called 'Technology Acceptance Model' (TAM) as discussed by Davies (1986), Frambach and Schillewaert (2002) and Rogers (2003), for example. This user approach shows a large variety of obstacles blocking widespread acceptance of PSS in planning practice. The main bottlenecks are the following: (i) the lack of awareness of the existence of PSS and the purposes for which they can be used; (ii) the lack of experience with PSS, which leaves users unaware of the benefits of PSS and the conditions under which they can be used; and (iii) the low intention to start using PSS among possible users. The list of other high scoring bottlenecks includes insufficient user friendliness and usefulness, the absence of the required organizational facilitators and social influences and, moreover, data quality and accessibility problems.

The transfer approach explains usage of PSS in planning practice based on characteristics of the diffusion of PSS within planning practice. It is concerned with the process of moving from an innovation context to a practical context, through the acceptance by individuals, groups and organizations. Diffusion is envisaged as a process that takes the innovation from the system developers towards widespread usage in practice over various levels of aggregation (Rogers 1995). Studies adopting the diffusion approach show that diffusion of PSS in planning organizations is more likely to start bottom-up than top-down since geo-information specialists are more likely to spot PSS developments and adopt them as they emerge from the working environment. Nonetheless, lack of opportunity for innovation and personal characteristics often cause geo-information specialists at the bottom of the organization to be unable to transfer the technology from the external environment and bring it to the attention of top management. In addition, these individuals are not able to bring PSS to the attention of planners since geo-information specialists themselves are often unable to communicate effectively with spatial planners and set up cooperation networks. Innovative ideas are also poorly diffused due to differences in appreciation of PSS between individual geo-information specialists and others within the organization. Thus, for example, geo-information specialists often encounter a discrepancy between planners' questions and their offers. From this summary of bottlenecks, it is possible to derive some lessons for future action (Fig. 1.2).

As far as instruments are concerned, there is a desire and an intention to improve the instrumental quality. However, those in practice willing to use PSS are often unable to improve the instrument quality of PSS on their own. Nonetheless, if they cooperate with researchers and system developers, they can at least contribute to PSS quality improvement. This requires that the demands expressed by potential users should be taken much more seriously than has been the case thus far. With regard to the users, much more awareness of and subsequent experience with PSS is needed. For those willing to start using PSS in planning practice, it is recommended that they actively spread the news of the existence of PSS and their potentials through the appropriate communications channels. Furthermore, they should make the PSS message better suited to the receivers. Good examples of applications and best practice will help to overcome some of the bottlenecks associated with lack of awareness and experience. As for transfer bottlenecks, a distinction can be made between diffusion towards planning organizations and diffusion within planning organizations. Some intermediate actors can fulfill the role in the diffusion towards planning organization. Of these actors, governmental research agencies

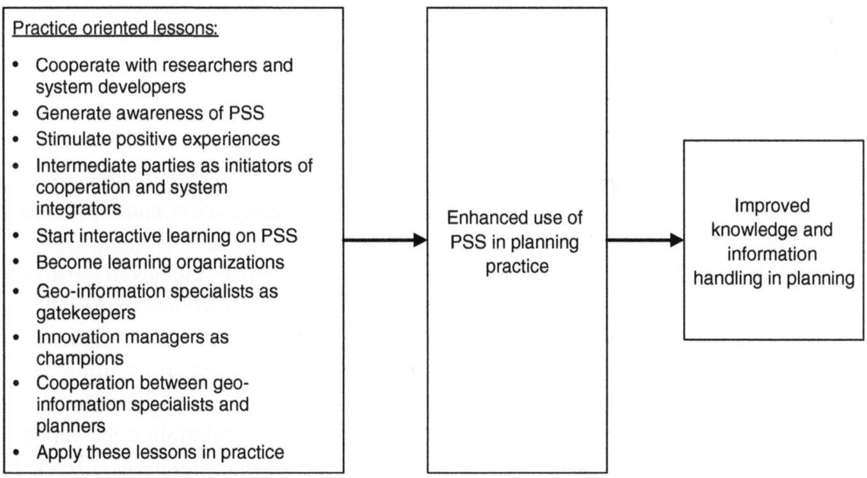

Source: Vonk (2006)

Fig. 1.2 Lessons to enhance usage of PSS to improve knowledge and information handling in planning

and consultant organizations usually have greater knowledge of and accessibility to planning practice than scientists working within universities. They are expected to be capable of getting the actors of the PSS innovation network working together to engage in cooperative development. In terms of diffusion within planning organizations, managers of planning organizations are advised to adopt the management paradigm of the 'learning organization' (Senge 1990) and adopt knowledge management strategies (Nonaka and Takeuchi 1995). Managing information technology adoption and implementation in complex environments is challenging. The adoption of the managerial paradigm of the learning organization can change this because it stimulates the flow of knowledge towards and within organizations, thereby stimulating innovation, acceptance and diffusion of PSS. This allows geo-information specialists to function as gatekeepers, matching innovations in the organizational environment with internal demands, and subsequently, innovation managers can then function as champions. This last role entails bringing the PSS innovation further into the organization towards utilization of the opportunities that PSS have to offer. To achieve this innovation, these champions need to convince planners of the use of PSS in their daily practices and persuade other managers to decide on adoption or at least provide room for some experimentation.

In conclusion, the instrumental quality of PSS, the awareness and experiences of potential users, and the transfer of these instruments towards and within planning organizations, all need improvement to overcome the bottlenecks that are known to exist and to fulfill the potential of the PSS to its fullest extent. Therein, spreading the news about good applications and best practice and improving cooperation (e.g. between geo-information specialists and planners) can be considered as key activities in what has to be undertaken to enhance PSS use in practice. Whilst best practice examples and cooperation are key, we must also acknowledge that the environment of planning practice, as well as the technological infrastructure and resources available to support planning are under continuous change and it is this dynamic which we consider in the next section.

1.3 Developments in PSS Context

In recent years, it seems that the speed and intensity of change continues to increase in virtually all aspects of economic, social and environmental development. Governments demand more evidence on which to base more effective policies, require better means of evaluating action or more strategic planning at different spatial scales, and encourage more stakeholder participation in major infrastructure projects as well as local community developments. Knowledge continues to accumulate as increasing amounts of information are extracted from ever-increasing volumes of data. Developments in technology are a key driving force in relation to the capture, storage, manipulation, analysis, generation, display and interpretation of data. In this section of the chapter, we explore some of these developments insofar as they impact on PSS.

1.3.1 Participation in Planning Processes

There is an increasingly widely held view that encouraging citizens and stakeholders to participate in planning and decision making is an essential ingredient for healthy and democratic development. Whilst many developed countries have seen a growing disinterest in local and national politics and extensive disillusion with planning in a capitalist environment, many governments have recognized that public involvement in the process of planning has potential value in renewing community interest and rekindling trust. Furthermore, there is no doubt that public participation has had a longstanding presence in the planning process but the trend in recent years is for governments, sometimes in the face of community breakdown, to develop new ways of getting the public actively involved in deciding the future of the places where they live, work and play. It can be argued that a major cultural shift is required to an ethos in which citizens feel empowered to create an environment over which they have some control. Moving to a system in which planning is not left entirely to a cadre of professionals but which involves citizens and other stakeholders presents a whole range of problems, yet collaboration involving a wide range of organizations and individuals is seen as making much better use of the resources and expertise that are available and therefore creating a higher potential for success. Collaborative problem solving and collective decision making are seen as a means to resolve confrontation at 'grass roots' and generate solutions that are more tailored to what local communities require. In many countries, of course, participation and collaboration in the planning process are required by law.

It is therefore not surprising that, over the past decade, considerable attention has focused on what has been termed Participatory Geographic Information Systems (P-GIS) or Public Participation Geographic Information Systems (PP-GIS) or Participatory Planning Support Systems (P-PSS) or Public Participatory Planning Support Systems (PP-PSS) for use in planning contexts where there is an explicit geographical or spatial dimension. Each type of system may have distinctive hallmarks but all of them have been developed to support democratic decision making and enable the opinions of local people to be channeled into proposals that will ultimately lead to an improvement of their own livelihoods. Consequently, these systems aim to facilitate greater stakeholder or public involvement in decision making, enhance effective communication and understanding, and monitor the impacts of policies and management plans on local communities more effectively.

When digital spatial data and software are made available to community groups and individuals using specially designed P-GIS or P-PSS tools, this type of activity is frequently referred to as PP-GIS or PP-PSS since it involves the participation of local people. Moreover, there is a trend towards increasing use of the internet to experiment with the development of online PP-GIS or PP-PSS, particularly in North America (Ventura *et al.* 2003; Al-Kodmany 2003) and in Europe (Carver *et al.* 1999; Kingston *et al.* 2003). These approaches offer huge potential for enhanced participation as online systems and web mapping software mature and as the public's experience of the internet grows in countries that have the required infrastructure. However, the systems and their use are not without certain shortcomings: they require significant technological investment in the first instance; they are not likely to be accepted by less technologically skilled participants; and there are a range of problems in representing people's opinions, beliefs, perceptions and value judgements in a PSS and then incorporating and using them in GIS together with more quantitative data. Whilst PP-GIS/PP-PSS is one of the areas embraced by the new wave of researchers promoting critical GIS (C-GIS) and examining how practices of GIS and mapping are fundamentally political, (Elwood 2005, 2006), we should acknowledge that the importance of a human-centred approach and the focus on human-computer interaction were major drivers towards the development of Collaborative Planning Systems (CPS) as outlined by Shiffer in the early 1990s (Shiffer 1992, 1995).

Public participation is one of the dimensions that have to be embraced by those public sector organisations charged with planning or policy making responsibilities. Different institutions carrying out different functions at national, regional or local levels have various requirements in terms of the data sets relevant to their functions and the technical tools that they need to process the data. Increasing importance is attached to providing the 'evidence base' to underpin policy decisions; there is increasing emphasis on performance of individuals, teams and institutions as well as policies; more comprehensive monitoring and auditing of processes is required as well as better evaluation of methods and outcomes; and public organisations are under increasing pressure to make their data sets more available and accessible. In the context of the last of the latter, Campagna and Delano (2004), for example, provide an interesting evaluation of the geographic information that public sector bodies provide on their websites. These issues raise concerns about the lack of awareness of what is best practice and the need for more and better trained staff with rising expectations as far as technology is concerned and how it can be applied. In the case of public participation in planning, this means an awareness of whether new forms of visualisation, animation, interactivity, customisation and prediction can be used in each particular planning context.

1.3.2 Information Provision and Data Integration

Virtually every sector of planning has seen an expansion in the volume of relevant information extracted from secondary sources whilst researchers strive to create and identify new primary quantitative and qualitative data sets that give fresh insights into their fields of interest. Large corporations are encouraged to be more transparent in their information holdings, to share data with others, to avoid the data silo mentality and to maximise the value of administrative information and data that they hold through linkage with other existing public or private data sets. The theme of data integration has been particularly important in the last decade both at an individual micro scale as well as at more aggregate macro scales, although there are frequently constraints on data integration at the micro level due to confidentiality issues. The use of administrative data in different planning contexts is also increasingly important in countries like the UK where the decadal Census of Population assumes huge importance in the absence of a population registration system. The 2001 Census in the UK is the most reliable and comprehensive source of socio-demographic data on small geographical areas throughout the country (Rees *et al.* 2002) but, as the years pass by and census data become more out-of-date, social commentators and researchers turn to alternative sources to establish changing trends and new social phenomena that planners must deal with. For example, administrative registers of births and deaths enable insights to be gained into the components of natural population change whereas re-registration data of National Health Service patients enable internal migration flows to be monitored and worker registration or national insurance data provide some indication of international immigration, a trend with economic and social implications that challenge policy makers and service providers across the country in both urban and rural environments.

Owing to the exponential increase in computing power and communication bandwidth, the past decade has also witnessed a spectacular growth in volume of data 'born digital'. This includes digital image data captured by sensors and satellites, but it also includes data generated by telephones and mobile phones as well as internet traffic, and has become an increasingly important focus of attention by social and computer scientists working on new e-social science initiatives.

One responsive trend to the proliferation of data sources and data sets has been the emergence, throughout the world, of national and international spatial data strategies linked to standards and integration. In many countries, there are various national databases and strategies that are relevant for planning. However, fragmentation of datasets and sources, gaps in data availability, lack of harmonisation between datasets at different geographical scales and duplication of information collection all make it difficult to identify, access and use data that are available. In response, at national and international levels, more and more initiatives are undertaken to tackle these problems. For instance at an international level the *Infrastructure for Spatial Information in Europe* (INSPIRE) initiative has been undertaken that aims to stimulate the creation of a European spatial information infrastructure that delivers to the users integrated spatial information services, allowing users to identify and access spatial or geographical information from a wide range of sources, from the local level to the global level, in an interoperable way for a variety of uses. INSPIRE is targeted at policy makers, planners and managers at European, national and local level and the citizens and their organisations. It has a geoportal at www.inspire-geoportal.eu/index.htm which is the gateway to geographic data and services, distributed around Europe, allowing users to search, view or, subject to access restrictions, download geographic data or use available services to derive information. Other examples of gateways to improve access to resources available on the web are ESRI's Geography Network (www.geographynetwork.com) the U.S. Bureau of the Census's American FactFinder (http://factfinder.census.gov/), the Social Science Information Gateway (SOSIG), an online catalogue of thousands of internet resources relevant to social science education and research, and the Census

Programme in the UK sponsored by the Economic and Social Research Council has a census portal (http://census.ac.uk/) that provides one stop shop access to all census resources used by the academic research community.

So, with increasing volumes of data becoming available, the need for standards, harmonisation and metadata becomes more and more essential. It is important for those who develop PSS in the public sector to be aware of the infrastructures that exist and the pressures to conform to national and international standards both in terms of the PSS inputs and outputs that might be used by others.

1.3.3 Technology Developments

The technology that underpins PSS is also developing rapidly and those who commission PSS are confronted with a host of important questions: Do we need a local or a national information system? Should the system be simple or more sophisticated? Should it be static or interactive? Should it contain tools for analysis as well as visualisation? Whilst the technology is getting easier to use it is also getting more sophisticated in what it can do. The impact of ICT and geo-technology on the visualisation of spatial data has been as remarkable and the pace of change is unrelenting with the new developments taking place in grid technologies, web mapping, interactive web services and advent of *Google Earth* and *Google Maps*.

These new developments have been made possible through the existence of the internet and the web using new languages such Hypertext Markup Language (HTML), the predominant markup language for the creation of web pages that provides a means to describe the structure of text-based information in a document by denoting certain text as headings, paragraphs, lists and so on, and by supplementing that text with interactive forms, embedded images and other objects. The Extensible Markup Language (XML), on the other hand, is a general-purpose markup language, classified as extensible because it allows users to define their own tags, whose primary purpose is to facilitate the sharing of data across different information systems, particularly via the internet. XML is a simplified subset of the Standard Generalized Markup Language (SGML), and is designed to be relatively human-legible. By adding semantic constraints, application languages can be implemented in XML that include XHTML, RSS, MathML, GraphML, Scalable Vector Graphics (SVG), MusicXML and many others. XML is recommended by the World Wide Web Consortium (W3C); it is a fee-free open standard. Geography Markup Language (GML) is an XML-based encoding standard for geographic information developed by the OpenGIS Consortium (OGC), designed to allow internet browsers the ability to view web-based mapping without additional components or viewers.

A major technological development is the evolution of the 'grid', known increasingly as 'e-infrastructure'. The grid comprises networked, inter-operable, scalable computational tools and services that make it possible to locate, access, share, aggregate and manipulate digital data seamlessly across the internet. Grid computing is distributed computing that involves a number of dispersed computer resources (platforms, hardware/software architectures, computer languages) that can be used to tackle large, time-consuming tasks, providing large amounts of computing power more economically than costly high-end computers and more effectively than the most powerful supercomputers. This has led to the development of open standards and the virtualizing of computing resources. There are computational grids which focus on computationally-intensive operations, data grids that involve sharing and management of large amounts of distributed data and equipment grids where a primary piece of equipment and data produced are used remotely. By linking digital processors, storage systems and software on a global scale, grid technology has the potential to transform computing from an individual and corporate activity into a general utility (Foster 2003). The grid offers potential for PSS that are designed to support 'big science' and involve major use of computing resources such as the simulation of mega-events like earthquakes or tsunamis or the modelling of major weather events like hurricanes, where the simulation task is too big for any single supercomputer. Grid computing also provides a multi-user environment and in this respect also has potential for collaborative PSS. Grid infrastructure is also available for video-conferencing. Considerable research progress has been made in providing grid-enabled versions of High Performance Computing (HPC) tools. In the USA, grid services are being developed by the National Science Foundation (NSF) Cyberinfrastructure project at the University of Chicago using their Social Informatics Grid (SID Grid) infrastructure. The SID Grid will be deployed as part of the larger TeraGrid, a suite of grid computing resources available to the scientific community at large (http://www.teragrid.org/).

In the UK, the National Centre for e-Social Science (NCeSS) has a research programme that aims to build upon grid technologies and tools developed by researchers in centres of expertise around the UK and apply them to the particular needs of the social science research community in order to generate new solutions to social science research problems. The NCeSS nodes include those at University College London where the overall aim is to provide grid-enabled virtual environments within which users are able to link spatial data about cities to GIS software to create *Geographic Virtual Urban Environments* (*GeoVUE*). *GeoVUE* will provide decision support for a range of users from academics and professionals involved in understanding cities, to planners and urban designers who require detailed socio-economic data in plan preparation. It will provide geographic information for a more general public involved in viewing problems and policies associated with the impact of change in cities. Demonstrator VUEs include 'MapTube', a system where users can link conventional map and socio-economic attribute data to open source spatial analysis software linked to GIS where the focus is on better scientific understanding of spatial patterns of deprivation, income distribution and demographic ageing in cities. Other projects include pollution monitoring, building on existing virtual cities work (Batty 2006; Batty and Hudson-Smith 2006), where air pollution monitoring and visualization are linked to impacts on resident population through the medium of 3D-GIS using real time sensing of air pollution, and constructing VUEs, building on 3D-GIS applications in central London (Hudson-Smith *et al.*

2005) and enabling small segments of the city to be modelled and then populated by users from the various geographic databases that are available, with a view to assessing the impact of urban planning proposals. Another NCeSS node is based at the University of Leeds where the objective is to develop representation of the entire UK population as individuals and households, together with a package of *Modelling and Simulation for e-Social Science* (*MoSeS*) tools which allow specific social research and policy issues to be addressed. Microsimulation techniques are used to create synthetic population of the whole of the UK. Variants of this approach have been developed in various different contexts (Ballas *et al.* 2005a,b; 2007a,b).

The 'Maptube' idea mentioned above involves the use of *Google Earth* and *Google Maps,* both of which have changed the nature of sharing geographical information. *Google Earth maps* the whole earth by superimposing images obtained from satellite imagery, aerial photography and GIS 3D globe. *Google Maps* is a free web mapping service application and technology that underpins many map-based services including the *Google Maps* website, Google Ride Finder and embedded maps on third-party websites via the *Google Maps* application programming interface (API). *Google Maps* provides high-resolution satellite images for most urban areas across the world which are at least a year old and in some cases date back to 2001. A large amount of JavaScript was used to create *Google Maps*. The term 'mash-up' refers to a new breed of web-based applications that allow the mixing of data from at least two different services from disparate web sites. A mash-up, for example, could overlay traffic data from one source on a map from Google. Tools are available to introduce custom location icons, location coordinates and metadata, and even custom map image sources into the *Google Maps* interface. It is possible to add your own set of locations, scripts and photographs to create memory maps or image annotation features. The *Google Maps* API was created by Google to facilitate developers integrating *Google Maps* into their web sites, with their own data points. Worldwide, more and more local authorities are using maps on their web sites and are increasingly using Google to create mash-ups of information rather maps from national mapping agencies. They are moving to *Google Maps* as the primary interface for casual use by public users, leaving GIS systems for more specialist users because this provides a better user-friendly interface which is easy to use, has integrated aerial imagery, is attractive and does not require training or large manuals. For example, a group of citizens in Brent Borough in Greater London has used the local authority service to create interactive running and cycling routes (Fig. 1.3) which allows the user to run the mouse over an elevation graph to see the corresponding location on the map.

The up and coming semantic web technology provides a common framework that allows data to be shared and reused across application, enterprise and community boundaries (Berners-Lee *et al.* 2001; Hendler *et al.* 2002; Hendler and De Roure 2004). It is a collaborative effort led by W3C with participation from a large number of researchers and industrial partners and is based on the Resource Description Framework (RDF) used to represent information and to exchange knowledge in the web. Web Ontology Language (WOL) is used to publish and share sets of terms called ontologies, supporting advanced web search, software agents and knowledge

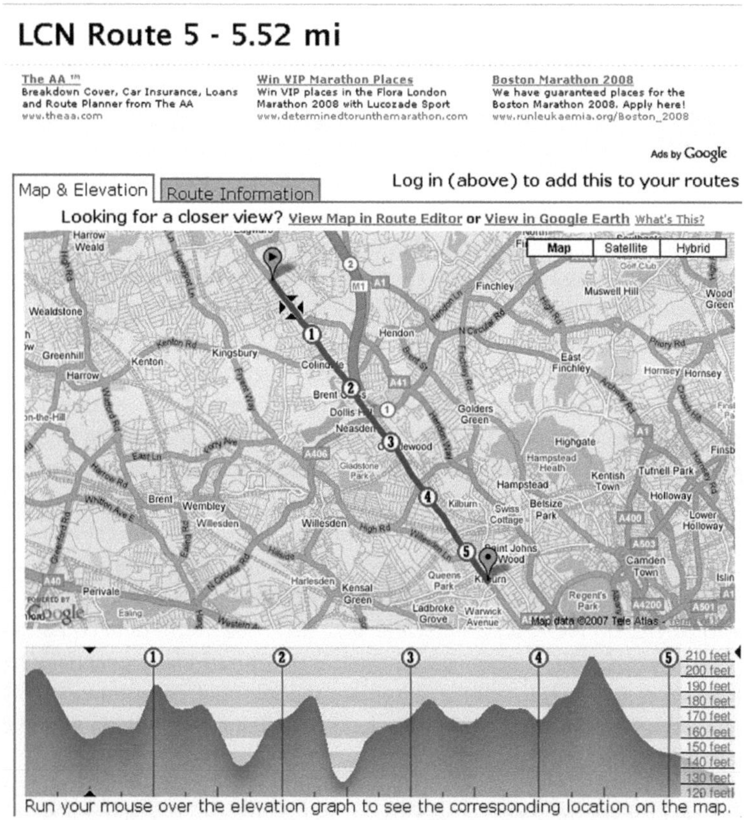

Source: http://www.runstoppable.com/routeoverview.php5?route_id=737044006

Fig. 1.3 Map view of Brent's runstoppable website (See also Plate 1 in the Colour Plate Section)

management. So, the semantic web is about common formats for integration and combination of data drawn from diverse sources, where the original web mainly concentrated on the interchange of documents. It is also about language for recording how the data relates to real-world objects that allow a person, or a machine, to start off in one database, and then move through an unending set of databases which are connected not by wires but by being about the same thing.

The sharing of information, data and resources is therefore a key factor in technology change in recent years and is responsible for driving the development of the semantic web and also the concept of a federated set of web services. Sharing applications over the internet with partners requires trust between two applications in different identity domains. Establishing this trust in user-machine interactions is challenging, and harder still in machine-to-machine environments. In order that a client application in one domain can request information from a web service in a different domain, the client will need to present proof of identity by presenting credentials trusted by the web service. The receiving service will need to be able to understand and evaluate these credentials to assess an identity's validity while also having evidence that the credentials were not tampered with or spoofed during transit. One challenge, therefore, is in finding a way to both federate identity and establish trust between machines in disparate identity domains.

In addition to the development in computing technologies associated with the web, there have also been significant advances in analysis and modelling techniques. One particular example is the *Epidemiological Simulation System* (*EpiSims*), the largest individual-based epidemiology simulation model ever created, built at Los Alamos National Laboratory with support from the U.S. Departments of Energy, Homeland Security, and Health and Human Services, with the purpose of providing an experimental testbed for analysing proposed responses to natural or intentionally caused disease outbreaks (Barrett *et al.* 2004; 2005a,b). *EpiSims* models the spread of disease in urban areas, allowing for the assessment of prevention, intervention, and response strategies by simulating the daily movements of synthetic individuals within an urban region. It allows the user to specify the effects in detail of a pathogen on a specific person, and to assign different effects to various people based on demographic characteristics. In conjunction with population mobility models it can represent behavioral reactions to an outbreak, including official interventions. Spatial micro simulation is one of a series of techniques used for spatial analysis, simulation modelling or prediction. Other new techniques include Geographically Weighted Regression (GWR) (Fotheringham *et al.* 2002), structural equations modelling (Smith *et al.* 2007), agent-based models optimised using genetic algorithms (Heppenstall *et al.* 2007), cellular automata models that incorporate fuzzy logic rules (Al-Ahmadi 2007) and other new land-use planning methods as reported in Koomen *et al.* (2007).

Some of the trends that have been identified in this section are exemplified in the chapters of the book that follow and which we now review in the final section of this introductory chapter.

1.4 Structure of the Rest of the Book

In producing an up-to-date overview of PSS, it is our primary aim to demonstrate that examples of best practice using proven methodologies are available in certain planning contexts but that new methods and approaches are being developed all the time. This compendium has been assembled with the ultimate goal of seeking to exchange proven knowledge, thoughtful insights and new experiences associated with the implementation, operation and evaluation of PSS among those people directly involved in planning. In so doing, our intention attempts to prevent repeated ‘reinvention the wheel’ and to emphasize the new and exciting developments in the application of geo-information technologies in diverse planning practices that are taking place.

The chapters have been divided into four parts. Part I is a collection of contributions that illustrate the application of existing PSS, demonstrating the range of useful functionalities. Part II contains chapters that focus on various PSS that have been developed in recent years for use in a variety of planning contexts, some of which have been implemented whilst others still are in prototype form. There is an emphasis on design issues, particularly when the systems have been constructed to fulfil a number of objectives, contain a wide range of data and incorporate a number of components. Part III contains chapters concerned with the development of a particular method into a planning support tool for analysis, evaluation or visualization and Part IV is comprised of a series of chapters that consider issues and processes of user engagement, stakeholder participation and the integration of analytical and participatory techniques.

1.4.1 Part I: Application and Assessment of Existing PSS

Chapter 2 is the first of three that constitute Part I in which several PSS are introduced that are likely to be known to many readers familiar with the PSS literature. *Deal* and *Pallathucheril* explain the functionality and operation of the *LEAM* model (Land-use Evolution and impact Assessment Model), a simulation model developed to forecast and evaluate land-use change over space and time that will enable planners and stakeholders to view and assess the future outcomes of decisions and policies before putting these into action. A range of applications have been performed with this model, some of which are described in detail, providing valuable lessons from which we can learn a great deal. Of particular note is the fact that *LEAM* evolves as an iterative process of data collection, model building, and dialogue in close cooperation with local planners, policy makers and stakeholders. It is argued that such a use-driven, embedded approach to local model development best suits an evolutionary process in which complex PSS will gain acceptance into practice by becoming an integral part of local and regional planning.

In the following chapter by *Nijs*, a model-based evaluation of a new Dutch National Spatial Strategy is presented, exemplifying the application of the so-called *Environment Explorer*, a land-use simulation model based on cellular automata with which *ex ante* assessments of proposed policy decisions can be performed. The model has been calibrated and validated initially before being applied to simulate new spatial developments. Thereafter, a probability map of future urban developments has been estimated for two scenarios using a Monte Carlo simulation approach. With the help of the scenarios, the objectives of the new Dutch National Spatial Strategy have been evaluated and potential problem areas identified. The study reveals that using the *Environment Explorer* to evaluate strategic spatial planning in the Netherlands provides detailed insights into the feasibility of the policy goals and potential problems arising in meeting those goals. Moreover, *Nijs* concludes that proper calibration and validation of the model is restricted – as is often the case with PSS – by the confined availability of data sets of appropriate quality. As a consequence and outcome, future research will be directed towards reducing the uncertainties in the results of the land-use model by taking advantage of better (remote sensing) data for monitoring change.

In the last chapter of Part I, *Pettit* and *Wyatt* present the outcome of an assessment of four different but relatively well-known PSS (*SLEUTH, What if? Google Earth* and *Preference Prediction*). These were tested for their functionality and capability in modelling, visualizing and evaluating a number of land-use change scenarios up until 2030. From this assessment it appears, perhaps not surprisingly, that each PSS has its strong and weak points but, especially in combined applications, they provide a remarkably useful toolbox for planners. *SLEUTH* and *What if?* function as instruments to build a range of future land-use scenarios and appear to be appropriate tools for improving the understanding of the policy implications of these scenarios. *Google Earth*, as an exponent of virtual globe products, provides a novel way to visualize landscape futures and thus helps to better engage planners, politicians and communities in discussing the outcomes of complex models. Finally, *Preference Prediction* is capable of performing *ex post* evaluation and helping to create more participatory and consensus-based planning practice.

1.4.2 Part II: Design, Development and Implementation of Recent PSS

In the first chapter of Part II, *Levine* introduces a Motor Vehicle Safety PSS which is able to address severe traffic safety problems. The PSS is developed and applied in the Houston area, Texas. The system provides tools for storing, analysing, and presenting crash data and produces information for safety reports, for identifying hazardous locations and for policy decision making on road improvements. The chapter explains the application of the well-known freeware programme, *CrimeStat,* to perform crime travel modelling and to calculate various statistics for measuring spatial distributions, hot-spot identification, risk analysis of incidents, and space-time interaction analysis. The studies show clearly that crash information and crash analysis should be combined with crash expert knowledge to provide valuable recommendations for arriving at safety improvement measures. In principle, the PSS is just a tool which, to become valuable, should be embedded in a more extensive analytical framework that goes beyond the data that has been collected and which addresses the behavioural issues involved in traffic safety. It shows that there are no simple engineering solutions, but that increased enforcement and public education is needed too.

In Chapter 6 by *Hahn, Kofalk, de Kok, Berlekamp* and *Evers*, a PSS for the Elbe river basin is introduced which consists of interactive tools for simulation, analysis and presentation and which is intended for use in exploring appropriate measures of effective and sustainable river basin management. To arrive at a functional PSS for strategic river basin planning, it is argued there is a need to understand the driving forces of the river basin and its dynamic behaviour, to simulate the combined effects of policy options and external effects, to assess the aggregated ecologic and socioeconomic impacts of potential measures and to communicate the goals and results to stakeholders. From these requirements, the chapter describes some methodological and practical lessons learnt during the development of the PSS, as well as some reflections on issues related to the application of the system.

In the next chapter, *Van Esch, Vos, Janssen* and *Engelen* present a PSS for reducing pollution emissions in the surface waters of Flanders, Belgium. In fact, the PSS helps the Flemish Environmental Agency to fulfill its day-to-day monitoring and reporting obligations regarding the pressures and impacts of point and diffuse emissions on surface waters *vis-à-vis* the Flemish, Belgian and European authorities. Essentially, the PSS is an accounting system, keeping track of the pollutants from a variety of sources towards their sinks in the surface waters. It enables the detailed representations of sectors responsible for the emission of harmful pollutants, their transport to treatment facilities and finally their discharge into the surface waters. In addition to this function of providing an up-to-date technical database, it enables the design and assessment of alternative policies and spatial – alongside technical – measures targeted at particular sectors, groups in society and/or regions and aimed at improving the quality of the surface waters in different river basins and administrative entities in Flanders. It supports 'what-if' analysis in an interactive context.

Van der Hoeven, Aerts, Van der Klis and *Koomen* introduce an integrated discussion support system with which policy makers can gain insights into the consequences of diverse flood risk management strategies for the Netherlands under the influence of climate change. Flood risk is determined by combining spatial land-use and hydrological information. Thus, use is made of a land-use model that operates under different future trends using socio-economic scenarios and climate information. Flood risk assessments are presented in both monetary and casualty terms. Both the construction of the system and its application to the Netherlands are discussed. The system aims to support the learning process of the users by facilitating discussion on the long-term adaptability of the Netherlands to flood risk. The system does not provide unambiguous answers on which management strategy is preferable but it does provide knowledge that adds to understanding of the impact of various flood risk management strategies, such as information about flood probabilities, potential damage, potential number of casualties, flood risk, as well as costs and benefits of the proposed safety strategies.

A spatial planning support system (SPSS) developed for the city of Bangalore in India is introduced in Chapter 9 by *Sudhira* and *Ramachandra*. The core of the system consists of an integration of spatial dynamics and agent-based land-use models. The system is dedicated to simulate different urban extension scenarios in order to get to grips with the patterns, processes, causes and consequences of urban sprawl. In particular, the set-up of the system and its user requirements are addressed, emphasising its utility as an effective tool for policy, planning and decision making. Moreover, its present drawbacks and the future intentions for development in the direction of a more web-based and participatory PSS are discussed.

In the last chapter of Part II, the *GRAS* system is introduced by *Pelizaro, Arentze* and *Timmermans*. This system can be considered a prototype SDSS for the planning, design and maintenance of urban green spaces. In short, *GRAS* is a GIS scenario-based, micro simulation multi-criteria DSS that incorporates a range of domain-specific models. The system is able to predict every individual's spatial-temporal 'green-space-choice' behaviour. It is capable of assisting every stage of the decision-making process, i.e. from the identification of a problem and the definition of multiple objectives to allowing the users to generate alternatives, and to the evaluation/assessment of alternatives. Its architecture and design are described, its constituting components explained and its first experiences of application in practice are documented.

1.4.3 Part III: New Methods for Planning Support

In the first chapter of Part III, *Johnson* and *Sieber* present a prototype PSS for the tourist sector based on the principle of agent-based modelling (ABM). With the help of the so-called *TourSim* model, experiments can be performed to simulate and visualize tourism planning scenarios under diverse policy changes. Examples of experimentation include discovering the effects of developing new tourism infrastructure (e.g. a new music festival) on tourist distribution, 'what if' questions concerning global tourism trends (e.g. monetary changes), and changes to tourism demand as a result of geopolitical or economic reasons (e.g. more domestic travel). The chapter demonstrates how *TourSim* works by applying it to the Canadian province of Nova Scotia where it is used to shows the effects of port of entry on tourist dispersion. Moreover, *TourSim* is evaluated in the context of three potential areas of adoption constraint: awareness of and experience with ABM; technological considerations; and overall fit with planning tasks. In a comparison with GIS adoption, it shows that the use of ABM in PSS holds great potential, but also is accompanied by significant hurdles that still have to be overcome.

Chapter 12 by *Gibin, Mateos, Petersen* and *Atkinson* describes a geographic visualization tool, named the *London Profiler*, for supporting public health service planning. It is built by University College London in cooperation with Southwark and Camden Primary Care Trusts and is based on an implementation of *Google Maps* APIs as a framework for geographical visualization of changing population characteristics. The system provides a frequently updatable picture of the London population at postcode unit level. With the help of *London Profiler*, it becomes possible to target public health initiatives and services to tackle health inequalities to specific population groups at risk or in need. Due to its flexible and inexpensive features, *Google Maps* APIs are considered to be a perfect platform for the development of these kind of future PSS.

In the following chapter, *Schaller, Gehrke* and *Strout* describe a new method involving geodatabase design and GIS processing modelling to support regional environmental planning procedures in a more effective way. The existing large geodatabases of the planning region of Bavaria, Germany, have been updated and restructured to be able to perform environmental modelling, sensitivity testing and site analysis for both decision support and scenario applications. With the help of these analyses, the authors are able to evaluate actual developments in land-use monitoring and predict future developments and their possible impacts on natural resources. The working of the new concept is illustrated with the help of some actual planning projects such as the urban growth model of the Munich region. Future research will entail, *inter alia,* the incorporation of the internet in the development of the new method.

In the last chapter of Part III, *Besio* and *Quadrelli* discuss knowledge bases they have built to support three environment and landscape planning processes to assess where and how to locate environmental systems. The first planning process relates to the definition of EU programme in the Liguria Region of Italy, the second to the preparation of a Cinque Terre National Park plan, and the last to strategic transport projects (new railways and motorways) in the Greater Genoa metropolitan area. PSS were built according to a cognitive procedure which organized data, processed information and produced knowledge at subsequent synthetic and interpretative levels, using geo-information technologies. These systems have been used to organize data for investigating many phenomena, the analysis of their numerous relationships to identify meaningful information, and the elaboration of synthetic models representing the environmental systems subjected to the planning process. The data and their meaning have been structured by making use of conceptualization and categorization procedures, whilst information processing has been carried out through qualitative, morphological and topological procedures. The experiences with the three planning processes are shown to be significant, not only in terms of the technological instrument adopted, but above all for the way in which it was used.

1.4.4 Part IV: Participation and Collaboration in PSS

Part IV contains the largest number of chapters, reflecting perhaps the relative importance of using PSS for participation in planning. In the first of seven chapters, *Lieske, Mullen* and *Hamerlinck* describe a participatory planning methodology in which planning support methods play an important role in order to arrive at comprehensive plans with high levels of community support. Therein, a distinction is made between planning support instruments (e.g. key pad polling, gaming) that were used to gather public input on issues, attitudes and values; and planning support systems (e.g. *CommunityViz*) which were used behind the scenes to integrate public values with geographic data and to evaluate citizen-generated development alternatives. Application of the participatory planning methodology has taken place in Albany County in Wyoming, USA, resulting in high levels of community support for the resulting plan and enhanced probabilities for plan adoption and implementation. Other applications of the methodology in different settings show similarly positive outcomes.

Chapter 16 by *Miller, Vogt, Nijnik, Brondizio* and *Fiorini* discusses the integration of analytical and participatory techniques and tools for planning the sustainable use of land resources and landscapes. It recognizes a need to provide a way of assessing the balance between the quality of the visual, ecological, cultural and production functions of the countryside. Two examples illustrate its findings. The first example considers land-use change in the Amazon, whilst the second discusses the socio-economic, ecological and visual aspects of land-use changes in a European landscape. Each example has employed the active involvement of stakeholders like landscape professionals and the public in the process of decision making. Stakeholder and public perspectives of landscape and land-use change were generated with the help of participatory techniques to enable these values, objectives and preferences to be incorporated into an analysis of options for future land uses. In short, the chapter presents a comparison of lessons learnt from the development and implementation of tools and methods for the assessment of scenarios of change in land use and landscape, in European and Brazilian contexts, in order to develop good practice in planning.

In the next chapter, by *Van Delden* and *Hagen-Zanker,* a methodology is presented for the linking of qualitative storylines to quantitative modelling in a participatory approach. In this integrative framework, a two-sided methodology is envisioned in which storylines and scenarios steer the model development, but also vice versa, with modelling outputs providing information and arguments to adjust the storylines and scenarios. The methodology is illustrated by using a case involving, on the one hand, a range of qualitative storylines about possible futures of Europe and, on the other, the application of a quantitative land-use model for three regions within Europe: the Netherlands, Estonia and northern Italy. As an important outcome, it is concluded that the integration in a participatory approach of qualitative scenario development and quantitative modelling seems to be a very promising direction, in which one method contributes substantially to the other and vice versa. Nevertheless, it is recognized that future work is needed to link both approaches in a more advanced way.

A new implementation of information and communication technologies is presented by *Soutter* and *Repetti* in Chapter 18 to support public participation. A system called *SMURF* (*System for Monitoring URban Functionalities*) was created for supporting participatory planning and management. The system consists of a geographic database and spatial indicators for viewing and sharing information, for editing information and for evaluating city development. After a review of the content and set-up of the system, the chapter describes two implementation examples, one in Thies, Senegal, and one in the Seychelles. From these and other applications of *SMURF*, a range of experiences was gained from which various conclusions and initiatives for further research have been derived. One important conclusion is that the process results in better knowledge for all participants and in a strong consensus about the diagnosis of the actual situation as well as clearer strategic objectives for local development.

In the chapter by *Kahila* and *Kyttä,* the so-called *SoftGIS* method is introduced and presented as a bridge between residents and urban planners. The *SoftGIS* method entails a range of methods, and their underlying theories, concepts and ideas that are dedicated to expose the knowledge that residents have of their own living environment, which can be included thereafter in formal urban planning procedures. *SoftGIS* methods reveal how the everyday lives of residents are organised, the kind of place-based positive and negative experiences residents have and how they behave in their physical environment. They therefore contribute to participation and collaboration of citizens in planning processes. Besides introducing different variations of the *SoftGIS* method like *softGISquality* and the *softGISchildren*, the approach is also evaluated in the context of current critical GIS discourse. Moreover, its theoretical basis is elaborated upon extensively and future plans for extension are considered.

In the penultimate chapter, *Chin* introduces the so-called *Mainport Planning Suite* (*MPS*) with which studio-based planning can be facilitated. The *MPS* has been designed and prototyped in close cooperation with practitioners in the Port of Rotterdam in the Netherlands. In short, the suite consists of software for the user-friendly presentation of geographic information in maps, also linked to a timeframe. Moreover, it consists of a sketchbook for the drawing up of ideas and the storage and presentation of these ideas. In addition, with the help of a so-called 'matchbox', an overview is presented of scores of alternatives on both quantitative and qualitative criteria, and finally, the outcomes of the matchbox can be graphically displayed. It is foreseen this toolbox can be applied too in quite different contexts and will support more and more upcoming needs.

Finally, in Chapter 21, the last in the sequence and in the book, *Carver, Watson, Waters, Matt, Gunderson* and *Davis* explain the participatory approaches developed to map landscape values for landscape and resource management, in particular in order to implement wildland fire-use plans in wilderness areas. In that framework, a tool and a methodology are developed with which fuzzy qualitative information can be gathered, stored and geographically presented. This qualitative information concerns the meanings that are ascribed by Flathead Indian Reservation residents to certain places within the Mission Mountains Tribal Wilderness in Montana, USA. This information is used in focus group discussions with forest managers about fuel treatments. The qualitative information is subsequently confronted with more crisp quantitative information, concerning land-use categories and planning activities. It is envisioned as being crucial that fire planners must understand how proposed actions interact with human meanings ascribed to the land and describe a prioritization process that addresses publicly perceived threats. The case study demonstrates that the approach is well suited to developing a better understanding of indigenous peoples' relationships with the land, and appears to be particularly useful for contrasting meanings attached to different classifications of land, articulating what is worth protecting and what it should be protected from, and showing how it is viewed by people from different cultures and/or stakeholder groupings. It is expected that the need for simultaneous handling of both qualitative and quantitative information will growth substantially (e.g. NIMBY effects, mental mapping).

1.5 Conclusion

The collection of chapters that we have introduced and which follow in the remainder of the book represents a cross-section of PSS development and experience around a decade after the concept of PSS originated and the first examples were formally identified. It is of course for you, the reader, to assess how the state-of-the-art has evolved during the last ten years but we hope that the contributions in this volume, by providing explanation of system components together with exemplification of applications and user experiences, will help you to structure your evaluation. We also hope to persuade you that PSS are innovative and exciting new tools whose wider application and adoption are likely to emerge with increasing regularity over the next ten years in different planning contexts across the world. Satisfactory progress will only take place if the obstacles that have been discussed in this chapter are confronted and overcome and if the planners are prepared to take advantage of some of the new technologies that this chapter has also reviewed.

References

Al-Ahmadi, K. (2007) Modelling urban growth using cellular automata and geographical information systems: a case study of Riyadh, Saudi Arabia, Unpublished PhD, School of Geography, University of Leeds, Leeds.

Al-Kodmany, A. (2003) Web-based tools and interfaces for participatory planning and design, In Geertman, S. and Stillwell, J. (eds.) *Planning Support Systems in Practice*, Springer, Berlin, pp. 65–86.

Ballas, D., Clarke, G., Dorling, D., Eyre, H., Rossiter, D. and Thomas, B. (2005a) SimBritain: a spatial microsimulation approach to population dynamics, *Population, Space and Place,* 11: 13–34.

Ballas, D., Clarke, G. and Wiemers, E. (2005b) Building a dynamic spatial microsimulation model for Ireland, *Population, Space and Place*, 11: 157–172.

Ballas, D., Clarke, G., Dorling, D. and Rossiter, D. (2007a) Using SimBritain to model the geographical impact of national Government policies, *Geographical Analysis*, 39(1): 44–77.

Ballas, D., Kingston, R., Stillwell, J. and Jin, J. (2007b) A microsimulation-based spatial decision support system, *Environment and Planning A*, 39: 2482–2499.

Barrett, C., Eubank, S., Anil Kumar, V.S. and Marathe, M. (2004) Understanding large-scale social and infrastructure networks: a simulation-based approach, *SIAM News* 4, May.

Barrett, C. Eubank, S. and Marathe, M. (2005a) Modeling and simulation of large biological, information, and socio-technical systems: an interaction-based approach, In Goldin, D., Smolka, S. and Wegner, P. (eds.) *Interactive Computing: A New Paradigm*, Springer Verlag, Berlin.

Barrett, C.L., Eubank, S.G. and Smith, J.P. (2005b) If smallpox strikes Portland..., *Scientific American*, 292(March): 54–61.

Batty, M. (1995) Planning support systems and the new logic of computation, *Regional Development Dialogue*, 16(1): 1–17.

Batty, M. (2006) Virtual London, In Heywood, I., Cornelius, S. and Carver, S. (eds.) *An Introduction to Geographical Information Systems*, Pearson Educational, Prentice Hall, Harlow, pp. 269–271.

Batty, M. and Hudson-Smith, A. (2006) Digital cornucopias: changing conceptions of the virtual city, *Environment and Planning B*, 33: 799–802.

Berners-Lee, T., Hendler, J. and Lassila, O. (2001) The semantic web, *Scientific American*, 279(5): 34–43.

Brail, R. and Klosterman, R. (eds.) (2001) *Planning Support Systems, Integrating Geographic Information Systems, Models and Visualization Tools*, ESRI Press, Redlands, California.

Carver, S., Blake, M., Turton, I. and Duke-Williams, O. (1999) Open spatial decision-making: evaluating the potential of the world wide web, In Kemp, Z. (ed.) *Innovations in GIS 4*, Taylor and Francis, London, pp. 267–278.

Campagna, M. and Delano, G. (2004) Evaluating geographic information provision within public administration websites, *Environment and Planning B*, 31: 21–37.

Clarke, M. (1990) Geographical information systems and model-based analysis: towards effective decision support systems, In Scholten, H. and Stillwell, J. (eds.) *Geographical Information Systems for Urban and Regional Planning*, Kluwer, Dordrecht, pp. 165–175.

Davies, F. (1986) *A Technology Acceptance Model For Empirically Testing New End-User Information Systems: Theory and Results*, Unpublished PhD Thesis, MIT, Cambridge Massachusetts.

Dishaw, M. and Strong, D. (1999) Extending the technology acceptance model with task-technology fit constructs. *Information & Management,* 36: 9–21.

Dishaw, M., Strong D. and Bandy, B. (2002) Extending the Task-Technology Fit model with self-efficacy constructs, *Eight Americas Conference on Information Systems, AMCIS*, Association for Information Systems, Dallas.

Elwood, S.A. (2005) GIS use in community planning: a multi-dimensional analysis of empowerment, In Fyfe, N. and Kenny, J. (eds.) *The Urban Geography Reader*, Routledge, London, pp. 400.

Elwood, S. (2006) Critical issues in participatory GIS: deconstructions, reconstructions, and new research directions, *Transactions in GIS*, 10(5): 693–708.

Foster, I. (2003) The Grid: computing without bounds, *Scientific American*, 228(4): 78–85.

Fotheringham, A.S., Brunsdon, C. and Charlton, M. (2002) *Geographically Weighted Regression: The Analysis of Spatially Varying Relationships*, Wiley, Chichester.

Frambach, R. and Schillewaert, N. (2002) Organizational innovation adoption; A multilevel framework of determinants and opportunities for future research, *Journal of Business Research*, 55: 163–176.

Geertman, S. and Stillwell, J. (2003) *Planning Support Systems in Practice*, Springer, Heidelberg.

Goodhue, D.L. (1995) Understanding user evaluations of information systems, *Management Science,* 41: 1827–1844.

Goodhue, D.L. and Thompson, R.L. (1995) Task-technology fit and individual performance, *MIS Quarterly,* 19: 213–235

Harris, R. and Batty, M. (1993) Locational models, geographic information and planning support systems, *Journal of Planning Education and Research*, 12(3): 184–198.

Hendler, J. and De Roure, D. (2004) E-science: the grid and the semantic web, *IEEE Intelligent Systems*, 19(1): 65–71.

Hendler, J., Berners-Lee, T. and Miller, E. (2002) Integrating applications on the Semantic Web, *Journal of the Institute of Electrical Engineers of Japan*, 122(10): 676–680.

Heppenstall, A.J., Evans, A.J. and Birkin, M.H. (2007) Genetic algorithm optimisation of a multi-agent system for simulating a retail market, *Environment and Planning B*, 34(6): 1051–1070.

Hudson-Smith, A., Evans, S. and Batty, M. (2005) Virtual London: from CAD to 3D-GIS to global digital cities, In Chiu, M-L. (ed.) *Insights of Digital Cities*, Archdata CAAD Talks 4, Taipei, pp. 75–92.

Kingston, R., Evans, A. and Carver, S. (2003) Public participation and online democracy, In Geertman, S. and Stillwell, J. (eds.) *Planning Support Systems in Practice*, Springer, Berlin, pp. 45–64.

Klosterman, R. (1997) Planning support systems: a new perspective on computer-aided planning, *Journal of Planning Education and Research*, 17(1): 45–54.

Klosterman, R.E. (1999) Guest editorial: New perspectives on planning support systems, *Environment and Planning B*, 26: 317–320.

Klosterman, R. and Pettit, C. (2005) Guest editorial: An update on planning support systems, *Environment and Planning B*, 32(4): 477–484.

Koomen, E., Stillwell, J., Bakema, A. and Scholten, H.J. (eds.) *Modelling Land-Use Change Progress and Applications*, Springer, Dordrecht.

Nonaka, I. and Takeuchi, H. (1995) *The Knowledge-Creating Company: How Japanese Companies Create the Dynamics of Innovation*, Oxford University Press, Oxford.

Rees, P., Martin, D. and Williamson, P. (eds.) (2002) *The Census Data System*, Wiley, London.

Rogers, E. (1995) *Diffusion of Innovations*, Free Press, New York.

Rogers, E. (2003) *Diffusion of Innovations*, Free Press, New York.
Senge, P. (1990) *The Fifth Discipline: The Art and Practice of the Learning Organization*, Currency Doubleday, New York.
Shiffer, M.J. (1992) Towards a collaborative planning system, *Environment and Planning B*, 19: 709–722.
Shiffer, M.J. (1995) Interactive multimedia planning support: moving forward from standalone systems to the World Wide Web, *Environment and Planning B*, 22: 649–664.
Smith, P.W.F., Berrington, A. and Sturgis, P. (2007) A comparison of graphical models and structural equation models for the analysis of longitudinal survey data, In Lynn, P. (ed.) *Methodology of Longitudinal Surveys*, Wiley, Chichester.
Ventura, S., Miller, T. and Barry, G. (2003) Community engagement in land-use planning through web-based technologies, In Geertman, S. and Stillwell, J. (eds.) *Planning Support Systems in Practice*, Springer, Berlin, pp. 87–98.
Vonk, G. (2006) *Improving Planning Support; The Use of Planning Support Systems for Spatial Planning*, The Netherlands Geographical Studies 340: pp. 126.
Vonk, G., Geertman, S. and Schot, P. (2005) Bottlenecks pp. blocking widespread usage of planning support systems, *Environment and Planning A*, 17: 909–924.
Vonk, G., Geertman, S. and Schot, P. (2007a) A SWOT analysis of planning support systems, *Environment and Planning A*, 39: 1699–1714.
Vonk, G., Geertman, S. and Schot, P. (2007b) New technologies stuck in old hierarchies; An analysis of diffusion of geo-information technologies in Dutch public organizations, *Public Administration Review*, 67: 745–756.

Part I
Application and Assessment of Existing PSS

Chapter 2
A Use-Driven Approach to Large-Scale Urban Modelling and Planning Support

Brian Deal and Varkki Pallathucheril

2.1 Introduction

Urban system simulation models are used to forecast and evaluate land-use change over space and time. Such simulation models offer planners and stakeholders an ability to view and assess the future outcomes of future policy alternatives before final decisions are made about implementation. This technology also offers the ability to improve our fundamental understanding of land-use transformation dynamics and the complex interplay between urban change and sustainable systems (Brail 2001; Deal 2001).

The literature on Planning Support Systems (PSS), and the large-scale urban modelling and simulation tools on which those systems are built, has largely focused on technical issues and system mechanics, especially the theoretical underpinnings, software architecture and tool functionality. That focus is understandable given the novelty, complexity and scale of such tools. However, several authors (Deal and Pallathucheril 2003; Pettit 2005) approach the topic from a qualitative angle, considering basic questions about the real-world relevance of specific PSS, such as:

- Are the tools useful, or even usable?
- Can they effectively support planning decisions and policy choices?
- Does the planning profession understand how to apply these tools?

The answers to such questions are important both for the shorter-term rate of adoption of these systems and for longer-term PSS acceptance within the planning and stakeholder communities.

The literature also conveys a sense that these systems and tools are being developed to meet functional requirements and use cases as conceived by the systems' developers. However, there is little evidence provided to indicate that deployment of these systems has informed development or that the use of these systems has

B. Deal (✉)
University of Illinois at Urbana-Champaign
Department of Urban and Regional Planning
611 Taft Drive, Champaign, Illinois 61820, USA
E-mail: deal@illinois.edu

S. Geertman, J. Stillwell (eds.), *Planning Support Systems Best Practice and New Methods*,
DOI 10.1007/978-1-4020-8952-7_2, © Springer Science+Business Media B.V. 2009

informed system development. In describing how a land-use model was operationalized for a regional PSS, Deal and Pallathucheril (forthcoming) suggest that embedding PSS development in real-world planning processes offers significant mutual benefits to both users and developers: the PSS informs the dialogue on regional planning while the application subjects the PSS to critical in-progress review and spurs developers toward further enhancements.

In this chapter, we elaborate on this idea of use-driven PSS development, which we distinguish from theory-driven development, based on our experience in developing and deploying *LEAM,* the *Land-use Evolution and impact Assessment Model.* First we briefly review the PSS literature and provide a succinct description of the *LEAM* process (Deal 2003; Deal and Pallathucheril 2007). Then we describe *LEAM*'s use-driven development by reviewing some of the policy-related questions that *LEAM* has been involved in testing. These policy reviews help to illustrate three key ideas about use-driven development:

- deploying a PSS during its nascent stages has significant benefits;
- addressing policy questions during system development is valuable; and
- developing a system can be as useful in planning as delivering the end product.

We conclude this chapter with a discussion of some of the difficulties inherent in use-driven PSS development.

2.2 Available PSS Tools

A growing set of PSS tools for simulating and evaluating large-scale urban dynamics have become available in recent years. Brail and Klosterman (2001) outlined the state of the art and described new approaches in PSS development that were, and continue to be, due in some measure to increased computational capabilities and availability of digital data. In their edited volume, the basic theoretical constructs and issues are examined by Batty, Harris, and Hopkins individually. A number of authors then describe a number of different PSS: *DRAM EMPAL* (Putman); *TRANUS* (de la Barra); *What if?* (Klosterman); *CUF, CUF II,* and *CURBA* (Landis); and *URBANSIM* (Waddell). A third section is devoted to issues of visualization.

To this body of knowledge, Geertman and Stillwell (2003) add a review of the role of PSS in planning with the intention of documenting best practices, promoting use of these tools and enhancing planning. They describe the evolution of PSS-like tools and identify spatially explicit tools (spatial decision support systems) as an important sub-category. The edited volume includes three discussions of planning processes: discussions of three tools used in each of a number of domains: participatory planning; strategic and land-use planning; and environmental planning. More recently, Klosterman (2005) has provided an update on the state of the art in a guest editorial in *Environment and Planning B* that attempts to define the types of PSS tools within a planning context in order to structure discussions of their utility. Klosterman describes categorizing PSS along two dimensions: by the planning task the model addresses; and the technique or approach it utilizes. Four techniques are recognized: (i) large-scale urban models – identified by their scale and the presence of spatial interaction markets; (ii) rule-based models – incorporate explicit decision rules that allow users to specify model behavior; (iii) state change models – use statistical methods to replicate geographic patterns without identifying underlying causes; and (iv) cellular automata (CA) – the least developed area of inquiry, combining rule-based and state-change dynamics over large-scale flat lattice grids. CA models provide an important basis for testing ideas and might find their way to useful applications in practice. At the time of this writing they appear extremely complex to build and operate and may be more useful as pieces coupled or integrated within other models (White and Engelen 1997; Wu and Martin 2002; Deal 2003).

This recent evolution of large-scale urban modelling towards dynamic spatial simulation systems contrasts somewhat with earlier work in spatial (and aspatial) reasoning systems (Kim *et al.* 1990). Knowledge-based reasoning systems are loosely based on a philosophical ideal of capturing the manner in which expert knowledge is applied to address complex planning problems. These systems are characterized by the use of multiple types of domain knowledge and complex domain models to support reasoning processes. This knowledge may include task and goal structures, various kinds of constraint, search control techniques and use of human expertise when necessary (Wilkin 2000). Almost two decades ago, Goel (1989) distinguished between case-based and model-based reasoning systems. In case-based reasoning, a body of known cases is assembled along with operations for inferring applicability to a particular situation; in model-based reasoning, inferences are drawn from a hierarchical abstraction of the system at hand. As with the present status of CA systems, future PSS work might employ integrated reasoning and dynamic simulation systems to help improve our ability to capture complex planning and implementation behaviors.

2.2.1 LEAM

LEAM is one approach to large-scale urban modelling and simulation that attempts to bridge the four categories described by Klosterman (2005). Similar to CA approaches, *LEAM* utilizes a structured lattice surface with state-change conditions that evolve over time. The *LEAM* lattice surface, however, is not flat and is shaped by biophysical factors (such as hydrology, soil, geology and land form), and socioeconomic factors associated with administrative boundaries, census spatial units and planning areas. As with state-change techniques, a probability is calculated for change of each cell from one land-use category to another. Unlike other state-change approaches, this probability is predicated on local interactions (e.g. the accessibility of the cell to a given attractor), global interactions (e.g. the state of the regional economy), and other mechanisms of causation (e.g. social forces). As with rule-based approaches, causal mechanisms are used to produce diverse planning scenarios. And akin to large-scale urban simulations, *LEAM* works at the regional scale with regional macro socioeconomic models combined within the *LEAM* modelling framework. Unlike other large-scale efforts, *LEAM* aggregates to the regional scale from a fine-scaled (30 × 30 m) resolution that includes cell-based micro models. This enables loose and tightly coupled linking with other models that might operate at a different spatial scale.

More detailed descriptions of the *LEAM* framework have been described elsewhere (Deal 2003; Deal and Pallathucheril 2007) and is not our intention to replicate these efforts. Rather, the intent here is to provide a very brief description of the basics of the model that will enable the reader to follow some of the reasoning behind its use-driven local applications. Fundamentally, the *LEAM* model consists of two major organizational parts: (i) a land-use change model (*LUC*) – defined by a dynamic set of sub-model drivers that describe the local causality of change and enable easy addition and removal of variables and the ability to play out 'what-if' scenarios; and (ii) impact assessment models that facilitate interpretation and analysis of land-use change depending on local interest and applicability – these help to assess 'so-what' questions and explicate the implications of a scenario. The need in planning and policy making to answer both 'what-if' and 'so-what' questions is a key basis for the *LEAM* framework.

In *LEAM*, the land-use transformation potential of individual cells is evaluated by explicitly quantifying the forces (drivers) that contribute to change. Understanding the causal mechanisms of change provide local decision makers an opportunity for testing policy and investment choices and are a critical component for completing scenario-planning exercises. Driver sub-models are locally dependent and derived through both analysis and local stakeholder interaction. An open architecture and modular design facilitates incorporation of additional local drivers needed to improve the explanatory power of the model.

A regional econometric, input-output model determines the regional demand for residential, commercial and open space land (Sarraf *et al.* 2005). Households and jobs are established and converted into land demand using sector-based economic and demographic analysis (in lieu of sub-regional constraints on demand to determine spatial allocation used in other approaches, such as Wu and Martin 2002). The estimated demand serves as a target for regional land allocation. Market variables increase or decrease development rates based on how well the regional demand targets were met or not met.

Simulated outcomes are described in graphs, charts, text and in map form and are used in engaging local dialogue and in analyzing the potential implications of the changes described. The environmental, economic and social system impacts of alternative scenarios can be modelled and tested (Deal and Schunk 2004). Scenario descriptions of alternative land-use policies, investments decisions, growth trends and unexpected events (among others) can be simulated, analyzed and compared for regional importance. *LEAM*'s visual and quantitative representation of each scenario's outcome provides both an intuitive means of understanding and a basis for analyzing the implications of potential decisions. These representations act as a catalyst for discussion and communal decision making.

For a particular region, *LEAM* evolves as an iterative process of data collection, model building, and dialogue. Local planners, policy makers and stakeholders provide feedback and input about the local salience and value of any given simulation. This feedback is gathered regularly and begins at project inception. It is used to more effectively capture the local condition, to provide a better local version of the tool and to inform local stakeholders about the tool and its uses. This form of use-driven modelling and system development, which takes place in very public forums, most distinguishes the *LEAM* approach. The authors believe, based on their experiences, that feedback and local dialogue are critical in the creation of useful PSS tools. Relying only on theory or the underlying mathematics does not tell a complete or even (at times) a compelling story about local conditions. Constant internal and external review and interaction are critical to informing both the modeller and the local stakeholders of modelled changes, improvements and scenario outcomes.

2.3 Use-Driven Simulation Model Development with *LEAM*

Application of a robust analysis tool like *LEAM* provides a rich source of data and information for a region. In applying *LEAM* in diverse contexts, we have found numerous occasions where this wealth of information has been used to specifically inform policy deliberations. In our use-driven, feedback-based process of model and system building, *LEAM* simulations inform policy deliberations. Policy deliberations call for more and better information, which in turn drives additional *LEAM* enhancements and refinements (that find their way to other policy deliberations). In this section, we draw on three examples to illustrate this mutually beneficial use-driven, feedback-based relationship. One example considers how visioning exercises can be more thoroughly grounded, another looks at assessing alternative transportation investment choices, and the third considers an environmental planning task. Each draws on *LEAM*'s inherent ability to quickly provide useful information and a forum for dialogue. In terms of model refinements, the first example required simulating alternative scenario futures, the second required loosely coupling *LEAM* with a separate model to inform deliberations, and the third required developing and refining a separate, but integrated model.

2.3.1 Grounding a Regional Vision

In 2001, the Northeastern Illinois Planning Commission (NIPC) embarked on an effort to engage the Chicago region in a bottom-up planning effort that culminated in the *2040 Regional Framework Plan*. An important aspect of this effort was *Common Ground*, a process to build regional consensus: 'Common Ground engaged a cross-section of people in the City of Chicago and the six-county region: residents, community leaders, public officials, business owners and planners at all levels. Nearly 4,000 people participated in 200 local and regional workshops and meetings across northeastern Illinois. These public meetings, combined with specialized work by a range of planning experts and elected officials, identified local and regional assets, needs, and challenges' (NIPC 2004: p. 6).

This multi-year effort produced an impressive set of 52 regional goals and associated objectives that addressed issues ranging from 'education to water supply, transportation to taxation' (NIPC 2004: p. 11). This diverse set included goals such as protecting natural resources, enhancing social equity and preserving economic competitiveness. A framework of centres, corridors, and green areas was formulated as the region's way of growing towards these goals. With this framework in place, the region's planners were faced with the need to ground their visions of the future in a manner that could be conveyed to the public in a tangible and meaningful way. To address this challenge, they commissioned two simulations of future land-use change using *LEAM*. A *Baseline* simulation played out current development patterns and trends into the future; a *Future Framework* simulation implemented the idea of centres, corridors and green spaces. Rather than the simulations themselves, it was expected that the comparison between the two would make clear the differences between the current trajectory of regional change and the alternative future envisaged by NIPC staff.

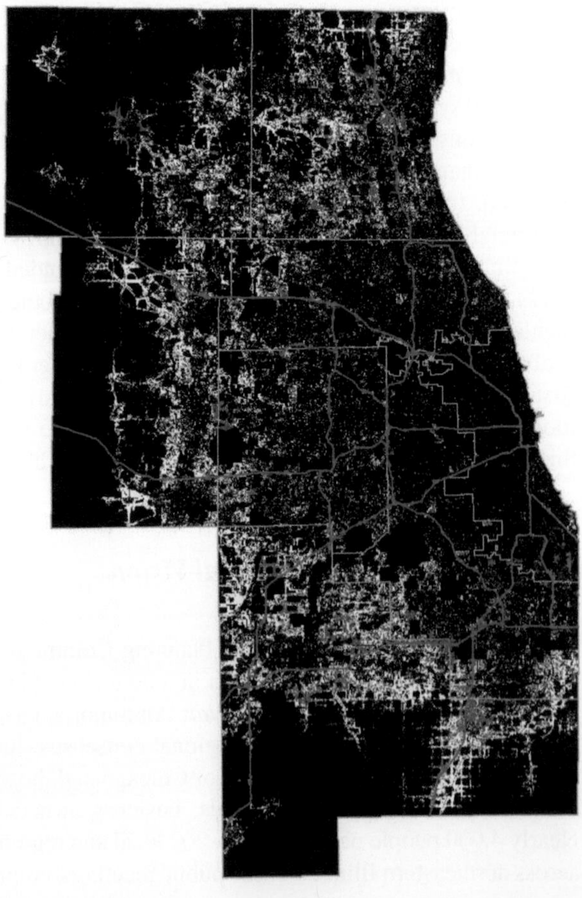

Fig. 2.1 Land-use change in the Chicago region: *Baseline* simulation (See also Plate 2 in the Colour Plate Section)

The typical *LEAM* protocol was followed. A preliminary set of simulations was created in about a month using a limited set of drivers and nationally available data sets. These simulations were reviewed with NIPC staff and some key problems with these simulations identified: for instance, development was occurring in unlikely places such as parks and recreation areas not identified in the national data. To remedy these kinds of problems, NIPC staff produced land-use and other data that better reflected current conditions. Model parameters were adjusted in *LEAM* to better reflect current development patterns in the Chicago region. A reasonable *Baseline* simulation was generated with these refinements to *LEAM*. Figure 2.1 is the map created to show the change in land use over the 40 year time period modelled. Figure 2.2 is a graph showing annual land-use change in each of the region's six counties; note that growth is neither linear nor necessarily sustained over time.

With the *Baseline* simulation as a reference, the *Future Framework* was simulated: places identified as population centres were made more attractive to residential development; places identified as jobs centres were made more attractive to industrial and commercial development; proposed transportation corridors made some parts of the region more attractive to future development than in the *Baseline* simulation; designated green areas were not available for development. Specific allocations to parts of the region were not made; rather various locations in the region had to compete for growth based on changes in the underlying drivers. This approach was developed through intense consultation with NIPC staff and others; it was not easily arrived at. Here too, periodic review with NIPC staff produced valuable insights: for instance, development was not occurring in brownfields as envisaged because the national data did not indicate that these areas were functionally obsolete.

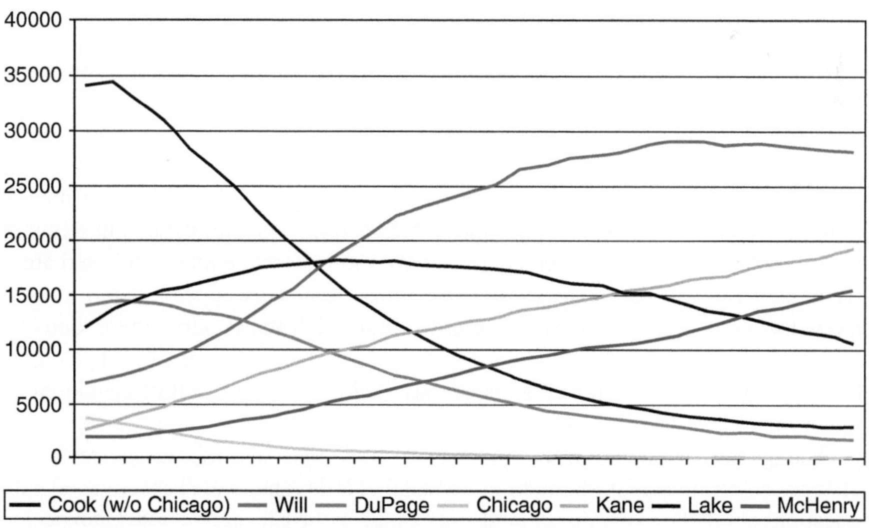

Fig. 2.2 Annual land consumption by county: *Baseline* simulation (See also Plate 3 in the Colour Plate Section)

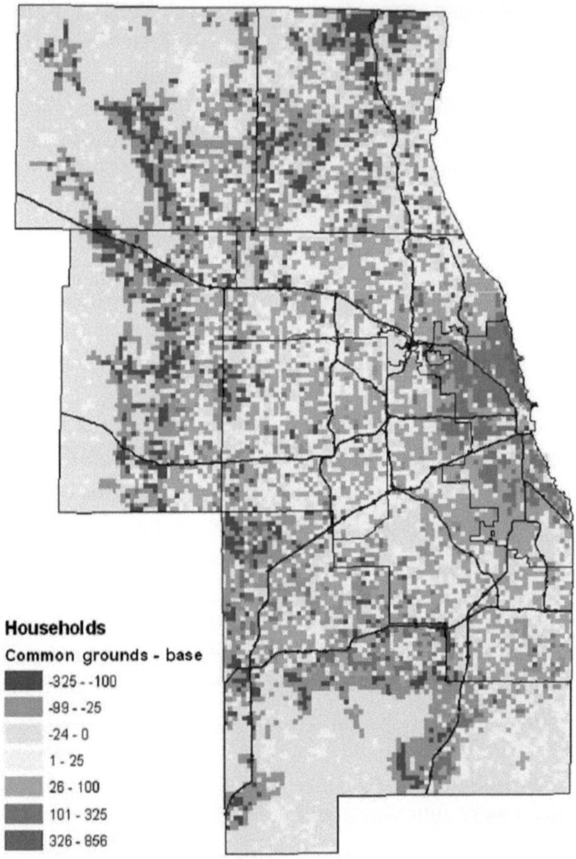

Fig. 2.3 Difference in household location between *Baseline* and *Future Framework* simulations (See also Plate 4 in the Colour Plate Section)

Comparing various aspects of the two simulations allowed planners to highlight some stark differences. Figure 2.3 shows differences between the two simulations in terms of location of new households. Green areas in the map have more *Baseline* households, while red areas have more *Future Framework* households. The *Future Framework* appears to pull residential development away from peripheral regions. Figure 2.4 shows the differences between the two simulations in terms of location of new jobs. Again, green areas have more *Baseline* jobs, while red areas have more *Future Framework* jobs. The *Future Framework* appears to pull jobs to a few locations in the inner ring of suburbs. Figure 2.5 shows agricultural land and unprotected open space lost in the two simulations. The difference varies by county, but all counties will see fewer acres of these lands lost as a result of development under the *Future Framework*.

The quick turnaround on the first set of simulations, and the weekly review of changes being effected in simulations, meant that NIPC staff were kept engaged with questions about the dynamics of land-use change in the Chicago region. Unexpected outcomes were not always attributable to shortcomings in the model or the underlying data. Rather, these outcomes challenged preconceived notions and facilitated

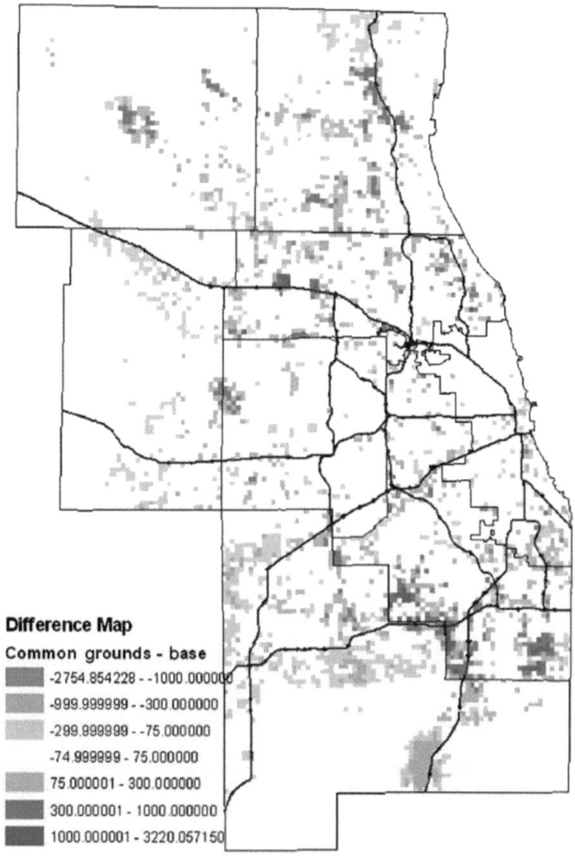

Fig. 2.4 Difference in job location between *Baseline* and *Future Framework* simulations

new insights about the region. For instance, most simulations showed more development in Lake County than NIPC staff expected. It emerged in discussions that Lake County has placed stringent restrictions on new development, and that was why the outcome appeared unexpected. The outcome brought to the surface some important

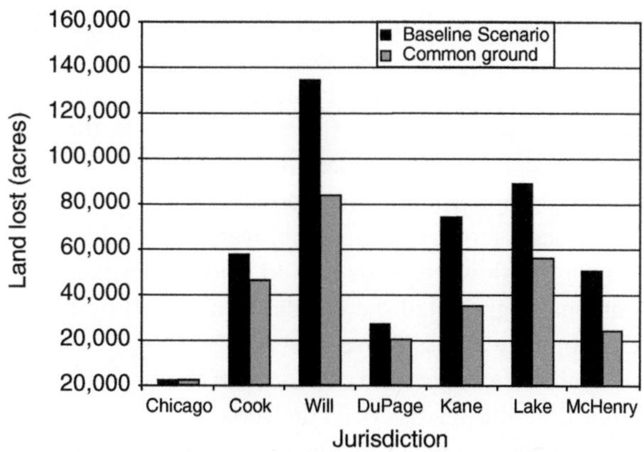

Fig. 2.5 Difference in consumption of agricultural land and open space by county

ideas that remained under the surface: Lake County is a very attractive place for development, and the restrictions placed in Lake County are pushing development to other locations in the region.

2.3.2 Assessing Transportation Investment Alternatives

The application of *LEAM* to the two-state, ten-county region around St. Louis, Missouri (MO), is a great example of a long-term, use-driven, embedded modelling effort with real implications for the process of planning in the region (Deal and Pallathucheril 2007). In 2003, the East-West Gateway Coordinating Council (EWGateway), the Metropolitan Planning Organization (MPO) and council of governments for the St. Louis region, began to use *LEAM* (in a version later called the *Blueprint Model*) as a platform for encouraging a regional dialogue on issues of economic development, social equity and environmental sustainability. Based on prior experience in the Peoria tri-county region (Deal and Pallathucheril 2003), instead of initializing the process with a lengthy model-building exercise, the initial focus was set on quickly producing a set of simulations. This quick-start process served two purposes: to quickly begin the process of engagement and build interest; and to collect information from the local stakeholders on the state of the local condition for model localization. These early simulations were subjected to public scrutiny in workshops, meetings and other public forums. Participants in these forums provided valuable insights into the dynamics of urban land-use change in the region and a direction for future modelling efforts. Conducted on an annual basis, they also provided an excellent platform for dialogue among participants.

One early critique of the preliminary *LEAM* simulations presented was aimed at the way in which new development was being distributed across the two sides of the Mississippi River – the Illinois side on the east, the Missouri side on the west. Preliminary simulations showed considerable new developments in Illinois relative to Missouri; at the same time, the central business district is in Missouri and has historically seen the bulk of new development. These simulations utilized posted travel speeds in some sub-models simulations were utilizing posted speeds and did not take into account the difficulty of crossing into the CBD. When observed, congested speeds were used to measure travel time (taking into account how traffic congestion makes portions of the region more or less attractive), simulated development shifted from the Illinois side to Missouri. A major factor was the effects of congestion on bridges and the approaches to them (bridges represent severe choke-points with very little opportunity for alternative routing). Figure 2.6 shows the change in an area's access to large-scale employers when bridges are crossed at (a) posted speeds versus (b) slower speeds due to traffic congestion. In the regional dialogue, this outcome highlighted the critical role played by bridges in the distribution of new development across the region.

To better connect the two sides of the region, the construction of a new Mississippi River bridge has been the subject of planning studies, preliminary design and environmental impact analysis for over 20 years. A concerted civic and political

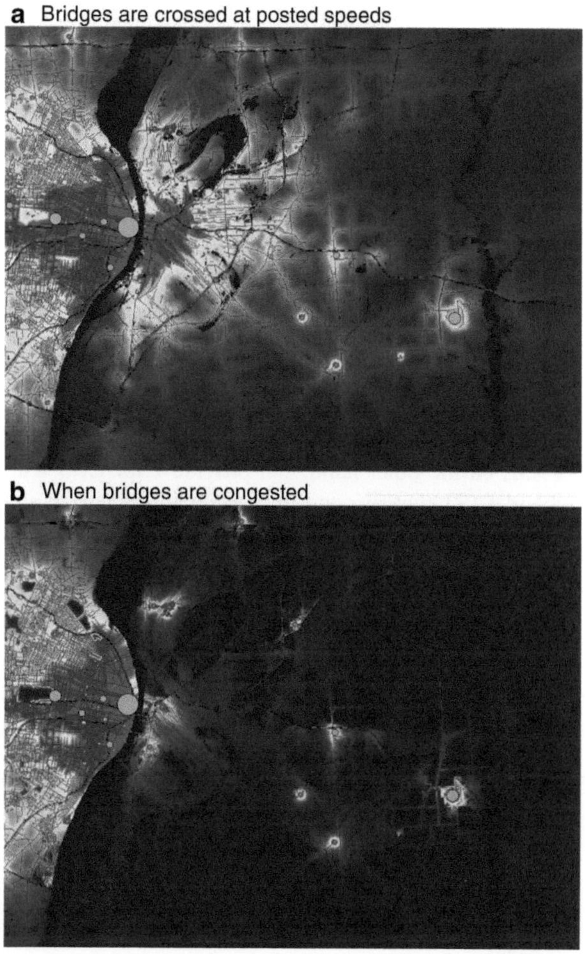

Fig. 2.6 The change in the attractiveness (red is high – blue is low) of large scale employers in the regions (*green dots*) when (a) bridges are crossed at posted speeds and (b) when they are congested (See also Plate 5 in the Colour Plate Section)

effort to secure earmarked federal funding was only partially successful. The resulting funding shortfall called into question the original bridge proposal and how it would be implemented. Alternatives considered included covering the shortfall with a toll and constructing less expensive alternatives such as enhancing the capacity of an existing bridge; there was no regional consensus on the way forward. Facing a stalemate on the issue, EWGateway took the lead and sought to inject an analytical basis into the regional debate. In order to do this, however, it became crucial to go beyond traditional cost-benefit analyses and to jointly simulate and analyze future transportation and land-use consequences of the different choices.

The desire for jointly modelling land use and transportation had emerged early in applying *LEAM* to this region. In the process of localization, future land-use changes were translated into changes in people and jobs across the region. The implications of these refinements presented EWGateway transportation planners and modellers with a more rigorous basis for determining future travel demand in the region. It allowed them to effortlessly and systematically assign future population, households and jobs to traffic analysis zones (TAZs), and to have this assignment reflect different future scenarios. The approach they previously employed evenly distributed new households and jobs across the region; *LEAM*-generated patterns were spatially varied and provided richer, more meaningful spatial data (in *LEAM* simulations, TAZs with new development can still lose population and jobs as a result of declining household size and increasing productivity).

With input from land use into the travel demand model in place, the next step was a loose coupling of *LEAM* and EWGateway's custom-built travel demand model, *TransEval*. Essentially, results from *TransEval* are used as inputs into *LEAM* which is run for several annual time steps, the resulting land use is the basis for socioeconomic inputs back into *TransEval*, and the process iterates until the simulation is complete. Each model helps address a limitation in the other. A travel demand model like *TransEval* can provide *LEAM* with indicators of travel conditions (congested speeds) that can continually change in response to changing land-use conditions. *LEAM*, on the other hand, can provide *TransEval* with changes in households and jobs in TAZs that respond to changes in the performance of the transportation network. This loose coupling is described in greater and more technical detail in Pallathucheril and Deal (2007).

Three simulations were created by coupling *LEAM* and *TransEval*. In all three simulations, *LEAM* was first run from 2000 to 2014, when construction of all the four alternatives is expected to be completed. At this point, the distribution of people and jobs from *LEAM* and different changes in the transportation network, representing each of the four alternatives being considered, were used as inputs in *TransEval*. Travel speeds on the transportation network estimated in *TransEval* as a consequence of these inputs showed differences across the alternatives. The resulting changes in travel speeds and the changes in demand for land (a function of the

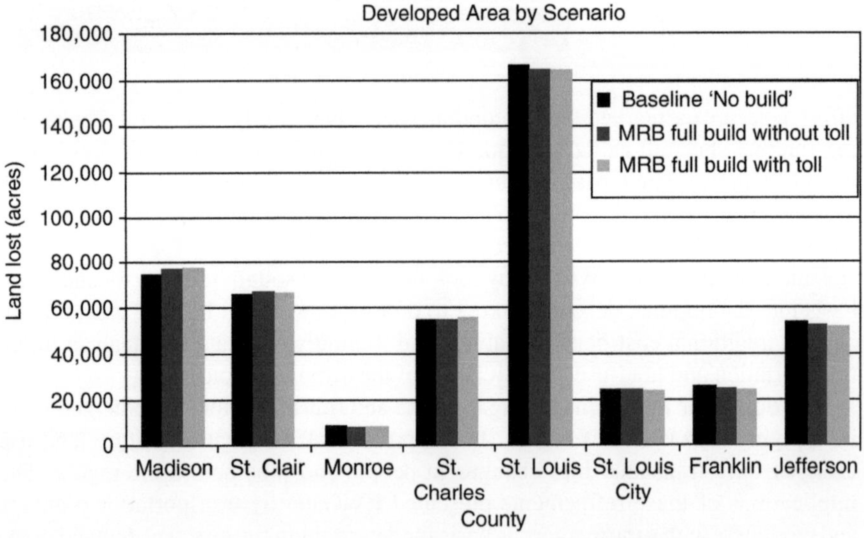

Fig. 2.7 Variations in land consumption, by county, across three simulations

impact of investment in each of the alternatives) were used as inputs in *LEAM* simulations of land-use change out to the year 2040. Differences between alternatives in terms of attractiveness of locations in terms of accessibility to jobs and urban amenities, and differences in the demand for land, produced different land-use outcomes in the year 2040.

The land-use, economic and transportation outcomes in the three simulations, and those of a baseline *No-Build* simulation, were the basis for informative comparisons. Figure 2.7 summarizes differences in land consumption by county among simulations associated with two alternatives (the original bridge design with and without imposing a toll) and the baseline *No-Build* simulation. Differences appear to be slight: building the bridge appears to slightly increase development in Madison and St. Clair counties and slightly decrease development in St. Louis and Jefferson counties; imposing a toll increases land development in St. Charles county. Figure 2.8 displays differences in land-use change between the *Full Build* and *No Build* simulations at a finer resolution; red cells see more growth in the *Full Build* simulation, green cells see more growth in the *No Build* simulation. The map presents a more complex set of differences and suggests that aggregating to the county level masks greater change: while building the bridge facilitates greater land development in the Illinois side of the region and takes away from development on the Missouri side of the river, there are significant differences in development at the local level.

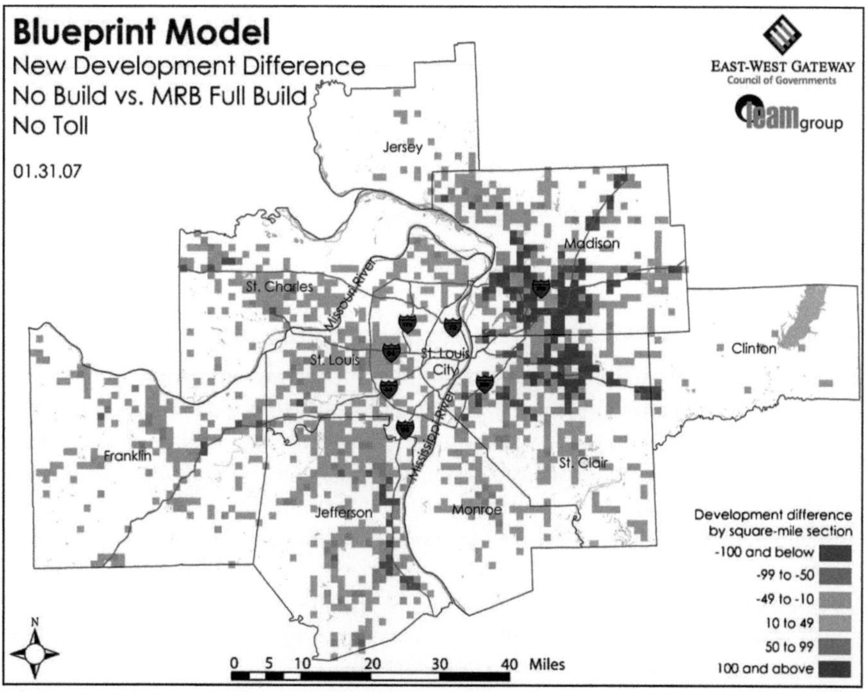

Fig. 2.8 Differences in land-use change between *Full Build* and *No Build* simulations (See also Plate 6 in the Colour Plate Section)

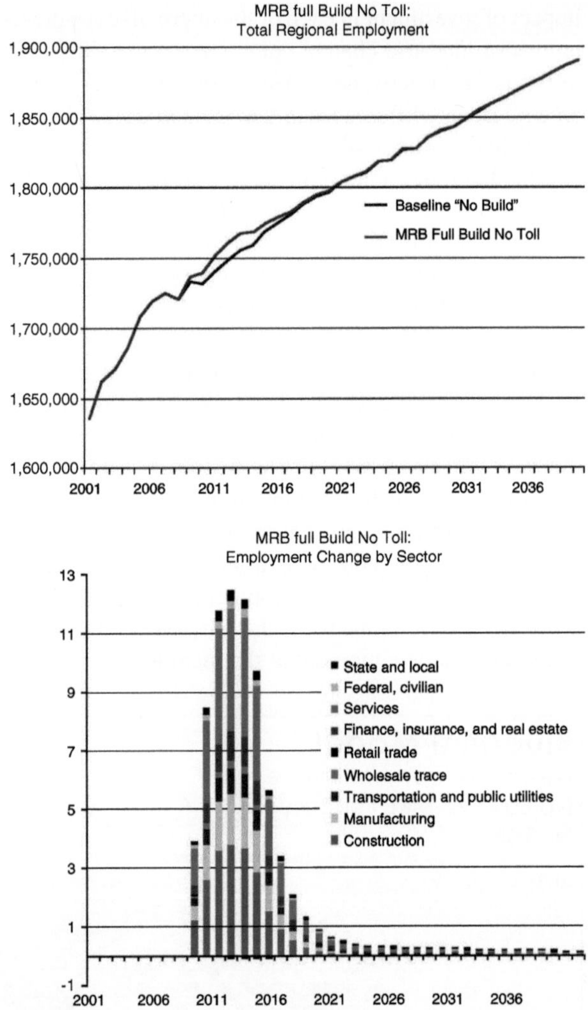

Fig. 2.9 Impact of the *Full Build* alternative on the regional economy

Differences in land-use change between the two simulations also result from greater amounts of land being consumed in the *Full Build* simulation as a result of expansion of the regional economy due to investment in bridge construction. This impact is summarized in Fig. 2.9. Beginning in 2009, the region will see an increase in jobs across different sectors as a result of bridge construction; after 2014, once the bridge is complete, the impact on the economy gradually attenuates.

Coupling the two models allows assessment of travel consequences of the different alternatives. Figure 2.10 presents a comparison of total time spent traveling from Missouri to Illinois (MOTOILVT) and Illinois to Missouri (ILTOMOVT); constructing the original bridge proposal and imposing a toll to cover the budget shortfall would increase travel times considerably.

As might be expected, discussions around these simulations and what to make of them were quite intense (and the regional dialogue is as yet unresolved). Some

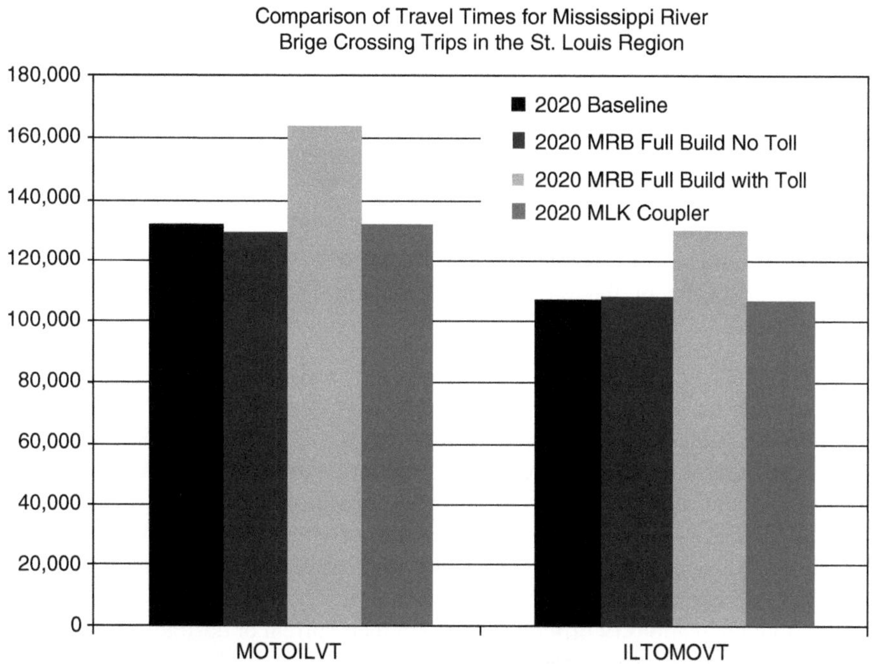

Fig. 2.10 Variations in total travel times across four simulations

outcomes appeared counter-intuitive; imposing a toll on the bridge increased total travel times in the region. Working through the complex interactions suggested a striking explanation; the toll was diverting traffic to the other bridges across the river that do not impose a toll, increasing congestion on these bridges, and increasing travel times. This explanation brought into question the wisdom of using a toll to cover the budget shortfall. There were other insights generated: patterns of land-use change are likely to change if additional river crossings are built; land-use policies and controls must be put in place in these areas to manage these impacts. Ultimately, however, only slight differences were uncovered even though the magnitude of the investment required for each of these alternatives is very different. This suggests that perhaps the lowest cost alternative is preferable, but it also suggests that demand-side tactics, such as investing in a better regional jobs-housing balance, might be more cost effective.

2.3.4 Planning Wildlife Corridors in an Urbanizing Region

The application of *LEAM* to the tri-county region around Peoria, IL, was one of the first involvements with a regional planning effort (Deal and Pallathucheril 2003). In 1999, regional planners from the three-county region surrounding Peoria and the Illinois Department of Natural Resources began studying the rapid land-use transformation in the region (Tri-County 2001). The concerns about transformation ranged from unplanned growth and natural resource depletion to low-density development and traffic congestion. The planners gathered spatial data, analyzed regional growth and apparent trends, and compared their data with historic spatial information. The group then generated explanations for the patterns that emerged from the analyses, and also developed general ideas about development impacts. However, a number of central questions remained unanswered for the group, namely:

- What should be done next with the information?
- What are some of the future impacts of current or alternative policies and decisions?

Through a collaborative process involving university researchers, state and local officials, and local stakeholder representatives, a version of *LEAM* was refined specifically for local application to the Peoria region. Scenarios were developed that described current and alternative policies and investment choices. Outcomes were used to visually examine and understand the spread of development in the region as well as probable environmental, social, and economic impacts of the different scenarios. An opportunity to leverage this work presented itself when the region sought to designate and protect wildlife corridors. The loss of habitat suitable for supporting populations of different species at their current or desirable levels is one particularly undesirable consequence of urban development. It is not so much that the amount of habitat is reduced; rather, it is change in both the amount and the spatial configuration of habitat. The same amount of habitat distributed across space in very small patches will likely support a smaller population.

LEAM simulations were used to study the threat from urban development to bobcat and wood thrush populations in the region. (The underlying technical basis for this work is available in Aurambout *et al.* 2005.) The advisory committee charged with directing the project considered the question of the species on which the study should focus. In the interest of keeping the number of species to a manageable number, the committee decided to focus on the bobcat and wood thrush populations. By choosing these two species the committee expected to cover the habitat suitable for supporting many of the other species found in the region. The threat to the bobcat and wood thrush populations from urban development was assessed in two steps. First, the amount and location of habitat suitable for each species was assessed using land-cover data from the year 2000. Second, this spatial distribution was assessed against land-use change as simulated up until 2030 using *LEAM*. Several *LEAM* simulations were used: three economic futures (business as usual, economic decline, and a high economic growth scenario), and high growth combined with different policies (agricultural land preservation, river bluff protection and limiting growth to areas currently served by infrastructure).

The highest decrease in suitable habitat, and the biggest impact on the bobcat population, occurs when high growth is combined with restricting future growth to areas currently served by infrastructure known as facility planning areas (FPAs). This is termed the *High Growth Contained* simulation. While the impact of high growth was only to be expected, the impact of restricting growth to FPAs was not.

Table 2.1 Total area of suitable core habitat (hectares)

Core habitat	Initial condition	Business as usual	Economic decline	Redirected	High growth Bluff protect	Contained
High home range	18,441	17,500	18,352	17,327	17,251	16,790
Average home range	24,086	23,701	23,979	23,459	23,350	22,931

Discussions of this surprising outcome with the advisory committee and local planners revealed that FPAs contain the largest forest patches most suitable as bobcat habitat, and brought into question the original basis for laying out FPAs.

The dynamic spatial model underlying this analysis allowed the sensitivity of these findings to be assessed. We looked in particular at the assumption of habitat requirement and considered two different assumptions: (i) high habitat requirements; and (ii) average habitat requirements. Table 2.1 describes the amount of suitable core habitat available across several simulations. As mentioned earlier, the *High Growth Contained* simulation results in the greatest reduction in core habitat. The underlying model developed for this analysis also allowed the population of breeding females to be estimated based on data concerning female bobcat territoriality. Table 2.2 shows that changes in land-use policies could potentially affect 10–15 per cent of the total potential bobcat population. Although a small sample set might skew the results, this may be a significant effect when dealing with threatened populations.

Figure 2.11 displays the areas of habitat lost in the *High Growth Contained* simulation. Lighter shaded areas on the map are habitat patches that remain after land-use change; darker shaded areas represent original habitat patch that is lost as a result of land development. Loss of habitat occurs in significant amounts to the north and south of the city of Peoria; there is a small patch of loss to the west of the city. In Fig. 2.12, we zoom in on the area north of the city: mid grey denotes bobcat habitat, lighter grey denotes existing developed land, and darker grey denotes land likely to develop in the *High Growth Contained* simulation.

Table 2.2 Potential number of bobcats hosted within home ranges (based on female territoriality)

Core habitat	Initial condition	Business as usual	Economic decline	Redirected	High growth Bluff protect	Contained
High home range	30	29	30	29	29	29
Average home range	35	34	34	33	33	33

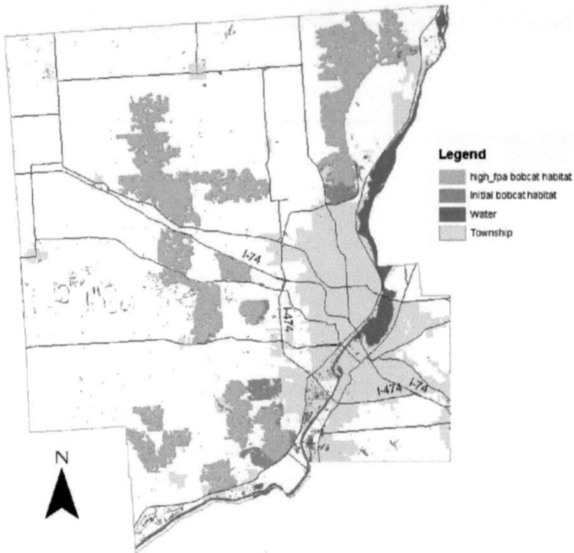

Fig. 2.11 Bobcat habitat lost (darker shaded areas)

Extending this analysis, we also overlaid habitat patches lost under all scenarios. In this way, we were able to identify patches that were likely to be lost in more than one simulation. Using this information, we were able to identify high-priority patches likely to be lost in all simulations as well as those that were threatened in fewer simulations. A number of patches that were not threatened in any simulation likely would require minimal attention. Using this information, conservation and land acquisition were prioritized in areas that are likely to be most threatened and would also contribute the most to connecting and expanding existing patches.

Future consumption of land for development reduces habitat suitable for supporting populations of native and/or threatened species. In some instances, the

Fig. 2.12 Detail of bobcat habitat lost

amount of habitat lost may not be as critical as where these losses occur. The spatial configuration of bobcat habitat is critical to its survival. High-quality patches, when surrounded by urban development that is difficult to traverse, might limit the genetic diversity of the local bobcat population. We were able to demonstrate to local stakeholders the value in using spatially explicit analyses. With *LEAM* simulations, they were able to visually see and participate in dialogue on potential policy levers and investments that might limit the fragmentation of bobcat habitat in the region. This led to discussions of implementing low-impact developments (LID) in specific parts of the region with protected corridors and greenways to enhance and connect high-quality habitat and limit its loss (Hopkins 2005).

2.4 Some Lessons Learned

Two of the key lessons learned in integrating *LEAM* with regional and local planning processes:

- *LEAM* is unlikely to succeed as a standardized, off-the-shelf product, and is mainly useful when customized as a localized application for the specific region of interest.
- Discussions about simulation results, including deficiencies, add considerable value to regional planning dialogue. Therefore, even 'quick and dirty' results can be very useful when planners, stakeholder representatives and decision makers work cooperatively to interpret them through the lens of expertise and local perspective.

Simulation results become an important part of the dialogue on transportation networks and other investment strategies. Simulation results offer a rich set of data that describe important relationships between growth pressure and investment choices. In places where growth pressure already exists, infrastructure investments can cause dramatic swings in growth (usually positive if not regulated). In places where the development pressures are not in place, infrastructure investments appear to have little economic growth benefits. Upon reflection this can be anecdotally surmised. If an interstate ramp is constructed in a rural area with low growth pressures, development from the investment may be limited to gas stations and other small, auto trip related developments. If an interstate ramp is constructed in a place with high growth pressures however, development as a result of the ramp investment can have large implications (perhaps like removing one's finger from a dike).

Using *LEAM* to test planning policy and investment decisions requires that local decision makers accept modelling, and more specifically the *LEAM* model, as a valid tool for understanding the local condition. The process of gaining acceptance and of becoming an integral part of local and regional planning evolves with time and familiarity (Deal and Pallathucheril 2007). A use-driven, embedded approach to local model development suits this evolutionary process. Our experiences in embedded modelling, as described in the policy questions presented, reveal some key lessons for operationalizing components of a regional PSS.

- There is value in embedding model and system development in real-world planning and policy making. Models are typically built by experts who work behind the scenes. By modelling in the public eye, the technical and analytical choices being made are subject to intense and valuable scrutiny that can foster improvements. Black-box approaches have limited value under these circumstances.
- The fundamental value of building models and PSS is that they foster thought experiments, the ability to test important investment and policy choices and discuss their potential future outcomes.
- There is a need to engage in the relevant dialogue among regional stakeholders very quickly. Even preliminary results, if they represent a tangible basis for discussion, not only enhance the regional dialogue but also help kick-start the building of a PSS.
- Some of the effects of embedding in planning and policy making may appear small but these are nonetheless important. An intervention may only reveal that further analysis is required, but doing so may slow the rush towards a poor choice.
- The process of model and system building is just as important as — if not more so than — the model itself. As a result, the precision or rigour of a model may be less of an issue than its ability to foster dialogue and produce useful results.
- There is value in being able to consider multiple futures rather than having to choose a single desirable future condition. We have argued this elsewhere (Deal and Pallathucheril 2007a). PSS should provide data and metrics that help stakeholders discern important differences among these multiple futures.

2.5 Conclusions and Future Efforts

Large-scale urban simulation models are being developed that enable planners and stakeholders to view and assess the future outcomes of current policies, and investment choices before they are put into action. But unlike our improving ability to create these complex planning support systems, the process of putting them into practice, gaining acceptance and becoming a part of local planning has been more difficult. We have argued here, through the use of examples from our experiences using *LEAM,* that a use-driven, embedded approach to local model development can help to speed up and improve this process. These applications revealed some key lessons for developing and implementing a regional PSS, including the fact that early and continued engagement and feedback are critical; and that embedding model development in real-world planning and policy questions is important for shaping tools that user groups will implement. We also found that the process of model building is just as important as the model itself because it provides insight and dialogue that may not otherwise take place; and that in an increasingly chaotic world, the ability to produce, analyze and discuss multiple futures is critical for effective decision making.

Better tools are needed to manage regional dynamics, not just as economic systems or static inventories of resources, but as complex systems that are part of regional and global networks (Campbell 1996). Effective decision making requires that we understand the systems to be administered and that we understand the future implications of our proposed strategies. We have attempted here to outline an approach for understanding the dynamics of urban systems and the potential implications of urban policy and investment management decisions. We described one modelling approach – *LEAM* – that utilizes technological advances in spatial simulation modelling to help improve a region's ability to make sound decisions. *LEAM* was intended to enable users to deal with stochastic influences and view the reported probable consequences of intended events in a scenario-based format that is comprehensible by local experts, decision makers and stakeholders.

Successful use of tools like *LEAM* requires that local decision makers accept the modelled outcomes as reasonable. The process of establishing 'reasonableness' and gaining acceptance unfolds over time. We believe that use-driven, embedded approaches will provide the basis for crafting a unique, local view of regional development problems, issues, and processes; that achievement, in turn, will be valuable both in gaining the necessary stakeholder acceptance and in promoting the effective implementation of policies based on consensus.

LEAM is a product of an academic laboratory. As such, a purpose of *LEAM* is to support the laboratory's underlying mission of advancing the science of modelling and the practice of planning. Ultimately, such advances must be verified in the 'laboratory of the real world', where planning decisions actually matter and have consequences for governments, regional planning authorities and their diverse constituents. That verification process necessarily involves ongoing relationships with end-users and repeated applications in multiple locations. The process is essential, but in itself it does not promote further advances in the state of the art. Lessons learned from real-world applications provide crucial input to further investigations and advances in the *LEAM* Laboratory. Progress may be encumbered when either phase of research and development is overemphasized.

In order to better exploit synergies between our basic research and applications, we have established an associated private-sector entity whose purpose is develop new, straightforward localized applications; provide technical support; and sustain long-term relationships with established *LEAM* users. At the same time, the purview of the *LEAM* Laboratory continues to be fundamental questions of urban land-use change and its implications. The laboratory concerns itself with advanced theoretical topics, emerging simulation modelling technologies, and new ideas about the potential of regional planning for helping to foresee and solve development problems before they become stubborn realities. This synergistic relationship requires considerable coordination and oversight, but it provides a positive and dynamic environment for future advances and problem solving.

Our embedded, use-driven approach to developing *LEAM* points to the need to address the following topics in future research:

- informing dynamic, continuous planning in other domains (such as watershed planning) based on the experience with embedding a PSS in regional planning;
- operationalizing and modelling the feedback from land-use change to regional economic change;

- developing a generic specification and framework for loosely coupling models based on the experience in coupling land-use and transportation models;
- supporting easy extraction of knowledge from burgeoning data sets that contain growing suites of land-use simulations of increasing complexity and resolution;
- leveraging the World Wide Web and other Internet technologies to democratize access to and enhance usability of the *LEAM* PSS; and
- deploying *LEAM* effectively on user desktops.

References

Aurambout, J.P., Endress, A.G. and Deal, B. (2005) A spatial model to estimate habitat fragmentation and its consequences on long-term persistence of animal populations, *Environmental Monitoring and Assessment,* 109(1–3): 199–225.

Batty, M. and Xie, Y. (1994) From cells to cities, *Environment and Planning B*, 21: 31–38.

Brail, R.K. and Klosterman, R.E. (2001) *Planning Support Systems: Integrating Geographic Information Systems, Models and Visualization Tools,* ESRI Press Redlands, California.

Choi, W. and Deal, B. (2008) Assessing hydrological impact of potential land use change through hydrological and land use change modeling for the Kishwaukee River Basin, *The Journal of Environmental Management,* 88(4): 1119–1130.

Deal, B. (2001) Ecological urban dynamics: the convergence of spatial modeling and sustainability, *The Journal of Building Research and Information*, 29(5): 381–393.

Deal, B. (2003) Sustainable land-use planning: the integration of process and technology, PhD Thesis, University of Illinois at Urbana-Champaign.

Deal, B. and Pallathucheril, V. (2003) The Land Evolution and impact Assessment Model (LEAM): will it play in Peoria? In *Proceedings of the 8th International Conference on Computers in Urban Planning and Urban Management*, Sendai, May 27–29.

Deal, B. and Pallathucheril, V. (2007) Developing and using scenarios, In Hopkins, L.D. and Zapata, M.A. (eds.) *Engaging the Future: Forecasts, Scenarios, Plans, and Projects*, Lincoln Institute for Land Policy, Cambridge Massachusetts, pp. 221–242.

Deal, B. and Schunk, D. (2004) Spatial dynamic modeling and urban land use transformation: a simulation approach to assessing the costs of urban sprawl, *The Journal of Ecological Economics*, 51(1–2): 79–95.

Deal, B. and Sun, Z. (2006) A spatially explicit urban simulation model: Landuse Evolution and Impact Assessment Model (LEAM), In Ruth, M. (ed.) *Smart Growth and Climate Change: Regional Development, Infrastructure and Adaptation*, Edward Elgar, Inc., Northampton, pp. 181–203.

Deal, B. Farrello, C., Lancaster, M., Kompare, T. and Hannon, B. (2000) A dynamic model of the spatial spread of an infectious disease, *Environmental Modeling and Assessment,* 5(1): 47–62.

Engelen, G., White, R., Uljee, I. and Drazan, P. (1995) Using cellular-automata for integrated modeling of socio-environmental systems, *Environmental Monitoring and Assessment*, 34: 203–214.

Geertman, S. and Stillwell, J. (2003) *Planning Support Systems in Practice*, Springer, New York.

Goel, A.K. (1989) Integration of case-based reasoning and model-based reasoning for adaptive design problem solving, PhD Thesis, The Ohio State University, Columbus.

Hopkins, L.D. (1999) Structure of a planning support system for urban development, *Environment and Planning B*, 26: 333–343.

Hopkins, E. (2005) Study identifies Peoria-Area wildlife corridors, *Peoria Journal Star*, September 12.

Kim, T.J., Wiggins, L.L. and Wright, J.R. (eds.) (1990) *Expert Systems: Applications to Urban Planning*, Springer-Verlag, New York.

Klosterman, R.E. (1999) The *What If?* collaborative planning support system, *Environment and Planning B*, 26: 393–408.

Pallathucheril, V. and Deal, B. (2007a) Coupled land use and transportation models: the LEAM/ TransEval experience in St Louis, MO, In *Proceedings of the 12th International Conference on Computers in Urban Planning and Urban Management*, Fas DeGuassu, Brazil, May 11–15.
Pallathucheril, V. and Deal, B. (2007b) Regional churning and land-use change, Paper presented at the 48th Annual Conference of the Association of Collegiate Schools of Planning, Milwaukee, Wisconsin, October.
Pettit, C.J. (2005) Use of a collaborative GIS-based planning-support system to assist in formulating a sustainable-development scenario for Hervey Bay, Australia. *Environment and Planning B*, 32: 523–545.
Sarraf, S., Pallathucheril, V.G., Donaghy, K. and Deal, B. (2005) Modeling the regional economy to drive land-use change models, Paper presented at the 46th Annual Conference of the Association of Collegiate Schools of Planning, Kansas City, Missouri, November.
Sun, Z. and Deal, B. (2006) Managing the dynamics of geographic information systems: the case of urban land use transformation in St Louis, MO, In Brady, S.R., Sinha, A.K. and Gunderson, L.C. (eds.) *Geoinformatics 2006: The U.S. Geological Survey Scientific Investigators Report*, USGS, Washington, DC.
Tri-County Regional Planning Commission. (2001) *The Peoria-Pekin Future Landscape Project*, State of Illinois, Department of Natural Resources Report, Springfield.
Waddell, P. (2002) *UrbanSim* – Modeling urban development for land use, transportation, and environmental planning, *Journal of the American Planning Association*, 68: 297–314.
Wang, Y., Choi, W. and Deal, B. (2005) Long-term impacts of land-use change on non-point source pollutant loads for the St. Louis Metropolitan Area, *The Journal of Environmental Management*, 35(2): 194–205.
White, R. and Engelen, G. (1997) Cellular automata as the basis of integrated dynamic regional modeling, *Environment and Planning B*, 24: 235–246.
Wilkins, D.E. and desJardins, M. (2000) A call for knowledge-based planning, In *Proceedings of AIPS Workshop on Analyzing and Exploiting Domain Knowledge for Efficient Planning*, Breckenridge, Colorado.
Wu, F. and Webster, C.J. (1998) Simulation of land development through the integration of cellular automata and multicriteria evaluation, *Environment and Planning B*, 25: 103–126.
Wu, F. and Martin, D. (2002) Urban expansion simulation of Southeast England using population surface modeling and cellular automata, *Environment and Planning A*, 34: 1855–1876.

Additional Reading

Albeverio, S., Andrey, D., Giordano, P. and Vancheri, A. (2008) *The Dynamics of Complex Urban Systems: An Interdisciplinary Approach,* Physica-Verlag, Heidelberg.
Batty, M. and Longley, P. (1994) *Fractal Cities: A Geometry of Form and Function,* Academic Press, San Diego.
Hopkins, L.D. and Zapata, M.A. (eds.) (2007) *Engaging the Future: Forecasts, Scenarios, Plans, and Projects*, Lincoln Institute, Cambridge, Massachusetts.
Maguire, D.J., Batty, M. and Goodchild, M.F. (2005) *GIS, Spatial Analysis, and Modeling,* ESRI Press, Redlands, California.
Ruth, M. (2007) *Smart Growth and Climate Change: Regional Development, Infrastructure and Adaptation*. Northampton, Massachusetts: Edward Elgar, Inc.
Torrens, P.M. and O'Sullivan, D. (2001) Cellular automata and urban simulation: Where do we go from here? *Environment and Planning B*, 28: 163–168.
Wolfram, S. (2002) *A New Kind of Science,* Wolfram Media, Champaign, IL.
Wu, F.L. and Webster, C.J. (2000) Simulating artificial cities in a GIS environment: urban growth under alternative regulation regimes, *International Journal of Geographical Information Science*, 14: 625–648.

Additional Readings

Chapter 3
Future Land-use in The Netherlands: Evaluation of the National Spatial Strategy

Ton C. M. de Nijs

3.1 Introduction

The Netherlands is a small but densely populated country with about 16 million people living in an area of 40,000 km^2. The 'Randstad' is a highly urbanized area of the country that comprises the major cities of Amsterdam, Rotterdam, Utrecht and The Hague, with a combined population of more than 5 million people. This conurbation is the economic heart of the Netherlands. Urban development threatens the natural and historical qualities of the landscape of the 'Green Heart', the central open area of this region. The first spatial plans for this area date from 1958 (RPD 1960; WWdL 1958), and these have been updated regularly since then (RPD 1966, 1977, 1988, 1994). New plans have been drawn up for the creation of about 2500 km^2 of forests and nature reserves by 2018 (LNV 2002; MNP 2002a). In addition, roughly 1500 km^2 of land will be needed for new residential and industrial areas by 2030 (ABF 2002). According to current policy, plans and trends, 10 per cent of the land use will change over the next 30 years. The National Spatial Strategy (VROM *et al.* 2005), drawn up by the present Dutch Government and prepared to provide a spatial policy framework to guide all of these developments, was adopted by Parliament at the end of 2005.

Roughly speaking, the National Spatial Strategy comprises three major categories of spatial planning policies. These concern:

- Restrictive areas, where further urbanisation will be curbed in the interests of a range of objectives; these areas include the nature reserves and protected areas and the risk zones around Schiphol Airport and other centres of hazardous activities;
- National Landscapes, where development is possible within these areas as long as the core qualities of the landscape are conserved or enhanced; within these

T. C. M. de Nijs (✉)
National Institute of Public Health & Environment
P.O. Box 1, 3720 BA Bilthoven, The Netherlands
E-mail: ton.de.nijs@rivm.nl

S. Geertman, J. Stillwell (eds.), *Planning Support Systems Best Practice and New Methods*,
DOI 10.1007/978-1-4020-8952-7_3, © Springer Science+Business Media B.V. 2009

landscapes, local authorities may only build new housing to meet the demand resulting from local population growth;
- Concentration areas, where new urban development will be clustered.

By order of the Dutch Ministry of Housing, Spatial Planning and Environment, the National Environmental Assessment Agency has evaluated the effects of this new National Spatial Strategy, giving particular attention to a number of questions. How will the National Spatial Strategy affect future land-use patterns in the Netherlands? Where, despite restrictions, is the pressure so high that new developments will ultimately be allowed to occur? What effect will the Strategy have on national landscapes? Will there be sufficient room in the concentration areas to accommodate all new developments? To answer these questions, future land-use developments have been simulated by the *Environment Explorer* in accordance with a 'Trend' scenario and a 'Sprawl' scenario (A. C. M. De Nijs *et al.* 2005; T. C. M. De Nijs *et al.* 2004). Limiting urban spatial sprawl is one of the main political topics in the Netherlands.

This chapter takes a closer look at the nature of the land-use model, at the Trend and Sprawl scenarios and at future land-use developments in general and restrictive areas, national landscapes and concentration areas in particular. The results of this study are discussed and conclusions are drawn that relate to the impacts of the policies, potential problems and the application of the land-use model itself. In closing, the effect of the study itself on the discussion of spatial developments in the Netherlands is discussed.

3.2 *Environment Explorer*

The *Environment Explorer* (A. C. M. De Nijs *et al.* 2001; T. C. M. De Nijs *et al.* 2004; Engelen *et al.* 2003; MNP 2005; White *et al.* 1997), a land-use simulation model, was used to simulate future land use in the Netherlands. This model attempts to plot regional spatial developments as accurately as possible on a land-use map with a grid size of 500 m. The model recognises 16 different land-use classes, including residential use, industrial use, offices, facilities, recreation, greenhouse horticulture, forest and nature conservation. The location of land-use change is determined by current land use, spatial policies, suitability, accessibility and the effect of current land uses in the neighbourhood (Fig. 3.1). The amount of land-use development is controlled in the scenarios. Zoning, suitability and accessibility maps have been compiled for each land-use category. The zoning map shows the areas where development is permitted and where it is prohibited. The suitability map shows how suitable each cell is for that particular land use. Accessibility by car and public transport is calculated by a dynamic traffic model in the *Environment Explorer*. This model computes changes in accessibility given the changes in land use, population, employment and infrastructure growth (RIKS 2002).

The neighbourhood potential is the influence exerted by the surrounding cells on the allocation of the various land uses. This neighbourhood potential is calculated

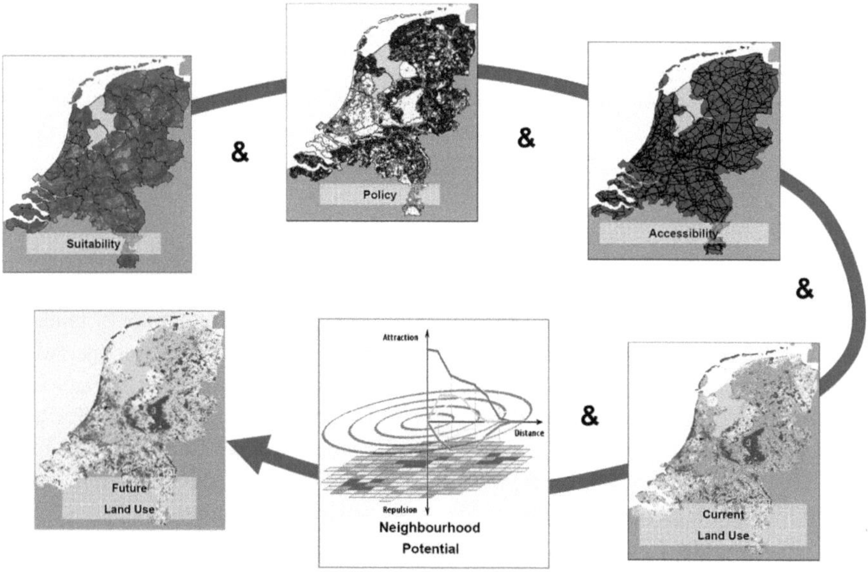

Fig. 3.1 The *Environment Explorer* simulates future land use as a function of the suitability of each area, the relevant spatial planning policies, accessibility, current land use and the influence of the neighbourhood

by a cellular automata (CA) model in which a set of transition rules describes how the various land uses interact – by either attracting or repelling each other. For example, over a short distance, airfields and industrial sites can have a negative impact on residential development, but the latter in turn is positively influenced by the proximity of greenspace. A set of transition rules determines the spatial interactions between each combination of land use over a number of distances. Where the various types of development will take place is relatively uncertain. This uncertainty is built into the allocation model in the form of a stochastic parameter, and it can be visualised in a probability map of future land use by running the model not once but 1,000 times in a Monte-Carlo simulation.

For the evaluation of the National Spatial Strategy, the spatial interactions between the various land uses in the *Environment Explorer* were calibrated according to the observed spatial developments over the period 1989–1996 (MNP 2005). Therefore, a semi-automatic calibration routine was added to the model with an 'error selection and correction' routine (Hagen-Zanker 2005b). The criterion to select the combination of land uses that gives the largest error is based on the Fuzzy Kappa statistic (Hagen 2003). The selected error is reduced by adjusting the transition rules, and errors that cannot be resolved are placed on a taboo list. This procedure iterates until the model error is minimised.

Analyses of the observed land-use changes in the period 1989–1996 revealed that 25 per cent of the changes can be considered to have been improbable. These unlikely land-use changes consist of changes from urban to agricultural uses or forest and nature conservation, and they can be attributed to changes in the defini-

Table 3.1 Results of the calibration and validation of the *Environment Explorer* based on the Fuzzy Kappa

	Environment Explorer	Random
Calibration 1989–1996	0.936	0.926
Validation 1996–2000	0.913	0.922

tions of the land-use classes over the years. For example, the grass bordering the roads is included in category roads in the map of 1989, while it is excluded from the map of 1996. In general, this is a major problem in the development of land-use models (Fang *et al.* 2006). The model was subsequently validated for the period 1996–2000, based on the Fuzzy Kappa. The results indicate that the *Environment Explorer* performed better during the calibration period than a random allocation model (Visser *et al.* 2006), whereas during the validation period it did not (Table 3.1). Two factors can mostly explain this performance: (1) the validation period was rather short; (2) further changes in the definitions of the land-use classes in the map of 2000. These latter changes have a number of effects, among which is a net decrease in residential area in the map of 2000 compared to the map of 1996, which can not be reproduced by the model.

The validation results over the short term are not so important; in contrast, the simulation results for the long term are very important. Nevertheless, it is impossible to verify these long-term simulations on observed land-use changes. These simulated land-use maps should be realistic, transparent and easy to explain, but it is not that easy to develop objective criteria to evaluate these characteristics. The first question that arises is just what are realistic land-use patterns. At the very least they should all have the same morphological characteristics. One very tight morphological constraint on models of urban growth is Zipf's law (Gabaix 1999; Gabaix *et al.* 2003; Zipf 1949). Zipf's law for cities is one of the most striking empirical facts in geography and economics. For most countries, the size distribution of cities strikingly fits a power law: the number of cities with populations greater than S is proportional to 1/S.

Therefore, we verified the long-term spatial developments of the model against the size distribution of urban clusters. The stochastic parameter in the *Environment Explorer*, which determines the number of new clusters over the long term, was used to accurately fine-tune this relation (Hagen-Zanker 2005a). The resulting size distributions for both the Trend and Sprawl scenarios in 2030 obey Zipf's Law (Fig. 3.2), and both approximately overlap with the city size distribution of observed land use in 1989 and 2000. The size distribution of the random allocation model does not obey Zipf's Law as a large number of small urban clusters have been allocated randomly on the map. The relative percentage of 50-ha clusters has increased from 60 per cent to nearly 80 per cent. In the long term, the *Environment Explorer* outperforms the random allocation model.

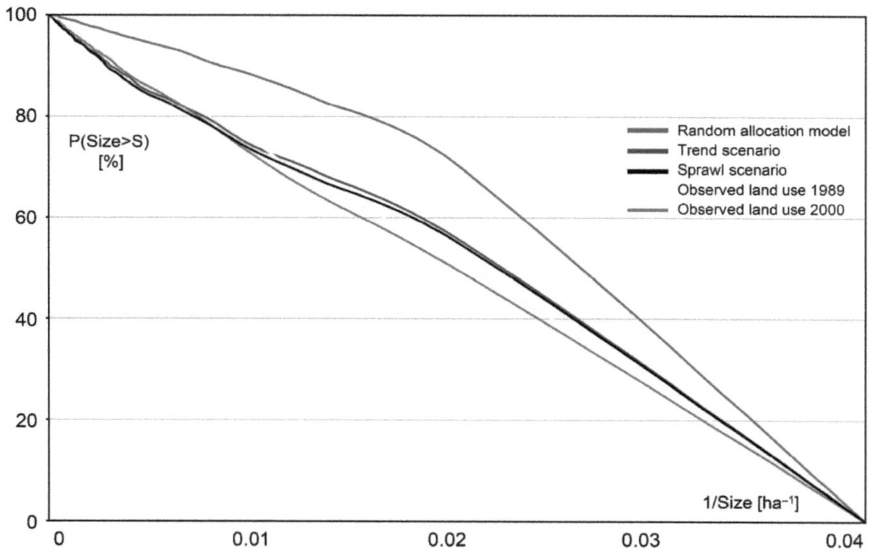

Fig. 3.2 Size distributions of urban clusters according to Zipf's Law for the observed land use in 1989 and 2000, the simulated land use according to the two scenarios and the random allocation model in 2030

3.3 The Trend and Sprawl Scenario

Although the names suggest otherwise, the Trend and Sprawl scenarios are almost identical. There is only one essential difference between both scenarios: smaller settlements will grow relatively faster in the Sprawl scenario than in the Trend scenario. Why and how this has been implemented will be described at the end of this section. Here, the common aspect will be discussed. Implementation of a scenario in the *Environment Explorer* includes the definition of:

- initial land use;
- amount of land-use developments;
- zoning, suitability and accessibility maps.

The spatial planning policies contained in the National Spatial Strategy have to be translated into either the input of these scenarios or the calibrated set of transition rules. In both scenarios, initial land use is based on the land-use/cover map from the *Bodemstatistiek 2000* (CBS 2003). Agricultural land use on this map is further broken down into 'grassland', 'arable' and 'other agricultural land' using a second land-cover map of the Netherlands, *Landgebruikskaart Nederland, 2000* (Alterra 2003). Residential land use is split into two categories based on the number of inhabitants: 'low-density population' and 'high-density population'.

The amount or quantity of land-use developments can be defined regionally in the scenarios of the *Environment Explorer*. For both scenarios, the growth of

Table 3.2 National targets per land-use category for the period 2000–2030

Land use	Target (ha)	Land use	Target (ha)
Residential	87,675		
Industrial	35,975	Recreation	17,625
Services	6,225	Forest	51,525
Social/Cultural	13,450	Natural grasslands	173,025
Greenhouse horticulture	5,425	Nature conservation	11,300

residential and employment uses is based on the so-called High Land-use Pressure Trend (HLPT) scenario (ABF 2002). The growth of recreational uses and greenhouse horticulture was derived by extrapolating the regional developments from 1989 to 2000. Regional developments in forest, nature and natural grassland areas are based on the 'Nature Reference Map 2020' (*Referentiebeeld Natuur 2020*) (Goetgeluk *et al.* 2000). Table 3.2 summarises the national targets for each land use for the period 2000–2030.

The same suitability maps have been used in both scenarios. These are based largely on a study (Verburg *et al.* 2004a, b) that shows that soil conditions are not important in determining residential, employment and recreational land uses. The suitability maps for forest and nature conservation were also taken from this study. The suitability for natural grassland is based equally on the suitability for nature conservation and the presence of grassland in the 'Land-use Map' of the Netherlands 2000. We assumed that suitability for agricultural uses is the highest in those areas where these uses are currently found. Areas under other agricultural uses are slightly less suitable, and areas under all other urban uses are not suitable at all for agricultural use. The suitability map for greenhouse horticulture was derived from a specific study 'Potential options for greenhouse development' (LEI-DLO 1997), which is based on the amount of light received per year and the proximity of distribution centres.

For each land use, the expansion locations and restrictions are defined in the zoning map. Expansion locations are locations where a specific land-use change has been planned in accordance with to local, regional or national policy plans. Restrictions are locations where a specific land use is not allowed to develop. The expansion locations for the various land uses are based on the plans contained in 'Plans in the Netherlands' (*Nederland in Plannen*) and the 'New Map of the Netherlands' (NIROV 2004). The expansion locations for forest, natural grassland and nature conservation marked on the zoning map were derived from the 'net boundaries' of the National Ecological Network 2003 (NEN). All of the restrictive policies in the National Spatial Strategy have been reproduced on the zoning map. The restrictions for each land use are listed in Table 3.3.

The policies for concentration areas and National Landscapes policies in the National Spatial Strategy, as described in the Introduction, are expected to have less effect on spatial development than the expansion locations and restricted

Table 3.3 Definition of restrictive areas per land use

Restrictive areas	Source	Residential, Services, Facilities	Industry, Greenhouse horticulture	Forest, Nature conservation
20-ke contour Schiphol 2004, noise disturbance contour	NR[1]: PKB 3	X		
Coastal Foundation Zone and Weak Spots	NR: PKB 4	X	X	
Net boundaries NEN 2003	NR: PKB 5	X	X	
Protected areas under the Birds and Habitats Directives and the Nature Conservancy Act	NR: PKB 6	X	X	
Existing and New Nature 2003	(MNP 2002b)	X	X	
Space for the rivers	(VenW 2005)	X	X	
National buffer zones/ Regional parks	(LNV 2002)	X	X	
Risk contours for companies requiring an external safety report	MNP (in preparation)	X		
Schiphol bird protection zone	(VenW 2004)			X

[1] NR: PKB, National Spatial Strategy: Spatial Planning Key Decision (VROM *et al.* 2005)

areas. In both scenarios, it is assumed that the impact of expansion locations and restrictive areas is 20-fold greater than that of the concentration areas and the National Landscapes policies. This figure is based on expert judgement. The road network, motorway entrances and exits and the locations of the stations are specified in the *Environment Explorer* to allow the Traffic module to calculate accessibility (RIKS 2002). Accessibility was calculated for the period 2000–2010 using the standard Dutch National Model System network of 1995 (HCG 1997). The assumption was made that after 2010 all extensions to the infrastructure will be completed, as planned in the Multiannual Programme for Infrastructure and Transport (VenW 2003), and that no further extensions will be made to the road network before 2030.

Up to this point, the Trend and Sprawl scenarios do not diverge. However, the policies in the National Spatial Strategy give the municipal and provincial councils more freedom to approve new developments in the National Landscapes than they had previously. Whereas the municipal councils in the Green Heart formerly had virtually no opportunities for expansion, they are now permitted to build new homes to meet demands resulting from local population growth. It is most likely that the smaller settlements in the Green Heart will grow relatively faster under this new policy than larger towns and cities outside this National Landscape. In time, this change in policy will affect the transition rule for housing development in the model. These rules have been calibrated on historical land-use changes. Therefore, to estimate the impact of this change in policy, we constructed a Sprawl scenario in which the transition rules for housing development were adjusted to allow the smaller settlements to grow slightly faster than they do in the Trend scenario. The effect of this change in policy and the size of the adjustment to the neighbourhood rules are difficult to determine. However, the actual difference between the rules should be minor. Therefore, the effect of this adjustment on urban clusters sizes in 2030 has been verified against Zipf's Law.

3.4 Results

3.4.1 Urbanisation Probabilities in 2030

Figure 3.3 depicts the urbanisation probabilities in 2030 according the Trend scenario. The urban area comprises the following land-use categories in the *Environment Explorer*: housing, employment, recreation, sports fields and greenhouse horticulture complexes. The map shows that a number of urban areas will probably expand towards each other and eventually merge. In the Randstad conurbation, the open spaces between The Hague and Rotterdam will close. The city of Utrecht will expand to the south, merging with several small towns. New development around Amsterdam will be more dispersed.

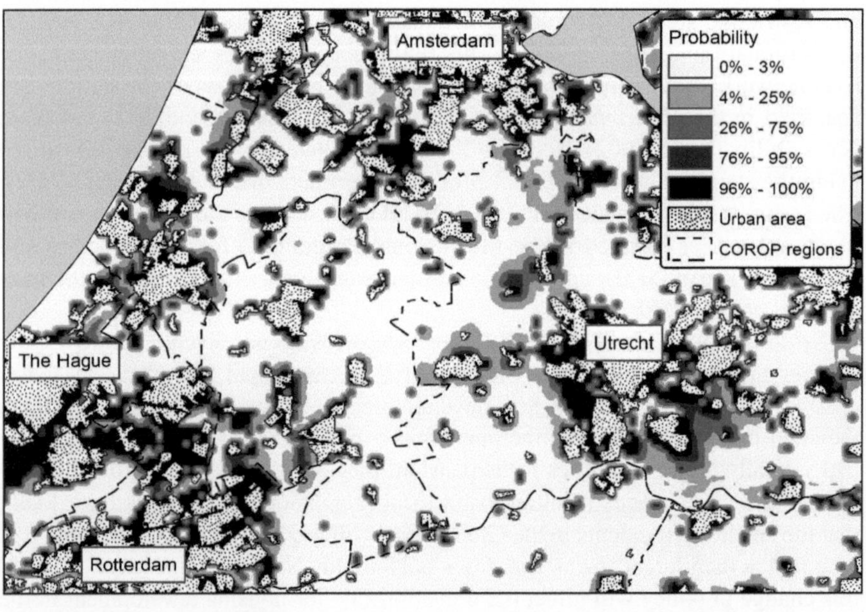

Fig. 3.3 Urbanisation probabilities in the Randstad conurbation in 2030 under the Trend scenario

In this scenario, the expansion of residential and employment land uses is heavily determined by the restrictive areas, including, among others, nature reserves and protected areas, regional parks and land in the floodplains reserved for flood control, water retention and habitat development. Developments in the concentration areas are given a preference over other areas. The National Landscapes are kept free of new housing and employment land as much as possible.

3.4.2 *Pressures on Land in the Restrictive Areas*

The Trend scenario was used to determine the relative pressure on land in the restrictive areas. These restrictive areas are only built on if all other available land within the COROP region has already been developed. COROP regions are the same as the NUTS Level 3 regions. The 40 regions are sub-divisions of the provinces in the Netherlands, each consisting of a central town or city and catchment area. Only if growth exceeds the amount of available land will these last open areas be developed. The probability of urban development in the restrictive areas was mapped for all urban restrictions, with the exception of the 20 km contour around Schiphol. This safety zone around Schiphol airport does not exclude industrial development and greenhouse horticulture.

There appears to be little pressure for urban development in the restrictive areas (Fig. 3.4), and only in the Randstad is the pressure on land so high that there is a chance of urbanisation in restrictive areas. New urban development may occur in the restrictive areas north of The Hague, around Amsterdam and in the vicinity of small towns in the nature conservation area to the north-east of Utrecht.

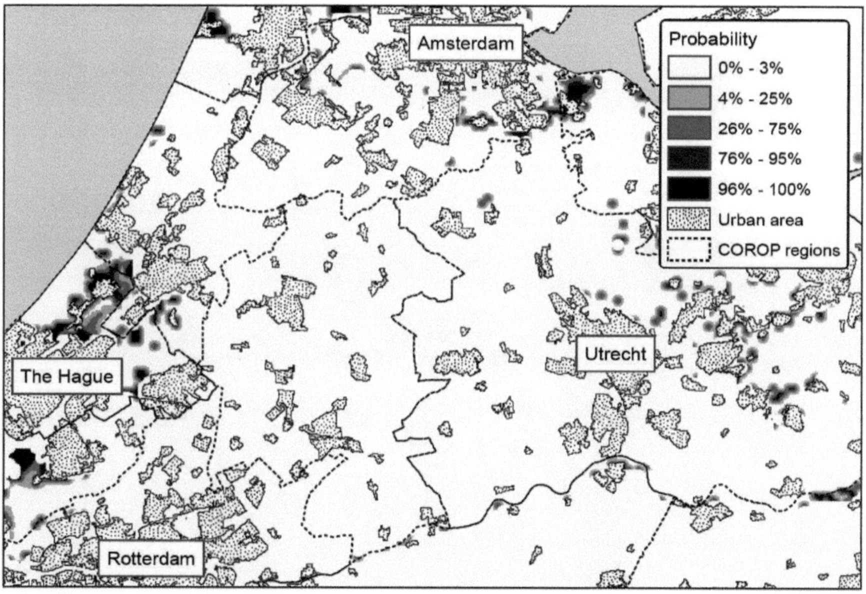

Fig. 3.4 Probability of housing development in the restrictive areas

3.4.3 Urban Development in the National Landscapes

Figure 3.5 shows the differences in urbanisation probabilities between the Trend and Sprawl scenarios in the Green Heart, one of the National Landscapes mostly threatened by urbanisation. Here, the probability of urbanisation is slightly higher under the Sprawl scenario, particularly in the eastern part of the Green Heart. The probability of urbanisation decreases in the area south-east of Utrecht and increases at the western side of the city. The areas of different grey shading indicate sites where the probability of urbanisation under the Trend scenario is higher; the black areas indicate where the urbanisation probability under the Sprawl scenario is higher. In the Sprawl scenario new residential development is less concentrated around the main cities, but it will be located more often near small villages in the Green Heart.

3.4.4 Space in the Concentration Areas

One of the main principles of the National Spatial Strategy is to locate new urban development in the concentration areas. The National Spatial Strategy states that the 'concentration percentage' in these areas should – at the very least – remain the same, but at the same time it does not clearly define concentration percentage. In addition, land also has to be reserved for water, nature and landscape conserva-

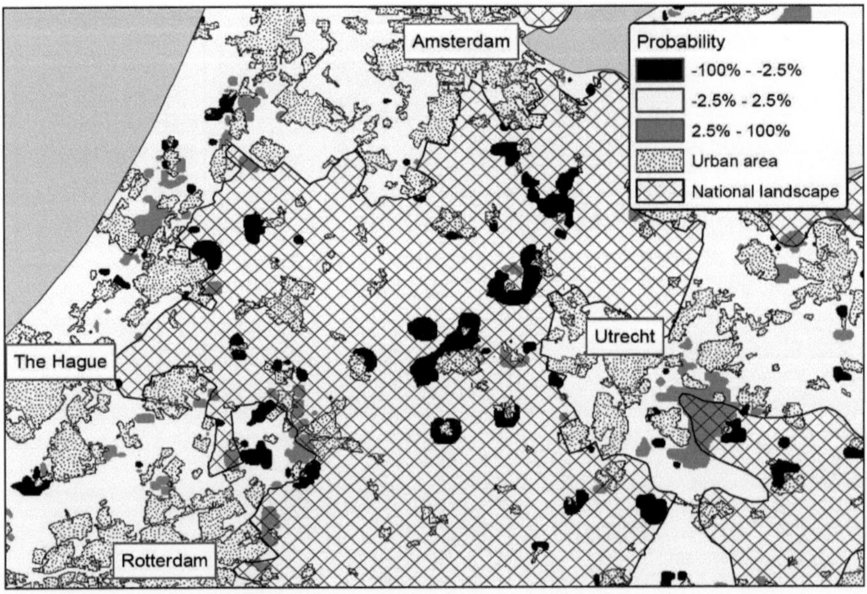

Fig. 3.5 Difference in urbanisation probabilities between the Trend and Sprawl scenarios

tion, recreation areas, sports fields and agriculture. In this analysis the concentration percentage (P) is taken to be the percentage of the urban area that is located in the concentration area ($A_{\text{Urban, Concentration Area}}$) relative to the total urban area in the province ($A_{\text{Urban, Province}}$):

$$P = \frac{A_{Urban,Concentration\,Area}}{A_{Urban,Province}} * 100 \tag{3.1}$$

Urban areas in this definition include residential and employment land, facilities, parks and greenspace, sports fields, recreation areas, roads, railways and greenhouse horticulture. It is questionable whether all spatial developments can be accommodated within these concentration areas. Therefore, for each province, the available, required and developed land has been determined for 2030. The available land is the land which will still be available for new urban developments; it is defined as the agricultural land and building sites in the land-use map of 2000, barring all physical and spatial restrictions. The required land in 2030 will be the land needed to maintain – at the very least – the same concentration percentage. It is estimated by multiplying the growth of urban land per province, in accordance with the scenario, by the concentration percentage in 2000; as such, it is the amount of land that will be needed in the concentration areas for the development of urban land uses. The developed land is the land which will most probably be developed in 2030. The developed land in 2030 was calculated from the urbanisation probability in 2030 according to the Trend scenario.

On the basis of this analysis, there is not enough land available in the concentration areas in the Randstad and in Limburg to accommodate all new developments (Table 3.4). Accordingly, the concentration percentage in these provinces will decline sharply in the future. The differences between the available, required and (probably) developed land provides some indication of the policy inputs needed to cluster new urban development in the concentration areas.

3.5 Conclusions and Future Directions

For the *ex ante* evaluation of the National Spatial Strategy, future land use was simulated by the *Environment Explorer*. Observed development trends were used to comprehensively calibrate and validate the *Environment Explorer* for short-term projections. The *Environment Explorer* outperformed a random allocation model in the calibration period, but not over the validation period. Land-use models generally perform poorly in validation studies when used for large-scale applications Pontius *et al.* (2008), while their performance in local-scale applications, such as at the city level, appears to be better (Hagoort 2006). The calibration and validation of land-use models are impeded by classification and aggregation errors in land-use maps (Fang *et al.* 2006; Pontius *et al.* 2004; Pontius and Spencer 2005). A novel methodology has been applied to verify long-term projections. The distribution of urban cluster sizes in future land-use maps was verified using Zipf's Law. The calibrated model was used to estimate the urbanisation probabilities in 2030 for a Trend and a Sprawl

Table 3.4 Available, required and developed land (ha) in the concentration areas, and the concentration percentages in 2000 and 2030 under the Trend scenario

	Available land	Required land	Developed land	Concentration per cent 2000	Concentration per cent 2030
Groningen	8,400	989	942	21.0	20.0
Drenthe	5,319	758	682	10.3	9.2
Overijssel	15,543	2,665	2,596	26.9	26.2
Gelderland	15,028	4,782	5,456	19.1	21.8
Utrecht	10,734	8,838	8,045	57.9	52.7
Noord-Holland	9,113	10,357	7,966	60.4	46.5
Zuid-Holland	24,931	17,195	16,202	73.0	68.8
Noord-Brabant	33,561	11,646	11,174	43.0	41.3
Limburg	5,634	6,491	4,444	36.2	24.8
Flevoland	6,012	2,255	2,991	28.8	38.2

There are no concentration areas in the provinces of Friesland and Zeeland

scenario using Monte-Carlo simulations. The effects of policy on the restrictive areas, the National Landscapes and the concentration areas were examined.

The probability of urbanisation in 2030 is highest near existing urban areas and is influenced by spatial planning policy. Urban development in restrictive areas is avoided as much as possible. Only in the Randstad is the pressure on land so high that it is likely that urban development will take place in the restrictive areas. Urbanisation in the National Landscapes will probably increase as a result, particularly in the Randstad and Limburg where development pressures are the highest. There is sufficient space in most of the concentration areas to accommodate all new developments. Given the scenario assumptions, there will be insufficient land in the provinces of Noord-Holland and Limburg to accommodate all new urban developments in the future while retaining enough land for water, nature and landscape conservation. This analysis identifies those provinces for which problems may be expected over the long term with implementing national spatial planning policies.

This evaluation of the National Spatial Strategy using the *Environment Explorer* provides detailed insights into the feasibility of the policy goals and potential problems. The study shows how land-use models can be used to determine the effects of spatial planning policies on future spatial development. Even so, the results of such modelling studies are inherently uncertain. The Monte-Carlo simulation enables the large uncertainties surrounding the location of developments to be visualized through the use of urbanisation probabilities. However, the uncertainty in the growth of land use and the effectiveness of current policies has not yet been incorporated in the simulated scenarios. These results provide policymakers with the opportunity to take potential problems into account beforehand. Monitoring programmes have been developed, especially those focussing on the simulated hot spots of urban developments in restricted areas. Additional measures can be defined to circumvent potential problems in the national landscapes and concentration areas. Moreover, the detailed spatial results have made it possible to determine various – potentially adverse – effects of these new policy plans; for example, to assess future noise levels and risk contours near airports in the Netherlands (Dassen 2005).

For this study, the *Environment Explorer* was calibrated for a relatively short period, given the slow pace of spatial development. Calibration on a consistent dataset over a longer time-span could improve simulation results. Moreover, a major part of the land-use changes in the calibration period was dubious: urban areas changing to forest, nature conservation or agricultural uses. Therefore, future research should be directed towards reducing the uncertainties in the results of the land-use model by taking advantage of better monitoring data. For example, to circumvent the classification problems in land-use maps, land-use models should utilise the primary remote sensing data instead of the interpreted results in the land-use maps. Moreover, all information available in subsequent remote sensing images (time-series) should be used to identify current land-use developments and develop algorithms to simulate future land-use developments. In terms of the allocation algorithm, the optimal spatial and temporal resolution needs to be determined. How well does the same model perform at different spatial or temporal scales? To what extent is it possible to aggregate land-use categories? How well does the same model distinguish two or twenty land-use categories. Finally, to measure these differences in model performance, it is of utmost importance to develop a set of sensitive and appropriate indicators.

As this study was first published in 2005 it is possible to look back on its impact in the Netherlands. Shortly after publication, the results of this study were presented to the relevant policymakers at the Ministry of Housing, Spatial Planning and Environment. The conclusions were quoted by various stakeholder groups in the Netherlands. One of these, the Netherlands Society for Nature and Environment (SNM 2005), published an article in its newsletter calling on the Dutch population to stop further urbanization. In general, the results of this study have been widely accepted. In *Revolutionary Future for Housing Construction*, Hugo Priemus (2005) explicitly supports the results of the study and states his opposition to the popular idea of a *'transformation of the Randstad'*, as suggested by planners and administrators. The results of this study have been used in several regional scenario studies in the Netherlands (Dijk 2007; SafeCoast 2005; Witmond *et al.* 2006). It is difficult to determine whether this study influenced policy, but the results have been discussed widely among scientists, stakeholders and policymakers.

References

ABF (2002) Ruimtevraag wonen, werken en voorzieningen herberekend. Verkenning 2000–2030 voor deel 3 van de Vijfde Nota Ruimtelijke Ordening, *No. R 2001–0097LM,* ABF Onderzoek en Informatie, Delft.

Alterra (2003) *Landelijk Grondgebruiksbestand LGN4 2000,* Alterra, Wageningen.

CBS (2003) *Bodemstatistiek 2000,* Central Bureau of Statistics, The Hague.

Dassen, A.G. (2005) *Evaluatie Regelgeving Burgerluchthavens en Militaire Luchthavens, No. 500047002,* Milieu- en Natuurplanbureau, Bilthoven.

De Nijs, A.C.M., Engelen, G., White, R. and Delden, H.v. (2001) De LeefOmgevings Verkenner. Technische Documentatie, *No. RIVM-rapport 408505007,* Rijksinstituut voor Volksgezondheid en Milieu, Bilthoven.

De Nijs, A.C.M., Kuiper, R. and Crommentuin, L.E.M. (2005) Het landgebruik in 2030: Een projectie van de Nota Ruimte, *No. Rapport 711931010,* Rijksinstituut voor Volksgezondheid en Milieu, Bilthoven.

De Nijs, T.C.M., De Niet, R. and Crommentuijn, L. (2004) Constructing land-use maps of the Netherlands in 2030, *Journal of Environmental Management,* 72(1–2): 35–42.

Dijk, v.T. (2007) Scale dependent synergy between risk management and open space preservation, Paper presented at the AESOP 2007 International Conference 'Planning for the Risk Society'.

Engelen, G., White, R. and de Nijs, T.C.M. (2003) The Environment Explorer: spatial support system for integrated assessment of socio-economic and environmental Policies in the Netherlands, *Integrated Assessment,* 4(2): 97–105.

Fang, S.F., Gertner, G., Wang, G.X. and Anderson, A. (2006) The impact of misclassification in land use maps in the prediction of landscape dynamics, *Landscape Ecology,* 21(2), 233–242.

Gabaix, X. (1999) Zipf's law for cities: an explanation, *Quarterly Journal of Economics,* 114(3): 739–767.

Gabaix, X. and Ioannides, Y.M. (2003) The Evolution of city size distributions, In H. a. Thisse, H. (ed.) *Handbook of Urban and Regional Economics. Volume IV: Cities and Geography,* North-Holland Publishing Company, Amsterdam, pp. 2341–2378.

Goetgeluk, R.W., Louter, P.J., Beurden, J.A.M.B.-v., Waals, J.F.M.v.d. and Geurs, K.T. (2000) Wonen en werken ruimtelijk verkend. Waar wonen en werken we in 2020 volgens een compacte inrichtingsvariant voor de Vijfde Nota Ruimtelijke Ordening? *No. RIVM-rapport 711931001,* Rijksinstituut voor Volksgezondheid en Milieu, Bilthoven.

Hagen-Zanker, A., Straatman, B. and Uljee, I. (2005) Further developments of a fuzzy set map comparison approach. Further developments of a fuzzy set map comparison approach, *International Journal of Geographical Information Science,* 19(7): 769–785.

Hagen-Zanker, A., van Loon, J., Maas, A., Straatman, B., de Nijs, T.C.M. and Engelen, G. (2005) Measuring performance of land use models. An evaluation framework for the calibration and validation of integrated land use models featuring cellular automata, Paper presented at the European Colloquium on Theoretical and Quantitative Geography, Tomar, September 9–13.

Hagen, A. (2003) Fuzzy set approach to assessing similarity of categorical maps, *International Journal for Geographical Information Science,* 17(3): 235–249.

Hagoort, M.J. (2006) *The Neighborhood Rules: Land-use Interactions, Urban Dynamics, and Cellular Automata Modelling,* Utrecht.

HCG (1997) *LMS 5.0 Deel 1–7,* Hague Consulting Group, The Hague.

LEI-DLO (1997). *Kansen voor Kassen,* LEI-DLO, The Hague.

LNV (2002) *Structuurschema Groene Ruimte 2. Samen werken aan groen Nederland,* Ministerie van Landbouw, Natuurbeheer en Visserij, The Hague.

MNP (2002a) *Natuurverkenning 2,* RIVM, Bilthoven.

MNP (2002b) *Natuurverkenning 2,* Kluwer, Alphen aan den Rijn.

MNP (2005) Kalibratie en validatie van de LeefOmgevingsVerkenner, *No. Rapportnr. 550016006,* MNP, Bilthoven.

NIROV (2004) *De nieuwe Kaart van Nederland,* NIROV, The Hague.

Pontius, R.G., Boersma, W., Castella, J.C., Kok, K., Veldkamp, A. and Verburg, P.H. (2008) Comparing the input, output, and validation maps for several models of land change, *Annals of Regional Science,* 42(1): 11–37.

Pontius, R.G., Shusas, E. and Mceachern, M. (2004) Detecting important categorical land changes while accounting for persistence, *Agriculture Ecosystems and Environment,* 101(2–3): 251–268.

Pontius, R.G. and Spencer, J. (2005) Uncertainty in extrapolations of predictive land-change models, *Environment and Planning B,* 32(2): 211–230.

Priemus, H. (2005) Wat staat ons allemaal te wachten? Revolutionaire toekomst voor de woningbouw, *Building Business,* 7.

RIKS (2002) *Verkeer in de LeefOmgevingsVerkenner: Ontwikkeling en bouw van een verkeersmodule in de LOV,* RIKS, Maastricht.
RPD (1960) *Nota Inzake de Ruimtelijke Ordening in Nederland,* The Hague.
RPD (1966) *Tweede Nota over de Ruimtelijke Ordening in Nederland,* The Hague.
RPD (1977) *Derde Nota over de Ruimtelijke Ordening. Deel 2: Verstedelijkingsnota,* The Hague.
RPD (1988) *Vierde Nota over de Ruimtelijke Ordening. Eenheid in Verscheidenheid,* The Hague.
RPD (1994) *Vierde Nota over de Ruimtelijke Ordening Extra,* The Hague.
SafeCoast (2005) *Inventory of Spatial and Economical Scenarios Available in the North Sea Region:* WL I Delft Hydraulics & Grontmij.
SNM (2005) RUIMTE - Verstedelijking gaat steeds meer ten koste van natuur, recreatie en water, *Nieuwsbrief.*
VenW (2003) *Meerjarenprogramma Infrastruktuur en Transport,* VenW, The Hague.
VenW (2004) *Wijziging Luchthavenverkeerbesluit Schiphol en Wijziging Luchthavenindelingbesluit Schiphol,* Ministerie van Verkeer en Waterstaat - Luchtvaart, The Hague.
VenW (2005) *Planologische Kernbeslissing Ruimte voor de Rivier Deel 3, Kabinetsstandpunt,* Rijkswaterstaat, The Hague.
Verburg, P.H., de Nijs, T.C.M., Ritsema van Eck, J., Visser, H. and Jong, K.d. (2004) A method to analyse neighbourhood characteristics of land use patterns, *Computers, Environment and Urban Systems,* 28: 667–690.
Verburg, P.H., Van Eck, J.R.R., De Nijs, T.C.M., Dijst, M.J. and Schot, P. (2004). Determinants of land-use change patterns in the Netherlands, *Environment and Planning B,* 31(1): 125–150.
Visser, H. and De Nijs, A.C.M. (2006) The Map Comparison Kit, *Environmental Modelling and Software,* 21: 346–358.
VROM, LNV, VenW and EZ (2005) *Nota Ruimte,* The Hague.
White, R. and Engelen, G. (1997) Cellular automata as the basis of integrated dynamic regional modelling, *Environment and Planning B,* 24(2): 235–246.
Witmond, B., Vos, W., Voogt, S. and Vervoort, K. (2006) *Wetlands in het IJsselmeer,* Ecorys BV, Rotterdam.
WWdL (1958) *De ontwikkeling van het westen des lands,* Werkcommissie Westen de Lands van de Rijksdienst voor het Nationale Plan, The Hague.
Zipf, G.K. (1949) *Human Behaviour and the Principle of Least Effort,* Addison-Wesley, Cambridge, MA.

Additional Reading

Dendoncker, N., Rounsevell, M. and Bogaert, P. (2007) Spatial analysis and modelling of land use distributions in Belgium, *Computers, Environment and Urban Systems,* 31(2): 188–205.
Jantz, C.A., Goetz, S.J. and Shelley, M.K. (2004) Using the sleuth urban growth model to simulate the impacts of future policy scenarios on urban land use in the Baltimore-Washington metropolitan area, *Environment and Planning B,* 31(2): 251–271.
Koomen, E., Stillwell, J., Bakema, A. and Scholten, H.J. (2007) *Modelling Land-Use Change Progress and Applications,* Springer, Dordrecht.
Manson, S.M. (2007) Challenges in evaluating models of geographic complexity, *Environment and Planning B,* 34(2): 245–260.
Torrens, P.M. (2006) Simulating sprawl, *Annals of the Association of American Geographers,* 96(2): 248–275.

Chapter 4
A Planning Support System Toolkit Approach for Formulating and Evaluating Land-use Change Scenarios

Christopher J. Pettit and Ray Wyatt

4.1 Introduction

Distributing land uses across space is the central problem addressed by the urban and regional planning discipline. Millions of words have been written about it, and there has been significant investment made by various planning agencies around the world trying to do it better. Nevertheless, from the time of the late-nineteenth century origins of town and country planning until around the mid-twentieth century, land-use planning was usually done in a non-analytical way, by architects and engineers who were simply content to tidy up the environment by such means as physically separating incompatible land uses and ensuring efficient servicing of them with properly distributed pipes, wires and transport arteries.

After the Second World War, however, the nature of urban and regional planning changed forever. This was due to at least two phenomena. Firstly, urban and regional planning came to be seen as a field where practitioners could actually see and touch social change, and so many other sorts of professions became involved in the discipline as it became more socially and less physically oriented. Secondly, the Second World War had seen the development of computers, and it was not long before they were applied to the urban and regional planning discipline in a bid to better manage its new found complexity. The problem was that early computers were difficult to manage. Indeed, the old adage that computers are frustrating because they do what you tell them rather than what you want, still holds true today. Nor was the theory of urban and regional planning up to the task of making the massive, computerized models of urban and regional growth plausible. The result was a huge disillusionment with large-scale urban analysis and modelling, and it is commonly regarded as having begun with the publication

C. J. Pettit (✉)
Landscape Systems – Spatial Sciences
Department of Primary Industries
Victoria, Australia
E-mail: christopher.pettit@dpi.vic.gov.au

S. Geertman, J. Stillwell (eds.), *Planning Support Systems Best Practice and New Methods*,
DOI 10.1007/978-1-4020-8952-7_4, © Springer Science+Business Media B.V. 2009

of Douglas Lee's article entitled *Requiem for large-scale urban models*, as far back as 1973 (Lee 1973).

The result is that today the use of computers in the urban and regional planning discipline is qualitatively different. It is now far less ambitious in its predictions of urban trends and patterns, which is a far safer way to go than adopting the recommendations of the large-scale urban models of the past. A more cautious, exploratory approach is less likely to draw false conclusions about the future. In other words, rather than run programs which unsuccessfully attempt to solve the whole of the intractable problem of spatial land-use planning, today's computer programs are quite content to simply model the most predictable consequences of different land-use pattern aspirations and assumptions. They proceed in a rather simple and factual way rather than in a fashion that was driven by the initial excitement of newly discovered computing machines and interesting, but mostly non-validated, socioeconomic theory.

Conditions are certainly ideal for the new approach. The advent of cheap and fast laptop computers, plus the Internet, has brought an unprecedented amount of information to the fingertips of not only the planners but also the stakeholders within the community. Hence the first part of the twenty-first century has seen urban and regional planning become much more participatory and data-driven: a new development which has nowhere near yet run its full course. One of its spin offs is the rise of computer-based planning support systems (PSS) in the early twenty-first century (Geertman and Stillwell 2003; Brail and Klosterman 2001), four of which will be applied and evaluated in this chapter, including *SLEUTH, What if?™, Google Earth*, and *Preference Prediction*. To do so we will use, as a case study application, a typical, rural-urban periphery municipality within a typical developed-world metropolis: the Mitchell Shire on the northern outskirts of Melbourne, Australia. There is nothing particularly out of the ordinary about this Shire, so it should serve as a convenient vehicle for generating findings that will probably be applicable to many other peri-urban regions around the world as well.

The first PSS to be considered will be the well known *SLEUTH* software, which has been used in a number of cities on several continents. Its origins were in California, and it simulates suburban and hinterland development by assuming that growth is dependent on natural tendencies towards such things as infill development, extra-urban estate generation and road-based community expansion. The second PSS, the *What If?™* software, is more flexible, having been designed to require the inputs of local Shire planners and stakeholders in order to specify the various importance levels of factors that influence urban development. Such weightings actually serve to specify what sorts of future community the participants would like to see evolve. The third PSS is *Google Earth*, which arguably falls into the visualisation PSS category (Klosterman and Pettit 2005). Virtual globe products such as *Google Earth* can be considered to be PSS because they enable planners to view, interrogate and potentially design alternative landscape futures. They are powerful planning, engagement and communication tools that can add value to the planning process and they are a likely extension to the already existing sketch planning psaradigm of PSS tools like *SketchGIS* (Geertman 2002). In any case, *Google Earth* is useful for conveniently showing the similarities and contrasts between different land-use patterns generated by two of our land-use modelling PSS (*SLEUTH* and *What if?*™). It is in the absolute vanguard of current trends which are seeing the traditional Geographic Information (GI) Science discipline morph into the disciplines of cybermapping, geo-visualization and neography. Finally, the fourth PSS is not spatial at all. It evaluates scenarios using its knowledge of decision-making priorities that it has learned from previous users of the system. It is known as the *Preference Prediction* software, and its function is to predict which (spatial) scenarios are likely to be favour by what sorts of people, thereby adding 'anticipation of community responses' to the PSS' repertoire.

4.2 The Mitchell Shire Case Study

The Shire of Mitchell is situated along the Hume Highway in the peri-urban hinterland north of the city of Melbourne, as shown in Fig. 4.1. The Shire comprises approximately 543,000 hectares of land with a mix of urban and agricultural land uses. Although the Shire is situated outside the Melbourne 2030 plan's urban expansion boundary, it is still experiencing significant population growth, with an expected population increase from approximately 30,000 people to 60,000 people up until 2030. This growth is expected to occur particularly in the southern parts of the Shire, in and around the townships of Wallan and Kilmore, due to spill over of housing demand in Melbourne itself. Such impeding urban growth is currently placing land-use pressures on existing agricultural land, which supports a range of agricultural practices including cropping, grazing, intensive agriculture, irrigated agriculture, plantation forestry, production forestry and natural vegetation grazing.

Fig. 4.1 The location of Mitchell Shire in South Eastern Australia (See also Plate 7 in the Colour Plate Section)

4.3 The *SLEUTH* Software

Originally developed by Keith Clarke at the University of California, Santa Barbara, *SLEUTH* randomly selects different pixels, examines neighboring pixels and uses the latter's pattern to decide whether or not to change the selected pixel from non-urban to urban, or from urban to non-urban, or leave it in the same state. As such, it is a type of cellular automata (CA) model, and it has been applied in at least 32 locations around the world (Clarke *et al.* 2006), including Atlanta in the U.S. (Yang and Lo 2003), Lisbon and Porto in Portugal (Silva and Clarke 2002), Chiang Mai in Thailand, Taipei in Taiwan (Sangawongse *et al.* 2005) and Santa Barbara in California itself (Herold *et al.* 2003). Moreover, it has been used in New York to underpin an air pollution-prediction model (Solecki and Oliveri 2004).

However, whereas it is usual to base underlying CA equations on theory, assumption or apparent fit with perceived urban mechanisms, *SLEUTH* simply calibrates them to past urban evolution. It massages the coefficients within its equations, by testing each possible combination of them, to find that combination which best replicates observed change within the urban area between two points in time. This makes it extremely computer intensive, which has prompted a number of attempts to speed it up by using a genetic algorithm-based search, details of which can be found both in Goldstein (2005) and in documentation by the New York Climate and Health Project (2006). Note also that it is difficult to know how to evaluate the performance of different coefficient combinations, which is why *SLEUTH*'s calibration routines actually use up to 12 alternative measures of best fit. Different researchers have different opinions as to what constitutes the best measure, and so there has been little consistency in the way the model has been applied around the world. This has prompted Dietzel and Clarke (2007) to run a comparative study, using Kohonen self-organizing maps, in which *SLEUTH* was applied to two contrived growth patterns, 'Coalescence' and 'Diffusion', and to a fictitious city, 'Demo City'. They then tested which measures of best fit tended to duplicate which others, to eventually produce what they claim is the best measure, the so-called Optimal *SLEUTH* Metric (OSM). The latter is a multiplying together of just seven of the 12 original measures.

It also needs to be remembered that *SLEUTH* is strictly a California derived model that reflects U.S. type growth mechanisms which may, but probably do not accurately simulate growth mechanisms within other parts of the world. Clarke *et al.* (2006) have begun to test this by mapping how large or small the generated coefficients become in 15 U.S., two Mexican, two Portuguese, one Dutch and one Cameroonian implementation of *SLEUTH*. The larger the coefficient the more important that parameter would seem to be in explaining urban growth within that part of the world, but this research still has a long way to go.

Nevertheless, we applied *SLEUTH* to Mitchell Shire by first using the *ArcMap* GIS to clip the various files that the model requires to make them all apply to the same area. We then used the GIS to convert the files to GRID (raster) format and then convert the raster files into TIFF files. Finally, we made the latter readable by

SLEUTH by using the *GIMP* software. It was then a matter of running *SLEUTH* on a Linux server, an experience which can involve climbing up a steep learning curve for those who have not done it before. We should observe, finally, that many researchers will never believe *SLEUTH*'s prediction because the model is always calibrated on the basis of past historical data that pertains only to the city being simulated. Cities change, and everyone believes that urban dynamic forces are altering continuously, everywhere, and so the only thing that can be said with certainty is that *SLEUTH*'s predictions will always be wrong; as are all forecasts. Hence if there were some way to alter *SLEUTH*'s coefficients so that they reflect anticipated trends and new forces that are expected to come into play within the foreseeable, local future, users might feel more confident about the model's efficacy. Yet the lack of any precise theory about how to make such alterations means that users will alter *SLEUTH*'s calibrated coefficients at their peril. It is for this reason that we now turn to an alternative model which is set up to make users feel safer whenever they tamper with parameter coefficients.

4.4 The *What If?*™ Software

What if?™ is a collaborative GIS-based PSS created by Klosterman (1999). *What if?*™ is a commercial off-the-shelf software package which is supported by *What if?*™ Inc (http://www.what-if-pss.com/). The PSS has been created using ESRI's *Map Objects* application programming interface (API) and the GIS proprietary shape file format to input and output scenario information. *What if?*™ has been applied in 15 counties by 48 end users. Some published use cases of *What if?*™ include: (i) simulating land-use change in Seoul's built-up area in Korea (Kim 2004); (ii) developing land-use change scenarios in the coastal Shire of Hervey Bay, Australia (Pettit 2005); and (iii) exploring growth strategies in the MidWest of the United States (Klosterman *et al.* 2006). All of these applications have applied using *What if?*™ Version 1. For our case study application, we apply *What if?*™ Version 2. We begin by first describing the functionality of the PSS before reporting on the results.

What if?™ comprises of two standalone programs, the first being *What if?*™ *Setup*. Before applying *What if?*™, substantial pre-processing of spatial (biophysical) and aspatial (socioeconomic) data is required. This needs to be undertaken within a GIS through the application of a number of geoprocessing functions, such as merge, intersect and overlay. This results in the formulation of *What if?*™ union file, which is essentially a geodatabase comprising all decision factors which will be used by the PSS in generating land-use change scenarios. Once all the data has been pre-processed the PSS user will populate the *What if?*™ *Setup* program by defining a number of parameters including: type of analysis to be undertaken, types of land use and associated land uses, suitability factors, factor types and value ranges, population and employment information, defined allocation controls, display layers, projection years, analysis areas and measurement units. The *What if?*™ *Setup* program provides a simple to use graphic user interface (GUI) with drop-down menus and selection boxes for customising and setting up a land-use planning project space.

The *What if?*™ program is where local and expert knowledge ultimately obtained from experts (scientists), planners, politicians and the community (soft data) can be combined with hard data in order to realise possible future land-use change scenarios. The *What if?*™ program comprises three modules (i) land suitability, (ii) land demand and (iii) land allocation. Users need to sequentially populate each of the modules with data before progressing to the next module. This is because the land allocation scenarios are driven by a combination of both the land suitability and demand modules. The land suitability module is based on McHarg's (1969) sieve overlay approach. Essentially, users decide which factors are most appropriate for determining land suitability for a particular area. For the Shire of Mitchell case study, these factors included erosion, flood, slope, soil, salinity, wildfire, landscape significance, vegetation protection, heritage protection and proximity to highways, railway stations and rivers. A comprehensive process was undertaken with planners for the Shire to decide upon the most appropriate factors and the value of these factors. Further details and results on the planner engagement process is the focus of ongoing research.

The *What if?*™ suitability module also requires users to specify which land use can be converted to what. For example, is it permissible for urban residential land to be converted to commercial activities or can cropping land be converted to rural or urban residential? The permissible land-use conversions have been varied for the Shire of Mitchell case study in order to generate two land-suitability scenarios. The first scenario captures the land-suitability factors combined with planning metrics and allows all agricultural land to potentially be converted to urban land uses such as residential, industrial and commercial (Fig. 4.2). The second scenario captures the same land-suitability factors and planning metrics but prohibits any potential agricultural land from being converted to an urban land use with a resulting suitability map illustrated in Fig. 4.3.

The second *What if?*™ module calculates land-use demand. The demand module is driven by both population and employment data. Population trends determine how much additional land is required to accommodate future residential housing. Employment trends determine how much additional land is required to service support industry and business sectors. For the Mitchell Shire case study, population projections have been obtained from the Victorian Department of Sustainability and Environment (DSE) population projection unit and from a private business, ID Consultancy, as commissioned by the Mitchell Shire. Employment projection information has been sourced by the Australian Bureau of Statistics (ABS) using the Australian New Zealand Standard Industry Classification (ANZSIC) industry data. For the purposes of this research, one demand scenario has been created using the population forecasts produced by ID Consultancy and employment trends derived by the ABS.

The final *What if?*™ module determines land allocation. The land allocation module utilises the suitability layers produced in the suitability module to determine land supply. The allocation module combines this information with the land-use demand calculated in the demand module. Also, the allocation module can combine

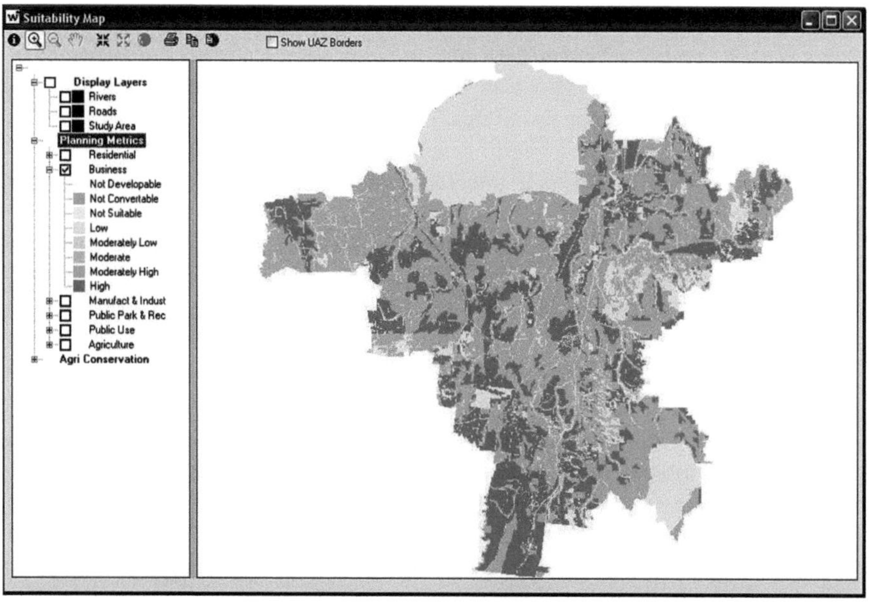

Fig. 4.2 Land suitability for business purposes with all agricultural land use convertible: a developer-centric, suitability-based scenario (See also Plate 8 in the Colour Plate Section)

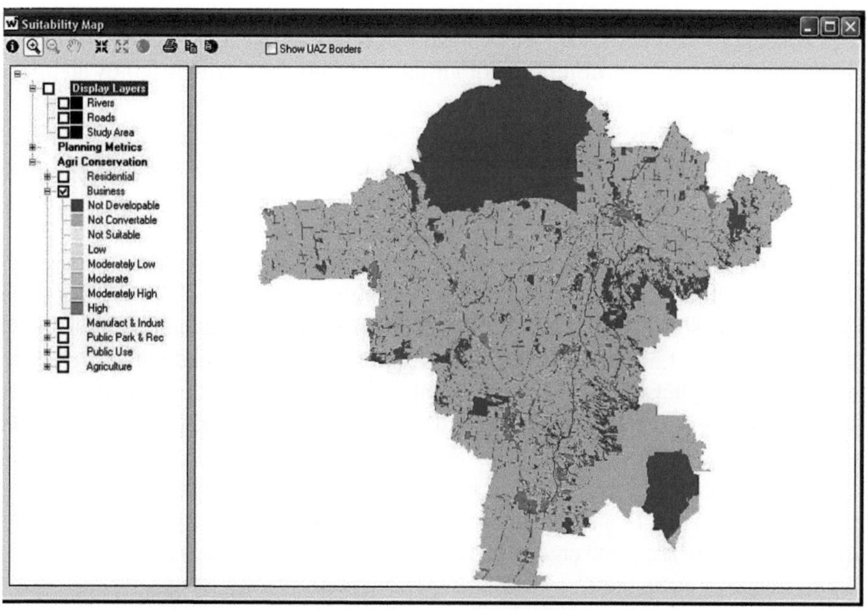

Fig. 4.3 Land suitability for business purposes with no agricultural land use convertible: a conservation-oriented scenario (See also Plate 9 in the Colour Plate Section)

any zoning or infrastructure requirements in order to guide the allocation of future urban land. A number of preliminary land-use allocation scenarios have been generated for the Mitchell Shire, However, when using the existing zoning schemes as a planning control, it has been found this only accommodated urban growth up until 2011. As our timeframe for scenario generation is 2030, we have therefore applied no planning control parameters in generating the developer-centric and agricultural preservation scenarios evaluated in the next section of this chapter.

Applying *What if?*™ 2.0 in generating two alternative scenarios has resulted in some interesting factual and critical experiences which are worth noting. Firstly, whilst *What if?*™ is based on a simple multi-criteria analysis (MCA) approach which is not complex in nature, it can be considered a complicated PSS. Conceptually, the system endeavors to bring together a significant number of variables and parameters and combine these into one tightly coupled decision support environment. It is no mean feat to bring all of this information together and undertake pre-processing. In fact, it could be estimated that 80 per cent of the time spent in developing these scenarios was in data pre-processing and refining the data information structure. Also, the time taken in working through system glitches should not be underestimated. Further evaluation of the *What if*™ PSS is given later in the chapter.

4.5 Using *Google Earth* to Visually Compare Land-use Scenarios

The concept of 'Digital Earth' was coined by Al Gore in the early 1990s. In recent times, there has been a proliferation of virtual globe products which deploy an information technology framework for realising the concept of Digital Earth. Such virtual globe products include: *Google Earth*, Microsoft's *Virtual Earth*, and NASA's *Worldwind*. For an overview of virtual globe technologies, see Aurambout *et al.* (2008). Virtual globes offer a user-friendly way to spatially explore and visualise geographic content and provide a powerful means for sharing spatial information. In our research, we have used *Google Earth* to display and visually compare the results of the three land-use change scenarios for the Shire of Mitchell. The resultant GIF files from *SLEUTH* and *What if?*™ shape files have been exported into Zipped Keyhole Markup Language (KMZ) formats so that the three land-use change scenarios for the Mitchell Shire can be visually analysed via the same spatial viewer software. *Google Earth* provided an excellent viewing platform to communicate to Mitchell Shire Council councilors and planners the results of the *SLEUTH* and *What if?*™ PSS products. However, it is worth noting that importing GIS information into *Google Earth* and developing custom legends is not as easy as one might think. Also, there are some challenges in ensuring KMZ files sizes are small enough not to freeze or crash the *Google Earth* viewing platform.

In a workshop environment, the models that are visualised using *Google Earth* stimulated debate about future land-use issues, including the preservation of good

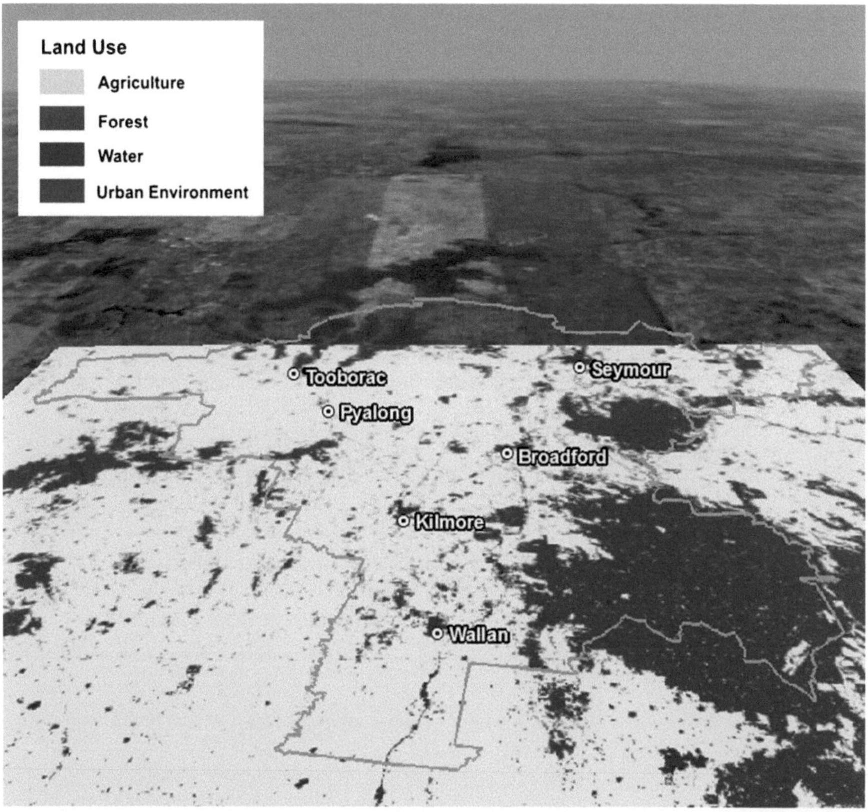

Fig. 4.4 The *SLEUTH*-generated, 'business as usual' (Trend) scenario for Mitchell Shire in 2030

quality agricultural land and development around key urban centres such as Wallan and Kilmore. *SLEUTH*'s 'business as usual' scenario is depicted in Fig. 4.4. As *SLEUTH* produces yearly output, a temporal animation of the simulated land-use change was produced for the period 2001–2030. This provided an effective means of communicating the expected incremental land-use changes within the Shire.

The 'business as usual' (Trend) scenario shows:

- most of future growth will occur along the main transport routes;
- growth is more likely to occur around existing urban centres such as Wallan, Kilmore, Broadford and Seymour, all of which will grow more rapidly than other centres such as Pyalong and Tooborac;
- both Melbourne and Mitchell urban centres will continue to encroach and exert pressure on farmland sandwiched between the two urban areas;
- the demand for urban land will continue to compete with other land uses in the Hume Corridor;
- Mitchell Shire will eventually join with the Melbourne urban area; and

- broadacre and livestock land uses will gradually be converted to other land uses such as industrial, residential, and intensive farming.

The *What if?*™ generated, 'developer-centric' scenario is depicted in Fig. 4.5. This scenario shows significant additional growth occurring within the Shire up until 2030. It is based on the high population projections produced by ID Consulting who suggest that the population will almost double from 31,594 in 2001 to 61,503 in 2031. Significant ribbon development is allocated along the Hume Highway in South Wallan around Beveridge and south of Broadford. There is also significant expansion of the existing centres of Kilmore and Broadford, along with some modest expansion around the township of Seymour.

The *What if?*™ generated 'agricultural conservation' (Conservation) scenario is depicted in Fig. 4.6. Surprisingly, this scenario involves a similar spatial allocation

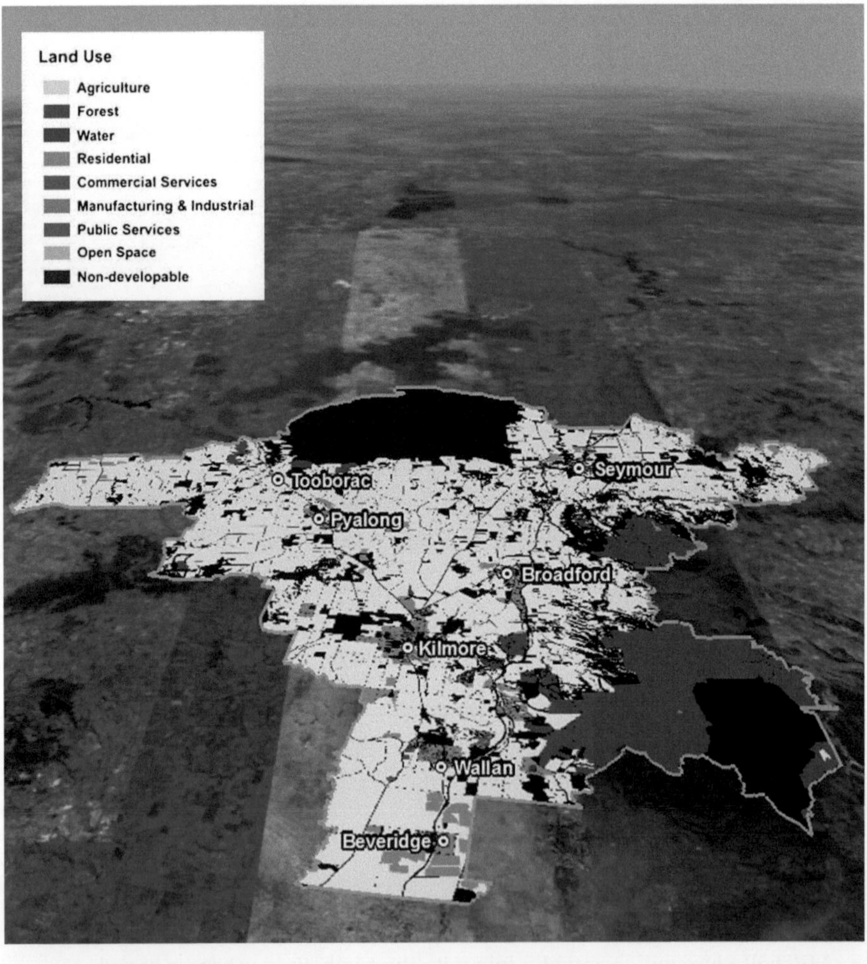

Fig. 4.5 The *What if?*™-generated, developer-centric (Suitability) scenario for Mitchell Shire in 2030 (See also Plate 10 in the Colour Plate Section)

Fig. 4.6 The *What if?*™-generated, agriculture protection (Conservation) scenario for Mitchell Shire in 2030 (See also Plate 11 in the Colour Plate Section)

of future urban growth to that within the 'developer centric' (Suitability) scenario. However, there is substantial additional urban growth south of Wallan in the Suitability scenario that is not apparent in the Conservation scenario. This is because such land is currently used for agricultural purposes and is therefore to be conserved. More generally, residual growth tends to be consolidated around the townships of Kilmore, Broadford and Payalong, which are all situated in the east of the Shire.

4.6 The *Preference Prediction* Software

In real-world practice, choosing which scenario to implement is usually done badly, and there are at least two reasons for this. Firstly, maps of alternative scenarios seldom provide a genuine feel for what it would be like to live under them, and so it is difficult to evaluate the scenarios realistically. Secondly, although those empowered to choose might feel that they understand the attitudes of the community towards each scenario, they actually do not, at least in any formal sense. Accordingly, our final contribution is to take spatial decision support one step further by boosting knowledge about how well, or otherwise, each scenario will be received within the community. More specifically, our fourth PSS predicts how different sorts of people – the old, the young, female, male, educated, non-educated and so on, will score each alternative scenario for utility. The program was formerly called *Strategizer* (Wyatt 2004) but its name has been changed to *Preference Prediction* because this more accurately describes what it actually does. A summary of how it works is to be given elsewhere (Wyatt 2008), so here we present only a brief description.

Basically, the software has so far been used to underpin 17 community-based, stakeholder workshops within a diverse range of problem domains ranging from designing a new ambulance headquarters right through to deciding where to locate a new Masonic temple, developing a new youth prison and choosing the best future planning strategy for a local government lobby group. The total number of workshop participants now numbers 288 and each person, before being allowed to apply the *Preference Prediction* software to their workshop's problem, was required to work through at least three typical choice problems that have been left on the system by past users. This involved scoring each such problem's alternative strategies on nine key, policy-evaluation criteria and then, after the software wiped the computer screen clean, scoring each of the alternatives for overall utility.

Since all such activity is recorded by the system, along with each user's age, sex and so forth, the software is able to plot the relationships between each criterion's scores and option/scenario merit. It is such 'learned' relationships – between criterion score and scenario merit, which enable the system to predict, once criterion scores have been agreed upon, any scenario's merit according to different sorts of people. The nine key, evaluation criteria comprise concepts like timeliness, effectiveness and responsiveness as shown below in Fig. 4.8. To demonstrate how they are used in this chapter, the authors have given scores (between –5 and +5) on the nine criteria for the three scenarios of Trend, Suitability and Conservation, as shown in the partial screen dump from the *Preference Prediction* software in Fig. 4.7.

The *Preference Prediction* software then combined these scores with its knowledge of the relationship between criterion scores and scenario merit, which it had learned from its past users, to predict each scenario's merit according to each group of people. For example, Fig. 4.8 shows the predicted merit scores for our three scenarios according to the total collective ('Everyone') of all the 288 people who have used the *Preference Prediction* software so far. They would probably think that the Conservation scenario is best because it has a desirability score of 1.6, Suitability is next with a score of 1.4 and the Trend scenario is last with a score of 0.8 (within a possible range of -5 to +5).

Note that the left hand sides of Fig. 4.8a and b suggest where each scenario performs well and where it performs badly. To understand this, look at the face charts in Fig. 4.8a. The top face shows the emphasis that the chosen group of advisors ('Everyone') places on each of the nine criteria – emphasis being proportional to the

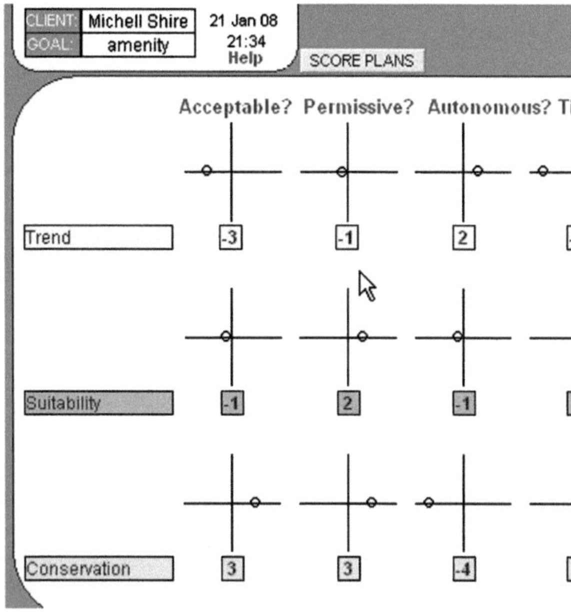

Fig. 4.7 Scores for our three scenarios on the *Preference Prediction* software's nine key criteria

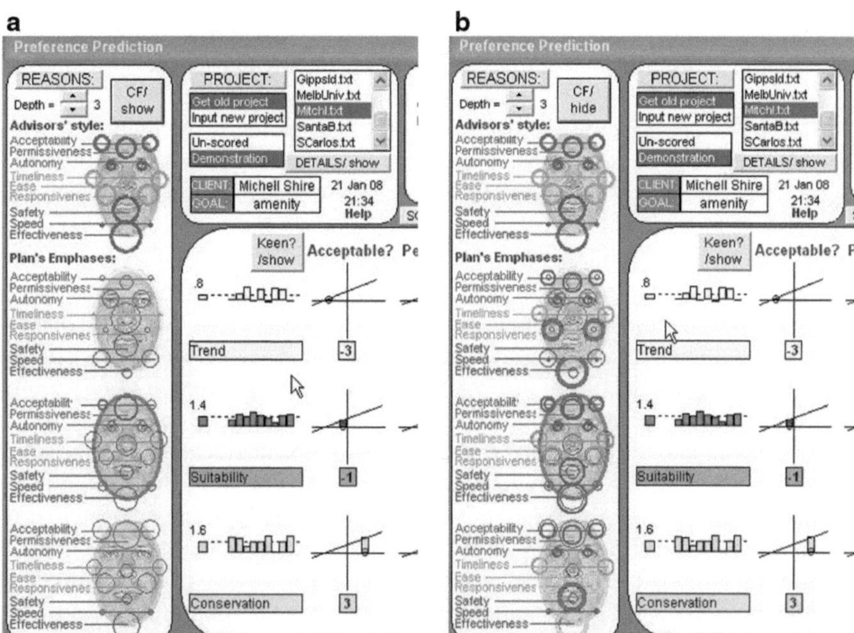

Fig. 4.8 Scores for utility that are likely to be assigned to our three scenarios by all of the past software users taken together (See also Plate 12 in the Colour Plate Section)

size of the circle(s) representing that criterion. Hence our advisors appear to place the most emphasis on effectiveness (the neck), safety (the mouth) and responsiveness (the cheeks). The other face charts show how each scenario has scored on the nine criteria – the higher the score the bigger the circle representing that criterion. It is likely, therefore, that the scenario whose face looks the most like the top face, the one that scores well on the advisors' emphasized criteria or the one whose face looks to be the most 'related' to the top face, will be regarded as the best scenario. Now the bottom face, for the Conservation scenario, looks to be the most similar to the top face and, sure enough, the advisors give it the top score of 1.6.

Figure 4.8b, obtained by clicking the 'CF/ show' button, tries to explain this in greater detail by highlighting each emphasized criterion which each scenario scored highly or lowly on. For example, the top-scoring option, the Conservation scenario, although it scored poorly for safety (the mouth), scored strongly for effectiveness (the neck) and for responsiveness (the cheeks). The Suitability scenario did not do as well because it did not score remarkably on any of the emphasized criteria at all. Finally, the Trend scenario scored poorly because although it scored well for safety, it scored lowly for the other two important criteria, effectiveness and responsiveness. It follows, therefore, that one way to make the Trend scenario more desired by our advisors would be to modify it so that its effectiveness and its responsiveness are somehow improved.

Yet different sub-groups of users will have different decision-making emphases, which will produce a different face chart at the top left of Fig. 4.8a and b if a new group of decision makers is selected by the user, and so this new group's scenario merit scores will be different. Indeed, the software is able to print out merit scores according to all groups of people, as shown in Fig. 4.9, with each advisory group's highest-scoring scenario being shown in bold font.

Hence we can instantly see that although Conservation is the highest scoring scenario according to most advisory groups, certain groups of people are likely to favour other alternatives. For example, tradespersons, councilors and those with just a primary school education (there is almost certainly some overlap between these groups) are likely to favour the Trend scenario, and the Suitability scenario is likely to be the most popular (marginally) according to managers.

Some of these results are thought provoking, for example, why should managers favour the Suitability scenario rather than Conservation scenario? This constitutes one of the greatest strengths of the *Preference Prediction* software. It is very useful as a discussion starter at stakeholder workshops – discussion that frequently leads to more insight into the merits and drawbacks of each scenario, along with possible improvements to them which might make them more palatable to different groups.

Note that the *Preference Prediction* software has recently been upgraded to take into account more sub-groups of users, and Fig. 4.10 shows the opinions of such sub-groups derived from a recent workshop held in Australia amongst landscape planning professionals. Here again Conservation is most often nominated as the best scenario, with Suitability second and Trend, once again, a distant last.

Yet some groups will favour the Trend option – people with one child, people born in the Middle East and Muslims. Moreover, several groups will probably favour the

RATINGS ACCORDING TO EACH GROUP

ACCORDING TO:	Trend	Suitability	Conservatior	ACCORDING TO:	Trend	Suitability	Conservatior
Everyone	.77	1.38	**1.57**				
11-20	1.57	2.04	**2.17**	21-30	.69	1.32	**1.55**
31-40	.31	.84	**.95**	41-50	.49	.96	**1.07**
51-60	.01	.52	**.68**	Males	.67	1.20	**1.36**
Females	.86	1.54	**1.76**	Non-parents	.80	1.41	**1.60**
1 child	.61	1.17	**1.42**	2 children	.43	.96	**1.05**
3 children	1.13	1.40	**1.55**	4 children	.11	1.00	**1.15**
5 children	.04	.15	**.24**	Primary ed.	**1.42**	1.39	1.34
Secondary ed.	1.09	1.66	**1.81**	Tertiary ed.	.64	1.26	**1.45**
Professionals	.40	.83	**.89**	Managers	.82	**1.24**	1.24
Administrators	.96	1.68	**1.93**	Salespersons	-.33	.43	**.69**
Tradespersons	**-.08**	-.18	-.29	Students	.82	1.47	**1.67**
Others	-.09	.53	.59	Cluster 1	.93	1.60	**1.82**
Cluster 2	.77	1.35	**1.53**	Cluster 3	.42	1.11	**1.29**
Cluster 4	.61	1.26	**1.53**	Cluster 100	.80	1.42	**1.63**
Powerful stake.	1.22	1.72	**1.82**	Legitimate stake.	1.01	1.66	**1.83**
Urgent stake.	.97	1.40	**1.49**	Bystanders	.55	1.02	**1.15**
Councillors	**.35**	.19	.12	Planners	.69	1.09	**1.34**
Geographers	.61	1.27	**1.44**	Non-geographers	.78	1.40	**1.59**
< 31	.82	1.45	**1.64**	> 30	.29	.80	**.93**
Degreed males	.49	1.01	**1.15**	Undegree males	1.17	1.62	**1.78**
Degreed females	.83	1.47	**1.71**	Undegreed	.89	1.59	**1.75**
Groups' ave.	.66	**1.18**	**1.32**				

Fig. 4.9 Utility scores for the three scenarios according to each sub-group of past software users

Suitability scenario – males, people aged 40–54, non-Anglican Christians, workers in extractive industries, members of a household comprising a couple plus children, home buyers, parents with two children, residents who have not moved during the

RATINGS ACCORDING TO EACH GROUP

ACCORDING TO:	Trend	Suitability	Conservatior	ACCORDING TO:	Trend	Suitability	Conservatior
Everyone	.58	1.20	**1.24**				
Males	.57	**1.01**	.91	Females	.21	1.01	**1.23**
18-29	-.61	.05	**.32**	30-39	1.27	1.85	**1.99**
40-54	.52	**.96**	.88	No children	-.14	1.02	**1.38**
1 child	**.90**	.57	.28	2 children	.55	**1.13**	1.12
in Australia	.58	1.20	**1.24**	M.Easterns	**.90**	.57	.28
Asians	-.20	.23	**.44**	Australians	.58	1.19	**1.21**
Others	.49	1.09	**1.19**	Anglicans	-.61	.05	**.32**
O. christians	.52	**.96**	.88	Muslims	**.90**	.57	.28
O. non-christians	-.20	.23	**.44**	No religion	1.12	2.00	**2.25**
Moved in 5 yrs.	.51	1.19	**1.27**	Stayed in 5 yrs.	.66	**.88**	.69
Extract. industry	.20	**.61**	.53	Other ind.	.80	1.44	**1.64**
Professionals	.69	1.28	**1.30**	Other occ.	-.61	.05	**.32**
Uni. graduates	.38	.93	**.94**	Cert./Dip. holders	1.08	1.73	**1.74**
Sing. pers.	-.20	.23	**.44**	Couple - children	.35	1.06	**1.09**
Couple + children	1.42	**1.57**	1.38	Hm. owners	.66	**.88**	.69
Hm. buyers	.20	**.61**	.53	Pve. renters	.84	1.59	**1.85**
Reg. comp. users	.58	1.20	**1.24**	$600-$1500/week	.58	1.20	**1.24**
Groups' ave.	.46	**.95**	**.99**				

Fig. 4.10 Utility scores for the three scenarios according to all sub-groups within a recent workshop amongst landscape planning professionals

last five years, and home owners. Again, there is likely to be considerable overlap between these groups. Note that this workshop did not consider the Mitchell Shire problem at all; it only contributed its relationships between criterion scores and scenario merit, and such relationships were combined with the scores shown above in Fig. 4.7 to produce Fig. 4.10.

Given the small number of people at this particular workshop, one needs to be careful not to make any definitive conclusions about the attitudes of whole cultural groups. The spirit of our software is to help start a discussion about the merits and de-merits of scenarios, not to end it. Note also that, as perceptive readers will have already realized, the updated sub-group categories used by the current *Preference Prediction* software correspond to Australian Bureau of Statistics (ABS) Census categories. Hence, now that we have an idea of such sub-groups' preferences, we can move towards mapping those geographical areas which will probably support the different scenarios – simply by mapping census data.

Finally it should be remembered that once we have scenarios' scores on the nine criteria, we can predict their merit on behalf of still other groups who have, also, never heard of Mitchell Shire and its alternative futures. For example, taking the scores above, plus the relationships between criterion scores and scenario merit that were found to exist amongst some transport planning postgraduates in Brazil who have used our package, the *Preference Prediction* software predicts that this group will overwhelmingly favour the Conservation scenario, although those aged over 30 will favour the Suitability scenario. Moreover, using the same relationships learned from those Italian students of architecture and urban planning who have used our software, the prediction is that all sub-groups will favour Conservation, except that the Trend scenario will be preferred by those who said they only have a primary school level of education. Hence our findings show signs of being replicated across cultures and across disciplines.

Yet any system will always give biased results so long as it bases its predictions only what has been learned from a sample of people. This is why the *Preference Prediction* software needs to be always run in conjunction with a stakeholder workshop that represents the immediate (Mitchell Shire) community – rather than simply implemented hypothetically, for demonstration purposes, as we have done here. Grounding the PSS in a two- or three-day workshop ensures some consensus for the criterion scores that we have merely floated above, in Fig. 4.7. It also stimulates deeper and more informed discussion about probable community rankings of, and justifications for, the alternative scenarios.

4.7 Evaluation of PSS Software Packages

In this section we provide a brief evaluation of the fours PSS we have deployed. There are a number of reports and papers which categorise, summarise and evaluate a wide number of PSS products (EPA 2000; Brail and Klosterman 2001; Geertman and Stillwell 2003; Klosterman and Pettit 2005). However, none of this analysis

Table 4.1 Comparisons between our four PSS software packages

Attributes	*SLEUTH*	*What if?™*	*Google Earth*	*Preference Predictor*
Customisable	Open program access to API	Closed program set-up module	Not an open API, but KML files are customisable	Closed program setup module
Cost	Free	Low cost license 1 Seat = $US2950	No cost – low costs	Under development
Type of model	Non-deterministic loosely coupled cellular automata	Deterministic loosely coupled multi-criteria analysis	3D landscape visualisation model	Artificial intelligence-based, evaluation system
Model driver	Historic land use	Population projections	Spatial data layers	Past users' planning behaviour
Model calibration	12 measures of best fit	No calibration or sensitivity analysis of MCA procedure	Visual calibration of accuracy using local knowledge	Learns decision priorities from past workshop attendees/users
Time intensive	High	High	Low	Low
Planner perceptions	Black box 'business as usual' scenario	Semi transparent user driven scenarios	Powerful communication tool	Completely transparent
Output format	GI Raster	Shapefile vector	KML, and KMZ (extensible)	Graphics of scenario preferences and reasons for them
Output resolution	20 m	Land parcel	Dependent on input data	User defined
Land use aggregation	Highly aggregated: urban vs. non-urban	Disaggregated up to 100 land use types	User defined	User defined
Computational requirements	High power computing	Desktop computing	Desktop computing	Desktop computing

has been undertaken using a scientific methodology applied to a consistent study area. In our research, we have applied four PSS tools to one selected study area for a consistent period of 2001–2030 and the evaluation of each PSS packages is documented in Table 4.1.

SLEUTH requires significant data inputs including four snapshots of historical, remotely sensed imagery. These data can be difficult to obtain and it takes a long time to process before they are even fed into the model. Also, significant computing resources are required to calibrate the *SLEUTH* model and it does not allow the user to safely tamper with its calibrated coefficients in order to better reflect current local circumstances – users are stuck with just the historically derived coefficients. On the good side, however, the *SLEUTH* model has the advantages of being open source and so at least the sophisticated modeller is potentially able to manipulate its basic structure. Moreover, *SLEUTH* can actually reveal emergent behavior because of its usage of the latest modelling algorithms including cellular automata.

In many respects the *What if?*™ PSS is opposite to *SLEUTH*, even though both are used to simulate the upshots of land-use scenarios. *What if?*™ can be described as a series of simple deterministic models which allow the user to interlink a number of socio-economic and biophysical variables to create a supporting geodatabase environment. *What if?*™ is a collaborative platform where the user group extends beyond just the modellers to include planners, councilors and ultimately the community. Users can enter suitability factor weightings and decide upon population projects scenarios and whether to include planning overlays or other planning controls. In real time, scenarios can be processed and presented back for further discussion.

Unfortunately, however, *What if?*™ is a commercial software package supported by a small developer team and there can be considerable frustration when encountering bugs and a time delay in getting these fixed. As the software is not open source, it is necessary to go back to *What if?*™ Inc. for a remedy. Another disadvantage of not being able to access the source code is the additional functionality cannot be built in by the model developer. The perceptions of the planners of *What if?*™ are that even though there are simple deterministic equations underlying land suitability, demand and allocation it takes a lot of time to understand how the model works and thus it is considered a complicated PSS tool. Once this is achieved, confidence in the model dramatically increases and the formulation and exploration of various *What if?*™ scenarios can commence.

The third software we have deployed is the virtual globe *Google Earth* product. *Google Earth* offers a no/low cost spatial information viewing platform, which has been downloaded by more than 100 million users world wide. *Google Earth* is intuitive to navigate and thus instills user confidence and it is particularly suitable as a spatial viewer for non-GIS experts. In *Google Earth,* spatial modelling outputs in a range of formats, including shapefile and other GIS formats can be exported to be viewed and shared with a wider audience. After we used the *Google Earth* viewer to display the outputs from the *SLEUTH* and *What if?*™ PSS packages, initial feedback from planners and councillors was extremely positive. Users could view the temporal and spatial changes expected in the landscape, and by using the layer transparency slider bar in *Google Earth* users can visually compare the existing land uses with future land uses. Unfortunately *Google Earth* is not open source so potential customisation is limited and it can be quite time consuming to produce even a simple customised legend. Also, the existing imagery supplied by Google does not include a time stamp, so it is difficult to ascertain what the 'current' land use practices are. This provides a substantial challenge when comparing current and future scenarios. Moreover, *Google Earth* struggles when rendering relatively small graphic outputs, particularly Keyhole Markup Language (KML) or KMZ files larger than 20 megabytes.

The final PSS we deployed was the *Preference Prediction* software, which has been used to evaluate the three land-use change scenarios. The software can theoretically support both top down and participatory planning processes, but in our application we have not engaged planners to score the scenarios on the nine criteria. That is, due to time constraints, we derived such scores using expert modeller opinion based on the authors' understanding of the study area. The *Preference Prediction* software is not a spatial tool. Rather, it evaluates scenarios from a holistic perspective of performance on a range of generic criteria. As such, there is no spatial output but a number of graphs and comparative statistics instead. One of its limitations is its 'training' population which, although it includes ordinary people from a number of community groups, is still too biased towards university students to constitute a truly representative sample of the population at large. On the other hand, its strength is that it can used in real time to provide instant feedback to end users, which ultimately would include planners, politicians and the community.

4.8 Conclusions

In this chapter we have applied four PSS tools and evaluated how well they can inform planners about future urban growth impacts within peri-urban areas. Specifically, we used two simulation model-based PSS tools, *SLEUTH* and *What if?*™ to develop three future land-use scenarios for the Mitchell Shire in 2030:

- business as usual (Trend);
- developer-centric growth (Suitability); and
- agricultural conservation (Conservation).

We then used *Google Earth* as a communication tool to enable local planners and councillors to visually explore each of the scenarios. This resulted in lively debate around some of the key issues facing the Shire, such as urban development pressures, the immigration of people from within and beyond the Shire, and zoning policies' impact on hobby farmers, primary producers and land developers. Finally, we demonstrated how to evaluate the three scenarios using the *Preference Prediction* software, and concluded that the Suitability scenario is likely to be the most popular one.

Whereas much previous research into PSS has focused on the development of a particular PSS tool, we have undertaken a more novel approach by applying and evaluating a number of PSS. As with climate change modelling, where a suite of models are run and trends and convergence are analysed to better understand future-based scenarios, we sought better understanding of landscape futures by taking a 'shot gun' approach. No person or model can predict the future; all that we can aim for is some simple understanding, based on a few predetermined factors, of what the likely implications of particular actions might be.

We conclude that both *SLEUTH* and *What if?*™ constitute useful tools for better understanding some of the policy implications for urbanising landscapes. Moreover, the recent proliferation of virtual globe products, like *Google Earth*, provides a novel way of visualising landscape futures. We believe that the innovative use of virtual globe products will better engage planners, politicians and communities, thereby offering exciting opportunities for improved communication of complex and complicated models' results. Finally, we believe the *post-ante* evaluation of scenarios using PSS tools like the *Preference Prediction* software, which should always be applied in collaboration with planners and communities, will lead to more participatory and consensus-based planning practice.

This is ongoing research and further engagement with the Mitchell Shire Council is envisaged in order to obtain feedback from the planners and councillors about the three future land-use scenarios presented in this chapter. To date we have developed three scenarios with the local planners and have deployed the *Preference Prediction* software to evaluate the future land-use scenarios. An interesting future research topic would be to compare the evaluations made by the 'real' Mitchell Shire community to those made by the *Preference Prediction* software's training population(s). Finally, future research is required to further develop better *post-ante* evaluation tools for understanding the sensitivity of modelling inputs, outputs and likely policy impacts.

Acknowledgements We would like to thank Hemayet Hossain who collected considerable information about the *SLEUTH* model, and whose persistence amid difficult circumstances enabled him to successfully implement it within the Mitchell Shire. Thanks are also due to Amanda Keogh, for her help in developing the *What if?*™ scenarios, Jean-Philippe Aurambout for his assistance with *Google Earth* and Mitchell Shire planner Katie Rizzuto who contributed much valuable input when formulating the land use planning scenarios.

References

Aurambout, J-P., Pettit, C. and Lewis, H. (in press) Virtual globes, the next GIS? *Landscape Analysis and Visualisation.*

Brail, R.K. and Klosterman, R.E. (2001) *Planning Support Systems,* ESRI Press, Redlands, California.

Clarke, K.C., Gazulis, N., Dietzel, C.K. and Goldstein, N.C. (2006) A decade of SLEUTHing: lessons learned from applications of a cellular automaton land use change model, *International Journal of Geographic Information Science and Systems*. Retrieved October 25, 2007, from www.toxicologynetbase.com/ejournals/books/book_summary/toc.asp?id=5245.

Dietzel, C. and Clarke, K.C. (2007) Towards optimal calibration of the SLEUTH land use change model, *Transactions in GIS,* 11(1): 29–45.

EPA (2000) *Projecting Land-Use Change: A Summary of Models for Assessing the Effects of Community Growth and Change on Land-Use Patterns*, United States Environmental Protection Agency, Washington.

Geertman, S. (2002) Participatory planning and GIS: a PSS to bridge the gap, *Environment and Planning B,* 29: 21–35.

Geertman, S. and Stillwell, J. (2003) *Planning Support Systems in Practice*, Springer-Verlag, Berlin.

Goldstein, N.C. (2005) *Brains vs. brawn: comparing strategies for the calibration of a cellular automata-based urban growth model*, In Atkinson, P.M. *et al.* (eds.) *GeoDynamics*, Boca Raton, CRC Press, pp. 249–272.

Herold, M.N., Goldstein, N.C. and Clarke, K.C. (2003) The spatio-temporal form of urban growth: measurement, analysis and modelling, *Remote Sensing of Environment*, 86: 286–302.

Kim, J-W. (2004) A simulation of the growth of the seoul metropolitan's built-up area with a GIS-based PSS Model, *Journal of the Korean Planners Association*, 39(7): 69–84.

Klosterman, R.E. (1999) The What if?™ collaborative planning support system, *Environment and Planning*, 26: 393–408.

Klosterman, R. and Pettit, C. (2005) An update on planning support systems, *Environment and Planning B*, 32: 477–484.

Klosterman, R.E., Siebert, L., Hoque, M.A., Kim, J-W. and Parveen, A. (2006) What if?™ evaluation of growth management strategies for a declining region, *International Journal of Environmental Technology and Management*, 6(1–2): 79–95.

Lee, D.B. (1973) Requiem for large-scale models, *Journal of the American Institute of Planners*, 39: 163–178.

McHarg, I.L. (1969) *Design with Nature*, Doubleday-Natural History Press, New York.

New York Climate and Health Project (NYCHP) (2006) Retrieved September 22, 2006. http://pages.csam.montclair.edu/luca/Calibration/calibration.html.

Pettit, C.J. (2005) Use of a collaborative GIS-based planning support system to assist in formulating a sustainable-development scenario for Hervey Bay, Australia.

Sangawongse, S., Sun, C.H. and Tsai, B.W. (2005) *Urban Growth and Land Cover Change in Chiang Mai and Taipei: Results from the SLEUTH Model,* Paper presented at the MODSIM 05 Conference, Melbourne, Australia. Retrieved October 25, 2007, from www.mssanz.org.au/modsim05/authorsS-T.htm.

Silva, E.A. and Clarke, K.C. (2002) Calibration of the SLEUTH urban growth model for Lisbon and Porto, Portugal, *Computers, Environment and Urban Systems*, 26: 525–552.

Solecki, W.D. and Oliveri, C. (2004) Downscaling climate change scenarios in an urban land use change model, *Journal of Environmental Management*, 72: 105–115.

Wyatt, R. (2008) A Decision Aiding System (DAS) for predicting people's scenario preferences, In Pettit, C. (ed.) *Place and Purpose*, Springer-Verlag, Berlin.

Wyatt, R. (2004) *A New Method for Predicting Humans' Scores for Plans*, Paper presented at the *Eighth World Conference on Systematics, Cybernetics and Informatics* (SCI 2004), Orlando, 1: 177–182.

Yang, X. and Lo, C.P. (2003) Modelling urban growth and landscape changes in the Atlanta metropolitan area, *International Journal of Geographical Information Science*, 17(5): 463–488.

Additional Reading

Harris, B. (1989) Beyond geographical information systems: computers and the planning professional, *Journal of American Planning Association*, 55: 85–92.

Landis, J. (1995) Imagining land use futures: applying the California Urban Futures model, *Journal of American Planning Association*, 61(4): 438–456.

Pettit, C., Cartwright, W. and Berry, M. (2006) Geographical visualisation a participatory planning support tool for imagining landscape futures, *Applied GIS*, 2(3): 22.1–22.17.

Shiffer, M. (1992) Towards a collaborative planning system, *Environment and Planning B*, 19: 709–722.

Part II
Design, Development and Implementation of Recent PSS

Chapter 5
A Motor Vehicle Safety Planning Support System: The Houston Experience

Ned Levine

5.1 Introduction

Planning Support Systems (PSS) have long been used for motor vehicle safety in order to improve roadways or implement programs for drivers. Usually called *crash information systems*, traffic safety specialists have long assumed that improvements come about through timely information on motor vehicle crashes and analysis of that information. Frequently referred to as a 'data driven' methodology, the analysis of motor vehicle crashes is the basis upon which many, if not most, improvements to the traffic system have come about in the USA, Europe and elsewhere. In the last decade, the use of geographic information systems (GIS) has brought a much needed spatial dimension to crash analysis and widened the analytical and policy tools available for safety planners and traffic engineers.

In this chapter, a motor vehicle safety PSS that was developed in Houston, Texas for safety planning is described. The limitations of such a system are explained and the mechanisms for dovetailing information with expert opinion in order to address both the behavioural as well as the physical road characteristics affecting traffic safety are outlined.

5.2 A Major Public Health Problem

Fatalities and injuries from motor vehicle crashes are a major public health problem. In 2005, there were 43,443 deaths in the United States (U.S.) resulting from motor vehicle crashes (NHTSA 2007). While the fatality rate per 100 million vehicle miles traveled (VMT) has been decreasing consistently and is the lowest on record (1.45 per 100 million VMT), motor vehicle accidents are still a major source of death in

N. Levine (✉)
Ned Levine & Associates
Houston, Texas, USA
E-mail: ned@nedlevine.com

S. Geertman, J. Stillwell (eds.), *Planning Support Systems Best Practice and New Methods*,
DOI 10.1007/978-1-4020-8952-7_5

the U.S. and elsewhere (NHTSA 2005). In fact, in 2004 in the U.S., fatalities from motor vehicle crashes was the third leading cause of death for males and the sixth leading cause of death for females (DHHS 2006: Table 307). Fatalities from motor vehicle crashes are the leading cause of death for all ages from age 1 through to age 44 (DHHS 2006: Table 32). The comparison is for fatalities from 'unintentional injuries', which includes most deaths from injury and poisoning. The category excludes homicides (including legal intervention), suicides, deaths for which none of these categories can be determined and war deaths (NSC 2007: pp. 8–9). The National Safety Council estimates that motor vehicle crashes cost about $237.7 billion a year (in 2005 dollars) from medical treatment, wage and productivity losses, administrative expenses, property damage and employer costs (NSC 2007, p. 7). This is in addition to the pain and suffering experienced by motor vehicle injury victims and their friends and relatives. In short, motor vehicle crashes are a major public health problem.

On a per capita basis, fatal crash rates are higher in the U.S. than most other developed countries though this comparison does not consider actual vehicle exposure (Scotland 1998; Peden *et al.* 2002; NationMaster.com 2007). Other studies show very low fatality rates relative to vehicle exposure for the U.S. when compared to other countries (Jacobs 1986; DSA 2005; Best 2007).

Over time, safety has consistently improved. In 1950, the age-adjusted mortality rate from motor vehicle crashes was 24.6 per 100,000 persons while it was 15.2 in 2004 (DHHS 2006: Table 320). Improvements to roadways and vehicles, increased seat belt use and alcohol enforcement have been major factors underlying the decline with a lesser influence on behavioural changes from drivers (Baxter 2006). Unfortunately, the decline has leveled off since about 2000, suggesting that additional measures will be needed, more tailored to drivers than to roadways (Chakiris 2006).

5.3 Crash Analysis as a Basis for Road Safety Decision Making

5.3.1 Federal U.S. Safety Programs

In the U.S. and many other countries, crash analysis is required as a basis for making roadway improvements. The Highway Safety Act of 1966 (23 U.S.C. Chap. 4) and the Highway Safety Act of 1973 (23 U.S.C. 152) established a requirement to reduce injuries, deaths and property damage from motor vehicle crashes through the development or upgrading of traffic record systems, traffic engineering studies, the development of technical guides for States and local highway agencies, work zone safety projects, the encouragement of the use of safety belts and child safety seats, roadway safety public outreach campaigns, enforcement to reduce impaired drivers, enforcement programs to combat drivers who speed or drive impaired, and enforcement to reduce aggressive driving (FHWA 1999).

The 1973 law, in particular, required that each State conduct and systematically maintain a survey of all highways to identify hazardous locations that may constitute a danger to vehicles and to pedestrians. Priorities for the correction of these hazards are required and a schedule of projects for their improvement must be established. The law established a benefit-cost methodology for identifying safety project locations and for assigning priorities. The Act provided mandates for States and an earmarked funding source for safety improvements. States could not use the excuse of lack of funding to avoid having to improve the safety of the highways.

5.3.2 Methodology for Safety Improvements

There is a formal methodology for funding safety improvements using Federal funds. Known as the *Highway Safety Improvement Program* (HSIP), there are four steps that are required for receiving Federal aid to improve safety on the roadways (FHWA 1981). Briefly, they are:

- the maintaining of crash and highway information for all jurisdictions at the State level;
- the requirement that the State identify hazardous locations on the basis of crash experience or crash potential;
- the requirement that an engineering study be conducted of hazardous locations to develop safety improvement projects; and
- the requirement that priorities be established based on a benefit-cost methodology for identifying the potential reduction in crashes from an improvement compared to the cost of the improvement.

This methodology is common throughout the U.S. and, to some extent, is used throughout the world. However, given the decentralized nature of planning in the U.S., compared to most countries, the use of the HSIP methodology is at the discretion of local decision makers. An extensive discussion of this methodology as applied to a crash hot spot in the East End of Houston can be found in the following source (H-GAC 2004).

5.4 The Greater Houston Motor Vehicle Safety PSS

5.4.1 Safety in the Houston Region

Between 2001 and the present, a motor vehicle safety PSS was developed at the Houston-Galveston Area Council (H-GAC), the metropolitan planning organization for the greater Houston region. The Houston metropolitan area has a severe safety problem. Between 1999 and 2001 in the eight-county region, there were 252,241 serious crashes, an average of 84,080 a year. From these crashes, 1,882 persons were killed and 281,914 persons were injured. The crashes accounted for 26 per cent of all serious crashes in the State of Texas during that period compared to a 22 per cent share of the State's population and a 21 per cent share of the State's VMT (H-GAC 2007a).

The likelihood of a driver in the region being involved in a fatal or injury crash in the region was 36 per cent higher than the State of Texas average and 149 per cent higher than the U.S. average. The region leads the State of Texas in virtually every type of crash and leads the nation in alcohol-related fatalities per capita (NHTSA 2005).

5.4.2 Geo-Coding Motor Vehicle Crashes

Starting in 2001, a GIS-based database was established that identified all serious crashes in the region (Levine 2006a). The data were obtained from the Crash Records Bureau of the Texas Department of Public Safety, the agency vested with compiling crash information for every jurisdiction in the state of Texas (DPS 2007). Texas State law requires that all crashes involving fatalities, injuries, or property damage in excess of $1,000 be reported to the local police. The local police department will then send copies to the Texas Crash Records Bureau if it meets the criteria of having fatalities, injuries or serious property damage (defined as one or more vehicles being towed from the crash scene). Because of delays in releasing information, data were only obtained on those crashes occurring between 1998 and 2001.

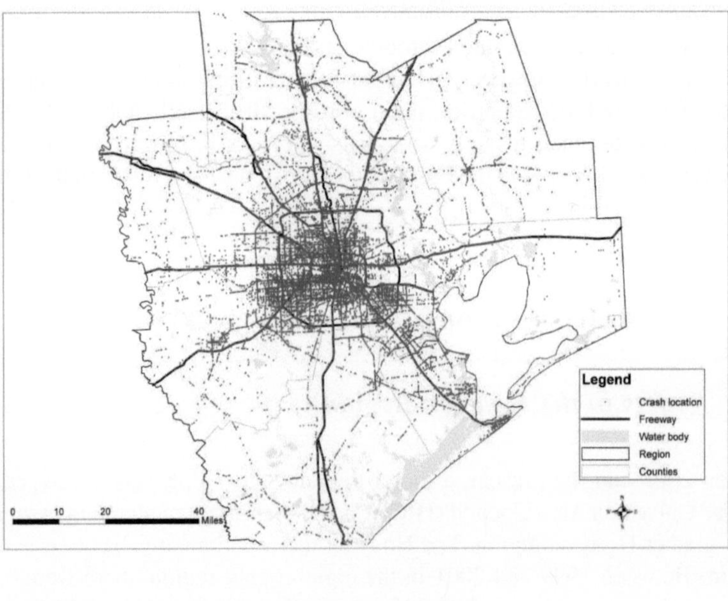

Fig. 5.1 Speeding crashes in the Greater Houston Region

The data were geo-coded according to a methodology that was developed initially in Honolulu (Levine *et al.* 1995a,b; Levine and Kim 1999). Approximately 82 per cent of the crashes were geo-coded with accuracy being around 91 per cent. Accuracy was tested by drawing a sample of 500 geo-coded crashes and examining the records to see whether they had been correctly located. Both matching and accuracy improved towards the centre of the region. For example, in the City of Houston, 90 per cent of the crashes were matched with 94 per cent accuracy. Once geo-coded, crashes or subsets of the crashes could be displayed. For example, Fig. 5.1 shows a map of all speeding crashes in the region.

5.4.3 A Data and Analysis-Driven PSS

Figure 5.2 illustrates a conceptual view of the motor vehicle safety PSS. At the root of the system is the geo-coded crash data. It is shown as two stages: the raw data collected from the Texas Crash Records Bureau and the geo-coding process that identifies the approximate location of the crash. The crashes can be then linked by a GIS to other information, such as roadway inventory, traffic volume information, land use data, EMS reports and even hospital reports. The aim is to widen the contextual information for understanding safety data in order to explore possible interventions.

Building on the basic spatial data and associated links is spatial analysis. Among these are hot spot analysis, crash risk analysis, spatial-temporal analysis (e.g. hot spots or crash risk by time of day) and visualization of crashes throughout the region; the latter is useful for a region-wide view of safety or a view of safety along a particular major arterial road. There is also an increasing interest in forecasting crashes based on expected mitigations. These analyses help identify high volume and high risk locations as well as place these locations in a broader context. The tools themselves may be part of a broader analytical focus. The analysis allows a number of applications, including routine reports that would be issued on a regular

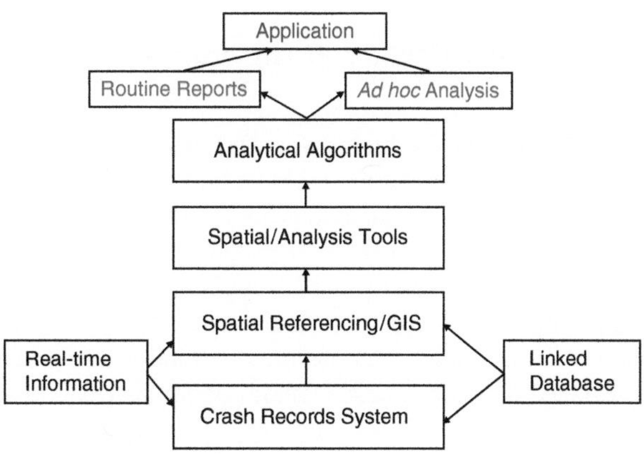

Fig. 5.2 A conceptual motor vehicle safety PSS

basis and *ad hoc* reports that are created to address a specific issue. The important point is that the spatial data and GIS are essential parts of the PSS, but the system also includes analytical tools and expert opinion.

5.5 Uses of the Greater Houston Motor Vehicle Safety PSS

The Greater Houston motor vehicle safety PSS has been used extensively in safety planning efforts. The following discusses three major applications of it.

5.5.1 Safety Reports

First, it was used for producing safety reports that were then placed on a safety web page (H-GAC 2007a). More than 40 reports were issued between 2001 and 2007. The most common type was an examination of safety along particular roadways. Typically, these were three to five page reports summarizing the number of crashes, their severity, major hot spots and crash risk. They were usually initiated by a request from a local government. A second type of report examined particular types of crashes, such as red light running crashes or pedestrian crashes. These reports have been widely distributed. Local governments, community groups and consulting organizations frequently use the reports as background material in framing policies or proposed actions. The reports also affected actual decisions made to improve safety. For example, in 2006, the City of Houston adopted photo enforcement for the running of red lights. The motor vehicle safety PSS was used to inform the City about the location of intersections with the most red light running crashes. The City used the information along with more recent data from their own database to identify 50 intersections throughout the city where red light running crashes were most likely to occur (H-GAC 2006; Stein and Dahnke 2006).

5.5.2 Identification of Hazardous Locations

Second, the motor vehicle safety PSS was used to identify specific locations where safety engineering studies could be implemented, following the HSIP methodology described above. An analytical tool used for this purpose was *CrimeStat* (Levine 2007). *CrimeStat* is a stand-alone program that was developed with funding from the National Institute of Justice and is available for free at http://www.icpsr.umich.edu/crimestat. The program interacts with most desktop GIS packages and calculates various statistics about the distribution of incidents, many of which can be displayed in a GIS. These include spatial distribution statistics, hot spot identification, risk analysis of incidents and space-time interaction statistics. There are also a

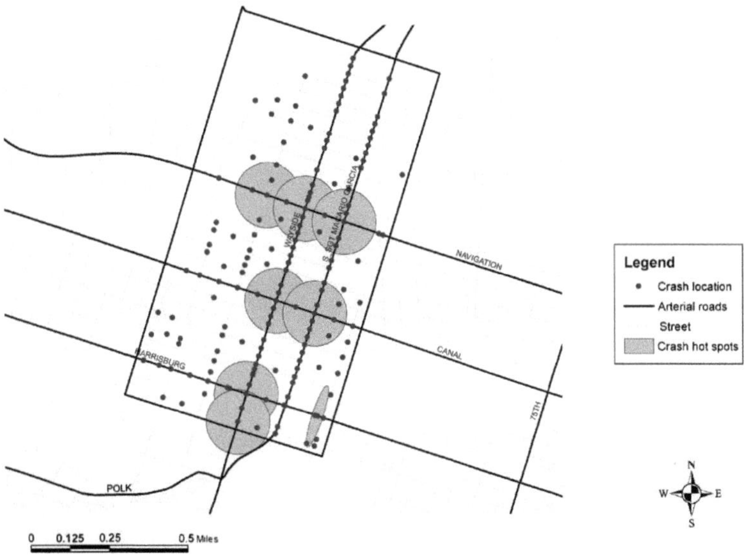

Fig. 5.3 Crash hot spots in East End of Houston

number of tools specific to law enforcement, including a full travel demand module for modelling crime travel. There are, of course, other programs that can be used for spatial analysis of crashes, but *CrimeStat* was developed with both crime and crash analysis in mind.

For crash analysis, mostly the hot spot routines were used to identify small areas with a high concentration of crashes. Figure 5.3 illustrates eight crash hot spots identified in the safety engineering study for the East End of Houston. The particular algorithm was a nearest neighbour hierarchical clustering routine that looks

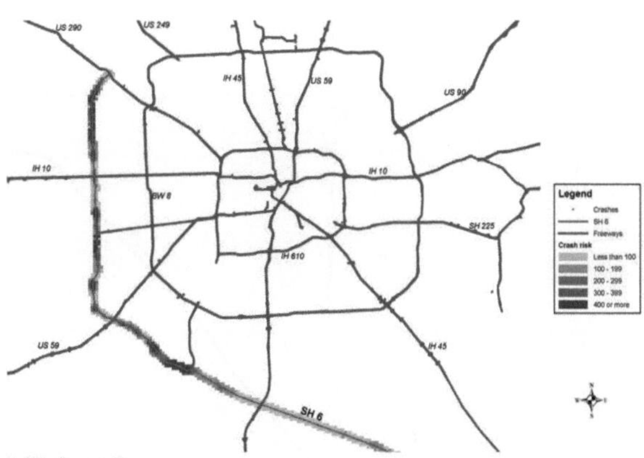

Fig. 5.4 Crash risk on State Highway 6, 1999-2001 (See also Plate 13 in the Colour Plate Section)

for pairs of points that are closer than a threshold and then groups these together in clusters (Levine 2007). The advantage of a hot spot approach over a frequency count of the number of crashes at each location is that it can include nearby crashes that both contribute to and are by-products of the hot spot. For example, in Fig. 5.3 the hot spots are centered on particular major intersections but also include crashes on the approaches to the intersections.

A second type of analysis that goes into the identification of hazardous locations is the identification of locations of high crash risk (crashes relative to traffic volume). When the volume of traffic approaches exceeds the capacity of the road to handle the traffic, crashes tend to occur. Thus, it is important to analyze crash risk, too, usually defined in units of crashes per 100 million VMT. Figure 5.4 shows crash risk along a major suburban arterial, State Highway 6. Again, the *CrimeStat* program was used to interpolate the crashes relative to the VMT. The particular algorithm interpolates both the number of crashes and the VMT to a fine grid using a smoothing algorithm. It then divides the interpolation of crashes by the interpolation of VMT (Levine 2007); the ratio is then multiplied by 100 million. Finally, Fig. 5.5 shows crash hot spots and crash risk along a segment of Westheimer Road, a major arterial in the Houston region. The map plots crashes per 100 million VMT and shows a difference between high crash locations, which typically are adjacent to the freeways because of the bottleneck of traffic from entry and exits ramps, and high crash risk locations some of which are away from the freeways and are often characterized by intersections with poor traffic control.

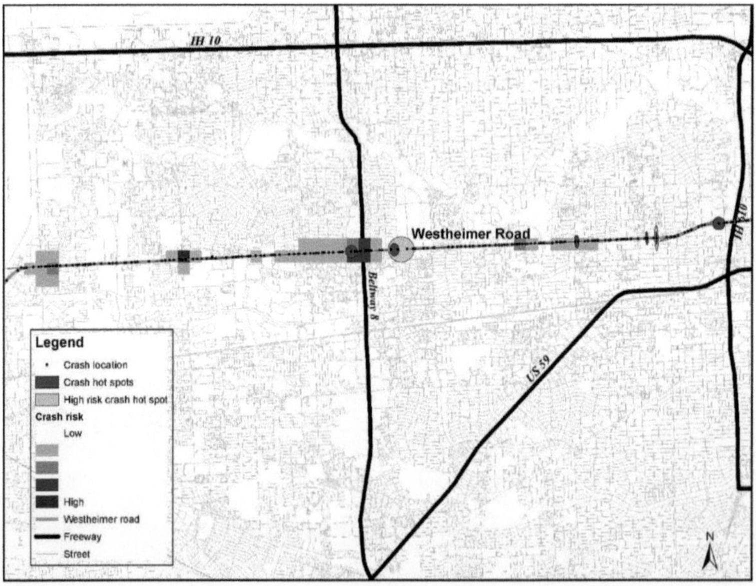

Fig. 5.5 Crash hot spots and crash risk on Westheimer Road (See also Plate 14 in the Colour Plate Section)

5.5.3 *Safety Engineering Projects*

Third, the motor vehicle safety PSS was used to develop safety engineering studies using hot spot analysis as the basis for selecting locations. The process involved establishing a partnership with local governments and with other relevant parties. Typically, discussions were held with the local government to establish a commitment, a funding stream, and a time line. Since the number of hot spots was almost always greater than the resources available, priority locations had to be established. Once agreed upon, there was still a need to establish the funding split between the regional agency and the local government and setting up advisory groups.

In the 2001–2007 time frame, five safety engineering studies were implemented. The first was a study of two intersections along a major arterial in Houston (Westheimer Road). The second was the East End safety study that involved examining all crashes within an area that was slightly less than one square mile. There were about 90 intersections within the study area, though about 10 accounted for most of the crashes. A third study involved studying five intersections in the City of Pasadena, a suburban community near Houston. A fourth study examined 13 intersections in the City of Galveston, an urbanized island about 50 miles from Houston, and a fifth study studied 12 intersections in the City of Sugar Land, another suburban city in the Houston region. These studies can be read on the H-GAC safety web page (H-GAC 2007a).

In all of these studies, the basic HSIP methodology discussed above was followed. The consultant first gathered all crash reports for a multi-year period. At the minimum, three years' worth of data were used. Second, *collision diagrams* were produced. These drawings indicate directionality and impact locations for the crashes. Additional data were obtained to provide contextual information about the crashes, in particular traffic counts, vehicle movements and signal timing. The consultant then examined contributing factors that are listed on the police report. Typically, these are behavioural factors such as driving too fast for the conditions, running a red light or changing lanes without regard for adjacent vehicles. The purpose is to establish a crash pattern that can then be mitigated. Fourth, the consultant then proposed various mitigation factors, which were then subjected to a formal benefit-to-cost analysis.

The methodology is aimed at identifying low cost improvements that will have substantial benefits. For example, at one intersection in the City of Houston, the consultant demonstrated that re-painting lane markings would reduce the number of crashes at that intersection by about 15 per cent with very little cost. At another intersection in the City of Houston, the proposal was to make a far left lane on a one-way street into a left turn only lane to minimize conflicts with vehicles in the second-to-the-left lane. At another intersection in the City of Sugar Land, it was proposed to create two dedicated left turn lanes because of the volume of traffic had grown beyond the capacity of the existing one left turn lane. At another intersection in the City of Galveston, crashes were occurring because vehicles were moving quickly to the left in order to make a left turn on to a major arterial; the consultants proposed putting lane signs much earlier on the approach to the intersection.

In short, this approach systematically mines the crash data in order to establish low cost improvements that can substantially reduce crashes. In practice, a benefit-to-cost ratio of 3:1 (meaning that the value of the benefits are at least three times the value of the costs) is considered the minimum criteria for choosing a particular measure. Many of the recommended measures had much higher benefit-to-cost ratios.

5.6 Advantages and Limitations of the HSIP Methodology

Even though the HSIP methodology has been around for more than 30 years, the Greater Houston motor vehicle safety PSS has allowed the metropolitan area to be scanned to identify all possible hazardous locations and to prioritize them. This is a major advance since the methodology was historically applied only when individual intersections or road segments were brought to the attention of transportation planners or engineers. The HSIP methodology, in turn, generally identifies low cost improvements, such as better signage, re-striping of lane markings, improved signal timing or re-location of lanes. If combined with routine maintenance, such a methodology can significantly reduce the number and severity of crashes on the road system. To a large extent, this has happened anyway as the roadway system has been improved over the decades. But, a more systematic integration of safety analysis with maintenance could provide a cost effective solution to safety improvement.

5.6.1 Ignoring of Behavioural Issues

The biggest disadvantage of this approach is that it is limited primarily to improvements on the road system and does not address the driver behaviour problems associated with crashes. Poor design of a roadway or inadequate traffic management of an intersection can contribute to crashes, but the crash is still a function of driver behaviour. For example, according to the Houston crash reports, 39 per cent of crashes occurring between 1999 and 2001 involved speeding; 20 per cent involved failing to yield the right of way; 19 per cent involved failing to stop at a red light or a stop sign; 7 per cent involved driving under the influence of alcohol or drugs; and 3 per cent involved following too close (H-GAC 2007b: p. 4). Almost 90 per cent of the crashes are caused by the behaviour of drivers, rather than defects in the roadways per se, though there may be interactions.

When the characteristics of drivers involved in crashes were examined, it was very clear that certain types of individuals were more likely to be involved in motor vehicle crashes than others. Teenage drivers were involved in 21 per cent of all crashes; their share of the driving age population is 9 per cent. Further, in the suburban areas of the Houston metropolitan region, the percentage of crashes involving teenagers was substantially higher than in the central city, typically in the 26–29 per cent range (H-GAC 2007b: p. 5). Males were more likely to be involved in serious crashes than females and there is a growing body of evidence suggesting that bad driving is a repetitive behaviour. For example, drivers who are convicted of drunk driving were more likely to have been involved in previous drunk driving behaviour (Roth 2006). Several studies have shown that drivers convicted of running red lights were more likely to have been convicted of previous red light running incidents (IIHS 2007; Schultz 2001).

Thus, the approach outlined above is too narrowly focused on the roadway system and not sufficiently on the driver. The roadway system is very important and improvements to it have been, and certainly will continue to be, important in improving safety. But, it is not the only factor.

5.7 Establishment of a Regional Safety Council

Consequently, in 2005, a decision was made to create a Regional Safety Council (RSC) in order to advise elected officials about actions that should be taken to improve safety. The RSC was made up of safety specialists from transportation, law enforcement, medicine, public health, trucking, insurance, business, safety advocacy, and safety research organizations. During its first year, the RSC set up individual committees to focus on four specific safety topics: (i) alcohol-related crashes; (ii) aggressive driving; (iii) freight safety; and (iv) safety information systems. At the end of the year, the RSC released a report that outlined major recommendations for improving safety in the region (H-GAC 2007b).

The Greater Houston motor vehicle safety PSS played a critical role in providing background information to the RSC. First, it provided basic facts on the extent of the problem and indicated what types of drivers were involved. Second, it provided locational information on the major hot spots and, third, it provided information that was used to estimate the cost of crashes. This approach provided a broader perspective on safety than the roadway-only orientation of the HSIP methodology. Planning and safety engineering was put into a framework of providing infrastructure that would facilitate education and enforcement efforts, rather than as a sole solution to safety problems. The recommendations focused on the interaction between driver behaviour and the roadway system. For example, one recommendation was to set up several safety corridors in the region in which enforcement, education and engineering would be focused. In each corridor, driver education programs would be targeted towards nearby high school students and would emphasize driving risk factors, particularly driving under the influence of alcohol (Wise 2006; Henk 2006). Engineering would be utilized to upgrade the intersections and segments along the corridor. In short, the safety corridor will involve the integrated contributions of engineering, education and enforcement in order to improve safety.

Another recommendation focused on distributing dynamic speed signs throughout the region to reduce speeding. If placed on roads with a sizeable number of crashes involving speeding, this technology would be a very cost effective way to improve safety. Several recommendations focused on reducing drunk driving. Ignition interlock devices are attached to the starter mechanism of vehicles. To start the engine, drivers have to blow into the device and be free of alcohol in their breath. Such devices are mandated by Texas law for those individuals convicted of a second or subsequent Driving While Intoxicated (DWI) offence at a high Blood Alcohol Content level of 0.15 or higher or for minors convicted of driving with any alcohol in their bloodstream. Nevertheless, only about 15 per cent of those persons eligible for such a condition were assigned it by their judges. The RSC recommended the expansion of this technology to the maximum and the development of educational efforts among the legal community to ensure enforcement of the law.

Another anti-DWI proposal recommended forbidding servers of alcohol in establishments from drinking while on the job (Copeland 2006; Davies 2006). Another measure addressed the 'intoxicants clause' in Texas law (and in many other States) that allows insurance companies to withhold payments for crashes in which there is drunk driving. One of the side consequences of this law is that hospital administrators are reluctant to order blood alcohol tests for motor vehicle crash victims who are brought into the emergency rooms even if it is clear that they were single drivers who had been drinking. The result is that many of these persons are treated and sent home without any legal actions being taken, only to repeat their behaviour later on.

Another anti-drunk driving recommendation was to identify and study both the hot spots where DWI crashes occurred and the residence location of drivers convicted of DWI. Most crash data that is released only includes information on the location of the crash, not where drivers live. But, some preliminary studies have indicated that drunk drivers may be concentrated in certain neighbourhoods with a sizeable number of bars and liquor stores and it may be possible to implement intervention programs in those neighborhoods (Canter *et al.* 2005). Other recommendations addressed truck safety and improving education to the trucking community, to elected officials, and to the public at large; Borchardt and Corder 2006; Curry 2006). Also, the RSC recommended improving data collection and the sharing of data across agencies (Benz 2006; Levine 2006b).

In short, the Regional Safety Council proposed a wide variety of different measures for improving safety. The motor vehicle safety PSS was important in providing information on the location and conditions associated with the crashes. But, the information needed to be placed in a larger, behavioural context in order to lead to decisions that have some chance of being effective.

5.8 Conclusions and Future Directions

The Greater Houston motor vehicle safety PSS has been successful in so far as it has identified safety problems in the region in a systematic way and has directed decisions towards improving safety, particularly at intersections where safety problems are severe. It was one of the first regional safety planning efforts in the country and has been highlighted in conference presentations and articles (Levine 2006a). Other regional transportation agencies in the country are adopting safety planning and the Houston system has been held up as a model (Washington *et al.* 2006).

There are key requirements to make such a program effective, though. First, the safety data need to be of high quality. Crash data contain lots of errors, primarily from poor data collection. Police officers are under extreme pressure to manage a motor vehicle crash, involving safely securing the road, ensuring that EMS vehicles come to the scene, making sure that the crashed vehicles are removed, overseeing the clean up after the crash and documenting the crash in a report. Filling out a crash form has the lowest priority in this sequence and is usually done after the crash scene has been cleared. That information is condensed and abbreviated is all too understandable. Nevertheless, the consequence is that there is incomplete information in reports which makes analysis difficult. There are a lot of efforts underway to improve the data collection process (NCHRP 2005; AECOM *et al.* 2002).

Second, the data needs to be timely. Unfortunately, at the time in which the Regional Safety Council met, the Texas Crash Records Bureau had released only data through 2001. The time delay meant that there was not current information about the state of safety, making it difficult to evaluate the effect of various mitigation measures. For example, in 2002, the Texas State Legislature implemented restrictions on teenage drivers that required a driver under 18 to be accompanied by a person 18 or over and that also restricted hours of driving. Without 2002 or 2003 data, however, it was unclear whether this legislation had been effective in reducing teenage crashes. Efforts are being addressed to improve the situation, not only in Texas but elsewhere in the country (Baxter 2006). The Federal Government provides funds for crash information systems and is accelerating its importance after years of neglect. In Texas, a new crash records management system is being implemented with accelerated processing of crash records and with all crash locations being geo-coded for use in a GIS; this will facilitate the use of crash data by regional planning agencies who can immediately map the crashes with their GIS programs. Prior to this, H-GAC had been the only regional agency in the State of Texas to have done this.

A third issue concerns the ability to link crash data with other information, particularly EMS records and hospital admission data. The importance lies in documenting the true costs of motor vehicle crashes as well as being able to evaluate the effects of certain policies (e.g. helmet use by motorcyclists, proper seatbelt use by young children). However, most medical providers are reluctant to share these data for a variety of reasons (concerns about privacy, fear of liability, competitiveness among medical providers). The National Highway Traffic Safety Administration pioneered efforts at linking crash data with medical records, but there is still a lot of resistance to doing this on a regular basis, especially at the local level (NHTSA 1997). A legal basis for data sharing needs to be established that will require actions at State or national levels.

Fourth, in addition to the data quality issues, there is the importance of analysis and spatial analysis in particular. Showing an elected official a map with thousands of dots on it representing the location of motor vehicle crashes is good for 'shock value', but can lead to inaction because the problem will be seen as overwhelming. On the other hand, showing a map with hot spots clearly identified can help focus on actions that can improve safety at those locations. In this case, the problem seems more contained and the hot spots point towards priorities that can be taken to improve safety. A statistical tool, such as *CrimeStat,* is invaluable in identifying high priority areas.

Fifth, and most important, political 'buy in' by local and State government agencies, elected officials, community groups and businesses is necessary in order to develop a framework for making decisions. Data and sophisticated analysis will not, in and of themselves, lead to decisions and implementation unless there is political support. In regards to traffic safety, funding is not that much of a problem because it can be shown that most safety improvements pay for themselves in a short time period. But, encouraging safety 'partnerships' of relevant players is essential for building support to allow decisions to be made to support safety.

Obviously, this depends on the type of decisions. Roadway improvements are generally not very controversial unless they involve major rebuilding of a roadway. However, imposing stricter enforcement or funding safety education programs will be more challenging as elected officials will find lots of reasons not to do anything. For example, the RSC's recommendation to identify the residence locations of drunk drivers involved in crashes was met by resistance by several elected officials who feared the consequences of such information on real estate values. Of course, other elected officials saw in this recommendation the need to address a serious public health problem; they were less concerned about the perception of their neighborhood and more on reducing the number of drunk drivers on the road. Not all the recommendations are accepted without controversy.

Further, some obvious measures are completely unacceptable politically. For instance, teenagers have a much higher likelihood of being involved in crashes than older age groups with the youngest drivers (aged 16 and 17) having much higher crash likelihoods than older teenagers (IIHS 2005). Because of this, many countries have established a minimum age requirement of 18 for a driver's license in order to reduce the likelihood of crashes. In the U.S., however, with very few exceptions, State legislatures are unwilling to impose a minimum age requirement of 18 for fear of alienating middle-class parents. Getting a driver's license at age 16 is considered a 'rite of passage' by U.S. teenagers and few elected officials want to be involved in advocating a policy that, while saving lives, would incur hostility by many of their parents. The consequence is that thousands of 16 and 17 year olds are involved in crashes each year, a condition that could be corrected with a change in policy. Frequently, recommendations which seem solid because they are based on evidence run up against other needs of elected officials. The RSC did not even bother to make such a recommendation in its first year report even though the majority of its members would have fully supported such an action.

This creates a dilemma for a traffic safety planning program. On the one hand, the public expects the roads to be safe. There is almost a universal acceptance of road safety and a commitment to improve safety. But, typically, safety is not a theme that organizes community groups. Most people perceive it as a ubiquitous problem and one not prone to easy fixes. A major exception is *Mothers Against Drunk Driving*, a national community-based organization that has been extremely effective in advocating for reductions in alcohol-related driving behaviour (MADD 2007).

In my experience, many elected officials see safety improvements as a 'high risk/low benefit' policy to support. If they have been in office, then they might become embarrassed by information that shows a high level of crashes in their districts. If they are new to their office, there is not an immediate political gain to be had from supporting safety. Further, identifying areas that are unsafe poses a potential risk to them because of possible backlash from their supporters and from business owners. For them, it is often a 'no win' proposition. Thus, in spite of the fact that safety improvements will generally pay for themselves, many elected officials avoid such issues.

On the other hand, creating a high visibility advisory body with specialists from a variety of fields (especially from medicine and law) can provide credibility and support for tough actions that need to be taken to reduce the number and severity of motor vehicle crashes. Such a body can provide extensive expertise allowing crash information and a substantial knowledge base of research on traffic safety to be used to make recommendations (NHTSA and GHSA 2006). It becomes more difficult for elected officials to avoid addressing safety issues when they have high visibility experts making recommendations. By and large, the recommendations of the RSC have been accepted by elected officials as it has provided a 'road map' for systematically improving safety in the region. It is still early in the process, but most of the recommendations are slowly being implemented.

In this effort, creating a safety planning support system, such as the Greater Houston motor vehicle safety PSS, can be an important tool in providing information that allows an advisory body to make recommendations based on knowledge and information.

But, it is only a tool. Such a system must be imbedded in a larger analytical framework that goes beyond the data that has been collected and which addresses the behavioural issues involved in traffic safety. Identifying hot spots and hazardous locations is but a first step to improving safety. Engineering improvements can be implemented at those locations to some extent. But, usually, increased enforcement is necessary for further improving safety. For example, lengthening the timing of a yellow light at a signalized intersection can reduce the number of persons who run the red light by allowing those vehicles to pass those who are just at the intersection when the light turns red. But, only enforcement can truly cut back on the more aggressive drivers who will run the red light from a substantial distance when the light turns red. There is not a simple engineering 'fix' for those types of drivers. Engineering improvements at intersections typically will reduce crashes by 20–30 per cent (H-GAC 2004). Further reductions will come about only through increased enforcement and public education.

Also, the information provided by a motor vehicle safety PSS needs to be interpreted to be meaningful, typically by people who have years of experience of working in safety fields. For example, the use of booster seats by children aged 5–9 is considered a high priority in the U.S. because fatality rates have dropped dramatically among 0–4 year olds but have decreased only slightly among 5–9 year olds. The reason is that, apparently, many parents put their children in adult seat belts as soon as they outgrow their infant seatbelts, not recognizing that a 5–9 year old child is not protected by an adult belt and could even be harmed by it. In Texas, the State Legislature has, to date, avoided requiring parents to use booster seat belts for fear of being 'too regulatory'. Thus, bodies such as the RSC and other safety organizations have been advocating for changes in the law. Using statistics on passenger fatality rates by age and, even, film footage showing how adult seatbelts do not protect young children, safety advocates will eventually get the Legislature to pass such a law. But, it will take time to convey this knowledge and explain why the current situation is not acceptable. In short, safety facts need to be interpreted and explained to decision makers in order to translate into actions that can improve safety.

Thus, the real PSS is a set of cumulative experiences about safety, a knowledge base of research about what works and what does not, and a timely and high quality crash information system that can provide data, analysis, and insights into the extent of the problem. It is the combination of information and experience that represents the greatest possibility for providing meaningful recommendations to improve traffic safety.

Future research needs to focus on three main issues. First is the obtaining of 'real time' information on the road system and on crashes. Many large cities have implemented camera systems for monitoring roadways with automated data collection on traffic volumes and speeds. When that information is combined with crash data, it will become possible to monitor crash risk on an ongoing basis and, thereby, refine further the identification of priorities, as well as allow much quicker emergency response.

A second priority is to better understand the connections between land use and crashes. Some relationships are obvious, such as a concentration of crashes in the central business district and in other commercial areas. But, there are more subtle relationships that need to be explored. A first one is in understanding where drivers who are involved in crashes live. Another is in understanding the relationship of crashes to drinking places and the potential for creating transit and para-transit services. A third is in understanding environments that encourage speeding. Predictive modelling might be important in this endeavour, but such models have to be used carefully. Predictive crash models are good for producing a system-wide model but are less useful for understanding specific locations. Traffic engineers have resorted to tables of expected crash reductions for particular engineering improvements (H-GAC 2004), but there are no such tables for modelling land use and social interventions. Research needs to be done to understand how modifying an environment (e.g. restricting access points around a shopping centre) or modifying the behaviour of participants in that environment (e.g. providing flexible para-transit services around areas with bars and nightclubs) can improve safety.

A third priority is to understand the legal impediments for enforcing drunk driving. Crashes from drunk driving are the most severe traffic safety problem, accounting for more than 40 per cent of all motor vehicle fatalities. The Federal Government and most States have enacted many measures to reduce this problem, including severe penalties for repeat behaviour, required treatment for teenagers and drivers convicted of multiple DWI offences, the requirement of ignition interlock devices and other measures. Yet, there are many legal obstacles that remain, including restrictions on sobriety testing, unwillingness to conduct autopsies on drivers who were killed and the insurance industry-supported intoxicants' laws. In this sense, the problem is not so much on the nature of the PSS, but on the implementation of measures that need to be taken to reduce the problem. As usual, social acceptance and politics in the broadest sense are the critical variables for implementing safety improvements. A motor vehicle safety PSS can be invaluable in identifying the problems and pointing towards certain solutions. But, it still takes legislative bodies having the courage to propose strong actions to improve safety that will lead to real changes.

References

AECOM Consulting Transportation Group, Bellomo-McGee and Levine, N. and Associates (2002) *Considering Safety in the Transportation Planning Process*, Federal Highway Administration, U.S. Department of Transportation, Washington. http://tmip.fhwa.dot.gov/clearinghouse/docs/safety/.

Baxter, J.R. (2006) Federal goals for safety planning, Presentation at the 1st Annual Regional *Safety Conference, Houston-Galveston Area Council, Houston. www.hgactrans.*org/safety/safetyconferenceslides.htm.

Benz, R. (2006) Proposed regional web-based crash analysis tools, Presentation at the 1st Annual Regional Safety Conference, Houston-Galveston Area Council, Houston. www.hgactrans.org/safety/safetyconferenceslides.htm.

Best, B. (2007) *Causes of Death*, Ben Best.com. www.benbest.com/lifeext/causes.html.

Borchardt, D. and Corder, S. (2006) Truck lane restrictions in Texas, Presentation at the 1st Annual Regional Safety Conference, Houston-Galveston Area Council, Houston. www.hgactrans.org/safety/safetyconferenceslides.htm.

Canter, P., Lu, W. and Levine, N. (2005) The use of crime travel demand modeling for drunk driver interdiction in Baltimore County, Presentation at the 8th Annual Crime Mapping Research Conference, Savanah, GA, National Institute of Justice, Washington, DC. http://www.ojp.usdoj.gov/nij/maps/savannah2005/papers/Canter.ppt.

Chakiris, G.S. (2006) National highway safety priorities, Presentation at the 1st Annual Regional Safety Conference, Houston-Galveston Area Council, Houston. http://www.hgactrans.org/safety/safetyconferenceslides.htm.

Copeland, K. (2006) Impaired driving: a serious problem for Texas and America, Presentation at the 1st Annual Regional Safety Conference, Houston-Galveston Area Council, Houston. http://www.hgactrans.org/safety/safetyconferenceslides.htm.

Curry, T.D. (2006) Improving freight safety enforcement efforts in the City of Houston, Presentation at the 1st Annual Regional Safety Conference, Houston-Galveston Area Council, Houston. http://www.hgactrans.org/safety/safetyconferenceslides.htm.

Davies, B.T. (2006) Impediments to efforts to reduce impaired driving in Texas, Presentation at the 1st Annual Regional Safety Conference, Houston-Galveston Area Council, Houston. http://www.hgactrans.org/safety/safetyconferenceslides.htm.

DHHS (2006) *Health, United States, 2006*, Centers for Disease Control and National Center for Health Statistics, U.S. Department of Health and Human Services. Washington. http://www.ncbi.nlm.nih.gov/books/bv.fcgi?rid=healthus04.TOC; http://www.ncbi.nlm.nih.gov/books/bv.fcgi?rid=healthus06.TOC.

DPS (2007) *Crash Records Bureau*, Department of Public Safety, State of Texas, Austin. http://www.txdps.state.tx.us/administration/driver_licensing_control/arb.htm.

DSA (2005) *Multi-Country Per Capita and VMT/VKmT Fatality Data for 2004*, Drive & Stay Alive, Inc, Williamsville. www.driveandstayalive.com/info%20section/statistics/stats-multi-country-percapita-2004.htm.

FHWA (1981) *Highway Safety Improvement Program (HSIP) User's Manual*, FHWA-TS-81-218, National Highway Institute, Federal Highway Administration, U.S. Department of Transportation, Washington.

FHWA (1999) *Conditions and Performance Report Appendix F – Federal Highway Safety Planning and Improvement Programs: State and Community Highway Safety Grants*, Federal Highway Administration, Washington. http://www.fhwa.dot.gov/////policy/1999cpr/ap_f/cpxf)4.htm.

Henk, R. (2006) Teens in the driving seat program, Presentation at the 1st Annual Regional Safety Conference, Houston-Galveston Area Council, Houston. http:/ www.hgactrans.org/safety/safetyconferenceslides.htm.

H-GAC (2004) *East End Safety Study February 2004*, Houston-Galveston Area Council, Houston. http://www.h-gac.com/taq/planning/default.aspx.

H-GAC (2006) *Facts about Red Light Running Crashes: 1999–2001*, Houston-Galveston Area Council, Houston. http://www.h-gac.com/taq/planning/default.aspx.

H-GAC (2007a) *Studies and Presentations*, Houston-Galveston Area Council, Houston. http://www.h-gac.com/taq/planning/default.aspx.

H-GAC (2007b) *State of Safety in the Region: Report of the Regional Safety Council for 2007*, Edited by Levine, N., Houston-Galveston Area Council, Houston. http://www.h-gac.com/taq/planning/default.aspx.

IIHS (2005) *Fatality Facts 2005: Teenagers*, Insurance Institute for Highway Safety, Arlington, VA, http://www.iihs.org/research/fatality_facts/teenagers.html.

IIHS(2007) *Q & A: Red Light Cameras*, Insurance Institute for Highway Safety, Arlington, VA. www.iihs.org/research/qanda/rlr.html.

Jacobs, G.D. (1986) Road accident fatality rates in developing countries a reappraisal, In PTRC Summer Annual Meeting, University of Sussex, 14–17 July 1986, Proceedings of Seminar H. London, PTRC Education and Research Services, pp. 107–119. http://www.transport-links.org/transport_links/filearea/publications/1_603_PA1164_1986.pdf.

Levine, N. (2006a) Houston, Texas, metropolitan traffic safety planning program, *Transportation Research Record*, 1969: 92–100.

Levine, N. (ed.) (2006b) Conducting hot spot analysis of motor vehicle crashes, Presentation at the 1st Annual Regional Safety Conference, Houston-Galveston Area Council, Houston. http://www.hgactrans.org/safety/safetyconferenceslides.htm.

Levine, N. (2007) *CrimeStat III: A Spatial Statistics Program for the Analysis of Crime Incident Locations* (version 3.1), Ned Levine & Associates, Houston, National Institute of Justice, Washington. http://www.icpsr.umich.edu/crimestat.

Levine, N. and Kim, K.E. (1999) The spatial location of motor vehicle accidents: a methodology for geocoding intersections, *Computers, Environment, and Urban Systems,* 22(6): 557–576.

Levine, N., Kim, K.E. and Nitz, L.H. (1995a) Spatial analysis of Honolulu motor vehicle crashes: I Spatial patterns, *Accident Analysis & Prevention*, 27(5): 663–674.

Levine, N., Kim, K.E. and Nitz, L.H. (1995b) Spatial analysis of Honolulu motor vehicle crashes: II Generators of crashes, *Accident Analysis & Prevention*, 27(5): 675–685.

MADD (2007) *Mothers Against Drunk Driving*. www.madd.org.

NationMaster.com (2007) *Health Statistics: Motor Vehicle Deaths (latest available) by Country*. http://www.nationmaster.com/graph/hea_mot_veh_deahealthmotorvehicledeaths.

NCHRP (2005) *Crash Records Systems: A Synthesis of Highway Practice*, National Cooperative Highway Research Program Synthesis 350, Transportation Research Board: Washington.

NHTSA (1997) *4313FA Crash Outcome Data Evaluation System (CODES)*, National Highway Traffic Safety Administration, U.S. Department of Transportation, Washington.http://www.nhtsa.dot.gov/search97cgi/s97_cgi.exe.

NHTSA (2005) *Traffic Safety Facts Annual Report 2005*, DOT-HS-810-631. http://www.nhtsa.gov/portal/nhtsa_static_file_downloader.jsp?file=/staticfiles/DOT/NHTSA/NCSA/Content/TSF/2006/TSF2005.pdf.

NHTSA(2007) *Comparison of Crash Fatalities by Gender and Year from 1996 to 2005*, Traffic Safety Facts, DOT-HS-810-780, National Highway Traffic Safety Administration, Washington. http://www.nhtsa.gov/portal/nhtsa_static_file_downloader.jsp?file=/staticfiles/DOT/NHTSA/NCSA/Content/RNotes/2007/810759.pdf.

NHTSA and GHSA (2006) *Countermeasures That Work: A Highway Safety Countermeasure Guide for State Highway Safety Offices,* National Highway Traffic Safety Administration and the Governor's Highway Safety Association, Washington.

NSC (2007) *Injury Facts: 2007 Edition*, National Safety Council, Itasca. http://www.nsc.org/lrs/statinfo/estcost.htm.

Peden, M., McGee, K. and Krug, E. (eds.) (2002) *Injury: A Leading Cause of the Global Burden of Disease,* World Health Organization, Geneva. http://www.who.int/violence_injury_prevention/publications/other_injury/injury/en.

Roth, R. (2006) Effective and ineffective laws to reduce drunk driving, Presentation at the 1st Annual Regional Safety Conference, Houston-Galveston Area Council, Houston. http://www.hgactrans.org/safety/safetyconferenceslides.htm.

Schultz, J. (2001) Gun and Go: analyzing red light running, *Quest 4(2): June*. http://www.odu.edu/ao/instadv/quest/gunandgo.html.

Scotland (1998) *Road Accidents Scotland 1998,* Scottish Executive, Edinburgh. http://www.scotland.gov.uk/library2/doc05/ras-10.htm

Stein, R. and Dahnke, R. (2006) Design evaluations of red light camera programs, Presentation at the 1st Annual Regional Safety Conference, Houston-Galveston Area Council, Houston. http://www.hgactrans.org/safety/safetyconferenceslides.htm.

Washington, S., Van Schalkwyk, I., Mitra, S., Meyer, M., Dumbaugh, E. and Zoll, M. (2006) *Incorporating Safety into Long-Range Transportation Planning,* NCHRP Report 546, National Cooperative Highway Research Program, Transportation Research Board, Washington. http://onlinepubs.trb.org/onlinepubs/nchrp/nchrp_rpt_546.pdf.

Wise, C. (2006) MADD's strategic direction and new campaign to eliminate drunk driving, Presentation at the 1st Annual Regional Safety Conference, Houston-Galveston Area Council, Houston. http://www.hgactrans.org/safety/safetyconferenceslides.htm.

Additional Reading

AECOM Consulting Transportation Group, Bellomo-McGee and Ned Levine & Associates (2002) *Considering Safety in the Transportation Planning Process*, Federal Highway Administration, U.S. Department of Transportation, Washington, DC. http://tmip.fhwa.dot.gov/clearinghouse/docs/safety/.

FHWA (2007) *Reducing Highway Fatalities*, Federal Highway Administration, U.S. Department of Transportation, Washington, DC. http://safety.fhwa.dot.gov/.

FHWA (2007b) *What is Transportation Safety Planning?* Federal Highway Administration, U.S. Department of Transportation, Washington, DC. http://www.fhwa.dot.gov/planning/SCP/.

FHWA/FTA (2005) *Safety Conscious Planning Peer Exchange*. Transportation Planning Capacity Building Program, Southeast Michigan Council of Government, the Federal Highway Administration and Federal Transit Administration. http://www.planning.dot.gov/Peer/Michigan/detroitSafety.htm.

H-GAC (2007) *State of Safety in the Region: Report of the Regional Safety Council for 2007*, Levine, N. (ed.), Houston-Galveston Area Council, Houston. http://www.h-gac.com/taq/planning/default.aspx.

NHTSA (2007) *2006 Traffic Safety Fact Sheets*, National Highway Traffic Safety Administration, U.S. Department of Transportation, Washington, DC. http://www.nhtsa.dot.gov/portal/site/nhtsa/menuitem.6a6eaf83cf719ad24ec86e10dba046a0/.

Chapter 6
Elbe DSS: A Planning Support System for Strategic River Basin Planning

Bernhard M. Hahn, Sebastian Kofalk, Jean-luc De Kok, Jürgen Berlekamp and Mariele Evers

6.1 Introduction

The German part of the Elbe River and its basin are characterized by multiple problems and objectives that call for strategic management based on an integrated approach. In August 2002, the region suffered a catastrophic flood with loss of lives and damage amounting to over 9 billion Euro. During the summer months, because of low flows, shipping along the river is problematic, which considerably reduces the economic transport capacity of the river. Several areas along the river act as a habitat for rare plant and animal species and have been designated as nature reserves. The output of diffuse and point sources of pollution in the river basin must be controlled in order to comply with standards of the *EU Water Framework Directive* (EU 2000).

Management actions taken as a part of implementing policy interact at different spatial and temporal scale levels, which complicate the identification of promising river management strategies. For example, the restoration of poorly maintained groynes to improve the navigation conditions may lead to undesirable changes in the hydrodynamic character and ecological vegetation patterns of the floodplains. Furthermore, long-term trends in socio-economic development and climate change can be expected to interact with measures. The diversity and structure of river-related research and planning organizations in Germany, the fact that the Elbe River cuts across six different federal states and that one third of the basin belongs to the Czech Republic, also complicate the formulation of promising river management strategies.

To deal with these problems, the German Federal Institute of Hydrology (Bundesanstalt für Gewässerkunde or BfG) took the initiative for a three-year project to develop the *Elbe DSS*, a pilot version of a planning support system (PSS) for strategic integrated management of the Elbe River and its basin. The *Elbe DSS*

B. M. Hahn (✉)
Research Institute of Knowledge Systems (RIKS)
PO Box 463,6200 AL Maastricht, The Netherlands
E-mail: bmhahn@riks.nl

S. Geertman, J. Stillwell (eds.), *Planning Support Systems Best Practice and New Methods*,
DOI 10.1007/ 978-1-4020-8952-7_6,© Springer Science+Business Media B.V. 2009

was completed in 2006 and includes measures and indicators for flood safety, water quality, navigation and vegetation ecology. The user can select from several demographic and climate scenarios and apply various combinations of management measures to explore their effects at different spatial scales ranging from the river basin to specific locations of interest along the river. Each measure can be configured interactively using a specialized graphical user interface that presents the underlying model parameters at an aggregation level appropriate for strategic decision making. All geo-information is accessible in a built-in map viewer/editor, which can also be used for interactive configuration of measures by general users.

In view of the problems with respect to the adoption of PSS that have been mentioned by Walker (2002) and Westmacott (2001), stakeholders and potential users were involved in the design of the system from the beginning. Potential user organizations, key problems and tentative measures were identified during a feasibility assessment that preceded the project. Scientific or so-called 'research' models are often data and location dependent and can be computationally demanding. This makes these models less suitable for application in an interactive instrument to be used during sessions with stakeholders where the instrument should be flexible enough to represent different management options or new political priorities as they emerge during discussion. Therefore, the *Elbe DSS* is based as much as possible on simple models, or simplifications of models that were too comprehensive for direct incorporation into the tool. Implementation of the DSS within organizations involved in river-basin management is currently being coordinated by the BfG through a series of testing and training workshops with users. The outcomes will be used to further refine the design in order to meet the expectations of the practitioners.

In Section 6.2, we provide a broad definition of *integrated special decision support systems* (ISDSS) as a specific type of PSS and we define how they can be used within the context of strategic river basin planning. A brief overview of the design methodology adopted for the development of the *Elbe DSS* is given in Section 6.3. In Section 6.4, we describe the key roles and interactions in a typical ISDSS project team. Knowledge integration is the main challenge of ISDSS development. We discuss our approach to knowledge integration in Section 6.5. In Section 6.6, we give an overview of the techniques used for technical model integration. In Section 6.7, we describe some aspects of the realization of the *Elbe DSS* prototype, and in Section 6.8 we report preliminary results of an evaluation of the prototype by users. Finally, in Section 6.9, we summarize our conclusions from the development of the *Elbe DSS* and indicate possible directions for further research.

6.2 DSS for River Basin Management: State of the Art

There is no generally accepted definition of what a decision support system (DSS) is. Table 6.1 shows which of the common DSS characteristics identified in Marakas (1998) are present in the *Elbe DSS*.

Table 6.1 Common DSS characteristics

Common DSS characteristic	*Elbe DSS*
Employed in semi-structured decision contexts	Yes
Intended to support rather than replace decision makers	Yes
Support for all phases of the decision-making process – for a model of the decision-making process, see Mintzberg (1976)	Yes
Focus on effectiveness of the decision-making process rather than efficiency	Yes
Is under control of the DSS user	Yes
Uses underlying data and models	Yes
Facilitates learning on the part of the decision maker	Yes
Is interactive and user friendly	Yes
Developed using an evolutionary, iterative process	Yes
Provides support for all levels of management	No
Can provide support for multiple independent or interdependent decisions	Yes
Provides support for individual, group and team-based decision-making contexts	Partially

Source: adapted from Marakas (1998).

Harris (1989) introduced the term planning support system (PSS) and defined it as an architecture for coupling a range of computer-based methods and models into an integrated system for supporting planning functions. DSS can be seen as part of PSS, when the scope of the DSS is more on the operational level. However, in the context of our view of strategic river-basin planning, we use the term PSS for a special type of DSS with distinguishing features as described in Table 6.2. PSS and policy support systems (PoSS) are developed to support the long-term, strategic planning process as opposed to management support systems (MSS), which are intended to support short-term decision making. An overview over the recent history and classification of PSS is given in Batty (2007) and policy support systems are defined in Kok (2007).

Table 6.2 Features of PSS as a subclass of DSS

Decision Support Systems (DSS)	
Management oriented DSS	Planning Support Systems (PSS) Policy Support Systems (PoSS)
Management oriented	Policy and planning oriented
Short-term/immediate	Long-term/strategic
Concrete topics (complexity due to technical limitations of the systems and methods used)	Abstract topics (complexity due to lack of knowledge and consensus on objectives)
Specialised sectoral context	Broad societal context
System based on GIS, database manipulations and (few) sectoral models	System based on complex integrated model(s)
Optimization oriented	Simulation and exploration oriented

Strategic river-basin planning may be considered as the attempt to control a complex and dynamic system of interdependent natural and human-driven processes. Due to these characteristics and the fact that experimenting with a real world system is not an option, information systems designed to support environmental planning are often model-centric. Model-centric PSS represent knowledge about the domain in formal models which can be used for analysis or simulation. Borshchev (2004, p. 1) describes a simulation model as *a set of rules (e.g. equations, flowcharts, state machines, cellular automata) that define how the system being modelled will change in the future, given its present state*. To accurately reflect the connectedness and interactions of the real world system, model-centric PSS typically couple several domain-specific models in an integrated system model (ISM). Each of the individual models then represents a process that is considered relevant within the planning and decision context. The system model is integrated, because it explicitly represents and simulates the interactions between the processes. Since, in the domain of environmental planning, most processes and variables of the ISM have a spatial dimension, we use the term *integrated spatial decision support system* (ISDSS) for a spatial PSS with an ISM as its core component. The basic assumption of the simulation approach is that the ISM mimics the behaviour of selected, relevant aspects of the real world system. Therefore, running simulations with the ISM enables the decision maker to explore a space of possible strategies prior to taking action.

The generic usage context of an ISDSS is shown in Fig. 6.1. Physical, ecological and socioeconomic processes are represented in the ISM as a network of coupled processes. The ISM embodies our current knowledge about the system at a level of abstraction that is appropriate for policy design and strategic planning. Apart from its autonomous behaviour, the state of the system is influenced by: (i) policy options; and (ii) external influences. Policy options represent variables of the system which are under human control and therefore are related to planning and decision making.

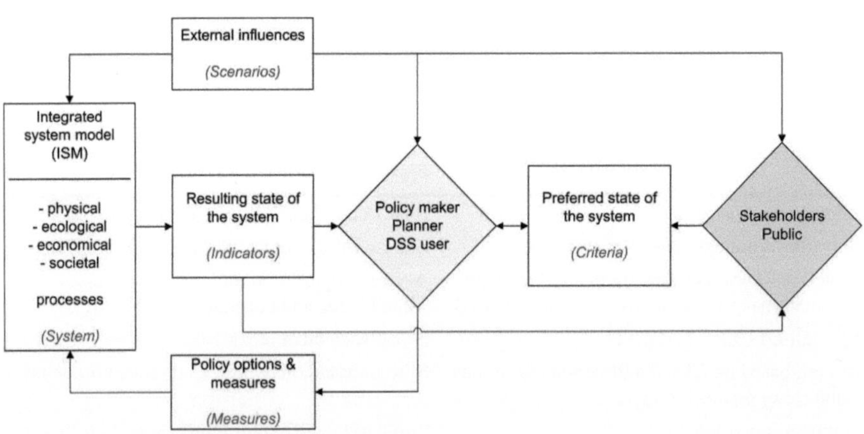

Fig. 6.1 Generic usage context of an ISDSS

External influences represent exogenous factors such as climate change. These influence the system, but are not under direct control of the decision maker, nor do these factors receive direct feedback from the system. In principle, the ISDSS user has access to all output variables of all processes represented in the ISM. However, policy formulation requires indicators that represent policy relevant aggregated information about the system. The relevance of such indicators depends on the policy context in which the system operates, and should be widely accepted among scientists and stakeholders. Indicator values can be analysed with respect to various criteria. Experimentation and exploration by hand or algorithmic optimization routines can be used to adapt the parameters of multiple combinations of measures in order to reach a close match between the resulting state of the system (indicators) and the preferred state of the system (criteria).

6.3 Design Methodology

The development of an ISDSS involves multiple disciplines and multiple roles and is an iterative process where the product is realized by interdisciplinary collaboration between scientists, modellers, domain specialists, IT specialists and end users (Fig. 6.2).

The aim is to develop a balanced product as represented by the position of the arrowhead in the centre of Fig. 6.2. To achieve this goal, ISDSS development ideally starts in the policy part of the triangle. Policy makers and planners should be involved in the project as equal partners from the start, or even better, be the initiators of the project. This increases the chance that the design of the system is driven by real world policy and planning requirements. Numerous projects have been initiated as part of a multi-disciplinary research project where the main motive for ISDSS development was to integrate the sectoral scientific results of other (sub)projects. In

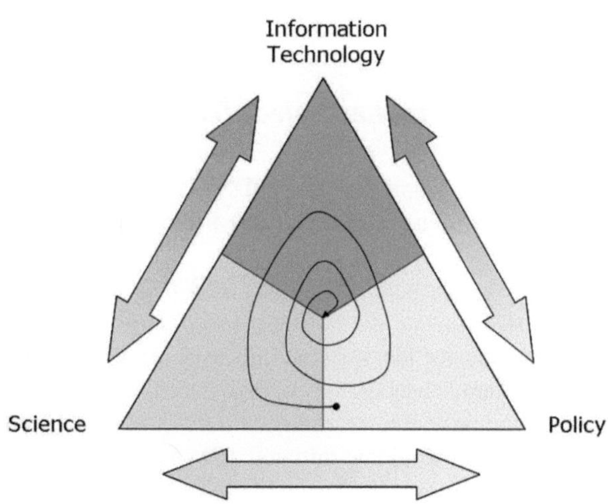

Fig. 6.2 Iterative ISDSS development for policy and planning support

general, projects set up in this way tend to serve the (legitimate) academic interests of the participating scientists more than addressing real policy and planning issues. Among the characteristics of the ISDSS products that come out of these so-called science-driven projects we find that:

- the ISDSS is based on weakly integrated comprehensive scientific models;
- models often do not reflect the scope of policy design questions: very detailed answers for questions in a very restricted context instead of sketchy answers in a broad context;
- complex user interfaces that expose the complexity of the underlying models to the extent possible;
- sluggish performance and weak technical integration due to insufficient software engineering quality of the individual models; and
- weak integration in the organisational context and existing IT infrastructure of the end user.

GIS-based DSS tend to be tools that analyse the past and the present but often lack the underlying integrated models necessary to explore how a complex system might evolve in the future. Therefore they tend to be more management oriented DSS that deal with operational and/or short-term tactical decision making rather than policy and planning oriented DSS, that deal with long-term strategic decision making.

Based on the lessons learnt from the *Elbe DSS* and other ISDSS development projects, we present here some elements of a design methodology for ISDSS which tries to avoid some of the problems discussed above with the aim of increasing the chance of developing a product that is perceived as relevant and useful for application to real world planning and policy problems. We will discuss three such elements of an improved ISDSS design methodology – project team operation, the integration of scientific knowledge and technical (software) integration.

6.4 The ISDSS Project Team

6.4.1 Roles in the Project Team

Successful development of an ISDSS is an interdisciplinary team effort. To achieve a balanced and useful product, the following key roles and responsibilities should be present in the project team.

End users provide the policy and usage context of the ISDSS. They define problems, delineate the business scope, and identify functional and organizational requirements for the system. End user representatives in the project team should consider themselves the 'problem owners'. They should have a clear role and set of tasks in the project team and should be involved from the start. In an ideal case, the ISDSS project would be initiated and sponsored by an end user organization.

Scientists provide the knowledge about natural and human-driven processes and their interactions. They define the spatial and temporal resolution, the scale and levels of detail of the individual models and the ISM. In addition to their deep domain specific knowledge, they should have a strong motivation for the integrated and interdisciplinary approach required for ISDSS development, without compromising the scientific integrity of their contributions.

IT specialists are responsible for system architecture, technical model integration, software design and implementation. They should have in-depth knowledge of state of the art software architectures and middleware technologies and a good vision on an appropriate technical design for the ISDSS. They should be highly flexible with respect to the formats, programming languages and environments they have to work with, and be able to manage a high level of complexity in a dynamic and iterative development process.

The *DSS architect* has primary responsibility for integration, communication and management of the development process. This person takes responsibility for integrating the work of the scientists and for controlling the quality of the ISM. As a generalist, this role bridges methodological and knowledge gaps between end users, scientists and IT specialists. In order to fulfil this role, the architect should possess a very good intuitive understanding of the application domain, the (policy and planning) problems posed, along with good communication skills.

6.4.2 Interactions in the Project Team

In the context of ISDSS development, the *interaction between scientists and end users* primarily deals with the issue of selecting policy relevant research findings, models and data. Developing solutions that fit the needs of policy makers and planners requires the scientists to shift their focus from process towards problem orientation – which processes need to be included and integrated for tackling particular problems? The processes should be represented at a level of detail and at spatial and temporal resolutions that are appropriate for the policy and planning problems at hand. For adequate representation, the integrated models often require that processes are linked across various spatial and temporal scales. This can mean that existing scientific models have to be adapted or reformulated.

Relying on (the predictions of) a computer based tool as a guide for policy formulation and planning requires wide acceptance and trust in the underlying integrated model and its individual components among the stakeholders. To achieve this, instead of being a black box, the ISM should be as transparent as possible. In the *Elbe DSS*, the documentation of the individual models as well as references to papers, calibration and validation reports are accessible to the user as part of a context sensitive online help system. Calibration, validation and uncertainty analysis are notoriously difficult and tedious for coupled (system) models. Therefore, it is important that scientists and end users develop a shared vision of how to tackle these tasks during the development phase. Providing analytical and visualization tools for uncertainty analysis can help to communicate the chosen approach.

In general, the *interaction between scientists and IT specialists* mainly focuses on the technical and, to a lesser extent, the scientific aspects of model integration. The goal of technical model integration is to develop a flexible ISDSS that allows upgrading each of its components, to keep it up to date with new research findings as well the changing needs of the policy context. The state-of-the-art approach to achieve this is to use an object-oriented software framework to construct the ISM from elementary building blocks, which pertain to a level of detail that is sufficiently high to be applied by scientists and domain experts. Most ISDSS development projects use existing models rather than developing new models from scratch. Existing models often have to be technically adapted in order to fit into the model integration framework. This often gives rise to a maintenance problem, because one might end up with the original (research) version and a separate 'DSS version' of a model. Keeping both versions in a synchronized state of development can be costly. The best way to avoid this is to make sure that scientists and software engineers collaborate at an early stage in the model development process.

The aim of the *interaction between IT specialists and end users* is to develop a user-friendly system that addresses the user's questions and seamlessly integrates in her/his existing IT infrastructure and organizational context. Here, user friendliness means that the system provides an ergonomic, transparent, interactive and responsive software environment, which enables the user to explore planning alternatives without being fully exposed to the complexity of the underlying ISM. Seamless integration in the users' existing IT infrastructure and organizational context is a key success factor for acceptance and adoption of the system. The preferred approach to integration on the technical level is using open standards and interface specifications (for example, the OpenGIS® standards and specifications developed by the Open Geospatial Consortium (OGC) and available at http://www.opengeospatial.org).

Embedding the system in the organizational context of the user is far more difficult. In the first place, this requires a thorough analysis of the user's organisations' work flows and business processes; an essential part of the requirements analysis phase of any ISDSS project. Furthermore, the interdisciplinary and integrated nature of the ISDSS may require that the task and business processes of the user's organisation change to some extent. Our tools shape our thoughts – our thoughts shape our tools. In the early phases of development, users often find it difficult to specify functional requirements for the ISDSS; once they have a first prototype, it will trigger their imagination about how it should be further developed in order to provide real added value for accomplishing their tasks. Involving end-user representatives during the entire development of the ISDSS, preferably as full project members, will enhance their feeling of ownership and control of the product and the changes it might inflict on their work processes. Especially in research driven ISDSS projects, the value of frequently interacting with and understanding the end users' changing organizational context can hardly be overestimated.

6.5 Knowledge Integration

Designing and implementing the ISM requires integrating knowledge, models and data from multiple disciplines. Since standardized methodologies for knowledge integration do not yet exist, this is often the most challenging element of ISDSS development. The structure of the ISM to a large extent determines the ISDSS functionality, e.g. which scenarios, policy options and indicators can be implemented? Knowledge integration, therefore, is an essential activity of the conceptual design phase of the ISDSS development process. Knowledge integration involves some scientific, technical and end-user related aspects of which we will give an overview here.

6.5.1 Scientific Aspects

How can knowledge from different disciplines be integrated in an ISDSS and how can we link models with different scientific paradigms? The starting point of model integration should be a detailed systems analysis based on user requirements, which identifies all relevant elements and cause-effect relationships that should be reflected by the integrated systems model. What are the model selection criteria and how can we then integrate these models?

A precondition for coupling models is that their interfaces match semantically and structurally. The spatial and temporal resolution and extent of the models must correspond, or appropriate aggregating and/or disaggregating transformations must be applied. In the case of the *Elbe DSS*, the spatial resolution ranges from the whole catchment (approximately 100,000 sqkm) to river sections of 2–3 km length, whereas the temporal resolution varies from decades to daily time steps. In addition, the modelling paradigm can also vary, because processes in the real world can be represented using different modelling approaches. In the natural system as an example, the runoff and infiltration of rainfall can be described using (partial) differential equations, simpler water budget approaches or stochastic approaches. Overall, the level of detail concerning spatial and temporal scale as well as the modelling paradigm should be selected based on adequacy for and appropriateness to the decision problem at hand. Applying the principle of Ockham's razor by choosing the simplest model that can describe a particular process of interest at the appropriate spatial and temporal resolution is often a good DSS design rationale. However, simple models are not always flexible enough and our DSS requirements are sometimes conflicting. So we often need to make a trade-off between flexibility and performance.

Although models can be linked technically as black boxes (see below), this is not recommended from a scientific point of view. To ensure robust and reliable behaviour of the integrated system, it is necessary to analyse model structure and information and data flows. Analysis of those model characteristics makes it possible to define adequate interfaces and ensure structural correctness of the integrated model. In some cases, it may become clear from model analysis that two models to be integrated share the same processes. In this case, it will be necessary to modify one or both models to ensure overall model consistency. Model selection also depends on input data requirements and run-time characteristics and can be supported by sensitivity and uncertainty analyses of the coupled models and DSS system as a whole. Models with excessive data needs and run-times longer than a few minutes are less appropriate for integration in an interactive PSS.

6.5.2 Technical Aspects

How can scientific models be technically integrated, without re-writing all software code, with maximum performance, user-friendliness and at low cost? Let us consider some general requirements for technical integration of models, most of which are also applicable to the ISM and ISDSS as a whole.

Flexibility is a key aspect of an instrument for strategic river-basin planning: an ISDSS is expected to have a long lifecycle. During its lifecycle, the requirements for the system will almost certainly change, data will need updating, models will evolve and scientific knowledge will change or advance. Therefore, the integrated model, as well as the system as a whole, should be as flexible, modular and as decomposable as possible. Individual models should remain accessible as self-contained software units, preferably at the binary level, but at least at the source code level.

Maintainability: The responsibility for maintenance and the potential further development of a model should remain with the original data/model provider and normally should not be transferred to the technical model integrator. It should always be possible to independently test a model before, and after it has been linked to other models in the ISM.

Extensibility: It should be possible to extend the ISM. Individual sub-models should communicate solely via interfaces in order to be exchangeable with as little effect as possible on other parts of the ISM.

Reusability: The development of an ISM may yield potentially reusable artefacts at various levels of abstraction including: (i) the conceptual system model; (ii) the integrated system model as a whole or parts of it; (iii) individual scientific models; (iv) source code; (v) compiled code; and (vi) user interfaces. The largest gain from reuse is obtained in the earliest stages of development cycle. Reusability is almost always formulated as a key requirement of any system development that consumes substantial resources. However, a generic and reusable design always takes considerable more effort compared to a design that only works in a narrow context (Hahn and Engelen 2000). This fact should be reflected in one way or another in the business model for the system. If the system is to be reused in several regions, the extra cost for providing this flexibility and generality (see below) should not be carried by the first client alone.

Generality: If fed with the appropriate data, ideally it should be possible to apply all models to other river basins or river networks. This is also a requirement for the ISDSS as a whole. Of course, the ISM can only be as generic as its least generic part. It is therefore important to avoid decreasing the generality of the ISM by integrating models with very specific locally bound data requirements if more generic models are available.

Accessibility: Often an ISDSS is designed to assist planning and decision-making processes for multiple users with different skill levels and backgrounds. The system design may reflect this by providing alternative user interfaces for different types of users, e.g. a system oriented user interface for the scientist/expert, or a task oriented user interface for practitioners. Thinking of participatory approaches to planning, it may be important to have location independent access to the system.

Performance: Since the primary use of an ISDSS is interactive exploration of policy designs and planning scenarios, the performance of the integrated model in terms of computing speed is a critical success factor. When interactively simulating a policy measure, the update of on-screen graphs, tables and dynamic maps ideally should not take much longer than a few seconds.

The main challenges of technical model integration are to meet the above mentioned requirements, to ensure software engineering quality for the ISM, whilst not imposing too much technical complexity or rigidity on the scientists and modellers participating in development. Knowledge, models and data which are selected to be integrated in an ISDSS, are expected to have a high level of scientific maturity. The ISM, therefore, typically integrates results from previous research projects rather than newly developed models. Integrating existing models and data requires the technical model integrator to deal with: (i) various programming languages; (ii) various compilers and interpreters; (iii) various run-time environments; (iv) various operating systems; (v) various concepts for data and memory management; (vi) open and closed systems; and (vii) a plethora of *ad hoc* data formats. If at all, the technical model integrator has only very limited control on the technical design and the form in which models will be delivered to the project. Geospatial and other data may be provided to the ISDSS project in a great variety proprietary and model specific *ad hoc* formats. The technical model integrator, together with the ISDSS architect, should select and/or define a small number of standard data formats for the project. Ideally, these formats should be open, widely used and precisely documented. Binary formats usually are more memory efficient and provide better read/write performance compared to ASCII based formats. However, they are almost impossible to access without exact documentation and specialized applications. When it cannot be avoided to develop a project or model-specific data format, this should be done on the basis of open and widely used meta-formats like XML.

Typical problems technical model integrators and DSS architects encounter when integrating existing scientific model code in an ISDSS include the following.

Architectural mismatch: Many scientific models are implemented in the 'read-compute-write' FORTRAN style of the 1970s, meaning they read their input data from a file and, after doing their computations, write the results to a file. Integrating such code in a modern event driven interactive application requires non-trivial changes in the control structures and application logic of the model.

Software interface: Software implementations of scientific models are usually designed as (stand alone) research tools. They are not designed as software components that can function as part of a larger system and they seldom feature a software interface that allows their execution to be controlled by another software component such as a simulation controller. To integrate such model code into the ISM, it needs to be 'wrapped' by an extra layer of software that implements the required interface. The worst case with respect to model integration is models that are developed within a closed modelling environment and can only be executed inside that environment. In such cases the only way to integrate the model is to re-implement it in an accessible environment.

Software engineering quality: In contrast to the quality of the scientific content, the software engineering quality of scientific model code is sometimes poor. When such code has to be integrated in an ISDSS, this can create performance, maintenance or usability related problems.

6.5.3 User Aspects

The user aspects of knowledge integration add the dimension of problem orientation to the scientific goal of improving our understanding of a complex system. We want to use our improved understanding of the system to inform management action, for example to better implement sustainability as framed in the EC Water Framework Directive (EU 2000). Therefore, knowledge integration in the context of DSS development should always start from policy, planning and management questions.

To find out more about these questions, and how a DSS could effectively help to solve them in a practical and adequate way, one of the authors conducted a broad evaluation with almost 200 experts in river basin management in five North Sea riparian countries (Evers, forthcoming). The evaluation confirmed that generally DSS are seen as a feasible means of supporting *integrated river-basin management* (IRBM). However, when searching for DSS requirements, one main issue is to find the tangible added value of a system – why and for what purpose is it worth investing financial and human resources into the development of a DSS for river-basin management? Based on the evaluation results four important reasons can be distinguished:

- compilation of data, information and knowledge from various areas (thematically and geographically) and quick access to it;
- participation, explanation and justification in the planning and decision-making process;
- support of planning and decision-making process by generating special insights, assessing impacts and measures and so forth; and

- handling of the complexity of IRBM and better understanding, including future perspectives.

Potential users were asked: 'What are key functionalities for a DSS from your perspectives?' The interlocutors considered: (i) building and exploring 'what-if' scenarios; (ii) assessing the diagnosis; and (iii) showing alternatives to meet management goals as the most important functionalities. This includes aspects such as analyzing and visualizing interrelations of land-water-ecosystems, cause-effect-interrelations, assessing scenarios and evaluation of management options (e.g. cost-effectiveness analysis of combined measures). Furthermore, goal seeking and visualisation of information were reported as important functionalities. The modelling aspect in DSS is generally greatly appreciated. However, live simulation is not necessarily required; using pre-calculated simulations is also an option. Another crucial factor for the success of a DSS is the way in which the DSS is implemented into the 'real' working environment of the user.

6.6 Techniques for Model Integration

The techniques used for technical model integration in the development of the *Elbe DSS* can be grouped into three categories – strong coupling, weak coupling and reimplementation.

Strong coupling includes techniques where an explicit software interface between model code and the simulation engine is created, which means that the simulation engine directly calls functions in the model code. Applying a strong coupling technique is possible with or without having access to the source code of the model. If the source code of the model is available and it is written in the same programming language as the ISDSS (C++ in case of the *Elbe DSS*), it can be integrated directly at the source code level. If the source code is available, but is written in a different programming language, this gap can often be bridged by applying special inter-language support libraries (e.g. Boost.Python, a C++ library which enables seamless interoperability between C++ and the Python programming language) or component middleware (e.g. COM, CORBA, .NET). In case the model code is only available as a binary component (e.g. dynamic link library or.NET component), strong coupling can still be applied. However, this requires the model code to feature the necessary software interfaces to connect it to the model controller of the ISDSS and possibly other model blocks in the ISM. Since this is unlikely for any model code that existed before the ISDSS started, having access to the source code gives the technical model integrator the most options by far.

The main advantage of the strong coupling approach is that, due to the lack of overhead and in memory access to data structures, it holds the potential to achieve high performance and good flexibility of the resulting ISM. The main disadvantage is that it creates complicated technical requirements for the model code, which may lead to costly modifications to the model code.

Weak coupling is the method of choice when strong coupling is not feasible due to technical, budgetary or organizational constraints. Instead of directly accessing in-memory data structures with weak coupling, the simulation engine uses the file system to communicate with a model. This approach is usually chosen when the model we want to integrate in the ISDSS is only available as an independent executable, which cannot be modified by the technical model integrator. In this case, the ISDSS first prepares an input file for the program that contains the model, and then issues an operating system call to start a separate process that runs the model program. After the model program has finished, the ISDSS reads the output file and applies the results to the internal data structures. The code in the ISM that controls the external model program takes the role of a proxy for the external model. It therefore features the necessary interfaces to connect to other model blocks in the ISM and the simulation engine.

The technical integration of the *GREAT-ER* and *HBV-D* models (Berlekamp *et al.* 2007) in the *Elbe DSS* are examples of the weak coupling approach with communication via the file system. Weak coupling techniques put very few requirements on the model we want to integrate. Therefore, their main advantage is that they allow the integration of models that otherwise could only be integrated by re-implementing them. The main drawback of weak coupling is that, due to slow file system operations, it often creates a performance bottleneck in the system. The dependence of the proxy on the external model program may also increase the complexity of maintenance of the system. A special form of weak coupling is 'asynchronous integration'. When a model is computationally too demanding to be integrated in an interactive application, one of the choices is to calculate some results offline and then integrate them as a scenario in the ISDSS. The climate scenarios in the *Elbe DSS* are an example of this. Asynchronous integration gives the ISDSS user access to the original model (e.g. via a web service) outside of the execution context of the ISM and thereby to add new scenarios.

Reimplementation of a model offers the most flexibility for the technical model integrator. In many cases, reimplementation yields the cleanest model code and the best possible performance. In the *Elbe DSS*, some of the flooding related models were delivered to the project as MATLAB® code (http://www.mathworks.com/products/matlab/) and then re-implemented in C++. In this particular case, the execution speed could be increased by a factor >100. For small models, reimplementation can be the most economical way to integrate the model in the ISDSS, because the code that would be necessary to 'wrap' the original model code can quickly become more complex than the model itself. The downside of the reimplementation approach is that it can make the maintenance of the code more complex. If the original author of the model decides to further develop the original model code instead of the reengineered code, two implementations of the same model have to be kept synchronized.

Table 6.3 shows for three models that were integrated in the *Elbe DSS* a comparison of the amount of code required for the model integration versus the model code itself. The last column in the table gives an estimation of the performance gain if the model code would be reengineered in C++.

Table 6.3 Proportion of model code versus code for model integration of selected models in the *Elbe DSS*

Model	Model code (# of lines)	Code for model integration (# of lines)	Potential performance gain with reengineering of model code in C++
MONERIS	(Excel model)	10,000	5x–10x
HBV-D	5,300	4,000	2x–5x
GREAT-ER	6,800	7,000	5x–20x
Total	–	21,000	–

6.7 Features of the *Elbe DSS*

6.7.1 Conceptual Modelling

A major goal of the feasibility assessment for the *Elbe DSS* was to clarify what is wanted and what is possible. Consultations with potential end users, together with a thorough analysis of results available from previous research projects on the Elbe River, yielded a description of the system scope and a preliminary set of functional requirements. On this basis, a *conceptual system model* (CSM) was developed in the first phase of the pilot project. The CSM identifies the processes, measures, indicators and external scenarios considered relevant for the policy context at hand, as well as the cause-effect relationships and information flows between these system elements. Depicted in a qualitative system diagram (Fig. 6.3; De Kok *et al.* 2008), the CSM proved to be a useful design artefact that: (i) effectively supported the communication between scientists; developers and stakeholders; (ii) formed the blueprint for the technical design of the system; and (iii) helped to monitor the progress of the conceptual system design during the development iterations of the project.

In the diagram shown in Fig. 6.3, system elements are connected by arrows which indicate cause-effect-relationships or information flows between the elements. The thick frames with the word 'System' represent the system boundary which separates system elements from their external inputs. External factors like external scenarios, measures and management objectives are linked to those system elements where they take effect. The CSM of the *Elbe DSS* is structured in four linked modules (catchment, river network, main channel, river section), each of them representing a specific spatial scale. Another useful technique to manage complexity is hierarchical composition, where a system element at one level may represent a full CSM at a lower level.

The qualitative system diagram is an effective communication tool for end-user and stakeholder participation. Early in the development process, particularly before any computer code is written, it shows what the ISDSS will represent and what it will not. In a collaborative effort between scientists and end users, the CSM is refined in several iterations, until mutual agreement is reached that the CSM: (i) meets the

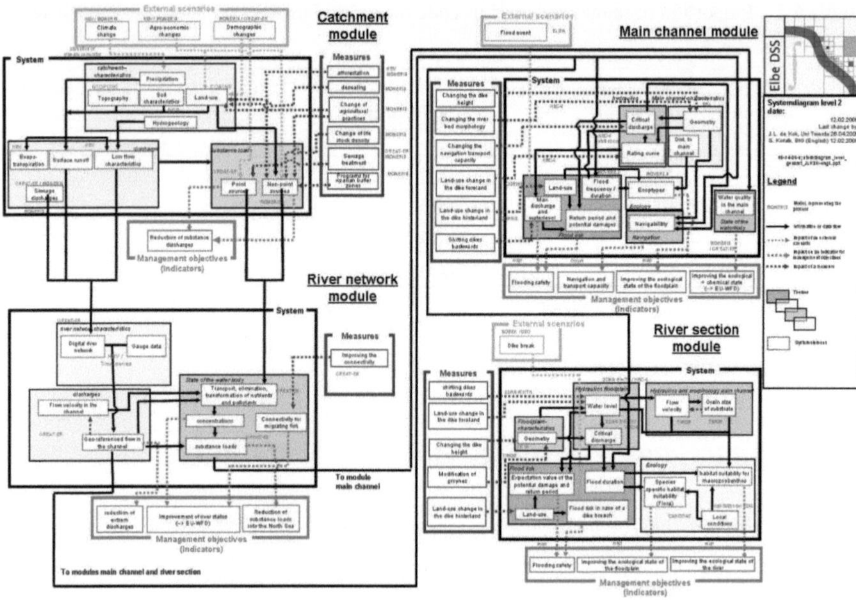

Fig. 6.3 Qualitative system diagram for the *Elbe DSS* (De Kok *et al.* 2008)

user requirements; and (ii) can be implemented with the available models and data. In the next step, models and data have to be associated with each system element. In the *Elbe DSS*, the qualitative system diagram has become part of the software as an interactive hierarchical diagram that acts as the central navigation instrument of the graphical user interface (Fig. 6.4). Scientists and modellers often prefer this system oriented type of interface; however, the results of the evaluation with end users (see below) indicated that they might be better served by a more sequential and task oriented approach to the user guidance. For example, the user interface of a river DSS may be based on a map showing the geography of the river basin.

6.7.2 *System Architecture*

In numerous projects, the authors have learnt to recognize generic components and architectural patterns that are common to many ISDSS. This has led RIKS to develop *Geonamica*®, an object-oriented application framework for ISDSS. While a decade ago the ISDSS developer had to start almost from scratch, nowadays mature application frameworks (e.g. *Geonamica*®, *SME* and others) can be utilized for time and cost efficient development of modular and flexible ISDSS applications. Fayad (1999) describes a framework as a reusable design of a system that describes how the system is decomposed into a set of interacting components. The

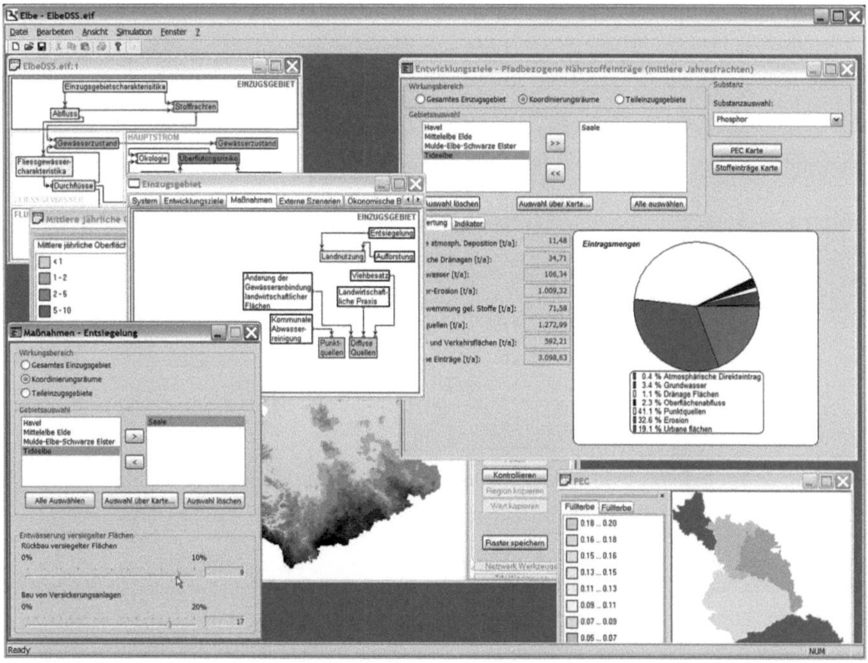

Fig. 6.4 Graphical user interface of the *Elbe DSS* (See also Plate 15 in the Colour Plate Section)

framework describes the component interfaces as well as the collaboration patterns between the components by means of special purpose textual (IDL, OCL) or graphical (UML) languages. Besides an abstract design specification, application frameworks may also provide a partial implementation in the form of a skeletal application that can be specialized to produce custom applications. Object-oriented frameworks are designed and implemented as a collection of collaborating abstract classes. They make use of the basic principles of object technology as described in Graham (1998): 'encapsulation' – data and processes are combined in objects and hidden behind an interface; 'polymorphism' – the ability to use the same expression to denote different operations; and 'inheritance' – implements the idea of classification and represents a special case of structural inter-relationship between a group of objects.

Figure 6.5 presents a simplified architectural overview of an ISDSS application based on the *Geonamica*® framework. *Geonamica*® has been used as the technological basis for the *Elbe DSS* and many other ISDSSs such as *RamCO, SImLucia* (Engelen *et al.* 1998a), *WadBOS* (Huizing 1998), *Environment Explorer* (Engelen *et al.* 1998b), *Modulus* (Engelen *et al.* 2000), *Murbandy* and *MedAction* (van Delden *et al.* 2007).

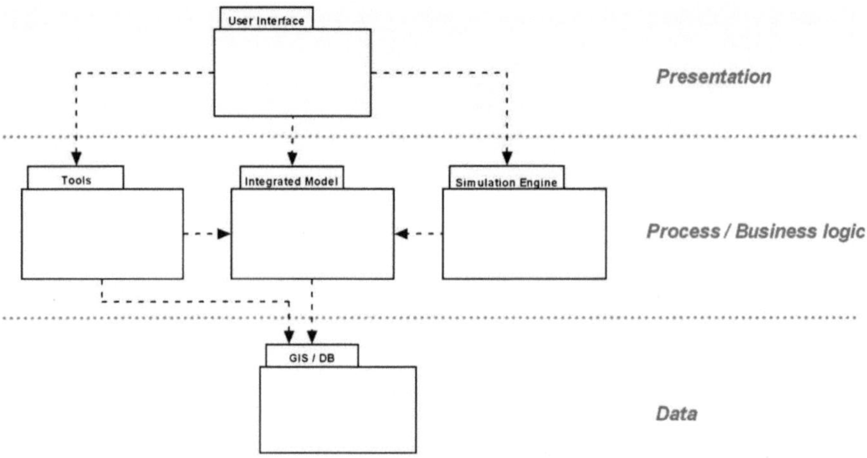

Fig. 6.5 Simplified reference architecture of a GEONAMICA® ISDSS application

6.7.3 *Iterative Development*

Requirements are seldom well understood at the beginning of an ISDSS project. Users usually start to recognize and articulate what they really want at the time when developers provide them with the first working prototypes of the product. From then on, requirements will never stop changing and evolving for the rest of the project lifetime and beyond. Acceptance of the fact that it is impossible to eliminate requirements' ambiguity and continuous change has lead to a shift from the classic linear lifecycle models (e.g. waterfall model) to more iterative and incremental approaches to ISDSS development. For the development of the *Elbe DSS*, we used the 'Evolutionary Delivery' lifecycle model as described in McConnell (1996). After an initial linear phase (top left-hand boxes in Fig. 6.6) that defines the core architecture of the system, it enters in an iterative and incremental cycle that elicits end-user feedback in joint workshops and delivers working prototypes at regular intervals.

This approach provided the development team with sufficient flexibility to adopt mid-course changes to the requirements (e.g. increased importance of flooding prevention functionality due to catastrophic events in 2002 or including policy options related to the recent climate change debate) and thereby increasing the relevance of the product for the potential end users.

6.8 Evaluation of the *Elbe DSS* Prototype

An important part of the ongoing requirements development process was the evaluation of the *Elbe DSS* prototype with end users. When the prototype became available by the end of 2005, an evaluation with users was carried out in close cooperation with BfG

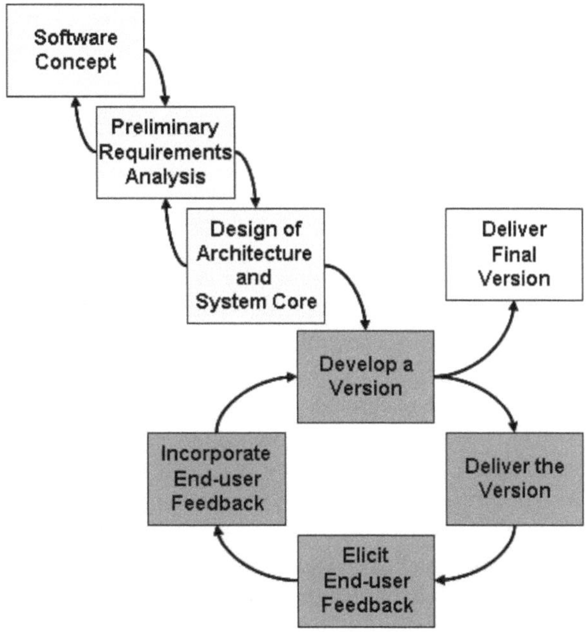

Fig. 6.6 'Evolutionary Delivery' lifecycle model (after McConnell 1996)

and the developer team. The *Elbe DSS* is one of the first systems realised which focuses on the complex field of IRBM for the strategic level of management. This evaluation served as a case study to analyse requirements for such different matters as system requirements, displayed system contents, data and models, decision and management support functionalities, performance, (sustainable) data management and maintenance. A detailed description of the *Elbe DSS* evaluation is given in Evers (forthcoming).

The evaluation was realised by means of user tests. As methodology, user tests with hands-on exercises, observations recorded in writing and questionnaires were undertaken. Furthermore, user test and system validation/verification were carried out both by a group of students and by one of the authors. During the winter term 2006–2007, an evaluation of the *Elbe DSS* prototype was undertaken with five different user groups who are concerned with IRBM on the strategic level. Test users were selected from the following organisations:

- a federal environmental institute;
- a sectoral management institute for coastal protection, water management and nature conservation in one of the Länder (federal states);
- a river-basin organisation of the River Elbe;
- an engineering company which is specialized in environmental planning; and
- a non-governmental organisation with focus on integrated water management.

The preliminary evaluation results can be summarised in the following points.
Key success/failure factors:

- Definition of purposes and requirement elicitation is a key issue. Thorough elicitation of requirements and users to define the purpose of the software is

essential rather than the software defining users' purpose. The system must have a positive effect on users' work (environment) and beyond, e.g. increase efficiency/quality.
- Definition of the diverse resources is crucial to assure continuous development, to ensure that enough data are available and to ensure maintenance of the system which is hardly ever finished.
- Active involvement of users, as already described above, is a key element of the development process.
- End user training sessions have to be provided.
- Prototyping of the system developed and tested by users involving realistic exercises. This gives tangible feedback before systems finalisation.
- The user-interface has to be oriented along users' technical backgrounds and should have integrated context-sensitive dialogues.
- The system has to be transparent (see above) with good documentation or library.
- The system should avoid fake accuracy and give information about the grade of correctness.
- It is important to consider the institutional and normative context of the institution where the software is supposed to be implemented. It has to be checked whether this institution has appropriate structure to use a PSS or can the system really depict the structure of that institution?
- Integration of evaluation during the development process is also an important point for successful implementation (see also next paragraph).

Knowledge integration: The elements of IRBM (despite groundwater management) are well integrated and the data pool is of good quality. The realisation meets the scientific requirements concerning models and methodologies.

User interface: A striking fact is that usage is complicated. The user gets lost easily and does not know what she or he can or has to do.

Decision-making process: The features of scenario simulation are highly appreciated. For example, the external scenarios with climate change and land-use change were considered to be very useful. Furthermore, the possibility of economic analyses is very welcome. Nevertheless, the usage of the scenarios was perceived as complicated. Shortcomings are evident in features like assessment, comparison of alternatives and optimisation of measures. These features are important for the decision-making process.

Transparency: This factor is regarded as very important. On the one hand, its realization can be regarded as very good as the test user assesses the scientific quality and data sources as good enough. On the other hand, the description of the content and the processes when operating the system was criticised. After the tests, all users were generally interested in the system.

Individual problems/aspects: The greatest interest in using the system was demonstrated by the members of the non-governmental organisations (NGOs), the federal environmental agency and the engineering company. The representatives of the institutions which are directly involved with operational management in IRBM, such as the regional water management authority and the river basin community, doubted whether they would actually use the *Elbe DSS* since their daily management demands are more of sectoral than integrative character.

Feedback by test persons concerning potential users of the Elbe DSS: was named with engineering companies (4), regional and federal authorities (3), a university, a river basin organisation, a water board, a sewage plant constructer and (diverse) institutions in nature conservation, agriculture, flood management and water management (each one nomination). This result uncovers a gap between the potential users as foreseen by the developers and the test users' assessment.

Against the background of the preliminary evaluation results, the question as to whether an ISDSS can provide useful support for IRBM can be answered affirmatively. The perceived added value can be summarized as the functionalities: developing a better insight of cause-effect-relations and getting a 'bigger picture' of relevant issues, relationships and optional measures.

For a generic *IRBM DSS* we can identify the following types of organisations as the most likely potential users:

- federal institutions which operate on a meta-management level;
- NGOs which focus on IRBM and are interested in gaining better insight into processes and optional management;
- engineering companies which want to generate specific information and data; and
- possibly universities to educate students in IRBM (this group did not take part in the DSS evaluation but one of the authors has practical experience in this field).

6.8.1 Future Trends

A standalone PC application is not the optimal system architecture for all ISDSS applications. The models and data in an ISDSS are often delivered by many different institutions. Knowledge management may become less complex, when these institutions remain in control and take the responsibility for the maintenance of their contributions to the ISDSS. Computationally demanding models are often better run on dedicated computers, which are optimally configured for this task. Keeping a tightly integrated PC application synchronized with quickly changing functional and organizational requirements for the ISDSS can be complex and costly. In view of participatory approaches to planning and policy design, the accessibility of the ISDSS for a wide class of users becomes a key issue. A service oriented architecture, where the ISDSS is an online application orchestrated from various web services offered by the knowledge and data providers, may in some cases be more flexible and lower the costs.

Environmental data that are used in an ISDSS often are monitored at regular intervals by ground-based observation networks or remote sensing instruments. In case of the *Elbe DSS*, integrating these data involved a lot of handwork by scientists and GIS specialists. For future ISDSS applications, especially when intended to support management oriented tasks, we envision enhanced integration with, and more direct access to environmental monitoring systems.

6.9 Conclusions and Future Directions

The experience with the design of the *Elbe DSS* leads us to the following recommendations. Firstly, both the design process itself and the practical application require a balance between users requirements, the quality of the scientific content, and sound software engineering. During the project the emphasis can temporarily pertain to a single aspect, but the final product should be based on a balance between all three aspects (De Kok *et al.* 2008). The key to achieve this goal is to start from real policy questions and to put together a multidisciplinary development team with an open communication culture, consisting of end-users, scientists and IT-experts as equal partners.

Secondly, effective communication between the developers and with stakeholders concerning the state of progress and possibilities of a tool is difficult when the tool is still being developed. A qualitative system diagram can be very useful in this respect. Furthermore the role of an 'architect', with knowledge integration and communication management as main responsibilities, turned out to be an important success factor for ISDSS development. The architect is a generalist and a true integrator, taking responsibility for linking the work of the modellers, assuring quality control of the mathematical core (model base) of the DSS and bridging the methodological and knowledge gaps between end users, modellers and IT specialists.

Thirdly, flexibility to cope with different measures, priorities or unexpected conditions is essential for relevance and acceptance of a tool. This calls for the application of relatively simple and flexible models. A modular system architecture, based on an object-oriented application framework helps to achieve a system that can be maintained, extended and adapted for future needs at relatively low costs.

The institutional embedding can be furthered by early anticipation in the implementation phase and providing key stakeholders with responsibilities for the design and financing, including maintenance.

Trans-national data harmonization and the development of metadata standards for the domain of strategic river basin planning are prerequisites for effective and cost efficient knowledge integration and ISDSS development. The fact that in the development of the *Elbe DSS*, nearly any model used its own *ad hoc* data format clearly showed how much needs to be done in this area. A lot of code needed to be developed just to translate between different representations of basically the same concept (e.g. a river network).

How to technically handle and, perhaps even more important, communicate uncertainty related issues in complex integrated models is still an open question. Solutions to this question could help to increase the acceptance of ISDSS as a tool for real world planning and policy making.

Finally, the interdisciplinary, integrated and holistic approach that underlies ISDSS often is ahead of the actual planning and policy making practice, where responsibilities and power are still organized in the classical sectors with only limited collaboration between sectors. Policy directives like the EC Water Framework Directive (EU 2000) and examples like the *Elbe DSS* may trigger further research into integrated methods for policy design, strategic planning and management of our natural resources.

References

Batty, M. (2007) Planning support systems: progress, predictions and speculations on the shape of things to come, Paper presented to the Seminar on *Planning Support for Urban and Regional Analysis*, Lincoln Institute of Land Policy, Cambridge, MA.

Berlekamp, J., Graf, N., Hess, O., Lautenbach, S., Reimer, S. and Matthies, M. (2007) Integration of MONERIS and GREAT-ER in the Decision Support System for the German Elbe River Basin, *Environmental Modelling and Software*, 22(2): 239–247.

Borshchev, A. and Filippov, A. (2004) From system dynamics and discrete event to practical agent based modelling: reasons, techniques, tools, in *Proceedings of the 22nd International Conference of the System Dynamics Society*, 25–29 July, Oxford.

Campagna, M., De Montis, A. and Deplano, G. (2006) PSS design: a general framework perspective, *International Journal of Environmental Technology and Management*, 6(1/2): 163–179.

De Kok, J.L., Kofalk, S., Berlekamp, J., Hahn, B. and Wind, H. (2008) From design to application of a decision-support system for integrated river-basin management, *Water Resources Management* (available online). http://dx.doi.org/10.1007/s11269-008-9352-7

Delden (van), H., Luja, P. and Engelen, G. (2007) Integration of multi-scale dynamic spatial models of socio-economic and physical processes for river basin management, *Environmental Modelling and Software*, 22(2): 223–238.

Denzer, R. (2005) Generic integration of environmental decision support systems – state-of-the-art, *Environmental Modelling & Software*, 20(10): 1217–1223.

Engelen, G., White, R. and Uljee, I. (1998a) Vulnerability Assessment of Low-Lying Coastal Areas and Small Islands to Climate Change and Sea Level Rise – Phase 2: Case Study St. Lucia, Report to the United Nations Environment Program, Caribbean Regional Coordinating Unit, Kingston, Jamaica, RIKS publication, Maastricht.

Engelen, G., White, R. and Uljee, I. (1998b) De Leefomgevingsverkenner. 'Proof of Concept' – Versie van een Integral Model voor het berekenen van het Leefomgevingskapitaal in Nederland, RIVM publication, Bilthoven.

Engelen, G., Meulen, Van Der M., Hahn, B., Mulligan, M., Reaney, S., Oxley, T., Mata-Porras, M., Blatsou, C., Kahrimanis, S., Giannoupoulos, P., Mazzoleni, S., Coppola, A., McIntosh, B. and Winder, N. (2000) *MODULUS: A Spatial Modelling Tool for Integrated Environmental Decision-making*, Final Report, Submitted to the European Union Directorate General XII Environment (IV) Framework, Brussels.

Evers, M. (forthcoming) Decision Support Systems for Integrated River Basin Management – Requirements for appropriate tools and structures for a comprehensive planning approach, Dissertation at the Faculty of Architecture and Landscape, Leibniz University of Hanover.

EU (2000) Directive 2000/60/EC of the European parliament and of the council of 23 October 2000. Establishing a framework for community action in the field of water policy, *Official Journal of the European Communities*, L327, 1–72. http://europa.eu.int/comm/environment/water/water-framework/index_en.html

Fayad, M.E., Schmidt, D.C. and Johnson, R.E. (1999) *Building Application Frameworks: Object-Oriented Foundations of Framework Design*, Wiley Computer Publishing, New York.

Graham, I. (1998) *Requirements Engineering and Rapid Development*, Addison-Wesley Longman, Harlow.

Hahn, B. and Engelen, G. (2000) Concepts of DSS systems, in German Federal Institute of Hydrology, *Decision Support Systems (DSS) for River Basin Management*, Koblenz, pp. 9–44.

Harris, B. (1989) Beyond geographic information systems: computers and the planning professional, *Journal of the American Planning Association*, 55(4): 85–90.

Huizing J.J., Engelen G., Ven van de K., Pothof, I. and Uljee I. (1998) *WadBOS: een prototype van een kennissysteem voor beleidsanalyse van de Waddenzee: eindrapport*, Rijkswaterstaat Directie Noord-Nederland, Harlem.

Kok, K. and van Delden, H. (2007) Combining two approaches of integrated scenario development to combat desertification in the Guadalentin watershed, Spain, *Environment and Planning B, 34 (in press).*

Marakas, G.M. (1998) *Decision Support Systems*, Prentice Hall, Upper Saddle River, New Jersey.

McConnell, S. (1996) *Rapid Development*, Microsoft Press, Redmond.

Mintzberg, H., Raisinghani, D. and Théorêt, A. (1976) The structure of "unstructured" decision processes, *Administrative Science Quarterly*, 21: 246–275.

Walker, D.H. (2002) Decision support, learning and rural resource management, *Agricultural Systems*, 73: 113–127.

Westmacott, S. (2001) Developing decision support systems for integrated coastal management in the tropics: Is the ICM decision-making environment too complex for the development of a useable and useful DSS? *Journal of Environmental Management*, 62: 55–74.

Additional Reading

Costanza, R. and Voinov, A. (eds.) (2003) *Landscape Simulation Modelling: A Spatially Explicit, Dynamic Approach*, Springer-Verlag, New York.

Seppelt, R. (2003) *Computer-Based Environmental Management*, Wiley-Verlag GmbH & Company KgaA, Weinheim.

Chapter 7
The *Emission Inventory Water*: A Planning Support System for Reducing Pollution Emissions in the Surface Waters of Flanders

Leen Van Esch, Greta Vos, Liliane Janssen and Guy Engelen

7.1 Introduction

The *Emission Inventory Water (EIW)* is an important policy instrument enabling a better insight into the pressures and impacts of point and diffuse emissions caused by different sectors on the surface waters of Flanders in Belgium. It is a spatially explicit computer-based information system developed to support the Flemish Environmental Agency (VMM) in its monitoring and reporting obligations *vis-à-vis* the Flemish, Belgian and European authorities. In addition to this function of providing an up-to-date technical database, it enables the design and assessment of alternative policies and spatially explicit (alongside technical) measures targeted at particular sectors and aimed at improving the quality of the surface waters in different river basins and administrative entities of Flanders. It supports to that effect 'what if' analyses in an interactive context. Such exercises are propelled by the obligation imposed on the Member States by the European Union to treat all waste waters by 2012. It is, therefore, an important tool in helping to meet water quality standards as well as evaluating and prioritizing measures (Crouzet and Bogestrand 1999; Fuchs *et al.* 2002; emissieregistratie.nederland).

The core of the EIW is an advanced MS Excel/Visual Basic module (Microsoft 2008) intimately linked to an *ArcGIS* module (ESRI 2008). The latter was only recently developed as part of the new version of the *EIW-PSS* (Engelen *et al.* 2006) delivered as an operational prototype in December 2006. As far as its GIS module is concerned, specific attention went into the emissions of metals caused by the building sector. In particular, the heavy metals lead, zinc, copper and, to a lesser degree, aluminium, chromium and nickel released by corrosion in the outer shell of the buildings as well as in the plumbing systems, are considered. Literature research and

L. Van Esch (✉)
VITO, Vlaamse Instelling voor Technologisch Onderzoek
(Flemish Institute for Technological Research)
Boeretang 200, 2400 Mol, Belgium
E-mail: leen.vanesch@vito.be

S. Geertman, J. Stillwell (eds.), *Planning Support Systems Best Practice and New Methods*,
DOI 10.1007/978-1-4020-8952-7_7,

measurement campaigns were set up to gather the required technical data alongside the development of the GIS methodology and implementation of the PSS. Various outputs of the PSS were calibrated and validated against independently measured data. The capabilities of the new version, the explicit geographical nature of its analysis and outputs, as well as its direct usefulness for policy and planning support, were the main criteria for continuing its development with the aim to incorporate more sectors and pollutants. Thus, ongoing work focuses on the generalization and enhanced automation of the methodology, the increased user-friendliness of the system, the documentation and training of technicians in the usage of the system for reporting, policy and planning tasks.

This chapter will emphasize the geographical and planning aspects of the *EIW*. It does not dwell on the large amount of work done to analyse the characteristics of the building sector in Flanders, to categorize and document the building components used and to estimate the potential corrosion of the latter. Interested readers are referred to the chapters on the subject by De Cuyper and/or Dinne from the Belgian Building Research Institute in the original report (Engelen *et al.* 2006).

7.2 The *Emission Inventory Water*: Basic Concepts

Essentially, the *EIW-PSS* is an accounting system, keeping track of the pollutants from a variety of sources towards their sinks in surface waters. Emissions are calculated on a sector by sector basis. Sectors are economic activities, such as the agricultural sector and chemical industry, but comprise also stocks such as buildings and road infrastructure.

Fundamental principles underlying the *EIW* are *conservation of mass* and *material flow analysis*. The latter analyses the flow of materials through a system. In the *EIW-PSS*, the materials are the wide variety of potential pollutants and the system is the Flanders region, looked at as a network consisting of nodes and links with pollutants flowing from node to node. Attention is paid in particular to chemical compounds known to be dangerous for humans and the environment: phosphates, heavy metals, Tributyltin (TBT), Polycyclic Aromatic Hydrocarbons (PAHs), *et cetera.* Mass is conserved, hence the amount of material flowing into a node has to leave it either through *collection* (in a piping system, for example), *accumulation* (in biomass, for example) or *removal* (by means of a treatment plant, for example). The mass of material leaving a node can be distinguished in function of its final destination: air, soil or water. In the *EIW-PSS*, only the destination water is currently considered.

Figure 7.1 is the material flow scheme of the *EIW-PSS*. Each box represents a node in the network and each arrow is a quantified flow. The material flow scheme is generic and applies to each sector separately. Moreover, in the *EIW*, emissions are dealt with on a compound by compound basis. Hence, the scheme applies to each chemical compound separately and in total. For example, when applied to the 'Source' building sector, node 'Sewerage' consists of the gross emissions from all

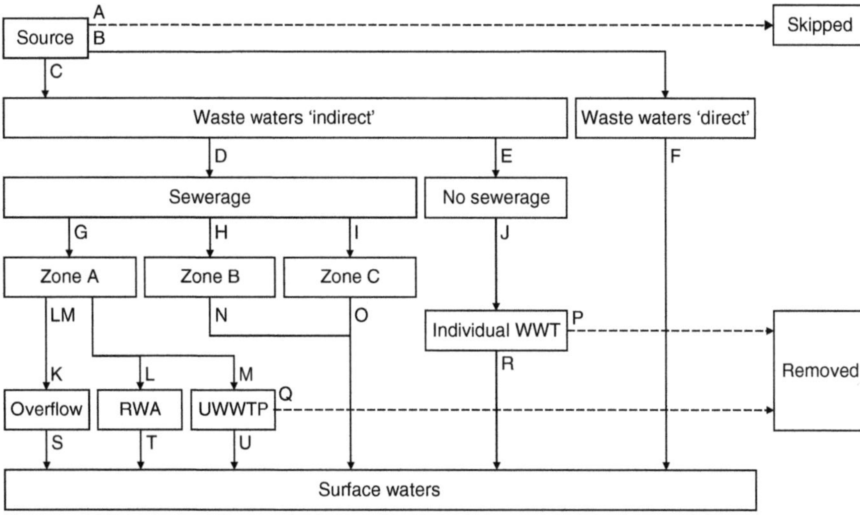

Fig. 7.1 Material flow scheme used in the *EIW-PSS* (After Syncera Water B.V. 2005)

buildings in Flanders connected to a sewage collection system. In contrast, node 'No sewerage' are the gross emissions from all buildings located in the part of the territory not serviced by a sewage collection system. Nodes 'Zone A', 'Zone B' and 'Zone C' consist of emissions from all buildings in those parts of the territory serviced by a sewage collection system equipped with an operational *Urban Waste Water Treatment Plant* (UWWTP) ('Zone A') respectively planned to be connected to an operational UWWTP ('Zone B'), and, not planned to be connected to an operational UWWTP ('Zone C'). The latter are predominantly industrial or commercial buildings occupied by sectors legally required to treat their own waste waters. The node 'Waste waters 'direct'' are gross emissions not transported through a sewage collection system but running off the surface and ending up without treatment in the surface waters.

In order to calculate the total emissions of a particular chemical compound in a sector, the so-called *emission factors (EF)* potentially causing the emission are itemized and quantified and the *emission explanatory variables (EEV)* are estimated. For example, in the building sector and for the emissions of heavy metals, emission factors are, for each building type, the building components in the outer shell of the buildings that are in contact with the air and precipitation including, for example, different types of gutters, metal hooks and pins to fix tiles on the roof, sheets of lead used at the interface between the walls, chimneys and the roof, *et cetera.* Emission factors are also components used in the piping systems of the building through which drinking or waste water is channelled, hence in contact with water and subject to corrosion. Finally, components of the fences, and other infrastructure used on the premises are considered. For each emission factor, its emission per metal and per unit length, surface or volume is determined. The amount of material emitted by each component of a building is finally calculated by multiplying, as in eq. (7.1), the emission factor with the so-called *emission explanatory variable (*EEV*)*, which is the typical length, surface or volume of the component used in a typical building of a given type. The *gross emission value* for each building type *t* (GEV*t*) is obtained after summing over all its components *i* as follows:

$$\mathrm{GEV}_t = \sum_i (\mathrm{EF}_i \times \mathrm{EEV}_i) \tag{7.1}$$

7.3 EIW's Spatial Components

From the beginning (Syncera Water B.V. 2005), the *EIW* was developed as a spatial PSS. It features 439 so-called *Water Treatment Areas* (WTAs). WTAs represent geographically contiguous areas which cover the full Flemish territory. Part of the territory of a WTA belongs to 'Zone A', 'Zone B', 'Zone C' and/or the 'No sewerage' node. In particular, 213 WTAs are serviced by UWWTPs (situation in 2002) meaning that part of their territory belongs to Zone A. A proportion of the waste waters generated in the 'No sewerage' node (this is the case for the sanitary waters) are treated in the so-called *individual Waste Water Treatment* (iWWT) *units,* which are small scale treatment installations operating at the level of the individual house, farm, office, social, commercial or industrial entity.

For each WTA, the gross emissions per sector and per metal are calculated. In the original version of the *EIW-PSS*, this was mostly done on the basis of coefficients estimating the share of activities, buildings, infrastructure, *et cetera,* to be attributed to the WTA. Coefficients also estimate a per unit pollution for each type of activity, type of building, infrastructure element, *et cetera.* Taking into consideration the technical characteristics of each UWWTP, the net emissions discharged into the surface waters per WTA were calculated. Potentials for policy, planning and management exercises with the system were plenty, but mainly limited to assessing changes in the technical coefficients representing the generation of pollutants and/or their removal. The approach is a simplification of complex processes, but is adequate for a first estimate and for carrying out the monitoring and reporting tasks required by the different authorities.

From the calibration and validation of the original *EIW-PSS* (Syncera Water B.V. 2005), it was concluded that a more detailed approach was required to achieve a better match between the calculated output of the system and the concentrations of particular chemical compounds measured at the nodes. Thus, it was decided to develop a new version leaving more room for detail in the representation of the sectors, the itemization and quantification of the *EF* and *EEV*, but also in the geographical capabilities of the system. The former improvements require a deeper scientific and technological insight into the individual sectors and the activities represented, the components generating emissions, and the physical and chemical processes responsible for the emissions. The latter improvements are much more generic in approach and add a new dimension to the *EIW* and its usage as a PSS.

In the next section, a closer look is taken at the spatial component of the new version of the *EIW-PSS*. Although it is designed to be generically applicable, for the sake of clarity, the discussion focuses on the emissions of heavy metals from the building sector, which was implemented as a first case. Building sector refers here to all buildings for residential, commercial and social activities. It also includes the housing and office buildings of small industrial production units and farms, but excludes industrial installations. Industrial installations are relevant to the analysis of the industrial and/or agricultural sector. Building sector also includes materials used for fencing off the terrains surrounding the buildings as well as garden objects.

7.4 GIS Analysis in Three Steps

The spatial module of the new version of the *EIW-PSS* is implemented in *ArcGIS* Software (ESRI 2008). It operates on data and GIS layers available from VMM and the Flemish or Belgian Government and is much more advanced in its spatial capabilities than the original version. It allocates the main sources of pollution to a raster grid, incorporates the detailed representation of the collection systems canalising the pollutants to UWWTPs; generates gross (prior to treatment) and net (after treatment) emission maps; and, produces accounts for catchment and administrative entities at different hierarchical levels. It addresses the problem in three steps carried out sequentially:

7.4.1 Step 1: Mapping the Building Sector

Gross Emission Values (GEV, see eq. (7.1)) are calculated for prototypical building types consisting of classes of buildings with a typical function, morphology and location in space and typically composed of a sub-set of building components. Based on the data available, eight main building types are used:

- terraced home;
- semi-detached house;
- detached house, farm or castle;
- apartment building;
- commercial building;
- industrial building;
- shed, garage and other small building including greenhouse; and
- other building.

Flanders has vector-based geo-referenced cadastral data which enables locating each individual building and categorizing its precise function in great detail. However, these data are not publicly accessible as they contain confidential information. Furthermore, the availability of the kind of detailed data is unique to the building sector only. As the *EIW* procedure should be generic and applicable to other sectors,

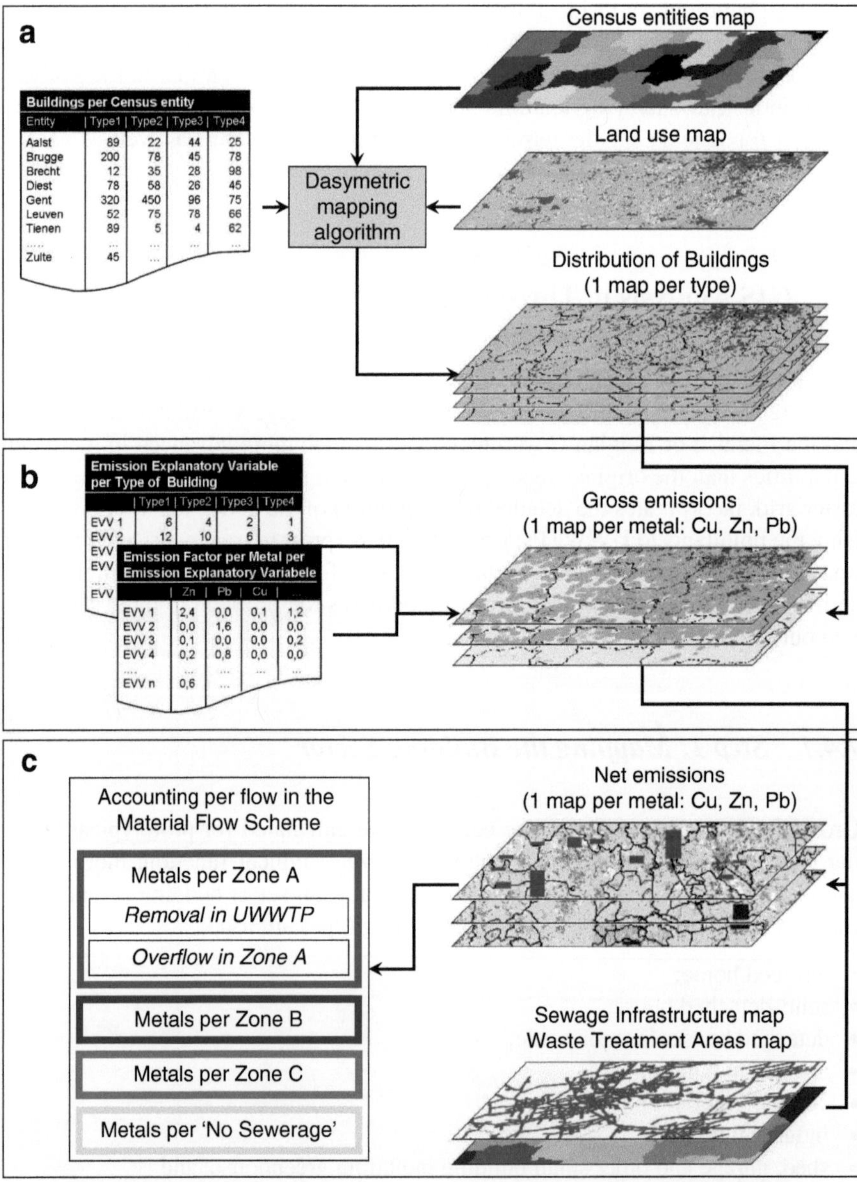

Fig. 7.2 Schematic representation of the three main steps of the GIS procedure: spatial allocation of the building sector (a), determination of gross emissions (b), transport and determination of net emissions (c)

without such detailed information, the vector data were not used. Instead a more generic raster-based approach was adopted on the basis of dasymetric mapping (Mennis 2003; Eicher and Brewer 2001; Gallego *et al.* 2001). An algorithm nearly identical to that of Mennis (Fig. 7.2a) is applied.

In comparison with choropleth mapping, dasymetric mapping is a technique enabling the cartographic representation of a spatial feature for which data are available at the level of relatively arbitrary spatial units, reflecting more realistically the precise spatial location of the feature within those spatial units. To that effect, use is made of ancillary (cartographic) information representative of the location of the feature. In the given case, the spatial feature concerns the number of buildings per type reported per municipality and/or statistical sector (see below). The ancillary data is derived from the land-use map of Flanders at the 15 m resolution available from the Agency for Geographical Information Flanders (AGIV) (www.agiv.be), which is the official government portal for geo-referenced data sets of Flanders.

The distribution of the building types is produced, at a 60 m resolution. This is in part a pragmatic decision enabling a straightforward aggregation of the original 15 m cells, but it is also selected because Flemish environmental legislation states that buildings within a 50 m buffer from a sewage collection pipe should be connected. When overlaying the detailed location of the sewage system (in Step 3 of the procedure), it is thus possible to distinguish between grid cells that should and should not be connected. The re-sampling of the 15 m land-use map was carried out by means of the *SpatAggr* tool developed by RIKS bv (www.RIKS.nl). This tool enables the spatial aggregation of the original land-use map ensuring the conservation of total area per land-use class. Failure to conserve total area in the various land-use categories in Flanders with its disperse building patterns would result in important errors. Prior to the re-sampling, the original 20 land uses were grouped into nine classes most relevant for allocating the building types.

The data relative to the distribution of the buildings were obtained from the Belgian Federal Statistics (FOD Financiën en FOD Economie, KMO, Middenstand en Energie). For all building types, information is available at the municipal level (308 in total for Flanders). However, for the three housing types and the apartment buildings, more detailed information is also available at the level of the so-called 'statistical sectors', which are sub-divisions of the municipalities (approximately 9,500 in Flanders). Next, the dasymetric mapping procedure is applied for each individual building type. The most detailed information available per building type is used to that effect. During calibration and validation of the resulting distributions, the amount of buildings estimated in a cell is compared with reality. Based on a sample of a few hundred cells, the coefficients of the algorithm were adjusted to obtain a best-fit. Figure 7.3 shows the result of the dasymetric mapping prior (Fig. 7.3b) and after (Fig. 7.3c) the calibration for the central part of Antwerp. Figure 7.3a is the land-use map for the same area with the borders of the statistical sectors superimposed. The figure demonstrates the much improved density pattern after calibration, which removed the high and unrealistic peak values. The result of Step 1 is a set of eight maps representing the distribution of the eight building types at the 60 m resolution.

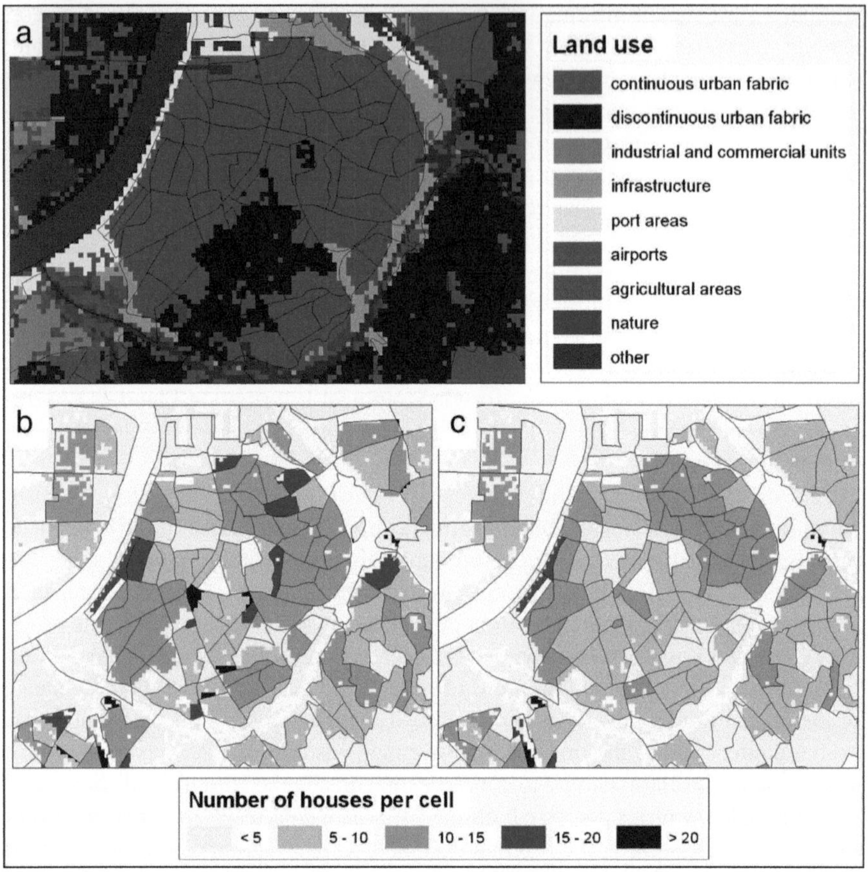

Fig. 7.3 Land use in the centre of Antwerp (a), distribution of housing as the result of the dasymetric mapping before (b) and after (c) calibration (See also Plate 16 in the Colour Plate Section)

7.4.2 Step 2: Detailed Localization of the GEVs Per Type of Building and Pollutant

Once the distribution of each building type is known, the next step consists in calculating the gross emissions generated in situ (Fig. 7.2b). To that effect, an important piece of research led to the *EEV*: a database of components used in buildings now and in the recent past (+/- 30 years back), their typical presence in the eight building types and the amount of emissions of metals generated on a yearly basis. Information is gathered for the following metals: copper (Cu), zinc (Zn), lead (Pb), aluminium (Al), chromium (Cr) and nickel (Ni). Sufficient information to perform all steps of the analysis was only available for Cu, Zn and Pb.

A literature research complemented with a survey among building practitioners and major producers and distributors of zinc and copper components revealed the typical configuration and composition of buildings pertaining to the three housing types, the apartment buildings and the commercial buildings of different ages. Regional differences in the building practice and tradition and associated choice of components were taken into consideration. For example, it is known that plumbers in the Antwerp region have traditionally used more copper and less galvanized components in the drinking water circuits than their colleagues in the rest of Flanders.

For the buildings pertaining to the categories 'other buildings' and 'industrial buildings', the approach taken differs from the one explained above. 'Other buildings' represents 4.3 per cent of all buildings in Flanders. They include: churches, schools and sports facilities, and hence vary considerably in function, size and age. To a lesser extent, the same applies for the category 'Industrial buildings'. An attempt was made to define a prototypical building for each of the two categories as well as to define their composition and contribution in terms of emissions. The approach consisted in gathering information from samples analyzed by means of the high resolution remote sensing images available for Flanders in *Google Earth,* followed by field visits to the selected buildings. This analysis revealed the size, roof type, materials used in the roof, metal building components on the roof, attached to the wall and/or on the premises. In total, 105 buildings, pertaining to the major sub-categories (schools, churches, hospitals, shopping malls, traditional industrial buildings, modern warehouses, *et cetera*) were thus analyzed. Both prototypes were finally defined as an average of the sampled buildings weighted according to their importance in the building sector belonging to each category.

Based on the above, the contribution to the emission of Cu, Zn and Pb for each prototypical building type was estimated. Next, these coefficients are multiplied on a cell-by-cell basis with the number of buildings to result in the gross emission map per building type and metal (Fig. 7.2b). The sum over all building types leads to the gross emission per metal at the 60 m resolution (Fig 7.4). The latter maps are input for the next step dealing with the transport of the metals from source to sink.

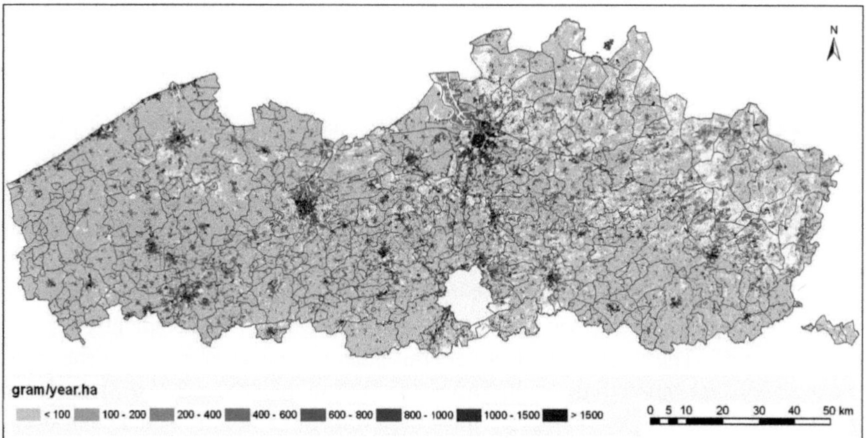

Fig. 7.4 Zinc: gross emissions (See also Plate 17 in the Colour Plate Section)

7.4.3 Step 3: Transporting the Emissions from Source to Sink

In the *EIW-PSS*, the pollutants running off the surface or infiltrating into the ground water to end up directly in the surface waters (node: 'Waste waters 'direct'') are currently estimated per sub-catchment on the basis of coefficients obtained from hydrological models external to the system. This will change in the future when such models will be incorporated (see Section 7.8). Of direct interest to VMM are the pollutants physically transported and treated by means of the existing and planned sewage infrastructure (node: 'Sewerage'). In the GIS component, this is implemented by overlaying the gross emissions with the vector layer representing the highly detailed sewage collection system of Flanders (Fig. 7.2c). For every pipe in the system, a buffer with a width of 50 m is created. As mentioned before, the 50 m buffer is an important criterion in the Flemish environmental legislation. Pipes and buffers are labelled according to the node they pertain to: 'Zone A', 'Zone B', 'Zone C'. Emissions in the cells falling entirely or partly within a buffer are assigned to it and summed. A special case applies when buffers intersect clusters of cells categorized as industrial, airport, harbour or inner city on the land-use map. All cells belonging to such clusters are assigned to the buffer and summed whether or not they fall within the buffer. This reflects the fact that most if not all such clusters feature a secondary piping system which may or may not be known to VMM and thus may or may not be incorporated in the network representation but should be.

Depending on the type of buffer and associated pipe, emissions are transported to the surface waters directly (node: 'No sewerage') or indirectly (node: 'Sewerage', and, 'Zone A', 'Zone B' and 'Zone C' fed by it). In fact, a reasonably limited number of cells deliver their waste waters and emissions to a sewage system to be transported to a UWWTP. This matches reality very well and is confirmed in the validation of the *EIW-PSS* carried out at this level. The application of the *EIW-PSS* to the building sector revealed that 57 per cent of the waste waters generated by the population of Flanders is treated in an UWWTP ('Zone A') and that 78 per cent of the population is connected to a sewage system ('Zone A' + 'Zone B' + 'Zone C'). These numbers match very closely those used by VMM and based on alternative sources (respectively 60 per cent and 80 per cent).

For waste waters delivered to the UWWTP, technical coefficients are applied to calculate the amounts of metal removed as part of the treatment. These are available in the *EIW* on a plant-by-plant basis and reflect specific situations, treatment technologies applied and efficiencies attained. Other technical coefficients represent physical pre-treatment (node: 'RWA') and storm weather overflow (node: 'Overflow'). After treatment the waters with the remainder of the pollutants are discharged in surface waters at the outlet point of the plant. This is typically a single cell in the GIS. The total amount of net emissions for the service area of the treatment plant is attributed to this cell. This result is available in the *EIW-PSS* as a map per metal, which is, by and large, the most important output of its spatial component (Fig. 7.5).

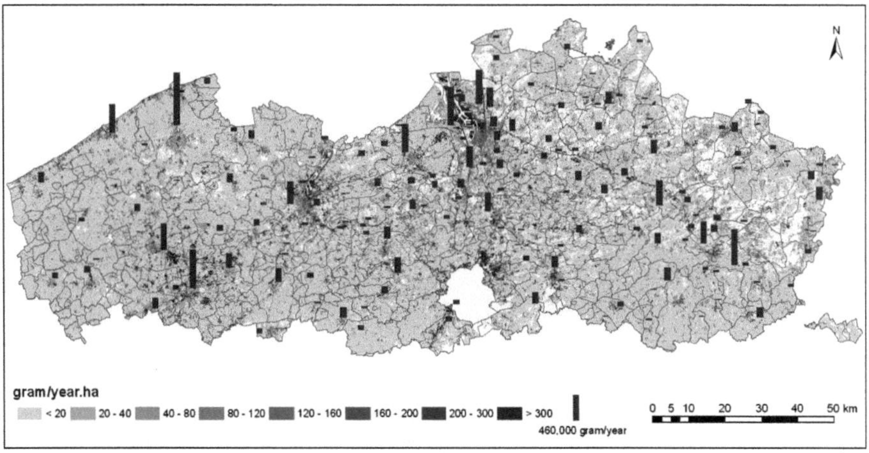

Fig. 7.5 Zinc: net emissions (See also Plate 18 in the Colour Plate Section)

7.5 Some Results

Not discussed in the previous sections is the fact that gross and net emissions are calculated separately for the metals corroded in the outer shell of the buildings, hence feeding the surface run-off, and the sanitary waters originating in the plumbing circuits. As a result, the *EIW-PSS* provides maps for both the surface and sanitary waters for both gross and net emissions. The reason for distinguishing one from the other is to assess planning and policy measures related to the decoupling of waste waters. In Flanders, like other regions with high proportions of hard surfaced areas, decoupling is a growing trend in the context of integrated water management as it serves multiple purposes simultaneously. It promotes secondary transport systems for the waters, typically precipitation, running-off at the surface and canalized in a separate piping system or in open-air ditches. Before all, decoupling stabilizes the quantity and quality of waste waters delivered to the UWWTP by means of the sewage system, limits sewage treatment to the more heavily polluted sanitary waters, and hence increases the efficiency and cost effectiveness of the UWWTP. But, it also limits overflow and the 'leakage' of waste waters in case of storms and other cases of heavy precipitation; enables replenishment of the groundwater, and limits the frequency and severity of floods because of the increased capacity to buffer the surface run-off.

In addition to the various maps discussed in the previous section, the *EIW-PSS* generates synthetic output at the level of each node and flow in the material flow diagram. It enables assessing the efficiency of the waste water treatment in operation and the potential effects of planning and management measures targeted at particular nodes. For example, Fig. 7.6, taken from the *EIW-PSS*, reveals that only a limited amount of pollutants from the building sector end-up in the surface waters without treatment. Finalising the waste water treatment plants in Zone B (already

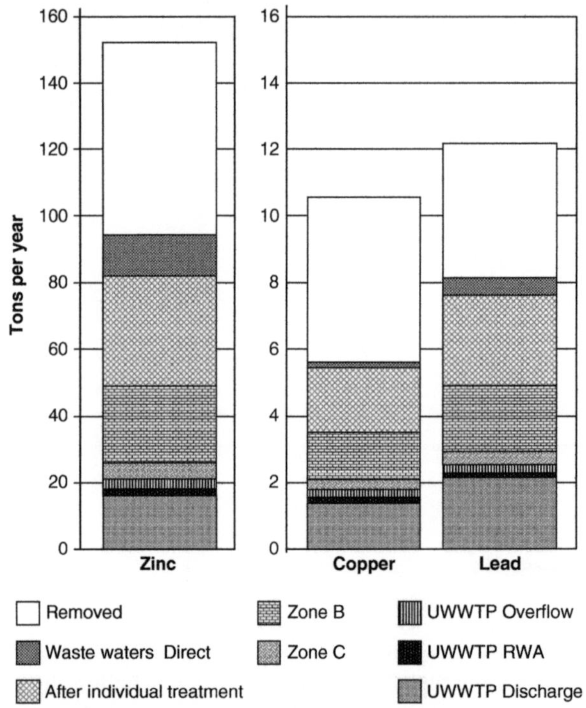

Fig. 7.6 Gross emissions (complete bar, including Removed) and net emissions of zinc (left), copper (middle) and lead (right) calculated per node in the *EIW-PSS*

equipped with a sewage system) is important, however, more could possibly be gained from more or improved iWWT units.

The final and intermediate results of the system have undergone calibration and validation. Some of this has been discussed at the level of the respective steps. For example, the validity of the EEV have been tested and confirmed by the Belgian Building Research Institute as part of a field campaign measuring the amounts of metals corroding from roofs, gutters and fences of different sorts. Also, the EEV have been compared against the quantities of copper, zinc and lead components sold to the building industry in Flanders yearly. The final results of the *EIW-PSS*, the gross and net emissions, have been validated against the loads in the influents and effluents of the UWWTP. However, comparisons between calculated and measured are difficult as the results of the model are for one sector only, while the influent of the UWWTPs concern the sum of all sectors. Past applications of the original *EIW-PSS* to all sectors revealed more metals in the influent than calculated. For example for arsenic, the measured load is close to ten times the load calculated. Consequently, a large proportion of the metals in the influent remained unexplained. The current exercise bridged part of that gap and attributed much of the unexplained loads of metals to the building sector. This is certainly so for lead. The new version of the *EIW-PSS* attributes 68 per cent (6 ton/year) of all lead in the waste waters to the building sector (rather than 19 per cent or 1 ton/year in the original *EIW*). For zinc and copper, higher loads are calculated too. The building sector represents no less than 71 per cent (72 ton/year) of the zinc in the waste waters and 40 per cent (6 ton/year) of the copper. In the original *EIW-PSS*, this was 56 per cent and 27 per cent, respectively. Thus the more in depth, geographical oriented bottom-up analysis was beneficial in providing more insights into the issue.

7.6 The *EIW-PSS*: Planning and Policy Examples

The *EIW-PSS* is intended for exercises that are more technical and sophisticated in nature than reported above. It is fair to state that these have not yet been fully explored at the time of writing, and that work is currently ongoing to uncover the full power of the system in support of various monitoring, reporting and planning tasks of VMM. Technical measures are associated with the particulars of the sectors represented, the stricter separation of industrial waste waters, the promotion of iWWT or characteristics of the UWWTP including efficiency, capacity, overflow discharge frequency, *et cetera.*

Already implemented is the possibility to assess emissions at the level of selected administrative and hydrological entities (Fig. 7.7). To that effect, the net and/or gross emission maps are overlaid with the boundaries of such entities at different hierarchical levels to generate sums per entity at each level. Thus VMM is now able to report results at the catchment and sub-catchment level as well as the municipal level, which is important to set targets and norms or meet preset criteria.

Management and planning options can be considered in a regional context too. For example, sub-catchments particularly hit by emissions and/or areas of particular natural value can be given a higher priority when it comes to improving the collection and treatment infrastructure. Similarly, areas densely populated or responsible for high net emissions, yet insufficiently equipped with sewage infrastructure and treatment facilities and/or not incorporated in investment schemes, can be easily delineated on the net emission maps. Totals can be generated and analyzed per spatial entity defined by the analyst. Precisely localized additional sewage infrastructure and treatment facilities can be inserted in the respective map layers as part of scenarios, and, the effects thereof on the surface waters can be simulated and quantified.

Ex ante simulation is another application. The effects of using materials generating less pollutants, as well as fundamental changes in the sectors, activities and hence the EF and EEV can be analyzed in depth. One example in the building sector is an ongoing evolution towards the usage of more plastics (replacing metal frames, pipes, gutters, *et cetera*) and better coatings. An evolution also towards more compact housing styles in new sub-divisions better planned for and equipped with sewage systems from their design onwards. Such analysis can be carried out at the regional level, hence assuming coefficients to change with time for the whole of Flanders, or very locally and targeted at a precise sub-division within a particular municipality and sub-catchment. Scenarios combining sets of measures can be

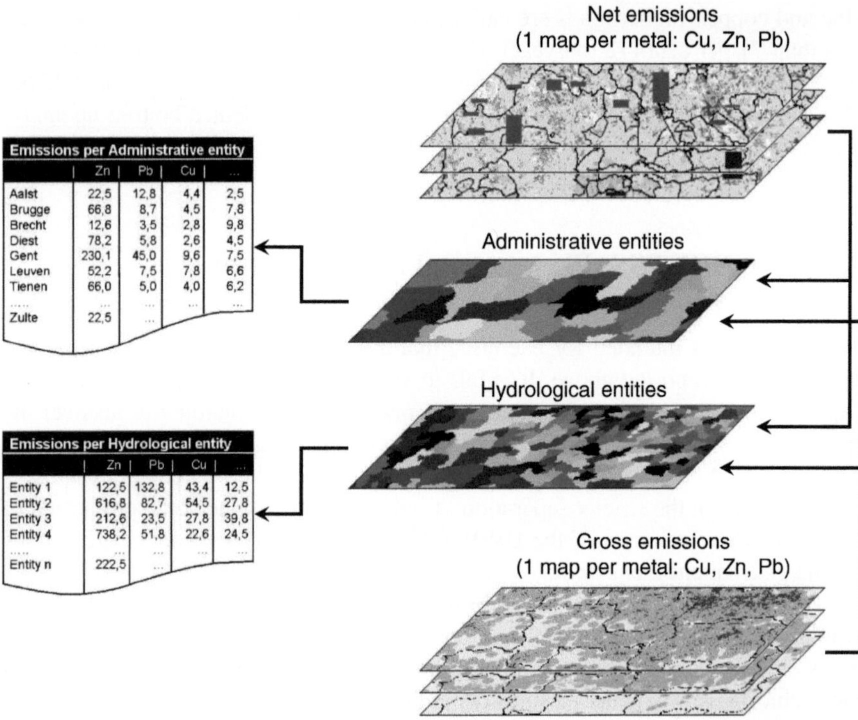

Fig. 7.7 Assessing spatial policy and management options with *EIW-PSS*: aggregating emissions per administrative and hydrological entities

composed, calculated and finally compared to inform the user about the effectiveness of an investment plan, a subsidy or an instrument embedded in a law or regulation. *Ex ante* assessments aimed at the longer term involve the analysis of scenarios featuring changed activity and land-use patterns in Flanders. Given the importance of the land-use map in the dasymetric mapping applied in Step 1 of the GIS component, there is little doubt that these may have important consequences in the output of the system.

Also, *ex post* and *trend* analysis with the *EIW-PSS* are of interest to the planner. A recently finished exercise (Engelen and Van Esch 2007) assessed the progress made with regards to the emissions of metals from the building sector in the period 1998 to 2005. The exercise emphasized improvements in the sewage system and treatment facilities as well as changes in the mix of buildings, their location in space and changes in the building materials used. The analysis carried out for the full territory also supports local analysis. For the sake of this section results are reported for lead and for one Water Treatment Area (WTA) only, namely Rotselaar. Figure 7.8 shows the location of the WTA in Flanders, the distribution of the buildings in 2005, and the sewage infrastructure associated with Zone A, Zone B and Zone C in 1998 and 2005. With a view to assess investment programs from the recent past and to provide input for choices and measures targeted at the

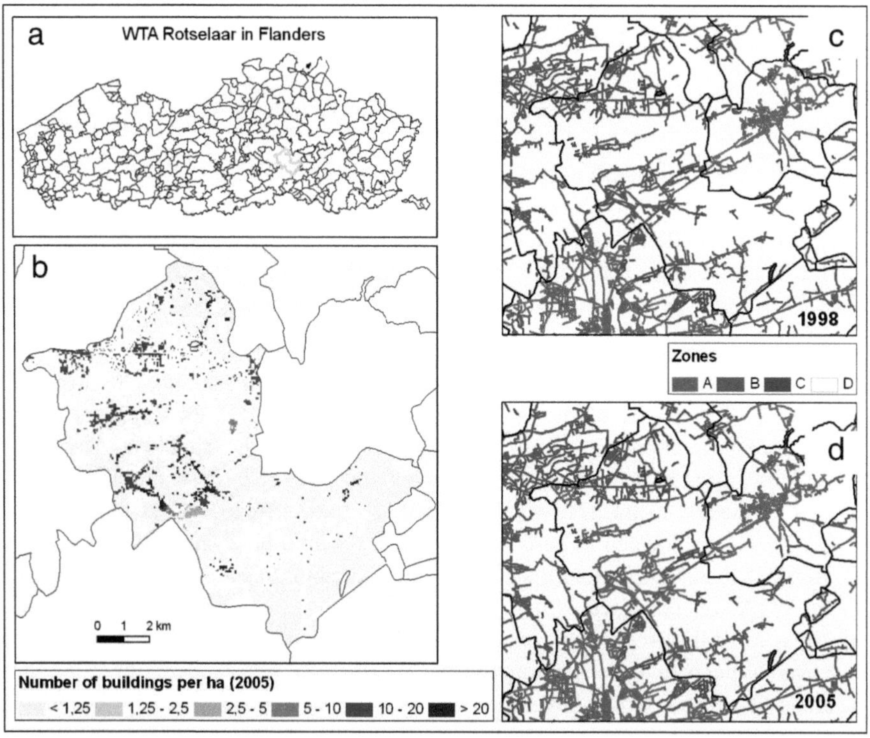

Fig. 7.8 Inputs for the trend analysis described for the Rotselaar WTA: location of the WTA in Flanders (a), distribution of buildings in the WTA (b), sewage infrastructure pertaining to Zone A, Zone B and Zone C in 1998 (c) and 2005 (d)

future, an indicator was developed representing in a synthetic manner the treatment efficiency achieved in the period concerned. The indicator is calculated as follows:

$$\text{Treatment efficiency} = \frac{\dfrac{\text{Net emission } \mathit{2005}}{\text{Gross emission } \mathit{2005}}}{\dfrac{\text{Net emission } \mathit{1998}}{\text{Gross emission } \mathit{1998}}} \tag{7.2}$$

Two conditions apply for the fraction applied in both the numerator and the denumerator:

$\dfrac{\text{Net emission}}{\text{Gross emission}} = 1$ applies when no treatment takes place. The discharge of metals in the surface waters is equal to the gross emissions.

$\frac{\text{Net emission}}{\text{Gross emission}} < 1$ applies when treatment takes place. The result of the fraction is the proportion of the gross emission not removed as a result of the treatment.

When eq. (7.2) is applied and the years 1998 and 2005 are considered *vis-à-vis* one another, the indicator enables us to assess, in relative terms, the advances made in the treatment efficiency in the period 1998–2005. Figure 7.9 shows the outcome of this spatially explicit indicator for Rotselaar WTA as it is presented in the *EIW-PSS*. Three types of areas result:

- The areas coloured in mid grey represent the 7 per cent of the territory with an improved treatment efficiency. For cells pertaining to these areas the outcome of eq. (7.2) is less than one. As can be verified in Fig. 7.8, the improvement in the treatment efficiency is mostly due to a change in status from Zone B to Zone A, meaning that the sewage system which was already largely in place in 1998 was connected to the UWWTP by 2005. The same figure shows that the mid grey areas are located in parts of the territory with a high number of buildings. Hence, substantially more than 7 per cent of the buildings and the population of the WTA are concerned.
- Areas left blank represent a status quo. The value of the indicator is one. They cover 16 per cent of the territory and are mostly areas pertaining to the Zone B or Zone C, hence in 2005 they still discharge untreated waste waters in the surface waters.
- Finally, the treatment efficiency of areas coloured in light grey and dark grey has degraded marginally (for areas in light grey) or slightly (for areas in dark grey). They represent 77 per cent of the territory of the WTA. They are those parts of

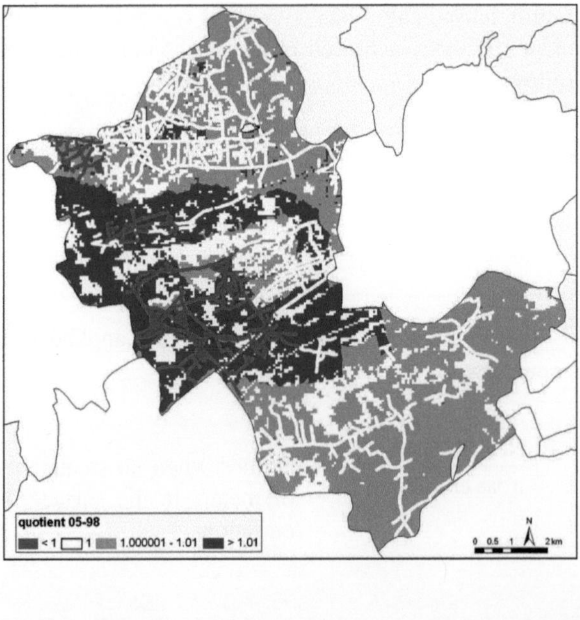

Fig. 7.9 Treatment efficiency indicator for Rotselaar WTA

the WTA not equipped with a sewage piping system. The reasons for the degradation are a complex of positive and negative terms. On the positive side of the balance is mainly the fact that the amount of emissions of lead in the sanitary waters has decreased because of the rejuvenated building sector in 2005 and the absence of lead in the plumbing systems of more recent buildings, houses in particular. On the negative side of the balance is the fact that the amount of buildings has generally increased in 2005, and, that there has been an evolution towards more semi-detached and detached houses. They have replaced buildings of all other types. They are generally bigger and thus emit more lead from their outer shell. Much of the water running off at the surface of the buildings is not collected and treated in the iWWT facilities that may be in place, hence end up in the surface waters directly.

Analyses of the kind described are supported by the *EIW-PSS*. They are very useful for the evaluation of investment programs from the recent past and provide input for choices and measures targeted for the future.

7.7 Discussion

The *EIW-PSS* is a system that has grown and will grow with new needs of its users; initially its tasks were very restricted to the very practical and legal requirements to monitor and report, but its availability generates more demand for more advanced applications. This is unlike many of the planning support systems (PSS) developed in the research community. In the latter case, end users are not in the driver's seat, hence their requirements are not necessarily initiating nor pushing forward the development. In the case of the *EIW-PSS*, the end user of the system is prominent as each new version of the system is used immediately for the day-to-day monitoring and reporting tasks. Hence, it has to work with the latest and best available data, has to be trustworthy and foolproof, and has to be calibrated and validated to the extent possible. Contrary to research oriented projects, the financing of the project is not necessarily the system per se, rather directed towards the outputs that the system should be able to generate. This puts a lot of stress on the development, the developers and the end users as strict deadlines and deliverables apply. Seemingly small technical or administrative details, such as the late availability of a new data set, or a map, may turn out to become major stumbling blocks in meeting such deadlines. Strict deliverables guarantee to keep the development on track and well within the realm of what is technically and practically feasible. However, this may also lead to a piecemeal and fragmented development which, in the longer run, may not be very efficient.

Lessons learnt in the past with the development of PSS (Engelen *et al.* 2002, 2003) show the many advantages of working on an advanced information system that has its place and role in the end user's organisation, right from the beginning. A system that is transparent and understood will be used by its intended end users in the organisation. Thus, it is a tool they are familiar with and feel confident about. In fact, it is an instrument that they stand for and will support when it comes to furthering its development and finding the necessary funding. PSS dropped into an organization from the outside very often lack such commitment and disappear as quickly as they have appeared despite their quality and inherent potential. Their role, therefore, becomes the general advancement of the domain in a scientific or technical sense rather than the real application of the instrument.

A word of warning is to be issued for those engaging in the kind of work. This warning relates to the generally difficult conditions in which PSS are developed in organizations using the tools and/or their components while they are under construction. For example, VMM is very well equipped with high resolution GIS data on most items related to environmental pollution in Flanders. It has highly qualified collaborators knowledgeable about GIS generally and its application to their domain. As far as this is concerned, VMM is a near to ideal partner for carrying out the kind of work. However, when it comes to integrating the available material in a new advanced product, one runs into the problem that use is made of materials that evolve constantly and rapidly. For example, in the *EIW-PSS*, the sewage piping systems keep on growing, new treatment plants come into operation, administrative maps of higher quality become available, *et cetera*. This material is available from different sources and incorporating it into a planning support system results in the typical mismatches known to the PSS community. In theory, all these problems have long been dealt with and solved. However, in a practical context like the one reported here, they may hamper the PSS development severely.

State of the art development of information systems assumes a design phase in which the intended system gets an appropriate design to carry out its intended tasks. A perfect design is theoretically possible, certainly if it concerns a new system to carry out a new series of tasks. However, organizations like VMM are given new tasks every day. Their information systems get updates and grow continuously and so do their databases. The pressure of due deliverables forces *ad hoc* solutions more often than wanted by any of the technicians involved. Yet, this is the reality in which PSS grow and evolve. More so, these will be the conditions in which more PSS will be born and grown in the future. VMM and many similar organizations are day after day better equipped to develop and use them; the human and technical resources are present. Their shifting tasks require more advanced analytic capabilities carried out with more responsibility and concern for the citizen. In densely populated regions such as Flanders, the straightforward and easy solutions to meet the environmental criteria put forward by the European Union among others have long been implemented: the low-hanging fruits have been picked. More sophisticated solutions have to be designed and have to be tested for their effectiveness, their cost efficiency and their equity in meeting the environmental criteria without jeopardizing the quality of life of the citizens nor the fair request for growth of the entrepreneurs propelling economic growth. There is an indisputable need and role here for PSS. It calls for flexible systems and appropriate development methodologies for building and upgrading them.

7.8 Conclusions and Further Research

In this chapter, we have presented the *Emission Inventory Water Planning Support Systems,* developed by the Flemish Environmental Agency. Initially this was for monitoring and reporting but its availability propels more and more advanced planning tasks. One and the other have been greatly enhanced by the new version which is equipped with a GIS component enabling an explicit geographical analysis at high resolution. This system is generic in its design, but has been applied to the building sector only.

The PSS enables the detailed representation of sectors responsible for the emission of harmful pollutants, their transport to treatment facilities and finally their discharge into the surface waters. It supports scenario, *ex ante* and *ex post* analyses of different kinds assuming technical changes, changes in the socio-economic structure of Flanders, the spatial configuration of Flanders, investment schemes, new legal requirements and criteria to be met, *et cetera*. Extensive analysis of the outcomes is facilitated by means of the map outputs (in its *ArcGIS* component) and the tabular and graphical outputs (in its *MS Excel* component). Configuring and running a scenario mainly consists of providing (consistent sets of) technical coefficients (to be changed interactively), statistical data, and map layers representing the location of activities and sectors as well as the waste water treatment infrastructure. Exercises of this kind can be carried out by the high level technicians of VMM. They may need help in their analysis from GIS technicians and domain experts preparing more advanced input for the system in the context of scenarios tried out.

The results of the new version have been validated to the extent possible. The tests carried out have been very convincing. In its application to the building sector, a much better estimate of the amount and proportion of the metals in the influents of the UWWTPs could be attributed to the sector. Ongoing work focuses on the generalization and enhanced automation of the methodology, the increased user-friendliness of the system, the documentation and training of technicians in the usage of the system.

Further research, currently specified in proposals, involves the extension of the *EIW-PSS* to all sectors and all emissions known to contribute to the pollution of the surface waters of Flanders. This is an ambitious endeavour that will take time and effort to be accomplished. Incorporating more pollution sources requires before all extending the computational framework. In addition to sources covering the entire territory, like the one discussed in this chapter, there are point sources and sources distributed along networks that need to be dealt with. Examples are respectively industrial facilities and transportation infrastructure. To address these, the *EIW-PSS* is to be extended with additional algorithms spatially allocating the kind of sources with similar levels of detail. Further, pollutants are not only transported via the sewage collecting system. Rather run-off and transport via small ditches, drainage networks and the groundwater is important for particular types of pollutants and/or sectors. This is certainly so for emissions caused by the agricultural sector. Again, the appropriate algorithms need to be developed and incorporated in the *EIW-PSS*.

The architecture of the resulting system will be one consisting of a library of modules, implementing the various algorithms. Depending on the precise characteristics of the source and the pathway of the pollutants, modules will be linked and executed in a sequential manner. Knowledge and modelling components to that effect will be derived from the literature as well as from state of the art applications. In addition, for each new sector incorporated, exhaustive literature research and measuring campaigns will be carried out to bring to the surface the information and data relative to the emission explanatory variables and the emission factors. This should result in a system as generic as possible, yet, one implementing each sector and pollutant of interest. At any particular point in time, the availability of information and resources will largely determine the level of depth that can be attained in representing sectors and their emissions. Such stepwise development will not hamper the usage of intermediate versions of the system for practical purposes as the methodology allows for the representation of different sectors at different levels of detail. It is in any case the only way in which an ambitious project of the kind can be implemented given the needs, the resources available, and the complexity of the tasks.

Acknowledgements The work presented in this article was carried out under two research contracts between the Flemish Environmental Agency and the Flemish Institute for Technological Research. The authors wish to thank Ir. K. De Cuyper and Ir. K. Dinne from the Belgian Building Research Institute for their work in the project leading to the estimation of the Emission Factors and the Emission Explanatory Variables. Thanks go also to the colleagues at VMM and VITO that participated actively in both projects.

References

Crouzet, P.H. and Bogestrand, J. (1999) A European Inventory of Emissions to Inland Waters – A First Proposal, *Technical Report nr 8*, EEA.

Eicher, C.L. and Brewer, C.A. (2001) Dasymetric mapping and areal interpolation: implementation and evaluation, *Cartography and Geographic Information Science*, 28(2): 125–138.

Engelen, G., Van Esch, L., Janssen, L., Van Holderbeke, M., Provoost, J., De Cuyper, K. and Dinne, K. (2006) *Kwantificering en verfijning van het aandeel van de diffuse bronnen op de emissies naar oppervlaktewater in het stedelijk (gerioleerd) gebied. Deel: emissies van metalen te wijten aan corrosie van bouwmaterialen*, Report 2006/IMS/R397, VITO, Mol.

Engelen, G. and Van Esch, L. (2007) *Evolutie van de emissies in water uit corrosie van bouwmaterialen aan de hand van de referentiejaren 1998, 2002 en 2005*, Report 2007/IMS/R/428, VITO, Mol.

Engelen, G. (2003) Desertification and land degradation in Mediterranean areas: from research to integrated policy making climate, In Giupponi, C. and Shechter, M. (eds.) *Climate Change in the Mediterranean. Socio-Economic Perspectives of Impacts, Vulnerability and Adaptation*, Edward Elgar, Cheltenham, pp. 159–195.

Engelen, G., Uljee, I. and Van de Ven, K. (2002) WadBOS: Integrating knowledge to support Policymaking in the Dutch Wadden sea, In Geertman, S. and Stillwell, J. (eds.) *Planning Support Systems in Practice*, Springer-Verlag, Berlin, pp. 513–537.

ESRI (2008) www.esri.com.

Fuchs, S., Scherer, U., Hillenbrand, T., Marscheider-Weidemann, F., Behrendt, H. and Opitz, D. (2002) *Emissions of Heavy Metals and Lindane into River Basins of Germany*, Research Report 200 22 233, UBA-FB, Umweltbundesamt (Federal Environmental Agency), Berlin.

Gallego, J. and Peedell, S. (2001) Using Corine land cover to map population density, in European Environment Agency, *Towards Agri-environmental Indicators, Integrating Statistical and Administrative Data with Land Cover Information*, Topic Report 6, pp. 94–105.

Mennis, J. (2003) Generating surface model of population using dasymetric mapping, *The Professional Geographer*, 55(1): 31–42.

Microsoft (2008) http://office.microsoft.com/en-us/excel/default.aspx.

Syncera Water, B.V. (2005) *Emissie-Inventaris Water*, Syncera Water B.V., Delft. Web-site emissieregistratie.nederland/methoderapporten.

Additional Reading

http://water.europa.eu/.

http://ec.europa.eu/environment/water/water-framework/index_en.html.

http://www.milieurapport.be/.

http://geography.wr.usgs.gov/science/dasymetric/index.htm.

Berlekamp, J., Lautenbach, S., Graf, N., Reimer, S. and Matthies, M. (2007) Integration of MONERIS and GREATER in the decision support system for the German Elbe river basin, *Environmental Modelling & Software*, 22(7): 239–247.

Langford, M. and Unwin, D.J. (1994) Generating and mapping population density surface within a geographical information system, *The Cartographic Journal*, 31: 21–26.

Marakas, G.M. (2002) Decision Support Systems in the 21st Century, Prentice Hall.

Mennis, J. (2002) Using geographic information systems to create and analyse statistical surface of population and risk for environmental justice analysis, *Social Science Quarterly*, 83(1): 281–297.

Wright, J.K. (1936) A method of mapping densities of population with Cape Cod as an example, *Geographical Review*, 26: 103–110.

Chapter 8
An Integrated Discussion Support System for New Dutch Flood Risk Management Strategies

Noor van der Hoeven, Jeroen Aerts, Hanneke van der Klis and Eric Koomen

8.1 Introduction

Water management in the low-lying delta of the Netherlands is a precarious planning issue that continuously needs adjustment because of the ever-changing conditions of the water system and the society that inhabits this risk-prone area. The recent concern for climate change has given rise to additional challenges for the sustainability of current water management strategies. These are related to the anticipated rise in sea level, the increase in river discharge and changes in the local precipitation patterns.

To support the discussion on future water management strategies, the development of a new information system has been commissioned by the Dutch Ministry of Transport, Public Works and Water Management. This 'discussion support system' presents an overview on the anticipated future developments in the coming 50–100 years, their impacts on flood risk and the water management options that are possibly needed to anticipate those impacts. Flood risk in this project is determined by combining spatial land-use information and hydrological information. This information is available from a land-use model that operates using a 100 by 100 m grid. Flood risk is then calculated under different future trends using socioeconomic scenarios and climate information.

Flood risk assessments in the system are presented in both monetary terms and possible number of casualties following the approach that is common in Dutch water management. This approach applies fixed mathematical functions that relate the possible inundation depth with the local occupation pattern (land use and number of residents). The system, furthermore, includes a wide range of management options, related to both technical solutions and physical planning strategies, and

E. Koomen (✉)
Department of Spatial Economics/SPINlab
Vrije Universiteit Amsterdam
De Boelelaan 1105
1081 HV Amsterdam
The Netherlands
E-mail: ekoomen@feweb.vu.nl

S. Geertman, J. Stillwell (eds.), *Planning Support Systems Best Practice and New Methods*,
DOI 10.1007/978-1-4020-8952-7_8,

also assesses their impact on flood risk. These assessments follow from a simulation of their impact on local land-use patterns and, furthermore, take the changed climate conditions into account.

The system thus provides insights into the complex interactions of changing physical conditions, socioeconomic scenarios and water management strategies. It helps water managers, politicians and stakeholders to explore different future scenarios and to discern the possible impact different management strategies have under these future conditions. This chapter will briefly describe the main components of the system and the way the different components are integrated. The graphical interface of the system will be explained and some characteristic outputs are shown. Past experiences and user feedback with the similar Planning Kit instrument (Van Schijndel 2005) are used to develop the system.

8.2 The Current Dutch Flood Risk Management Debate

Dutch hydraulic engineering still leads the world in flood safety. In governance, too, the Netherlands has a rich history in which policy makers, water managers and researchers have repeatedly acted in unison to cope with extreme weather conditions and the danger of flooding. However, new scientific knowledge calls for a critical assessment of the sustainability of the current safety policies in relation to long-term processes of climate change and land subsidence. In fact, it can be questioned whether the current investments in, for example, urban development, infrastructure and safety, are (cost) effective in the coming 50 – 100 years (Bouwer and Vellinga 2007).

Climate change will have a profound effect on flood risk in the Netherlands. The sea level along the Dutch coast is expected to rise approximately 80 cm by the year 2100. The consequences of sea-level rise in the Netherlands are exacerbated by the subsidence of land that may locally lower the ground level by 1 m in 2100. In addition, Middelkoop *et al.* (2001) estimate that peak flows of the Rhine River (one of the three major rivers in the Netherlands) may increase by about 5–8 per cent by the year 2050. This would imply that the current technical measures such as dikes are not meeting the legal safety standards, and additional adaptation measures are required.

The low-lying areas in the Netherlands are protected by a system of dikes and embankments along the main rivers and coastal areas. A so-called 'dike ring' is a geographical unit bounded by its flood protection system normally consisting of dikes. It is also a separate administrative unit under the Water Embankment Act 1996. According to the act, a dike ring area should be protected against floods by a system of primary embankments and each dike ring has been designed such that it meets a safety norm. These safety norms are based on the risks that were deemed acceptable in the mid 1950s, taking into account the lives and capital at risk. For example, a dike ring with a safety norm of 1:10,000 means that this dike ring has been designed such that it can withstand a flood that has a statistical probability of occurring once every 10,000 years. This expected return period of the flood water levels has been derived from extrapolation of historical records. There are 95 dike ring areas, each having

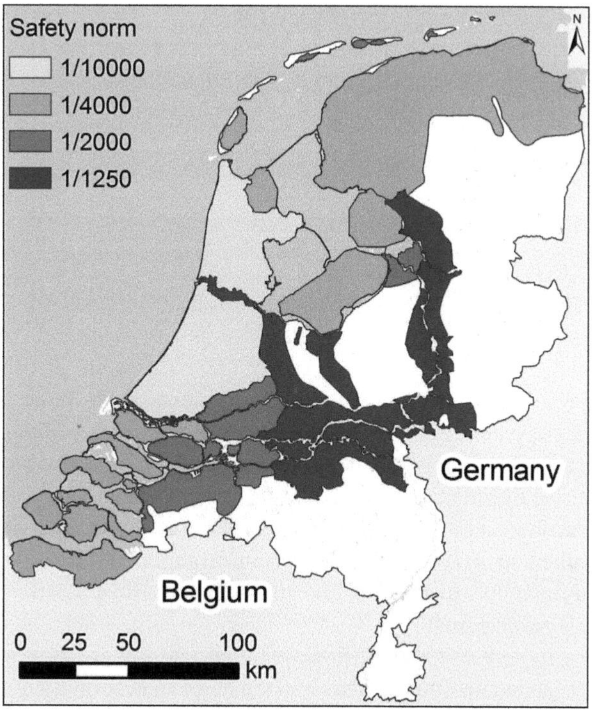

Fig. 8.1 Regional differences in safety norms in the Netherlands

different safety norms. The most important safety norm areas are shown in Fig. 8.1. In general, Dutch water management is rather technology driven and bounded by strict legislation that aims at guaranteeing a high protection level against flooding. This becomes clear from a comparison with the area of New Orleans, where the pre-Katrina safety norms of most levees were based on events occurring once in 100 years (Grossi and Muir-Wood 2006).

The Dutch government has already taken measures in response to certain aspects of long-term trends such as climate change and land-use change. Local flooding and changes to the river discharge regime are addressed by the 'Water Management for the 21st Century' policy (Commissie Waterbeheer 21e eeuw 2000). Adaptation to higher river discharges is dealt with in greater depth in the 'Space for the Rivers' programme (RvR 2006), whereas the 'Weak Links' project (V&W 2003) looks at the risks on the coast. The safety of the primary flood defences was analysed in the 'Safety of the Netherlands' project (DWW 2005).

Within this framework, we develop a method to carry out a flood risk study on long-term safety. The potential of the method is demonstrated through application in a prototype system. The questions to be answered include:

- What are the expected changes in climate and land use in the long term (50 – 100 years)?
- Which administrative, social and economic variables are important for protection against flooding?
- What flood safety strategies can be developed to deal with these changes?

The current project focuses on the following components: (i) scenario analysis; (ii) inventory of effective safety strategies; (iii) vulnerability assessment; (iv) appraisal through different evaluation methods; and (v) development of a discussion support system that is the topic of this chapter.

8.3 Building the System

The system is developed to support the *discussion* on the long-term adaptability of the Netherlands to flood risk. It aims to facilitate the learning of the user on the subject, instead of giving unambiguous answers on what management strategy is preferable. This is a significant difference with the more traditional *decision* support systems. The system, furthermore, aims to provide insight into the impact of various flood risk management strategies. For discussions on complex issues, like flood risk management, a discussion support system is a potentially powerful tool to facilitate a meaningful exchange of ideas. Advantages of such a system are amongst others are: (i) that it structures information and controls the amount of data a participant has to process; and (ii) that it develops a collective basis of knowledge for all participants.

Experience shows, however, that various decision and discussion support systems remain unused. This is often due to a lack of attention to the needs of the target group. From their survey of existing support systems in the Netherlands, Hooijer and De Bruijn (2005) distil a number of recommendations:

- The development of a system requires close contact with a real user group (wanting to solve a real problem) to prevent it from becoming an academic exercise. Furthermore, it requires a very clear statement of the problem and the solution limits;
- A system is only used when users have confidence in its results. It is thus important to calculate effects well, and make any remaining uncertainties clear;
- The system must be simple to use. This does not mean that the underlying calculations should be simple, but it implies that the user should not be bothered with their complexity;
- Effects of flood risk management strategies should be presented spatially, i.e. on maps, to make them more easily understandable for decision makers; and
- Widespread acceptance of a system can be obtained by enabling the parties involved to introduce their own knowledge to the tool. This requires a design in which it is easy to add new information.

A good example of a flood risk management support system that is actually used in decision making in the Netherlands is the Planning Kit (Van Schijndel 2005). Many of the requirements mentioned above stem from experience with this system. In the development of the discussion support system presented here, these recommendations have been taken to heart in order to meet the requirements of the target group of flood risk managers as closely as possible. This is mainly done through organising workshops in which users, experts and others are consulted. In this way, the experiences and knowledge of all these groups have been incorporated. In addition, the lessons learned from other models and systems such as the Planning Kit were recognised and integrated.

8.3.1 Target Group and Requirements

The target groups are both policy makers and more technical water managers involved in the discussion on long-term flood risk management and for those who prepare new management strategies. The system is not meant to serve as an analysis tool for modellers. Nor is it meant for those who are involved in short-term flood event management; for example it, does not support evacuation or emergency measures. The above-mentioned recommendations and types of end users have resulted in a list of requirements of the flood risk discussion support system, of which the main requirements are that:

- the system should be user friendly and quick;
- the system should be exciting enough to keep the single user's attention;
- the user should trust the information in the system and understand its reliability; and
- it should be relatively easy to add new strategies and scenarios to the system.

To design the system such that it fulfils these requirements attention is paid to the information presentation aspects of *order*, *focus*, *spatial representation* and *comparison*. The *order* in which the system presents the aspects related to flood risk management (e.g. probability, flood damage, potential casualties and costs) is such that it is understandable and logical to the end user. The *focus* of the user is set to the main purpose of flood risk management: to meet the safety levels now and in the future. Users are, therefore, asked to explicitly define the safety levels that are to be maintained. Furthermore, most information is *spatially represented* in maps covering the full Netherlands to provide an overview. Many flood risk related aspects are aggregated here to the regional level of safety norm areas as this the level at which safety is generally analysed. To make a *comparison* between various strategies and future scenarios, users can choose among maps, graphs and tables depending on their preference for visual or textual output. Furthermore, standard reports are produced containing a summary of the results. The system thus presents a wealth of information that is required to discuss the effectiveness and efficiency of different strategies.

Concerning the technical requirements of the system, the following aspects are relevant. The system only performs simple calculations. All complex simulations and analyses have been performed in advance, where the results are gathered in a database. For each scenario or strategy selected by the user, the relevant information is obtained from this database. This guarantees a quick system, in which the user receives the results of her or his choices within seconds. The technical design of the system, furthermore, allows the easy addition of more information.

The functionality of the system can also be extended relatively easily to incorporate the new ideas that are expected to come up once it is actually used. To obtain a system that will actually be used in practice, extensive attention is paid to the user-friendliness of the system. A user group, representing the target group, is involved in the development in order to fulfil the expectations of the target group and thus gain its confidence.

8.3.2 System Design

The different components in the system are simplified and shown in Fig. 8.2. Its main components include socioeconomic and climate change scenarios, strategies and moments of investment. The choices that are made for each of these components are evaluated in terms of different indicators related to the years 2040, 2100 and the long-term future. Two different shapes (rounded rectangles and hexagons) are used in the figure to denote the two types of input that determine the model results. The rounded rectangles represent variable elements whose values can be selected by the user. The hexagons, on the other hand, are pre-set, and depend on preceding choices. These elements have fixed values as is the case with, for example, the river discharge that depends on the selected sea-level rise.

As is illustrated in Fig. 8.2, first a choice has to be made between two socioeconomic scenarios. In addition, a climate change scenario must be chosen by specifying an anticipated sea-level rise. Based on this sea-level rise the values for river discharge, precipitation and surge level are set. The necessary socio-economic and climate information is taken from two recent national studies (CPB *et al.* 2006; Van den Hurk *et al.* 2006). When the socioeconomic and climate scenarios are specified, the user can vary between different safety strategies. Subsequently, an investment moment can be selected, thus introducing a time aspect in implementing management options. The output resulting from this set of choices is

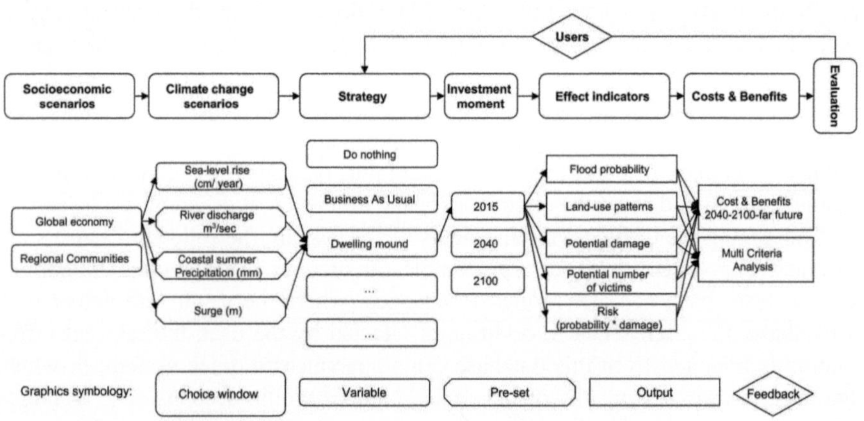

Fig. 8.2 Structural design of the flood risk management support system

presented graphically in a map showing the full Netherlands. These effect indicators represent flooding probabilities, land-use patterns, potential damage, possible number of casualties and resulting flood risks.

Finally, the user can evaluate the strategies and moments of investment by comparing them mutually and by comparing the effects of one strategy in several future scenarios. This comparison can be based on the effect indicators and estimated costs and benefits. Figure 8.3 illustrates the comparison of two maps with different content, according to the choices of the user. By incorporating user knowledge in the development of the relevant adaptation strategies, the relevance and usefulness of simulated developments are tested.

8.4 Land-use Simulations

The land-use patterns for the current situation (2015), the scenarios and the different strategies that can be chosen in the system are simulated with the *Land Use Scanner*. This is a GIS-based land-use model that simulates future land use based on the integration of sector-specific inputs from dedicated models (Dekkers and Koomen 2007;

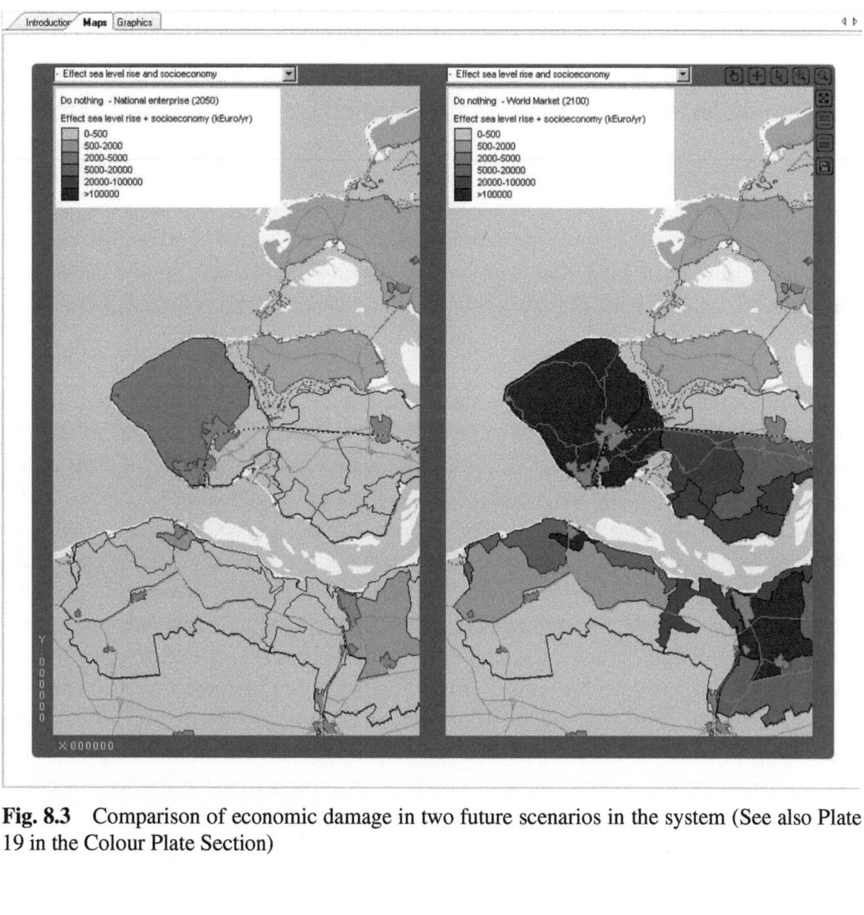

Fig. 8.3 Comparison of economic damage in two future scenarios in the system (See also Plate 19 in the Colour Plate Section)

Hilferink and Rietveld 1999). The model offers an integrated view on all types of land use, dealing with urban, natural and agricultural functions. It is grid-based and this application uses a 100 by 100 m grid to cover the terrestrial Netherlands with about 3.3 million cells. This resolution comes close to the actual size of building blocks and allows for the use of homogenous cells that only describe the dominant land use.

The model is based on demand-supply interaction for land, with sectors competing for allocation within suitability and policy constraints. The model employs a mathematical optimisation approach to allocate land to its optimal location given a set of boundary conditions related to the regional demand for land and the local suitability definition. The solution of this discrete allocation problem is considered optimal when the sum of all suitability values corresponding to the allocated land use is maximal. This allocation is constrained by two conditions: the overall demand for the land-use functions that is given in the initial claims and the total amount of land that is available for each function. The suitability maps are generated for all different land-use types based on location characteristics of the grid cells in terms of physical properties, operative policies and expected relations to nearby land-use functions. This suitability can be interpreted to represent the net benefits (benefits minus costs) of a particular land-use type in a particular cell. The higher the benefits (suitability) for that land-use type, the higher the probability that the cell will be used for that type. The economic rationale that motivates this choice behaviour resembles the actual functioning of the land market. For a more detailed description of the most recent model version and the calibration and validation of its new allocation algorithm, the reader is referred to another publication (Loonen and Koomen 2008).

Unlike many other land-use models, the objective of the *Land Use Scanner* is not to forecast the dimension of land-use change but rather to integrate and allocate future land-use claims from different sector-specific models. The land-use simulations integrate expert knowledge from various research institutes and disciplines and thus represent the best-educated guess regarding the possible spatial patterns. It should be noted, however, that the simulations are based on many assumptions about future developments. They can by no means be seen as exact predictions and should therefore not be treated like that.

In the system, several land-use simulations have been executed under different socioeconomic scenarios. These are derived from the study ‘Prosperity and quality of the living environment’ (CPB *et al.* 2006) that aimed to investigate long-term developments in prosperity and their effects on the physical environment. For this analysis, we selected the two most extreme socioeconomic scenarios: *Global Economy* and *Regional Communities*. The former combines globalisation with a focus on the economy and has the following main characteristics: high population and economic growth; global free trade without political integration; and no initiatives on international environmental agreements. The latter scenario has a focus on regional development and communal values and is characterised by: moderate population growth until 2015 and a slight decline thereafter; modest economic growth; trade blocks and taxes for protection of the environment; emphasis on national environmental policy and increased public environmental awareness. A more extensive discussion of the selected scenarios is provided elsewhere (Koomen *et al.* 2008; Riedijk *et al.* 2007).

The scenarios have been made spatially explicit with the help of several sector-specific models and a number of additional assumptions. Various specialized institutes have performed the development of the socioeconomic scenarios. The Netherlands Bureau for Economic Policy Analysis (CPB) and the ABF Research company have provided the expected amount of residential development (ABF 2006; CPB *et al.* 2006). CPB has delivered the demand for industrial and commercial land and office space (CPB 2002; CPB *et al.* 2006) and the Agricultural Economics Research Institute (LEI) has provided the projections for agricultural land-use changes (Helming 2005). The Netherlands Environmental Assessment Agency (MNP) subsequently inserted the regional spatial demands in the *Land Use Scanner* model together with a spatially explicit translation of the scenario assumptions into suitability maps. The land-use simulation starts by creating a 2015 land-use map from a 2000 base map taken from Statistics Netherlands (CBS 2002). In this step current, explicit land-use plans, mainly taken from the new map of the Netherlands survey (NIROV 2005), are included in the simulation to represent autonomous developments. Based on this base situation, simulations for 2040 are made for the two scenarios according to the scenario-specific assumptions and sector-related developments discussed in the following section.

Both scenarios indicate that the amount of land used for sectors such as residence, industry, nature and recreation increases at the expense of agriculture. Figure 8.4 depicts the local impact of these national changes and shows that the increase in residential land is more moderate in the *Regional Communities* scenarios. However, urban sprawl is expected to continue due to the continuing demand for rural residences and green, spacious urban housing. The extent of this sprawl is limited, however, because population growth is coming to a halt in all regions. The regional development in employment is strongly related to the provision of consumer services.

Based on the assumptions of the two socioeconomic scenarios and the limited literature available on scenarios for 2100, we also simulate land use for that year. Although we realise that the uncertainties related to such long-term simulations are enormous, they are included nonetheless to assess the potential effect of the projected strong increase in sea-level rise of 1.5 – 5 m.

Following the two basic scenarios, additional simulations have been made for different safety strategies. Two of these strategies are the *Business As Usual (BAU)* and the *Terpen* strategies. The *BAU* strategy contains standard water protection measures that will be prolonged until 2040. These measures include dike reinforcement, additional space for rivers, coastal supplement, replacement of flood barriers and other coastal defence works. The *Terpen* strategy contains a specific additional safety measure. The Dutch word *terpen* refers to dwelling mounds and these artificial elevations were common in the Middle Ages when dikes were absent or offered limited protection. In this strategy, all new housing development will take place at five metres above mean sea level on artificially elevated construction sites. As the surface level in parts of the western Netherlands lies up to 6 m below mean sea level, this strategy calls for enormous amounts of sand. This sand is available in the nearby, shallow North Sea and can be transported at reasonably low costs

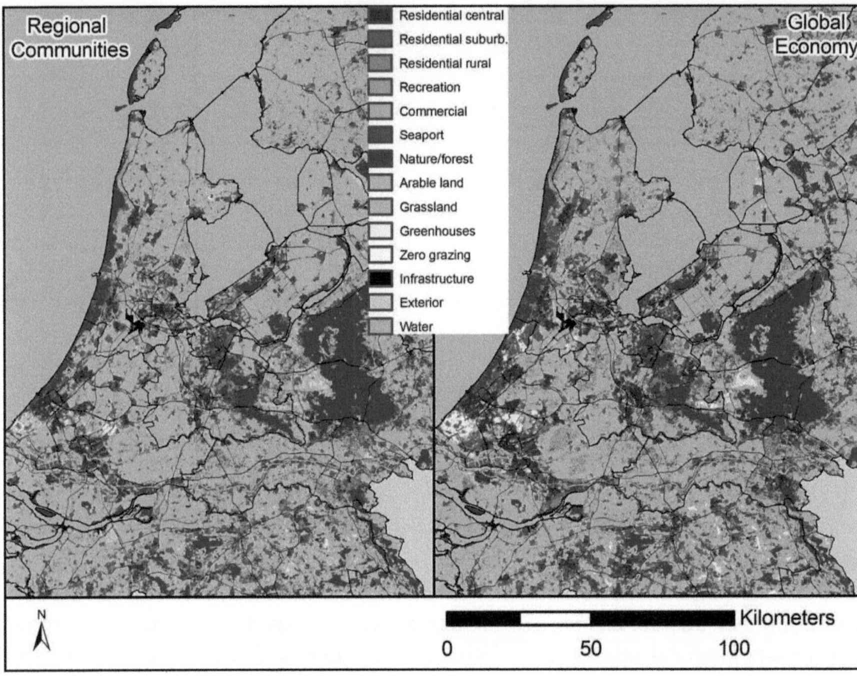

Fig. 8.4 Simulated future land use in 2040 for the *Regional Communities* and *Global Economy* scenarios (See also Plate 20 in the Colour Plate Section)

(Van der Meulen *et al.* 2007). Furthermore, flood barriers and, after 2040, dike rings will locally be reinforced to create a new integrated safety system that is interconnected with the many new *terpen*. This strategy is included in the system to test its potential as an alternative option to limit flood risk.

8.5 Impact Assessment of Flood Risk

The concept of risk encompasses both the probability and the consequences of a failure (Howard 2002). Generally, it is defined as the product of probability and consequences. In our case, these consequences consist of economic damage and number of casualties. The following aspects have to be considered when deciding upon a method to assess flood risk:

- the focus is on the long-term impacts, for which the inherent uncertainty in the circumstances is large;
- the incorporated strategies have a relatively global, explorative character and are not described in exact technical details; and
- consequently the uncertainty in flooding probability and the related long-term impact is large.

Therefore, the flood risk assessment used in this system is largely based on simple heuristics and expert knowledge. In the following sections, we will describe the method used to compute both the probabilities and the impacts of a flood in the Dutch safety or dike ring areas.

8.5.1 Flooding Probability

As discussed in Section 8.2, the protection level of Dutch water defences is expressed in safety norms that are related to the statistical return periods of specific flood water level. It is assumed that the system is designed to safely (i.e. without failure) withstand water levels below this design water level. Higher water levels will lead to severe overtopping and consequent failure of the flood defence system. As no other failure mechanisms are considered, the probability of failure of the flood defence system equals the probability of exceedence of its design water level.

Under the influence of the changing climate, the flooding probabilities are expected to increase as the sea level is thought to rise further, storms may become more severe and river discharges are likely to increase. From complex probabilistic analyses, it is known how flooding probability increases with increasing water levels, assuming that no measures are implemented. Of course, it is unlikely that no additional flood management strategies are implemented. The system shows the impact of this option, nonetheless, as it aims to provide policy makers with an opportunity to analyse the effectiveness of different strategies in terms of flooding probability and associated risk against a common reference point.

8.5.2 Flood Impact

The impact of flooding, both in terms of economic damage and in casualties, depends largely on the location of failure in the water defence and the hydraulic conditions of the flood. For example, sea floods have different impacts than river floods and the impact of a flood in an extensive deep polder differs from a comparable flood in an area where height differences may offer a rettfuge from flooding. Also the economic value and the number of inhabitants present in the flooded area determine the impact of the flood.

To estimate the effects of a flood, the so-called 'Damage scanner' has been developed within the *Land Use Scanner*. This instrument computes the economic damage and number of casualties per scenario and strategy. Per scenario, per strategy and per year (2015, 2040, 2100), the flood effects are computed by applying damage functions to maps describing water levels after flooding and land-use configurations. These functions relate economic damage and number of casualties to specific land-use types, considering the expected water levels. These results are aggregated per dike ring area to facilitate an easy comparison of the different strategies. As a reference for the flood risk computed for the different future situations, we use the expected situation in 2015. For this year, we assume the actual flooding probabilities to match the safety standards. Until that time, some major projects are being undertaken in the Netherlands to meet the standards (RvR 2006).

8.5.3 Uncertainties

Flood risk analysis involves many uncertainties. The input data are uncertain, the models involve uncertainties, the analyses are usually not complete and the simulations of future situations are inherently uncertain. However, these uncertainties should not prevent us from drawing conclusions. Although the exact figures may be wrong, the order of magnitude is probably right and the differences between alternatives become clear through a consistent comparison. In the analysis it is important to explain clearly which assumptions were made, where uncertainties are large and what effect they may have.

8.5.4 Current Experiences and Further Development

A first demonstration version of the discussion support system was presented to end users in a seminar late 2007 to derive feedback for further development. During the session most users claimed that the system clearly added value to the discussion on long-term climate change effects and flood risk. Many stakeholders expressed that the issue of climate change is complex as its effect and hence the solutions, are inherently uncertain. Apart from climate change, other issues were also deemed important such as land-use change and socioeconomic trends. Together, these issues pose huge dilemmas to many policy makers and experts in water management as to what policy would be most robust in the future. For this discussion, the system is considered to be important as it provides insight into the possible consequences of the combined changes in climate and society. It, furthermore, shows what impact possible management strategies may have. So, the facilitating role of the system has been clearly acknowledged. An issue mentioned for further development is the use of more detailed quantitative information. The current version uses straightforward hybrid methods and consequently the resolution of the results is coarse. Some experts expect that technology-oriented users may see this as a disadvantage of the system. Furthermore, they suggest making the methods and assumptions underlying the outcomes more explicit.

The current system is not exhaustive in its scope and applied methods, leaving room for further innovations in the long term. Three aspects of climate change are now integrated in the system: sea level, river discharge and precipitation. From these, the user is currently only allowed to vary sea-level rise. In the future, it could be interesting to also vary river discharge and precipitation and maybe even integrate other climate change aspects such as temperature, drought or storms. Information from other projects about climate change could supply the necessary information. Besides climate change, land subsidence is another important safety issue in the Netherlands. Isostatic changes and tilting are causing the Netherlands to sink by about 10 cm per century. Local subsidence within peaty polders caused by the oxidation of organic matter may even lower surface levels by about 1 m by 2100. Relative to the raised sea level, ground level can locally drop 1.5 m or more. Inclusion of this phenomenon would be a valuable addition to the system.

We currently follow a scenario approach where the 'uncertain future' is captured in a set of climate and socioeconomic scenarios that will or could influence flood safety and vulnerability to water damage in the Netherlands. An innovation in this respect could be the inclusion of discontinuities since most existing scenario studies take a linear outlook and thus produce a limited variety of scenarios for the future. Future research will pay more attention to identifying extremes, unanticipated events, trend breaks and feedback mechanisms. The challenge is to convert these discontinuities into spatially explicit changes in land use and related flood risk.

8.6 Conclusions and Future Directions

The last report of the Intergovernmental Panel on Climate Change (IPCC) shows that human activities lead to global warming and that extremes in the hydrological cycle, such as flooding and extreme rainfall, will be more frequent (IPCC 2007). Also other future developments will have an effect on flood risk and their management. For example, European and global political developments will have an increasing influence on Dutch flood safety policies. These include liberalisation and privatisation, decentralisation of responsibilities to regional government and an increased importance of private parties and public participation (Van den Brink *et al.* 2007). The Netherlands has, furthermore, become richer, and should therefore be more capable of defending itself against flooding. The drawback, though, is that the cumulative value of physical assets and the increased population mean that the risk—defined as the probability of flooding times the damage—has risen rapidly. It has recently also become clear that floods are much more likely to cause high numbers of casualties than most other disasters at, for example, industrial plants, airports and railway marshalling yards combined.

This chapter has demonstrated the development of a system that facilitates the discussion of different future trends, their impact on flood risk, and the effects of new flood safety strategies. The system follows a scenario approach where the 'uncertain future' is captured in a set of climate and socioeconomic scenarios that may influence flood safety and vulnerability to water damage in the Netherlands. The system is targeted at water managers that need to take action now to be safe in the future. This involves a lot of uncertainty, which is inherent in long-term flood risk assessments. There are simply too many options about how the socio-economic situation in the coming decades will develop or how the climate will change. Hence, the system will not predict exactly what will happen or tell us precisely what measurements we should take, but it is a learning instrument that gives more insights into what future scenarios are possible and which options exist in answering the possible changes. Not only will this lead to a better understanding but also to a more creative and integrative approach to climate change impact analysis.

Several possible methodological extensions to the system related to, for example, the inclusion of more variable climate indicators or land subsidence were discussed in the previous section. The addition of new safety strategies would also help to make it a more complete system dealing with future water safety. A different direction for future improvements of this and other systems for discussion support, however, is to offer such systems to different groups of users. The current system is, like many comparable instruments, intended for relatively well-informed experts. A challenge for future versions is to open the wealth of available information to non-expert users, such as interested citizens and local or regional stakeholders. This calls for serious improvements of, amongst others, the current user interface and the level of detail. To interest a younger user group it is worth exploring the option of introducing a game element to playfully familiarise them with the topic of flood risk management. In all these cases, however, the fundamental issue remains how to attractively translate complicated technical knowledge to a broad audience while doing justice to the uncertainties, choices and assumptions underlying the results.

Acknowledgements The authors are grateful to the Dutch Directorate for Public Works and Water Management and the Dutch research programmes 'Living with Water' and 'Climate changes Spatial Planning' for financing the *Attention for Safety* research project. Furthermore, we like to thank the Netherlands Environmental Assessment Agency (MNP) for providing the initial land-use simulations for the discussion support system.

References

ABF (2006) *Achtergrondrapport Bevolking,* ABF Research, Delft.

Bouwer, L.M. and Vellinga, P. (2007) On the flood risk in the Netherlands, Chapter 24 in Begum, S., Stive, M.J.F. and Hall, J.W. (eds.) *Flood Risk Management in Europe*, Springer, Doordrecht, pp. 469–484.

CBS (2002) *Productbeschrijving Bestand Bodemgebruik,* Centraal Bureau voor de Statistiek, Voorburg.

Commissie Waterbeheer 21e eeuw (2000) *Waterbeleid voor de 21e Eeeuw. Geef Water de Ruimte en de Aaandacht die het Verdient,* Advies van de Commissie Waterbeheer 21e eeuw, The Hague.

CPB (2002) *De BLM: Opzet en Recente Aanpassingen,* Centraal Planbureau, The Hague.

CPB, MNP and RPB (2006) Welvaart en Leefomgeving. Een Scenariostudie voor Nederland in 2040, *Centraal Planbureau,* Milieu- en Natuurplanbureau en Ruimtelijk Planbureau, The Hague.

Dekkers, J. and Koomen, E. (2007) Land-use simulation for water management: application of the Land Use Scanner model in two large-scale scenario-studies, In Koomen, E., Stillwell, J.,

Bakema, A. and Scholten, H.J. (eds.) *Modelling Land-Use Change: Progress and Applications*, Springer, Dordrecht, pp. 355–373.
DWW (2005) Flood Risks and Safety in the Netherlands (Floris); Floris study - Full report, *DWW-2006–014,* Ministerie van Verkeer en Waterstaat/ Dienst Weg en Waterbouwkunde, Delft.
Grossi, P. and Muir-Wood, R. (2006) Flood Risk in New Orleans. Implications for Future Management and Insurability, *Risk Management Solutions Inc.*, Newark, CA.
Helming, J. (2005) A model of Dutch agriculture based on Positive Mathematical Programming with regional and environmental applications, PhD thesis, Wageningen University, Wageningen.
Hilferink, M. and Rietveld, P. (1999) Land Use Scanner: An integrated GIS based model for long term projections of land use in urban and rural areas, *Journal of Geographical Systems*, 1(2): 155–177.
Hooijer, A. and De Bruijn, K. (2005) Development of a framework for long term flood risk management, *Draft Report Floodsite Project*, Waterloopkundig Laboratorium, Delft.
Howard, C.D.D. (2002) Sustainable development - risk and uncertainty, *Journal of Water Resources Planning and Management*, 128(5): 309–311.
IPCC (2007) Climate Change 2007 - Impacts, Adaptation and Vulnerability; Contribution of Working Group II to Parry, M., Canziano, O., Palutikof, J., Van der Linden, P. and Hanson, C. (eds.) *The Fourth Assessment Report of the IPCC*, Cambridge University Press, Cambridge.
Koomen, E., Loonen, W. and Hilferink, M. (2008) Climate-change adaptations in land-use planning; a scenario-based approach. In Bernard, L., Friis-Christensen, A. and Pundt, H. (eds.) *The European Information Society; Taking Geoinformation Science One Step Further.* Springer, Berlin, pp. 261–282.
Loonen, W. and Koomen, E. (2008) Calibration and validation of the Land Use Scanner allocation algorithms, Milieu- en Natuurplanbureau, Bilthoven.
Middelkoop, H., Daamen, K., Gellens, D., Grabs, W., Kwadijk, J.C.J., Lang, H., Parmet, B.W.A.H., Schädler, B. and Schulla, J. (2001) Impact of climate change on hydrological regimes and water resources management in the Rhine Basin, *Climate Change*, 49(1–2): 105–128.
NIROV (2005) *Nieuwe Kaart, Nieuwe Ruimte: Plannen voor Nederland in 2015 Nirov,* The Hague.
Riedijk, A., Van Wilgenburg, R., Koomen, E. and Borsboom-van Beurden, J. (2007) Integrated scenarios of socio-economic change, *SPINlab Research Memorandum SL–06*, Vrije Universiteit Amsterdam, Amsterdam.
RvR (2006) Planologische Kernbeslissing Deel 4; Ruimte voor de Rivier, Projectorganisatie Ruimte voor de Rivier, The Hague.
V&W (2003) Procesplan Zwakke Schakels in de Nederlandse Kust, Ministerie van Verkeer en Waterstaat, The Hague.
Van den Brink, A., Van Lammeren, R.J.A., Van de Velde, R.J. and Dane, S. (2007) *Geo-Visualisation for Participatory Spatial Planning in Europe*; Imaging the Future, Mansholt Publication Series Volume 3, Wageningen Academic Publishers, Wageningen.
Van den Hurk, B., Klein Tank, A., Lenderink, G., Van Oldenborgh, G.J., Katsman, C., Van den Brink, H., Keller, F., Bessembinder, J., Burgers, G., Komen, G., Hazeleger, W., Drijfhout, S. and Van Ulden, A. (2006) KNMI Climate Change Scenarios 2006 for the Netherlands, *KNMI Scientific Report WR 2006-01*, KNMI, De Bilt.
Van der Meulen, M.J., Van der Spek, A.J.F., De Lange, G., Gruijters, S.H.L.L., Van Gessel, S.F., Nguyen, B., Maljers, D., Schokker, J., Mulder, J.P.M. and Van der Krogt, R.A.A. (2007) Regional sediment deficits in the Dutch Lowlands: implications for long-term land-use options, *Journal of Soils Sediments*, 7(1): 9–16.
Van Schijndel, S.A.H. (2005) The Planning Kit, a decision making Tool for the Rhine branches, In Van Alphen, J., Van Beek, E. and Taal, M. (eds.) *Floods, from Defence to Management*, Taylor & Francis Group, London.

Additional Reading

Asselman N.E.M. and Jonkman, S.N. (2003) Consequences of floods: the development of a method to estimate the loss of life, *Delft Cluster Report DC1-233–237, Delft.*

Koomen, E., Stillwell, J., Bakema, A. and Scholten, H.J. (2007) *Modelling Land-Use Change; progress and applications*, GeoJournal Library, Vol. 90, Springer, Dordrecht.

Tol, R.S.J. and Langen, A. (2000) A concise history of Dutch river floods, *Climate Change*, 46: 357–369.

Chapter 9
A Spatial Planning Support System for Managing Bangalore's Urban Sprawl

H. S. Sudhira and T. V. Ramachandra

9.1 Urban Sprawl

Urban sprawl is the outgrowth along the periphery of cities and along highways. Although an accurate definition of urban sprawl may be debated, a consensus is that urban sprawl is characterized by an unplanned and uneven pattern of growth, driven by multitude of processes and leading to inefficient resource utilization. Urbanization in India has never been as rapid as it is in recent times. As one of the fastest growing economies in the world, India faces stiff challenges in managing the urban sprawl, while ensuring effective delivery of basic services in urban areas. The urban areas contribute significantly to the national economy (more than 50 per cent of GDP), while facing critical challenges in accessing basic services and necessary infrastructure, both social and economic. The overall rise in the population of the urban poor or the increase in travel times due to congestion along road networks are indicators of the effectiveness of planning and governance in assessing and catering for this demand. Agencies of governance at all levels: local bodies, state government and federal government, are facing the brunt of this rapid urban growth. It is imperative for planning and governance to facilitate, augment and service the requisite infrastructure over time systematically. Provision of infrastructure and assurance of the delivery of basic services cannot happen overnight and hence planning has to facilitate forecasting and service provision with appropriate financial mechanisms.

This chapter addresses the issue of urban sprawl in the Indian context with a focus on Bangalore and analyses the status of planning practice in India with a note on utility of spatial planning tools. As a synthesis of the prevailing situation, the ensuing section outlines the critical challenges for addressing sprawl. The subsequent section highlights the need for an integrated Spatial Planning Support System (SPSS), illustrating the evolution of a framework for SPSS, its development and

T. V. Ramachandra (✉)
Centre for Ecological Sciences and Centre for Sustainable Technologies
Indian Institute of Science
Bangalore,560 012, India
E-mail: cestvr@ces.iisc.ernet.in

S. Geertman, J. Stillwell (eds.), *Planning Support System Best Practice and New Methods*, 175
DOI 10.1007/978-1-4020-8952-7_9, © Springer Science+Business Media B.V. 2009

evaluation. The later section presents the policy analysis carried out using the SPSS, which offers insights into areas of concern. The chapter concludes by noting the drawbacks and challenges for future research for managing urban sprawl.

Earlier studies characterize urban sprawl (Barnes *et al.* 2001; Hurd *et al.* 2001; Epstein *et al.* 2002; Sudhira *et al.* 2004) using spatial metrics while highlighting the implications of sprawl on natural resources and how inefficient unplanned growth could be. Among the undesirable effects of sprawl are unplanned outgrowths which are neither aesthetic nor hygienic. Thus, there have been varied connotations to ascribe what constitutes sprawl. Galster *et al.* (2001) have addressed this issue as 'lost in semantic wilderness', by describing sprawl under six broad categories: by example that embodies characteristics of sprawl, such as Los Angeles; aesthetic judgment and general development pattern; cause of an externality; consequence or effect of independent variables; pattern of development; and process of development.

Extending the notion of urban sprawl articulated by Torrens and Alberti (2000), Galster *et al.* (2001) define it as a pattern of land-use in an urban agglomeration that exhibits low levels of some combination of eight distinct dimensions: density, continuity, concentration, clustering, centrality, nuclearity, mixed uses and proximity. Ascribing sprawl as a pattern of land-use alone does not throw light on the underlying processes, causes and hence consequences. In a developing country like India, where population density is high with significant urbanization rates, urban sprawl obviously cannot be characterized by pattern alone but processes, causes and their consequences. Hence, we suggest a modification to the definition of urban sprawl as the pattern of outgrowth emergent during the process of urban spatial expansion over time caused by certain externalities and a consequence of inadequate regional planning and governance. The pattern of urban sprawl is characterized by using spatial metrics on the extent of paved surface or built-up areas. The process of urban sprawl can be characterized by changes in the pattern over time, like the proportional increase in the built-up area leading to rapid urban spatial expansion. Analysing the causes of urban spatial expansion, the externalities can be modelled as agents in a geospatial environment like the location of jobs, housing, access to services and the level of economic activity. The interaction of agents can suggest the consequences on land-use changes. The sequence of *patterns*, *processes*, *causes* and *consequences* sets the research agenda.

9.2 Critical Challenges for Spatial PSS

The management of urban sprawl entails quantifying the pattern of sprawl and capturing the processes that involves analysis of causal driving factors. This requires understanding and visualising the consequences of policies applied by local planning and administrations to sprawl, like the lack of an effective public transport system with varying work-home distances, giving rise to independent motor vehicles and the resultant congestion and spatial expansion. Testing the effects of policy interventions on urban spatial expansion and mobility are some important questions into which a PSS could offer insights. Operational urban planning seeks answers for many such interventions. This necessitates integrated spatial PSS for managing sprawl. The framework of such PSS is discussed in the Section 9.4.

Noting the various studies and prevailing urban fabric conditions in India, it is found that lack of effective governance with operational systems and processes in local bodies of governance have resulted in unplanned and uncoordinated urban outgrowths. Urban governance requires an information system for keeping track of various services and functions of the urban local body. In the absence of any such systems, at the basic level, there is a strong and pressing need for an information system to cater for all these. In the next level, it becomes essential to build models based on the information systems involving simulation and analysis for specific urban contexts. Thereafter, there is a subsequent level involving the formulation of different strategy and policy options using the models and information systems. Thus, at the outset, there are three levels to address the problem of sprawl and to strengthen planning and decision making information systems, models and policies. Emphasizing this need, Sudhira *et al.* (2007) have discussed in detail the planning and management practices in India. With the interaction of key bureaucrats and concerned officials in Bangalore City Corporation, it is found that, on one hand, the prevailing policy and practices concerning planning and governance are in line with the State Government's aspirations, while on the other hand, there does not exist any other mechanisms or modes through which the Government could make this effective and plausible by being citizen centric. The key challenge for research is to encapsulate the spatial information coupled with models to facilitate policy analysis, evaluation and visualization of these consequences into a spatial PSS.

9.3 Planning in the Digital Age: GIS/Spatial Analysis and Planning Tools

Spatial tools, notably geographic information systems (GIS) for mapping and monitoring urban areas, have become extremely popular. Monitoring the spatial patterns of urban sprawl on a temporal scale can be undertaken using temporal remote sensing data acquired from space borne sensors. These help in inventorying, mapping and monitoring linear and radial growth patterns. In the recent past, the geospatial domain has seen significant developments in modelling urban systems using approaches ranging from operations research to system dynamics and agent-based models. Models of urban systems are essentially built to aid in planning for understanding, evaluating, visualising and deciding various interventions. Thus geospatial models have become an inseparable aspect of a PSS. In India, there are some attempts to address urban sprawl using geospatial tools (Jothimani 1997; Lata *et al.* 2001; Subudhi and Maithani 2001; Sudhira *et al.* 2004) and modelling (Subudhi and Maithani 2001).

Simulation tools based on the concepts of discrete-event system simulation approaches are being used extensively in recent times to capture and emulate urban systems and their dynamics. With the emergence of multi-agent systems from the domain of artificial intelligence, these are now being used to aid in simulation of urban systems. Another approach to model the urban dynamics is the system dynamics (SD) framework, which captures the system based on complexity involving dynamic relations represented by stocks and flows determined by various activity volumes in the city, which are synthesized from knowledge and observation. Operations research approaches and the SD framework have been applied quite rigorously in urban systems, and in the recent times, geospatial modelling aided by visualization has been very effective.

Globally, modelling urban sprawl dynamics has closely followed traditional urban growth modelling approaches. Subsequently, with the need to manage urban sprawl, modelling urban sprawl by understanding the nature of growth and its implications has been undertaken since 1960s. The key initial studies in developed countries based on traditional approaches of urban model building include Lowry (1967) in Batty and Torrens (2001), Walter (1975), Allen and Sanglier (1979), and Pumain *et al.* (1986). The traditional model building approach involved linking independent to statistically significant dependent variables in a linear or non-linear model. These models although used mostly for policy purposes, could not be used when the processes involved rule-based systems (Batty and Torrens 2001). Whilst urban development models were developed much earlier, modelling the dynamics of urban sprawl has been undertaken only recently (Batty *et al.* 1999; Torrens and Alberti 2000). Models developed using cellular automata (CA) and agent-based models would prove beneficial to pinpoint where sprawl takes place (including causal factors), which would help in effective visualization and understanding of the impacts of urban sprawl. Furthermore, to achieve an efficient simulation of urban sprawl, modelling has to be attempted in both spatial and non-spatial domains. Modelling urban sprawl in non-spatial domain is mainly by the application of statistical techniques while CA models and agent-based modelling are known to complement modelling in the spatial domain. The fusion of geospatial and agent-based models has been formalized as Geographic Automata Systems (GAS) by Benenson and Torrens (2004). These approaches and research in geospatial modelling towards a simulation framework can effectively complement a spatial PSS.

9.4 Integrated Spatial Planning Support Systems

For effectively managing the problem of urban sprawl; testing, building and visualizing different scenarios, it is imperative to have a robust spatial PSS which would not only aid in managing but also in planning, organizing, coordinating, monitoring and evaluating the system in question. Spatial PSS include instruments relating to geo-information technology that have been primarily developed to support different aspects of the planning process, including problem diagnosis, data collection, mining and extraction, spatial and temporal analysis, data modelling, visualization and display, scenario-building and projection, plan formulation and evaluation, report preparation, enhanced participation and collaborative decision making (Geertman and Stillwell 2004). Integration of different processes associated with the dynamics of sprawl phenomenon is required for addressing the problem of urban sprawl.

Currently there are few popular systems that try to emulate spatial PSS with an objective to make planning interactive and participatory. Among such existing PSS are *What-If?* (Klosterman 1999) and *RAMCO* (Uljee *et al.* 1999). Klosterman's *What-If?* system is an interactive GIS-based PSS that responds directly to both achieving the ideals of communicative rationality and traditional comprehensive land-use plans. It uses geographic data sets to support community-based efforts to evaluate the likely implications of alternative public policy choices. The package can be customized to a community's existing geographic data, concerns and desires, providing outputs in easy to understand maps and reports which can be used to support community-based collaborative planning efforts. Given a set of factors and factor weights for determining land suitability, the *What-if?* PSS generates projections for future land-use and subsequent land allocation can be based on user requirements. Although this system is claimed to be interactive, the dynamics of the factors and hence their interactions are less transparent, with only the results of a final land-use scenario obtained as output.

RAMCO (Rapid Assessment for Management of COastal zones) is a prototype information system for regional planning in a generic decision support environment for the management of coastal zones through the rapid assessment of problems (Uljee *et al.* 1999). The system was developed by integrating GIS, CA and SD. Subsequently, White and Engelen (2000), the developers of *RAMCO,* also support the integration of GIS, CA and SD with the usage of multi-agent systems for a high-resolution integrated modelling of spatial dynamics of urban and regional systems. This has currently set the standard of technology that can be used for achieving an integrated PSS.

UrbanSim and *OBEUS* are two other established frameworks with supporting packages for integrated modelling of urban systems. *UrbanSim* is implemented as a set of packages under Open Platform for Urban Simulation (OPUS) (Waddell 2002). This is fairly comprehensive in the sense that the framework integrates land-use, transportation, economic, demographic and environment variables. However, this framework does not support participatory simulations. The *OBEUS* (Object-Based Environment for Urban Systems) is more robust and represents an emerging trend to integrate various processes as agent-based models and to simulate them spatially, hence the term 'geo-simulation' (Benenson and Torrens 2004).

9.4.1 Framework for Spatial PSS

The framework for a planning and decision making process involving different phases of intelligence, design and decision/choice (Sharifi 2003) is depicted in Fig. 9.1. The intelligence phase is confined to defining, understanding and assessing the existing situation along with evolving appropriate metrics for quantifying

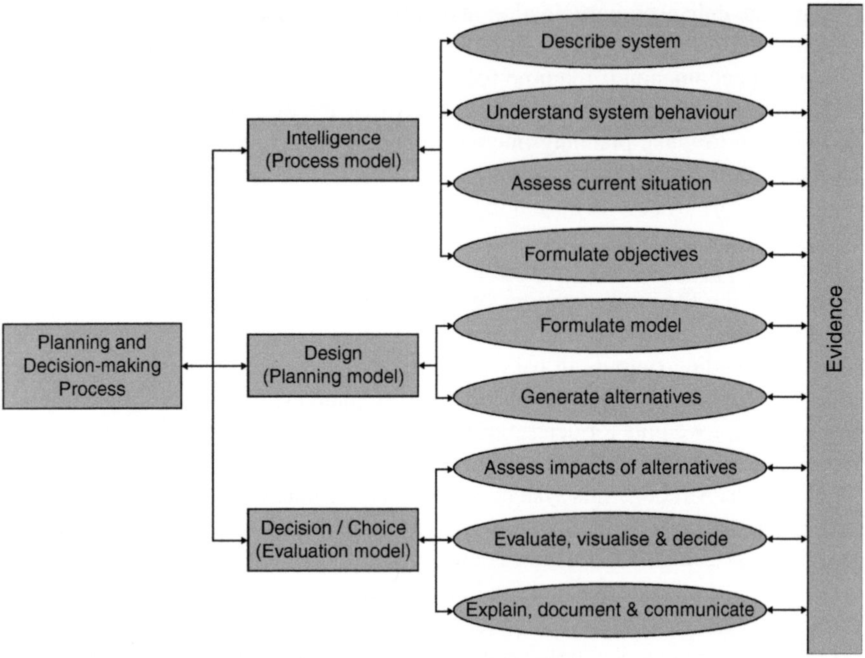

Fig. 9.1 The planning and decision-making process (Sharifi 2003)

urban sprawl. In the design phase, the dynamics of urban sprawl are captured and subsequently modelled. The design phase concludes with the generation of alternatives. In the decision/choice phase, the review and evaluation of the different policy options are undertaken to arrive at policy recommendations for managing and mitigating the urban sprawl.

9.4.2 Evolution of SPSS Through Integrated SD and Agent-Based Land-use Models

Keeping in line with the framework for planning and decision-making suggested by Sharifi (2003), a prototype of the SPSS was developed with the following four components: patterns, processes, causes and consequences. Accordingly, the evolution of the PSS is depicted in Fig. 9.2.

Modelling the dynamics of urban sprawl is attempted by a combination of system dynamics and agent based models over a geospatial domain (Sudhira and Ramachandra 2007). Initially, the land use of the region, Bangalore, was characterized using the satellite remote sensing data and the metrics for depicting the patterns were computed accordingly. A multi-stage classification process was employed resulting in classification of the entire landscape into four broad land use categories. In order to characterize sprawl, some of the metrics computed are: built up area, built-up

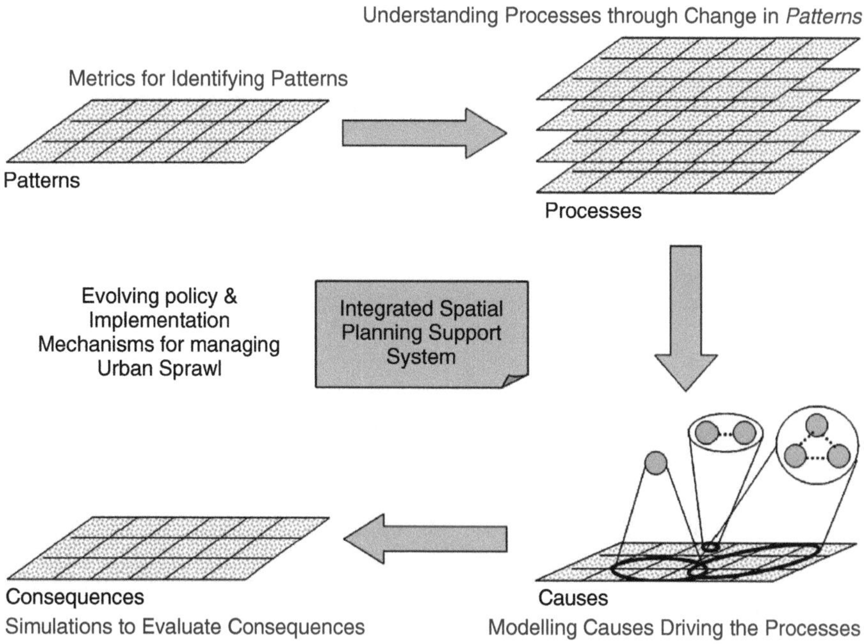

Fig. 9.2 Evolution of the PSS

density, patchiness (all spatial metrics), population and their densities. Apart from these metrics the percentage of different land-uses were also monitored. Further, to attempt modelling the processes and causes using system dynamics, variables were identified and depicted in a causal loop diagram.

The process of modelling in system dynamics begins with a clear statement of the problem, followed by gathering necessary data and identifying the key variables contributing to the problem. The variables are mapped on to what is referred to as causal loop diagrams (CLDs). With the gathered data and establishing relations with the variables amongst the variables are the stock flow diagrams created. It is possible to simulate the behaviour of the system through a stock flow diagram and plot the behaviour of key variables, test for different hypothesis and study the implications of specific interventions. The problem in question here is the pressure on land due to the extent of city outgrowth or the sprawl. The consequences of the interventions that can be attempted towards addressing this are analyzed through this analysis.

In Bangalore, one of the prominent measures to limit growth is by way of enforcing building height restrictions formalised through the floor area ratio (FAR). These restrictions are one of the popular measures of controlling the zoning of central and several parts of the city. Though the planners in Bangalore have already enforced such restrictions, this in turn has also resulted in increasing the extent of the city outgrowth. SPSS intends to explore the options of relaxations with the FAR to visualise the probable land-use patterns, which would be of interest for planners.

Accordingly, based on the analysis, the key variables identified were: population growth, economic activity, pressure for new housing and industrial areas, land-use zoning, available land, built-up area, level of services and building height restrictions. The typical approach to cause-effect modelling, the 'disjointed viewpoint' and 'linear, control viewpoint' of one cause leading to one effect, is replaced by a 'causal-loop, nonlinear feedback viewpoint' where multiple effects are the result of multiple factors (Sterman 2000), as also depicted in Fig. 9.3.

Analyzing the feedbacks generated – reinforcing and balancing, the resultant dynamics reveal that certain practices are actually proving counter-productive. The five feedbacks generated in this system are: (i) reinforcing feedbacks: housing demands (R1); industrial demands (R2); relocation (R3) and infrastructure provision (R4); and (ii) balancing feedback: outgrowth of the city (B1). The reinforcing feedbacks, R1 and R2, quite natural for most cities, set the demand for development of land into residential layouts/apartments or industrial estates, respectively. As the city grows, there is congestion in central areas forcing residents to relocate in the outskirts. This is captured by the reinforcing feedback, R3. The other reinforcing feedback, R4, for infrastructure provision suggests that augmenting infrastructure in a congested area can be counter-productive. Indeed this insight is well established by the classic Braess paradox (Braess 1968). This was demonstrated by adding a link within the network to ease congestion would be counter-intuitive in a congested road network. Reinforcing feedbacks often generate exponential growth and then collapse. Systems that are self-regulating or self-correcting are comprised of balancing loops. The only balancing feedback, B1, results in the outgrowth of the city through the process of land-use change limited by the availability of land for development. The other control variable, building height restrictions, suggests that it influences the extent of outgrowth and built-up areas negatively. Bertaud and Brueckner (2004) have shown that in an unrestrictive building height conditions the extent of the city spread is contained with higher densities in central areas.

Further, the causal loop diagram is translated into a stock-and-flow diagram with appropriate numbers for the variables and assumptions. An important assumption

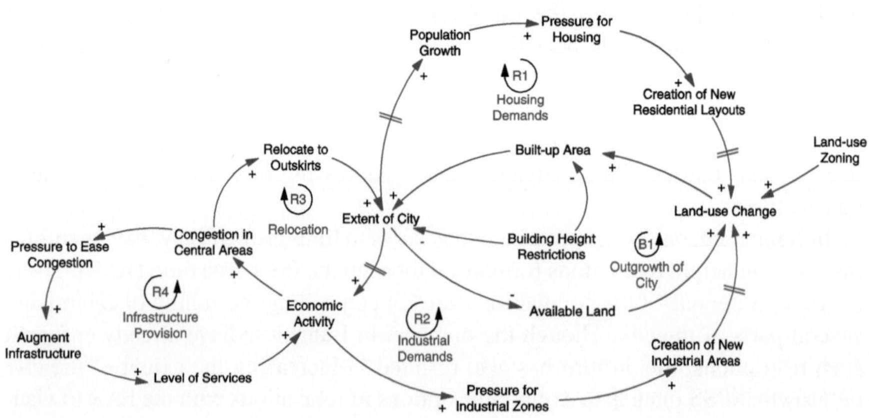

Fig. 9.3 Causal loop diagram for urban sprawl

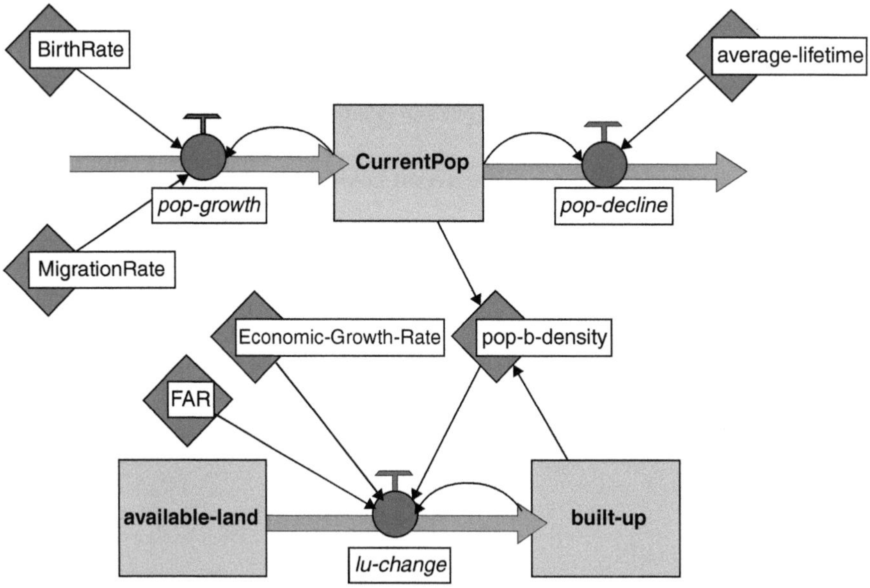

Fig. 9.4 Stock and flow diagram capturing population growth and land-use change

relating to the process of land-use change was that economic activity combined with an increase in the workforce (through migration) would lead to higher demand for land for development and hence drive the land-use change. It is observed that it is the market forces operating in the context of a market economy, which are influencing the process of land-use change. Thus the analysis of land-use change has to acknowledge the implications of prevailing market conditions and the supply and demand aspects of land. The stock-and-flow diagram capturing the dynamics of population growth and land-use change is depicted in Fig. 9.4. Essentially the stock-and-flow diagram sets the demand for new built-up areas.

9.4.3 Prototype Design and Implementation

9.4.3.1 Visualization Environment, Design and Implementation

The prototype of the spatial PSS has been developed as the *BangaloreSim* model, implemented using NetLogo (Wilensky 1999), an agent-based modelling environment developed by the Centre for Connected Learning and Computer Based Modelling, Northwestern University, USA which facilitates encapsulation of processes through rule-based procedures and offers adequate monitors and plots to visualise pattern, model the causes and evaluate the consequences through simulation. The pseudo code of the *BangaloreSim* model is presented in Fig. 9.5, which outlines the sequence of processes involved in the simulation.

9.4.3.2 Development of Prototype and Evaluation

Among the key criteria considered during the prototype development was enabling the user-interface for visualization of land-use along with the other requisite metrics in the common interface. Thus, the entire parameters are presented in a single interface with provisions to change views for different land-uses, sprawl metrics, transporta-

```
INITIALIZATION

Set Parameters -> User Defined or Set from earlier start-up
Import land-use data, elevation, transportation networks and CDP (Land-use
Policy)
Initialize System Dynamics setup
- Initialize for Population, Economic Growth Rate (City Development Prod-
  uct), Available Land

EACH TIME STEP

Check for Simulation End Time and Available Land [Stop run if exceeds
either of them]
Compute Demand for Land (Inputs from System Dynamics Model – Stock
& Flow Diagram)
- Population Growth Model Runs -> Current Population
- Current Population -> Population Density -> Land-use Change
- Land-use Change ~ Population Density, Available Land, Economic Growth
  Rate
- Land-use Change ~ Requirement for New Built-up
Change Land-use
- Obtains Built-up Demand from Land-use Change
- Site-Suitability
- Suitability for patches based on Distance from City Centre, Transportation
  Networks, Weightages for Certain Land-use based & LU Policy, and Prox-
  imity to Growth Pole
- Allocate Land
- Check for Built-up Demand [Stop run if exceeds current built-up or No
  Available Land]
- Check for Maximum Site Suitability in the Neighbourhood (3´3) and allo-
  cate Land-use to Built-up
Compute Metrics ~ Check for Impacts on Resources and Access to Services
Update Views and Draw Plots

END

Agents are Growth Pole centres influencing land-use in their neighbourhood
```

Fig. 9.5 *BangaloreSim* model pseudo code

tion networks, land-use policy (the comprehensive development plan) and suitability. Subsequently, for a spatial PSS to be effective, it would require the modeller/expert to control relevant parameters for generating scenarios simultaneously enabling dynamic interaction. Considering these key criteria, the prototype developed is depicted in Fig. 9.6. The base year for the model was considered as 2000 with the classified land-use map for the same year serving as the input.

The evaluation of the spatial PSS involved calibration of the suitability weights with mean floor area ratio and location of industrial estates as growth-poles. The FAR was set to 2.75 based on the average permissible ratio of the land-use zoning policy. Considering these model set-up parameters, the suitability weights for different factors like land-uses, distance from city centre and distance from transportation network were calibrated. The calibration for the weights was performed comparing the classified land-use and the simulated land-use for 2006.

In the earlier test simulations (before acknowledging growth poles) the land-use changes could not forecast the land-use for 2006 appropriately. There were other initiatives by the State Government like the creation of an outer ring road, which were captured in the system. The industrial estates were ascribed as a growth pole (GP) agent to capture these changes. The GP agent essentially boosts suitability of any particular land-use in the neighbourhood with the greater probability to change into built-up (see underlined procedure in the pseudo code in Fig. 9.5).

The simulation for year 2006 considering the influence of growth-pole agents and the mean FAR with the calibrated weights for suitability is presented in Fig. 9.7. The comparison of these outputs were also analyzed using the Map Comparison Kit 3 (Hagen-Zaker *et al.* 2006). The simulated land-use (Fig. 9.7a) was compared with the classified land-use of 2006 (Fig. 9.7b). The Kappa statistic of built-up areas was 0.413 and Kloc was 0.427 with Moran's I for the simulated and classified image being 0.643. The calibrated model was then used to perform policy analysis.

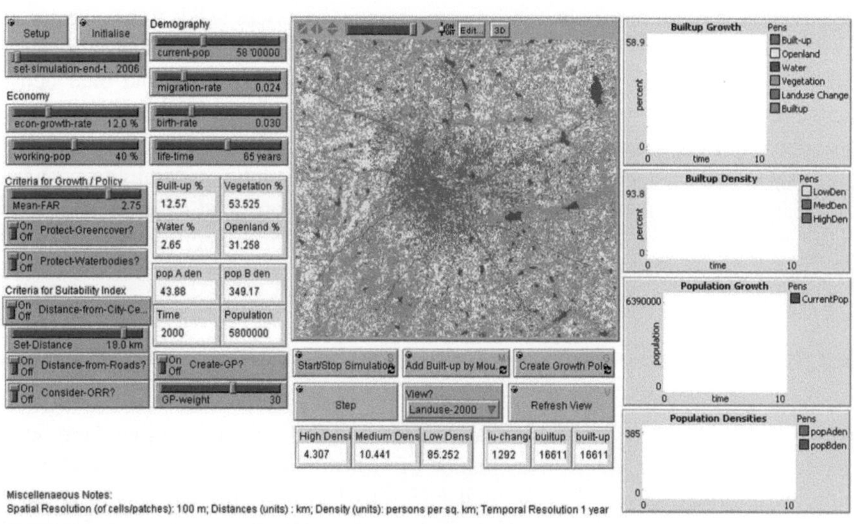

Fig. 9.6 Prototype of the SPSS (See also Plate 21 in the Colour Plate Section)

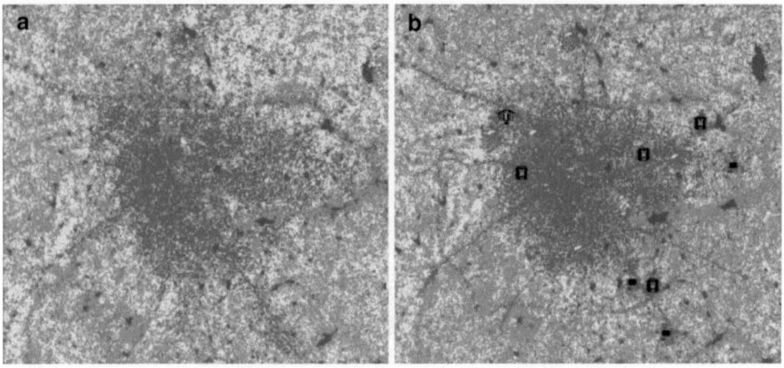

Fig. 9.7 Classified and simulated land-use of Bangalore for 2006. (a) Classified land-use (2006), (b) Simulated land-use (2006) (See also Plate 22 in the Colour Plate Section)

9.4.4 Policy Analysis

The scope of policy analysis is restricted for evaluating the impact of increasing floor area ratio on extent of sprawl up to year 2015. The effect on sprawl was analyzed based on the percentage built-up areas and built-up densities. The simulations revealed that higher FAR resulted in increase of built-up areas (386.6 km^2, 93.2 per cent) with high densities and the rate of built-up growth declining. However, during the duration of increase in density of built-up areas, the population 'b-density' (population over built-up area) depicted 'U' type behaviour. Accordingly, the population b-density decreased initially and increased after certain duration of time. This suggests that initially the expansion of the built-up area in spite of FAR continued and only after a threshold, the population chose to locate within the city's extent over locating in the outskirts. The increase in built-up areas with high densities suggests more congestion in certain parts of city. It is in these parts wherein infrastructure and services needs to be augmented to accommodate the forecasted population b-densities. The policy analysis revealed the probable effects of change in FAR on extent of built-up areas and their densities. The spatial PSS was effective in locating areas with higher densities of built-up area and requiring augmentation of infrastructure and services to relieve of impending congestion in these localities. While this exercise was useful in generating a scenario of urban land-use, there can be many more scenarios generated by undertaking the simulations for different policy settings to understand their respective consequences.

9.4.5 Drawbacks and Recommendations

An important distinction in the spatial PSS was the ability to incorporate the process that lead to land-use changes like creation of growth poles and studying the effect of outgrowth through the FAR. A spatial PSS thus facilitates the analysis of choices, visualization of such decisions and evaluation of certain policy decisions. However, the success of these will rest on the modellers and planners who can identify and capture these processes to realize the best possibilities from a PSS.

The scale of analysis affects the computational complexity of the system. Higher spatial scales increase the computational resources required to perform the simulations optimally and vice versa. In the current model, the spatial scale is set for 100 m x 100 m grid, with yearly time scales. Caution should also be exercised on the choice of scale based on the problem and processes that are modelled.

The research here is yet to validate the spatial PSS with the government functionaries. Validation of agent-based land-use models has been a contentious issue in recent times. However, recent work by Brown *et al.* (2005) has attempted to clarify this debate by acknowledging path dependence and bringing out the distinction of achieving predictive accuracy and process accuracy. Consequently, it is important that any spatial PSS should have greater process accuracy and be able to generate patterns resulting out of numerous processes. But planners and decision makers would always want some amount of predictive accuracy informing them what type/pattern of growth will emerge at which locations. Thus spatial PSS should be ideally achieving reasonable predictive and process accuracies. The process accuracy and predictive accuracy of this model is yet to be ascertained. However, the spatial PSS in its current state allows the modeller or experimenter to test for various options and evaluate the consequences.

The contention of the process accuracy can be either addressed by following a similar approach of Brown *et al.* (2005) or engaging in a focus group evaluation involving key bureaucrats, planners, policy makers and experts who would verify the processes captured in the SPSS and outcomes of the simulation in the SPSS. With such a focus group evaluation, it will also make it beneficial for SPSS to be adopted in their current planning practices.

9.5 Conclusions and Future Directions

Proper implementation of master/development plans is a critical aspect in regulating the development of urban areas. Although 1,200 master/development plans for important towns and cities have been prepared in India, their implementation so far has not been satisfactory due to a variety of reasons. City planning mainly addresses the preparation of land-use plans through zoning to cater for projected populations. However, civic authorities also need to plan for meeting the demand for infrastructure facilities and ensuring delivery of basic services. This is dismal in current planning practice since these are normally static master plans or development plans mostly addressing land-use issues. These plans are also less equipped to review and evaluate any policy decisions dynamically by visualising the potential implications of a policy directive and the regions of potential sprawl. It is therefore necessary to support administrators and planners by providing them with better understanding, methods and tools to tackle the problem of urban sprawl. Further, administrators and planners need to be informed of possible areas of sprawl to take corrective actions to mitigate the implications. The problem of urban sprawl is observed to be an outcome of improper planning, inadequate policies and lack of effective governance due to various reasons. The inability of the administration and planning machinery to visualise probable areas of sprawl and its growth is persistent with the lack of appropriate spatial information and indicators.

Thus, in the present context, with the escalating problem of urban sprawl, the evolution of a PSS in the form of the *BangaloreSim* model is the first step in this direction. The PSS aids in undertaking policy analysis for certain policy measures and its impact on urban land-use. The PSS was also effective in evaluating the influence of FAR on controlling urban sprawl. However, there are numerous challenges for incorporating the complex urban system in a dynamic modelling framework with which to support participatory decision making. The PSS can be used to regularly monitor and check the nature of sprawl for compliance with policy recommendations over time.

Most PSS allow analysis, evaluation and visualization only on standalone systems, wherein each stakeholder has to choose/decide the options on same system/ platform. This would suggest that all stakeholders have to meet physically to evaluate and decide. Moreover, such initiatives are not normal and very difficult to moderate. In this context, it becomes necessary for a parallel and distributed simulation framework to support PSS, so that all stakeholders and managers/administrators are able to interact, organize, plan, evaluate and decide through a network. Then the challenges are twofold: initially to integrate different models that are required to carry out the simulations, and then to synchronize the model inputs, feedbacks and outputs over space and time. Design of PSS with participatory simulation framework will largely contribute to its ready usage in urban planning and decision-making process.

The PSS is developed on an agent-based modelling environment as a process-based land-use model. The developments in PSS can significantly complement research in temporal geographic information science transforming the temporally static GIS to temporally dynamic systems. The exchange of geospatial data (import and export functions) is now possible for ASCII grid formats. In future, it is imperative that these systems are interoperable and support most of the standard geospatial formats. Provision of PSS over the internet can also imply that they may be accessible to everyone and can become pervasive. Such a web-based PSS can be a very effective tool for monitoring, planning and decision-making encouraging community participation. At the outset, the future development of the PSS can facilitate exchange of standard geospatial data formats and evolve as a web-based system.

Acknowledgements We thank Indian Institute of Science for financial and infrastructure support. The Global Land Cover Facility (GLCF), Institute for Advanced Computer Studies, University of Maryland, USA and the National Remote Sensing Agency (NRSA), Department of Space, Government of India is duly acknowledged for making available the remote sensing satellite data for the study.

References

Allen, P.M. and Sanglier, M. (1979) A dynamic model of urban growth: II, *Journal of Social and Biological Systems*, 2(4): 269–278.

Batty, M., Xie, Y. and Sun, Z. (1999) The dynamics of urban sprawl, *Working Paper Series, Paper 15*, Centre for Advanced Spatial Analysis, University College London, England.

Batty, M. and Torrens, P. (2001) Modelling complexity: the limits to prediction, *Paper 36, Working Paper Series,* Centre for Advanced Spatial Analysis, University College London.

Benenson, I. and Torrens, P.M. (2004) *Geosimulation: Automata-based Modelling of Urban Phenomena*, Wiley, Chichester.

Braess, D. (1968) Ãoeber ein paradoxon aus der verkehrsplanung, *Mathematical Methods of Operations Research*, 12(1): 258–268.

Brown, G.D., Page, S., Riolo, R., Zellner, M. and Rand, W. (2005) Path dependence and the validation of agent-based spatial models of land-use, *International Journal of Geographic Information Science*, 19(2): 153–174.

Galster, G., Hanson, R., Ratcliffe, M.R., Wolman, H., Coleman, S. and Freihage, J. (2001) Wrestling sprawl to the ground: defining and measuring an elusive concept, *Housing Policy Debate*, 12(4): 681–717.

Geertman, S. and Stillwell, J. (2004) Planning support systems: an inventory of current practice, *Computers, Environment and Urban Systems*, 28(4): 291–310.

Hagen-Zanker, A., Engelen, G., Hurkens, J., Vanhout, R. and Uljee, I. (2006) Map Comparison Kit 3, Research Institute of Knowledge Systems, Maastricht, from http://www.riks.nl/mck.

Jothimani, P. (1997) Operational urban sprawl monitoring using satellite remote sensing: excerpts from the studies of Ahmedabad, Vadodara and Surat, India, Paper presented at the 18th Asian Conference on Remote Sensing, Malaysia, 20–24 October.

Klosterman, R.E. (1999) The What If? collaborative support system, *Environment and Planning B*, 26: 393–408.

Lata, K.M., Sankar Rao, C.H., Krishna Prasad, V., Badrinath, K.V.S. and Raghavaswamy (2001) Measuring urban sprawl: a case study of Hyderabad, *GIS Development*, 5(12): 22–25.

Lowry, I.S. (1967) Seven models of urban development: a structural comparison, In Hemmens, G.C. (ed.) *Urban Development Models,* Special Report 97, Highway Research Board, Washington DC, pp. 121–163.

Pumain, D., Saint-Julien, T. and Sanders, L. (1986) Urban dynamics of some French cities, *European Journal of Operational Research*, 25(1): 3–10.

Sharifi, M.A. (2003) Integrated Planning and Decision Support Systems for Sustainable Water Resources Management: Concepts, Potentials and Limitations, Seminar on *Water Resources Management for Sustainable Agricultural Productivity*, organized by Asian Productivity Organization, Lahore.

Sterman, J.D. (2000) *Business Dynamics: Systems Thinking and Modelling for a Complex World*, Irwin McGraw-Hill, Boston.

Subudhi, A.P. and Maithani, S. (2001) Modelling urban sprawl and future population prediction using remote sensing and geographical information system techniques. In: Subudhi, A.P., Sokhi, B.S. and Roy, P.S. (eds.) *Remote Sensing and GIS Application in Urban and Regional Studies*, Human Settlement and Analysis Group, Indian Institute of Remote Sensing, Dehradun.

Sudhira, H.S. and Ramachandra, T.V. (2007) Modelling the dynamics of urban sprawl using system dynamics and agent-based models, Presentation at International Conference on Framing Land Use Dynamics II, University of Utrecht, 18–20 April.

Sudhira, H.S., Ramachandra, T.V. and Bala Subrahmanya M.H. (2007) Integrated spatial planning support systems for managing urban sprawl, Presentation at 10th International Conference on Computers in Urban Planning and Urban Management (CUPUM), Iguassu Falls, 11–13 July.

Sudhira, H.S., Ramachandra, T.V. and Jagadish, K.S. (2004) Urban sprawl: metrics, dynamics and modelling using GIS, *International Journal of Applied Earth Observation and Geo-Information*, 5(1):29–39.

Sudhira, H.S., Ramachandra, T.V., Raj, K.S. and Jagadish, K.S. (2003) Urban growth analysis using spatial and temporal data, *Photonirvachak, Journal of Indian Society of Remote Sensing*, 31(4): 299–311.

Torrens, P.M. and Alberti, M. (2000) Measuring sprawl, *Working Paper Series, Paper 27*, Centre for Advanced Spatial Analysis, University College, London.

Uljee, I., Engelen, G. and White, R. (1999) *Integral Assessment Module for Coastal Zone Management, RAMCO 2.0 User Guide*, Research Institute of Knowledge Systems BV, Maastricht.

United Nations (2004) *Executive Summary, World Urbanisation Prospects: The 2003 Revision*, Population Division, Department of Economic and Social Affairs, United Nations, New York.

Waddell, P. (2002) UrbanSim: Modelling urban development for land-use, transportation and environmental planning, *Journal of the American Planning Association*, 68(3): 297–314.

Walter, H. (1975) *Urban Systems Models*, Academic Press, USA.

White, R. and Engelen, G. (2000) High-resolution integrated modelling of the spatial dynamics of urban and regional system, *Computers, Environment and Urban Systems*, 24: 383–400.

Wilensky, U. (1999) *NetLogo*, Center for Connected Learning and Computer-Based Modelling, Northwestern University, Evanston, IL, USA, from http://ccl.northwestern.edu/netlogo.

Additional Reading

Batty, M. (2001) Models in planning: technological imperatives and changing roles, *International Journal of Applied Earth Observation and Geoinformation*, 3(3): 252–266.

Batty, M. and Torrens, P. M. (2005) Modelling and prediction in a complex world, *Futures*, 37: 745–766.

Couclelis, H. (2004) The construction of digital city, *Environmental and Planning B*, 31: 5–19.

Forrester, J. W. (1969) *Urban Dynamics*, Wright Allen Press, Cambridge, Massachusetts.

Gakenheimer, R. (2002) Planning, transportation and land use for cities in India, Workshop on Integrated Planning for Transportation, *Land Use and the Environment in Indian Cities*, Harvard University Centre for Environment and Administrative Staff College of India, Hyderabad.

Wegener, M. (1994) Operational urban models: state of the art, *Journal of the American Planning Association*, 60: 17–29.

Chapter 10
GRAS: A Spatial Decision Support System for Green Space Planning

Claudia Pelizaro, Theo Arentze and Harry Timmermans

10.1 Introduction

Green space is defined here as a vegetated public area in an urban setting that provides an open environment. Urban green spaces provide a number of benefits towards the overall goal of urban sustainability. Some of these benefits are more tangible than others (e.g. Heisler 1986; Randall *et al.* 2004). However, very little is known about the link between urban green space provision and human behaviour. There are surprisingly few published guidelines explaining how to assess the provision of green spaces on an intra-urban basis (e.g. Nicol and Blake 2000; Herzele and Wiedemann 2002). Projects on this field are in a premature stage, with current approaches just starting to explore the elaboration and application of choice experiments and contingent valuation techniques to explain the influence of green space on people's lives (Lawson 2000; Pozsgay and Bhat 2001; Bhat and Gossen 2002; Kemperman *et al.* 2005; Maat and de Vries 2006; Jim and Chen 2006). These are, however, isolated studies or *ad hoc* experiments rather than models or tools that can be explicitly used to support decisions on green space provision for public welfare.

Urban green spaces normally occupy large pieces of land, involve high maintenance costs and do not bring any direct monetary benefit. When the benefits of green spaces are indirect and intangible, sustainable funding will suffer in relation to other priorities in the urban context. With that in mind, we have developed a prototype Spatial Decision Support System (SDSS) to assist authorities to strategically enhance the supply of recreational green spaces (squares, parks, green corridors, waterfronts, *et cetera*) with the right type and variety of green spaces that optimise public welfare. The system has been given the acronym *GRAS*, the Dutch word for *grass*, which stands for GReenspace Assessment System. *GRAS*

C. Pelizaro (✉)
Landscape Systems
Future Farming Systems Research Division
Department of Primary Industries
32 Lincoln Square Nth
Carlton, VIC 3052, Australia
E-mail: claudia.pelizaro@dpi.vic.gov.au

S. Geertman, J. Stillwell (eds.), *Planning Support Systems Best Practice and New Methods*,
DOI 10.1007/978-1-4020-8952-7_10, © Springer Science+Business Media B.V. 2009

is a prototype SDSS proposed to underpin the planning, design and maintenance of recreational urban green space. By planning, we mean the process of forecasting or influencing the physical arrangement (location and size) of green spaces in the built environment. Green space design is related to the selection of features (facilities and attributes) within the physical locations to meet functional criteria. Examples of design considerations include the type of vegetation, provision of public transportation and facilities such as playgrounds, sport fields, picnic tables, toilets, *et cetera*. Maintenance encompasses the recurrent, periodic or scheduled work necessary to repair, prevent damage or sustain existing features in green spaces.

GRAS is a fully integrated, GIS scenario-based, microsimulation, multi-criteria evaluation system. The GIS-based scenario tool supports users in applying strategic thinking to the search for viable future options in green space portfolio within the built environment. A range of static to dynamic modelling methods is employed within the system to represent individuals' behaviours. In principle, these models operate on a '100 per cent sample' (i.e. the entire population) of individuals, which are synthesized from census data. Under the umbrella of the *Spatial Models Component,* spatial choice models traditionally used in the urban planning field can be applied. The *Spatial-Temporal Models Component* comprises an activity-based microsimulation model, the so-called *Aurora* (Joh *et al.* 2001; Joh 2004), which predicts space-time behaviour patterns at the individual level. It is based on the premise that the use of certain green space is not only a function of the attributes of the green space, socio-demographics and distance/travel time, but varies also as function of other activities that need to be conducted during the day and week and dynamic space-time constraints. Thus, this model provides considerably more detail than the traditional spatial choice models within the *Spatial Models Component*. A multi-criteria tool allows the comparison of current and future scenarios in light of multiple and possibly conflicting criteria.

GRAS's capabilities in supporting the planning, design and maintenance of urban green spaces were tested in a case study for the city of Eindhoven, Southern Netherlands. *GRAS*'s models and tools were used to identify and diagnose potential weaknesses and/or problems in the overall green space portfolio of the city. Based on a preliminary analysis/assessment of the current green space portfolio and planners' aspiration levels and goals for a future green space portfolio, changes were suggested and incorporated in the development of two new scenarios. These alternative green space scenarios were then evaluated and compared against each other using a range of performance indicators derived by the system. This exercise demonstrated *GRAS* to be a robust framework, capable of supporting every stage of the decision making process in green space planning and monitoring.

Although the main focus of this chapter is on the design of *GRAS* (i.e. the overall description of the program architecture, components, modules, interface and data infrastructure), we dedicate a section at the end of this chapter to outline the practical issues related to the system's application.

10.2 *GRAS* Architecture and Design

The traditional approach in developing SDSS or GIS-based DSS (Sprague 1980) for the urban planning domain uses a standard GIS package as a *DSS Generator* that is utilised to quickly and easily build the *Specific DSS* (Batty and Densham 1996; Keenan 1997; Yates and Bishop 1998; Wegener 2001; Booty *et al.* 2001; Geertman 2002; Randall *et al.* 2004; Yeh and Qiao 2005). Unlike the traditional approach, in this project we followed a more rigorous and flexible approach that used *DSS Tools* to develop the *Specific DSS*.

GRAS is a GIS windows-based application, written in Borland Builder C++ 5 (Borland™) and GIS functionalities were added using MapObjects 2.0 ActiveX Control (ESRI™). Rather than embedding less elaborate models within a comprehensive GIS, we embedded a limited range of GIS functions within a more elaborate modelling framework. Although the chosen approach has disadvantages, such as requiring advanced programming skills when changes in the system are required, this development method offers a number of advantages, such as: (i) overcoming limitations and overheads of standard GIS software packages; (ii) flexibility to include the domain analytical/simulation models desired; and (iii) full integration, where the users only intervene in the system to control the modelling process, not to conduct basic operations needed for data modelling and interchanges.

Three major components can be identified in *GRAS*: the Database Management System (DBMS); the Model Based Management System (MBMS); and the component for managing the interface between the user and the system, i.e. the Dialogue Generation and Management System (DGMS). The DGMS is left for future discussion. For now, it suffices to say that the DGMS was built upon a graphical user interface (GUI) to allow easy and intuitive manipulation of the system. The main user interface for *GRAS* (Fig. 10.1) is a GIS-based map window with icons, a pull-down menu and pop-up windows that can be manipulated by a computer mouse (and, to some extent, via a keyboard) to gather user's actions with the goal to act on the data (i.e. visualize a map, consult and process the data) and on the models (i.e. set parameters and run models to produce new information). The first two components are described in detail in the next sub-sections.

10.2.1 Data and the Database Management System

In line with trends in urban planning towards more disaggregated and realistic models, the type of models proposed for green space provision and monitoring require a fundamentally new organization of more detailed data based on a micro view of the urban environment. Spatial modelling in the straightjacket of the zonal systems is too aggregated to predict impacts on small-scale planning, for instance neighbourhood-scale planning (Wegener 2001).

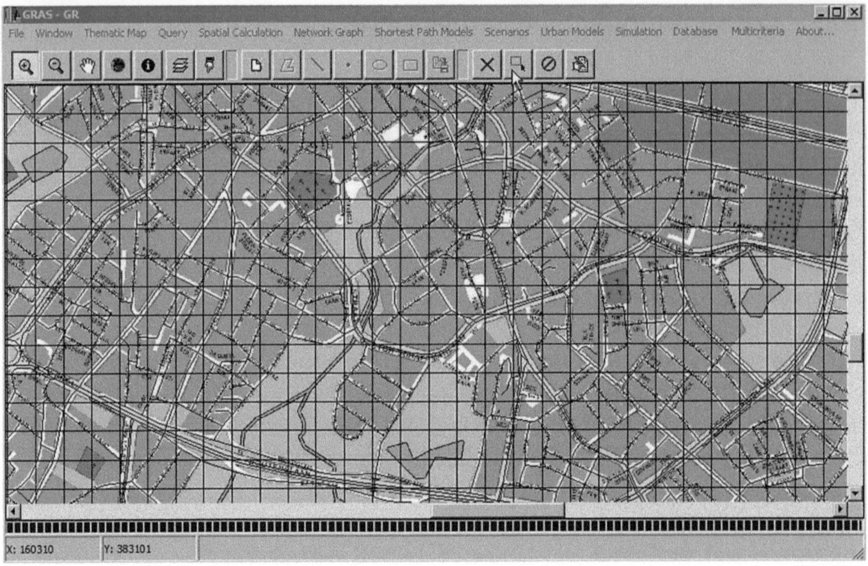

Fig. 10.1 *GRAS* main user interface

The move from macro to micro in complex spatial systems requires efficient data structures to overcome long data processing time and/or poor system performance (still an issue even with new technologies). The solution adopted to deal with such issues is to represent the study area as a cell system. Cells provide a stable representation of space, and are generally computationally more efficient to work with than zones, and, if defined on a sufficiently fine scale, will largely avoid the aggregation bias problems. On the other hand, they are very artificial constructs and their use can (perhaps) lead to an overlaid abstracted representation of space and spatial processes. In the limit, as the cell size goes to zero, cells become individual points in space and the spatial representation becomes fully continuous (Miller *et al.* 2004). A fully continuous representation of spatial distributions and processes is not a practical possibility at this time, at least for *GRAS*-type models. Hence, we use a cell system of $100\ \text{m} \times 100\ \text{m}$ to represent urban space. Spatially aggregate data of land use, green spaces, work establishments, urban zoning system/socio-demographics and postcode addresses must be disaggregated and combined into a singular spatial source defined as cells. This unique cell vector-based layer contains the baseline data of *GRAS*. Although the cell size is an arbitrary value that can be easily changed, we strongly recommend the dimension suggested. The cell resolution was experimentally adopted as the optimum size given limitations of memory and speed of personal computer technology and model sensibility and accuracy given spatial resolution.

In summary, there are many advantages of information storage following the cell structure, such as:

- much higher spatial resolution: spatial interactions between zones are established via networks linked only to the centroid of the zones. When zones are divided into

cells, interactions are established between cells via networks linked to the centroid of the cells;
- avoiding of serious methodological difficulties such as the modifiable areal unit problem and problems of spatial interpolation between incompatible zone systems (Openshaw 1984; Fotheringham and Wong 1991; Wegener 2001); and
- facilitation of data flow within the urban models improving computing time (processing time) and decreasing required computing power.

Technically, the DBMS is a database model built with Borland Database Engine and embedded with GIS functionalities from MapObjects (ESRI). Embedding GIS capabilities to the objects and relational tables of Borland, spatial and non-spatial data from different sources can be efficiently stored, retrieved, manipulated and processed. The full integration of these two technologies makes the DBMS the heart of the spatial and operational information system, allowing for communication and intermediate storage among models without user intervention.

10.2.2 The Model Based Management System (MBMS)

The MBMS is an integrated model environment built with a range of models designed and implemented to support the evaluation and assessment of the urban green space provision and monitoring. It incorporates various procedures that enable users and planners to easily create alternative scenarios and make decisions based on a number of analytical models for green space assessment. After opening the MBMS 'box', tools and models illustrated in Fig. 10.2 are found. The arrows in the figure represent the communication mechanism and data flow between models.

Models implemented within the system were developed using a conceptual framework of individual behaviour patterns within the urban built environment (Pelizaro 2005). Such a conceptual framework is a state-of-the-art modelling approach for urban planning and monitoring processes and implies that planners can predict the likely impact of their possible future plans and actions by observing and predicting individual behaviour in the urban environment in constant change. The MBMS has seven major fully-integrated modules, described hereafter.

10.2.2.1 The Scenario Management Module

As an instrument for strategic thinking and option search, scenarios are means to represent the future. Xiang and Clarke (2003) argue that scenario development tools are likely to contribute to an effective decision-making process if they perform well the two functions of bridging and stretching. A scenario bridges the process of modelling with that of planning. It is a cognitive apparatus that stretches people's thinking and broadens their views in planning. The authors claim that a scenario development tool may best perform the bridging and stretching functions if it (i) stimulates surprising and plausible scenarios creation, (ii) presents the information used in a

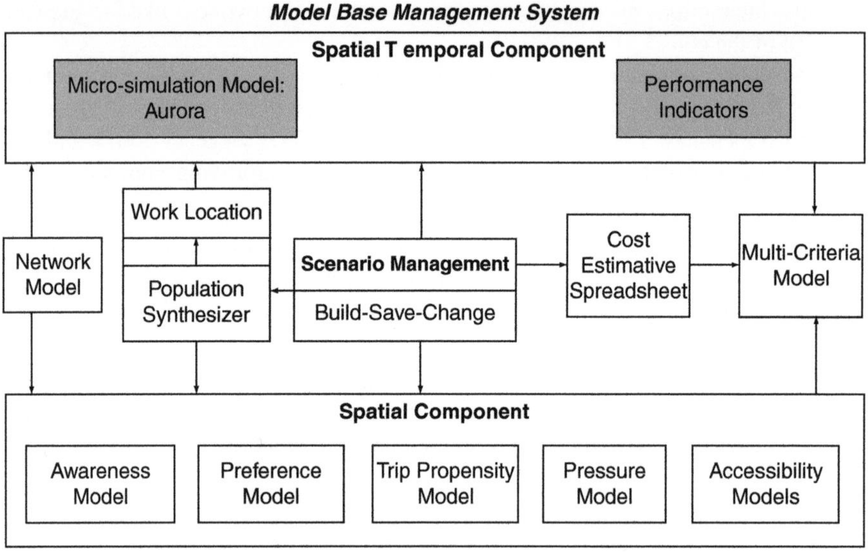

Fig. 10.2 Tools and models within *GRAS*' model based management system

vivid way and (iii) has a cognitively ergonomic design, that is, effectiveandsafe. Although we will not explore these issues here, it is worthwhile to mention that we have incorporated Xiang and Clarke's credentials in the design and implementation of the scenario tool in *GRAS* (Pelizaro 2005).

GRAS is operated around the concept of scenarios. A scenario is defined in terms of an appropriate description that represents the urban space/system in terms of land-use configuration, the transportation network, green space amenities, the zone system (socio-demographics) and work facilities. At the starting point, the user, having certain aspiration levels or goal states for the area being monitored, can modify the actual urban space. As suggested before (Arampatzis *et al.* 2004), this actual state, or 'status quo' or yet 'zero state' scenario, corresponds to the baseline situation, which is used as the basis for the creation of a new scenario.

The creation of a new scenario is carried out through interactive procedures supervised by the scenario management tool under the concept of 'themes'. In practice, the scenario development process involves a simple operation of changing the cell(s) attributes (data / information) of a cell-based map describing the urban space (which could be the baseline data or a scenario created previously), using the scenario themes strategy illustrated in Fig. 10.3a. Indeed, the target urban space to be modified in a new scenario is selected by the user on the map-based user interface, using the computer mouse. Then, the edit box window shown in Fig. 10.3a prompts, introducing the concept of scenario themes for data editing. For instance, suppose the user wishes to create a new scenario where a particular green space will be redesigned. The user then enables the scenario development tool by simply clicking on the appropriate icon of *GRAS*'s user interface. Then, the user selects any cell on the map that belongs to the green space to be modified. The window shown in

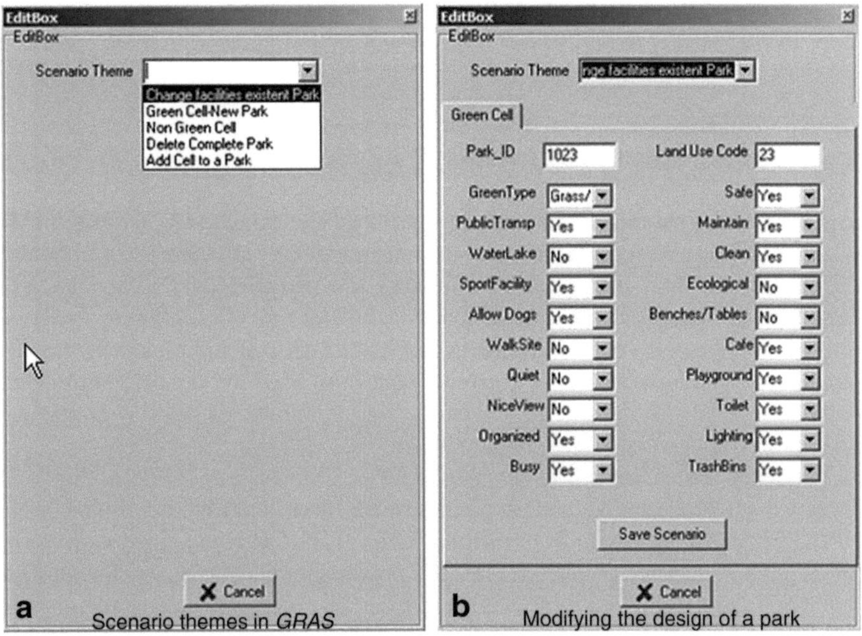

Fig. 10.3 Scenario management tool and scenario themes in *GRAS*

Fig. 10.3a prompts the user to choose the appropriate theme listed in the drop down box (in this case, 'change facilities of a green space'). The system refreshes the scenario edit box window as shown in Fig. 10.3b, displays the current description of the green space to be modified and sets the database to the edit mode. Yet through this interface window, the green space design can be modified by adding or removing the functional attributes as listed in Fig. 10.3b.

An important function of the concept around scenario themes is to restrict data access in order to guarantee data consistency. By ergonomically designing a scenario theme interface, only relevant data are made editable to users. As described in the example above, when the user redesigns an existing green space, only the data describing that green space attributes and facilities are made editable to the user. There are five themes on which to compose scenarios:

- Change the facilities of a green space: allows users to modify green space features (design), which will impact on individuals' perceptions/values.
- Green cell – new park: allows users to create a new green space on the area specified by the user on the map interface.
- Non-green cell: allows users to remove a parcel of land from an existing green space. The interface will then prompt the user to redefine the land-use and socio-demographic attributes for that parcel of land within the scenario being developed.
- Delete complete park: the selected green space (map user interface) is replaced with a new development, which is specified by the user in terms of land-use and socio-demographics attributes.

- Extend the area of a green space: increases the green space area with surrounded parcels of land specified by the user on the map interface.

10.2.2.2 The Population Synthesizer

The scenario population is defined by means of the Population Synthesizer. In *GRAS*, significant efforts were made towards adopting a disaggregated approach to system modelling, in which behaviour is the sum of individuals' behaviour of an urban system. The 'synthetic population model' takes care of creating a population imitation of the study area with demographics closely matching those of the real population. The information on the individual level used by the different models within *GRAS* is: age; gender; employment status in number of working hours; and household composition: with/without child(ren).

Following earlier work (Beckman *et al.* 1996; Bradley *et al.* 2001; Arentze and Timmermans 2000) the synthetic population model uses iterative proportional fitting (IPF) to extend a given sample of the population consistent with known statistics of the target population. To create a synthetic population, the Population Synthesizer requires the following types of data source: census tract data of the study area on the cell level and demographic data of a representative sample of the real population on the individual level.

The Population Synthesizer user interface allows creation of the entire population or only a percentage or even subgroups of the entire population, using age, gender, household composition and work status constraints (or a combination of these possibilities). The possibility of reproducing only segments of the population provides extra system capability to support decisions to benefit a particular group(s) of the target population.

10.2.2.3 The Network

The Network model plays an important role in the system. It estimates distances using information about the road network and Dijkstra's shortest path algorithm. Distances related to the movements of individuals in space and time will be given by this model and passed on to others system's models when required. Models within the 'Spatial Component' assume that individuals move in space looking for the shortest distance. In these cases, the shortest path is given by the shortest distance. On the dynamic level (spatio-temporal models), it is assumed that people seek to minimise travel times by choosing a route to a given destination. Hence, the shortest path is given by the minimum travel time.

10.2.2.4 The Spatial Models Component

The 'Spatial Models Component' consists of a family of discrete choice models to describe individuals' green space choice behaviour and a model to calculate spatial accessibility measures to urban green spaces. This component can be also called a '*Static Component*' in the sense that models do not consider the movements of the individuals in space and time. Individuals' green space choice set is restricted by some accessibility measure from each individual's doorstep. We estimated three discrete choice models with different degrees of complexity (and behavioural realism) capturing different aspects of the theoretical framework described in Pelizaro (2005). From these three discrete choice models, we derived four models to assess/evaluate green spaces: (i) an awareness model; (ii) a preference model; (iii) a trip making propensity model; and (iv) a pressure model.

The Awareness Model

The concept of awareness is defined here as the probability that individuals know a certain spatial choice alternative, i.e. green space. The model developed by Ponjé *et al.* (2005) and embedded in *GRAS* assumes that the awareness level of an urban green space is a function of (i) a set of relevant attributes of the green space, (ii) a set of relevant characteristics of individuals, and (iii) some measure of accessibility. The model estimates an average awareness level for each green space within the study area that is the sum of individuals' awareness of each particular green space. This quantitative measure will inform decision makers/planners on the most/least known green space in the study area. This measure could be used, for instance, as an indicator of maintenance budget management or improvement priorities.

The Preference Model

This model assumes choices are based on the utility individuals derive from green spaces, which allows them to rank green spaces in terms of preference. A preference scale was observed from actual individual stated choices or preferences under experimental conditions (Ponjé and Timmermans 2003). Based on that, utility and level of preference for a green space can be predicted as the accumulation of each individual's utility/preference for that green space. Figure 10.4 illustrates the user interface of this model in *GRAS*.

Notice there are three rules to define the choice set across individuals. Rule 1 defines the choice set for individual i located in cell z, as those green spaces $j \in J_i$, for which the distance d (calculated by the shortest path algorithm using the road network) between z and j is smaller than or equal to a threshold distance defined by the user via the user interface ('Maximum Distance fill-in-the-blank box'). Rule 2 uses the same principle, but allows the user to define different distance criteria for different types of green space, suggesting that individuals are willing to travel further to reach larger/more attractive green spaces. Rule 3, on the other hand, uses variable r defined by the user to delineate the choice set, where r is a constant that counts the number of alternative green spaces. In this rule, the closest r green spaces to the individual's doorstep are assumed to make up an individual's choice set.

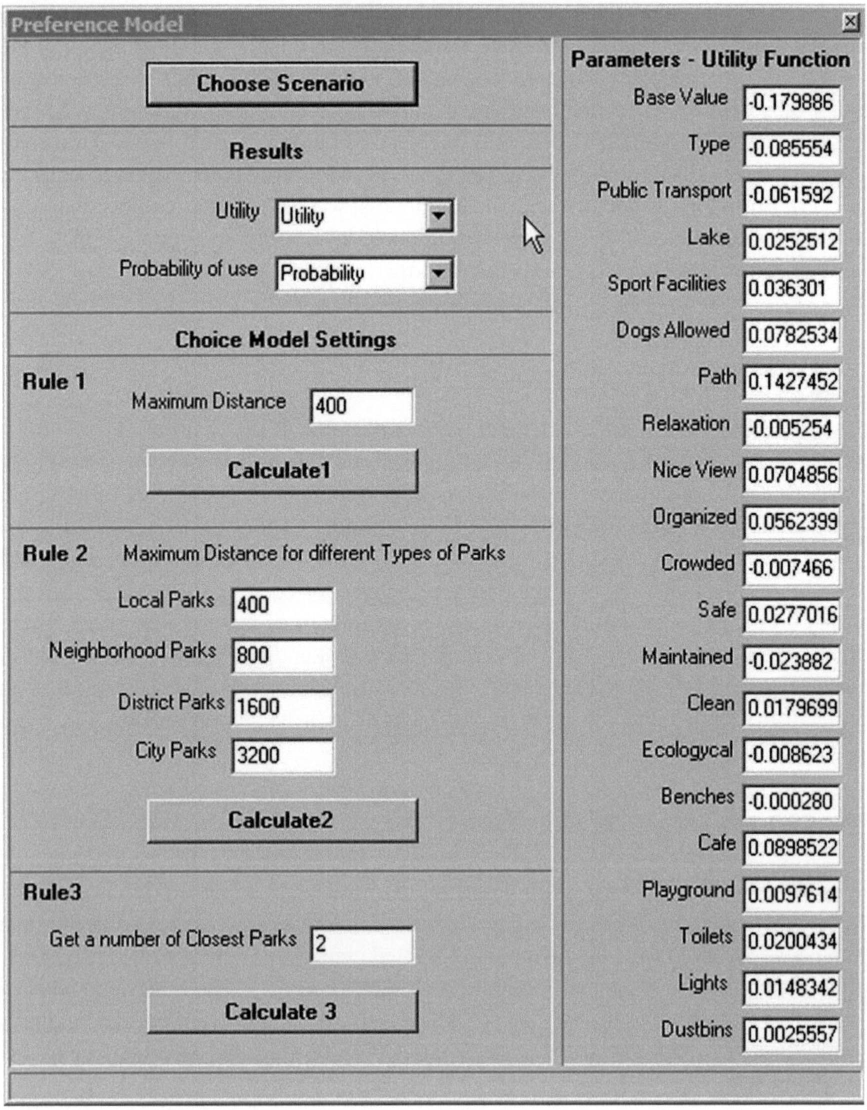

Fig. 10.4 Preference model user interface

The Trip Making Propensity Model

This model predicts the probability of green space usage, given the season of the year, the day of the week and the time of the day. To predict the allocation of trip frequencies across different types of urban green space, the model assumes that the choice of green spaces for recreational and leisure trips involves substitution. Therefore, besides individual socio-demographics and the specific attributes of each green space, the model tries to capture the effect of temporal variation (time of the day, day of the week and season of the year) on an individual's choice. For more information on the model calibration, see Kemperman *et al.* (2005). The individual's choice set is defined following the same approach as Rule 2 of the preference model. As output, green spaces are scored with the number of visits by individuals, given the season of the year, the day of the week and the time of the day.

The Pressure Model

The 'pressure' on green space usage is defined here as the number of individuals visiting a given green space per hectare of green space per day per season. Hence, the model focuses on informing administrators about the intensity of green space usage by season as an indicator of pressure. The frequencies with which individuals visit green spaces were estimated for the different seasons of the year using a survey data and the CHAID (Chi-square-Automatic-Interaction-Detection) technique (Pelizaro 2005). Then, the number of visits that an individual makes to each green space from the choice set is given by the number of visits that this individual makes to green spaces in a given season (result of the CHAID analysis) multiplied by the probability of visiting such a green space found by the trip propensity model. By repeating this procedure for each individual of the scenario population, an aggregate number of visits across the green spaces in the study area is derived. The aggregate number of visits to each green space is then divided by the number of days in the season and by the size of the green space (in hectare) resulting in a measure of green space pressure.

Accessibility Performance Measures

GRAS includes nine accessibility measures that refer to the level of access for green activity from any home within cell z to each of the green space locations j, given the distance dzj. Six of these measures are based on spatial interaction methods (Coelho and Wilson 1976; Dalvi 1978; Ben-Akiva and Lerman 1978; Martin and Williams 1992; Love and Lindquist 1995; Talen and Anselin 1998; Miller 1999). The so-called cumulative-opportunity measure evaluates accessibility with regard to the size of opportunities accessible within a certain travel distance from a given home location (Ingram 1971). The container measure counts the number of green spaces within a certain distance threshold defined by the user (Wachs and Kumagai 1973). The minimum distance measures the distance from home location to the closest green space facility. For all accessibility measures implemented, distance is calculated from an individual's doorstep to the green space entry point, using the road network.

10.2.2.5 The Spatial Temporal Component

The spatial temporal component is represented by an activity-based microsimulation model that predicts space-time behaviour patterns at the individual level.

This component is a simplified version of a model of activity (re)scheduling, named *Aurora* (Joh 2004). Our interest in applying the activity-based approach to the green space problem is to examine the participation of individuals in green activities. The ultimate goal is to understand and, thus, predict activity choice dynamics. This approach shifts the unit of analysis from green space facilities to individual green space activities participation. Spatial choice can then be viewed as, at least in principle, deciding which activities to conduct (activity participation choice), where (activity location or destination choice), when (choice of timing), for how long (duration choice) and the transport mode used (mode choice involved).

From an applied perspective, the problem of how individuals schedule their planned activities as a function of available time, transport mode, *et cetera*, is more relevant, especially, when the focus is on the examination of non-mandatory activities such as green space. Spatio-temporal constraints are critical in the sense that the temporal considerations on individual activity participation decisions (with more rigidity associated with work related activities compared to leisure activities) added to the spatial constraints imposed by the spatial distribution of opportunities for activity participation decisions may imply that an individual cannot be at a particular location at the right time to conduct a particular activity.

We remind the reader here that this chapter is not meant to include a detailed review of the model conceptualisation and operation (see Arentze *et al.* 2005; Pelizaro 2005; Joh 2004). As our focus is on the system as a whole, we concentrate on the model's outputs and green space assessment measures that can be derived from this component. The output generated by *Aurora* is the schedule of activities for every individual of the (synthetic) population for a given urban built environment (scenario). Individuals' schedules of activities provide a good source to derive several urban design performance indicators, based on the individuals' behavioural patterns. From this model, as implemented in *GRAS*, we can derive three performance indicators.

The so-called 'quality of life' is represented by the average utility of individuals' schedule. This is directly related to quality of life in the sense that, if facilities (green spaces, work, *et cetera*) are not well distributed in space, individuals must travel further. Because travel generally has a negative utility value (consumes time and costs money), the overall schedule utility will decline. Moreover, there will be less time left for individuals to conduct activities. Since the utility derived by individuals for conducting an activity is an increasing function of the time spent, less time available will once again reduce the total schedule utility. Similarly, if facilities are not very attractive, individuals may choose to travel further or simply to not conduct the activity at all. Either way, this will have a negative impact on the schedule utility (not conducting an activity returns 'zero' utility).

The second performance indicator is the 'average travel time to green spaces' and the third, is the 'average duration of green activities'. Visit duration patterns can be quite informative to authorities, regarding issues of green space attractiveness (design aspects), spatial location or a combination of both. A short period of time spent in certain green space can be an indication of poor attractiveness or good location (little traveling required). Therefore the impacts should be analysed within the urban context.

10.2.2.6 The Cost Estimation Spreadsheet

This tool calculates costs related to green space provision and maintenance. It requires users to input, via the user interface, elements of costs related to particular features and services provision as shown in Fig. 10.5. Maintenance costs are estimated on an annual basis, while provision costs are estimated as a fixed amount of lifetime investment. Costs are estimated per hectare of green space, allowing green space administrators to fairly compare expenses among green spaces of different dimensions (size). Together with the set of spatial (-temporal) behaviour models implemented in *GRAS*, the assessment of green space costs will allow decision makers and planners to trade off green space costs and benefits on a more quantitative basis. Decisions on the planning and design of green spaces will affect costs of provision and maintenance. Scenarios can then be compared not only in terms of the green space benefits to the community, but also in terms of the costs associated to the implementation of actions.

10.2.2.7 The Multi-Criteria Model

Performance indicators derived from the models described above are often measured in different scales and dimensions. Therefore, the multi-criteria evaluation (MCE) technique is used to rescale these indicators into a common measurement unit, allowing for the comparison of different scenarios in the light of multiple criteria and conflicting priorities. In *GRAS*, the Weighted Linear Combination (WLC) method (Voogd 1983) is used to combine the evaluation scores and weights to arrive at an overall scenario score. The criteria weights are assigned by the user via the user interface. To convert the various criteria scores into a common measurement unit, the interval standardization method (Voogd 1983) is applied. Hence, the score of each criterion will range from 0 to 1.

10.3 *GRAS* Application

In order to make *GRAS* operational, extensive data collection and preparation is required. *GRAS* strongly depends on data: spatial and non-spatial data is required for (i) description/representation of the study area, and (ii) model calibration. In the case study developed to test the system, the data required for the representation of the study area (outlined in Section 10.2.1) was gathered relatively easily from several governmental institutions. One exception however, was the data related to green space attributes (e.g. picnic facilities, sport field, *et cetera*), which was not available

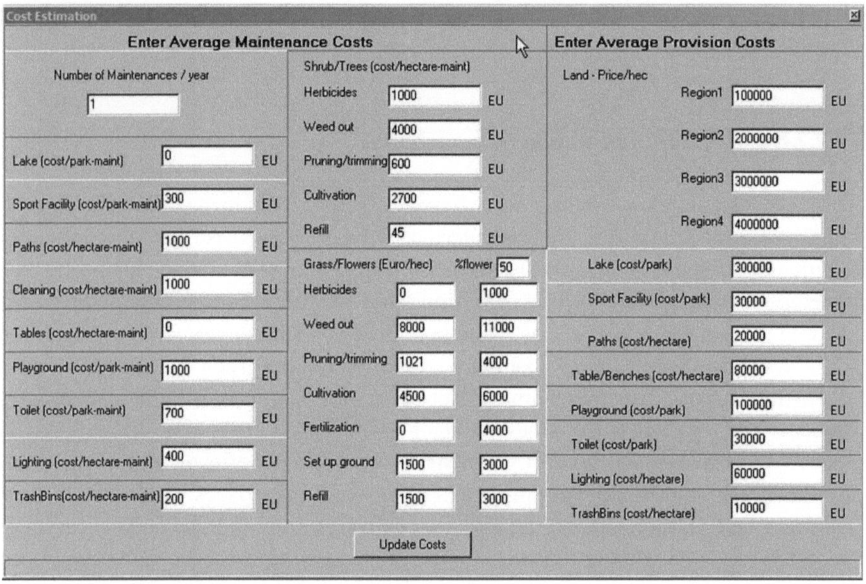

Fig. 10.5 Cost estimative spreadsheet user interface

and required additional fieldwork. The effort to gather this information and store it in a digital format was not trivial, considering the 1,600 green space areas within our study area.

A proper operational setting of the models within *GRAS* requires two sets of parameters to be estimated from empirical data. The first set is required for the static models and the second, for the activity-based model. Therefore, a multitude of survey and statistic procedures need to be performed, which can be highly complex in the case of the activity-based model. Data collection and model calibration are well documented elsewhere (Joh 2004; Pelizaro 2005).

To date, *GRAS* has been applied only once. A case study was developed in the city of Eindhoven, southern region of The Netherlands. In line with the local spatial planning strategy to improve green spaces in the city centre of Eindhoven, green space authorities were seeking answers to the following questions: (i) What would be the best available location to place a green space within the city centre area? (ii) What combination of facilities should it comprise to engender the best value perceived by the community; and (iii) Does perceived benefit pay off the economic investment (i.e. green space provision and maintenance costs)?

GRAS has shown to effectively support all phases of the decision-making process. First, *GRAS* models were used to evaluate/assess the current green space portfolio in the city centre of Eindhoven compared to the green space portfolio of the city of Eindhoven as a whole. This evaluation confirmed the speculation of a lack of green space within the city centre (measured as green space availability, i.e. area divided by the number of CBD's inhabitants). In addition, individuals living within the CBD were found to have relatively low measures of accessibility to green spaces and green spaces within the CBD were indeed under high usage pressure. The development of alternative solutions consisted of a search for suitable locations to build a green space and respective design (i.e. combination of facilities within the green space). The GIS-based user interface has shown to effectively support the location search process, although expert knowledge of the current urban design and expert judgement were essential to the development of possible scenarios.

On the design level, decision makers could simply include or remove facilities (or a combination of them) in the green space being created. Three different scenarios were created by variyng location and design. Models in *GRAS* were then used to assess each scenario based on 16 performance indicators derived in *GRAS*. Scenarios were then compared against each other by means of the multi-criteria analysis model.

Even though there is a lack of green space in the CBD, high costs associated to green space provision in the CBD would not compensate the small gain in individuals' perception. This conclusion surprised green space authorities who were already convinced that a new green space in the area would bring a great deal of benefit to the community. We have found that the small amount of land suitable/available to build a green space in the CBD and its high monetary value does not pay-off the small increase of individuals' perception of the green space portfolio.

10.4 Conclusion and Future Developments

In this chapter, the design of a prototype SDSS for the planning, design and maintenance of green space has been described. *GRAS* is a GIS, scenario-based, microsimulation, multi-criteria decision support system, with a range of domain-specific models inspired by a commonly accepted conceptual framework of spatial (-temporal) behaviour. The suggested system is capable of assisting every stage of the decision-making process, i.e. from the identification of a problem and the definition of (multiple) objective(s) to allowing the users to generate alternatives, and to the evaluation/assessment of alternatives. The system identifies interdependencies among planning, design and maintenance elements of the green space problem and integrates these three levels of decision effectively. *GRAS* was built using the full-integration approach, written in Borland C++ Builder 5 and GIS functionalities were embedded by MapObjects 2.0 ActiveX Control. As discussed in the chapter, the chosen approach provides a more flexible and elaborated development environment. One shortcoming, however, is that when available models in the system cannot fully meet users' needs, users may have difficulties in building and integrating other models themselves because *GRAS* does not provide procedures to enable models to be created and integrated.

GRAS data dependency is an important issue that deserves critical thought. The large amount of data and information needed to operate the system is not only a strength, but also a potential weakness. Obviously, the strength is related to the rich information provided for decision making. The weakness is the considerable effort required to collect the data to feed and calibrate the system and, later, to keep the database up to date.

An issue that needs to be addressed in future work is the further development of the *Aurora* model. Future elaborations may involve finer categorization of activities, especially related to recreation and leisure, which is relevant if the user wishes to have more detailed information about the kind of activities that can be conducted at the parks. The spatial representation in *GRAS* may also be a concern in future research. We did achieve a more disaggregated level of spatial representation than the traditional zone-based approach. Explorations indicated that the spatial representation of 100 m by 100 m cells is computationally efficient, but the cells sometimes appear rather artificial. The user interface can also be improved. To ensure optimal performance, usability testing must be addressed in future research.

References

Arampatzis A., Kiranoudis C., Scaloubacas P. and Assimacopoulos, D. (2004) A GIS-based decision support system for planning urban transportation policies, *European Journal of Operational Research*, 152: 465–475.

Arentze, T.A. and Timmermans, H.J.P. (2000) *ALBATROSS: A Learning-Based Transportation Oriented Simulation System*, European Institute of Retailing and Services Studies, The Hague.

Arentze, T.A., Pelizaro C. and Timmermans H.J.P. (2005) Implementation of a model of dynamic activity-travel rescheduling decisions: an agent-based micro-simulation framework, In *Proceedings of the Computers in Urban Planning and Urban Management Conference*, London, pp. 16.

Batty, M. and Densham P. (1996) Decision support, GIS, and urban planning, Retrieved November 12, 2004, from http://www.geog.ucl.ac.uk/~pdensham/SDSS/s_t_paper.html.

Beckman, R.J., Baggerly K.A. and McKay, M.D. (1996) Creating synthetic baseline populations, *Transportation Research A*, 30(6): 415–429.

Ben-Akiva, M. and Lerman S. (1978) Disaggregate travel and mobility-choice models and measures of accessibility, In Hensher, D. and Stopher, P. (eds.) *Behavioral Travel Modelling*, Croom Helm, New York, pp. 654–679.

Bhat, C.R. and Gossen, R. (2002) A mixed multinomial logit model analysis of weekend recreational episode type choice, *Working Paper*, Department of Civil Engineering, University of Texas, Austin.

Bradley M., Outwater M., Jonnalagadda N. and Ruiter, E. (2001) Estimation of an activity-based micro-simulation model for San Francisco, Paper presented at the 80th Annual Meeting of the Transportation Research Board, Washington, D.C.

Booty, W.G., Lam, D.C.L., Wong, I.W.S. and Siconolfi, P. (2001) Design and implementation of an environment decision support system, *Environmental Modelling & Software*, 16: 453–458.

Coelho, J.D. and Wilson, A.G. (1976) The optimum location and size of shopping centres, *Regional Studies*, 10: 413–421.

Dalvi, M.Q. (1978) Behavioural modelling, accessibility, mobility and need: concepts and measurement, In Hensher, D.A and Stopher, P.R. (eds.) *Behavioural Travel Modelling*, Croom Helm, New York, pp. 639–653.

Fotheringham, A.S. and Wong, D. (1991) The modifiable areal unit problem in multivariate statistical analysis, *Environment and Planning A*, 23: 1025–1044

Geertman S. (2002) Participatory planning and GIS: a PSS to bridge the gap, *Environmental and Planning B*, 29: 21–35.

Heisler G.M (1986) Energy saving with trees, *Journal of Arboriculture*, 12: 113–125.

Herzele, A. and Wiedemann, T. (2002) A monitoring tool for the provision of accessible and attractive urban green space, *Landscape and Urban Planning*, 63: 109–126.

Ingram, D.R. (1971) The concept of accessibility: a search for an operational form, *Regional Studies*, 5: 101–107.

Jim, C.Y. and Chen, W.Y. (2006) Recreation-amenity use and contingent valuation of urban green space in Guangzhou, China, *Landscape and Urban Planning*, 75: 81–96.

Joh, C.H., Arentze T.A. and Timmermans, H.J.P. (2001) Activity scheduling and rescheduling behavior, *GeoJournal*, 53: 359–372.

Joh, C.H. (2004) Measuring and Predicting Adaptation in Multidimensional Activity-Travel Patterns, Published Doctoral Thesis, Faculteit Bouwkunde, Capaciteitsgroep Stedebouw, Eindhoven University of Technology, Eindhoven.

Keenan P. (1997) Using a GIS as a DSS generator, *Working Paper MIS 95–9,* University College Dublin, Department of Management Information System, Dublin.

Kemperman, A.D.A.M., Ponjé, M.M.W. and Timmermans, H.J.P. (2005) Analyzing heterogeneity and substitution in trip making propensity to urban parks: A mixed logit model, *Tourism Analysis*, 10(3): 223–232.

Lawson, C. (2000, January) Understanding leisure travel/activity decisions: psst, want to know where to go for a good time? Paper presented at the 79th Annual Meeting of the Transportation Research Board, Washington, D.C.

Love, D. and Lindquist, P. (1995) The geographical accessibility of hospitals to the aged: a geographical informations systems analysis within Illinois, *Health Services Research*, 26: 629–651.

Maat, K. and de Vries, P. (2006) The influence of the residential environment on green-space travel: testing the compensation hypothesis, *Environment and Planning A*, 38(11): 2111–2127.

Martin, D. and Williams, H.C.W.L. (1992) Market area analysis and accessibility to primary health care centers, *Environment and Planning A*, 24: 1009–1019.

Miller, E.J., Hunt, J.D., Abraham, J.E. and Salvini, P.A. (2004) Microsimulation urban systems, *Computer, Environment and Urban Systems*, 28: 9–44.

Miller H. (1999) Measuring space-time accessibility benefits within transportation networks: basic theory and computational procedures, *Geographical Analysis*, 31: 1–26.

Nicol, C. and Blake R. (2000) Classification and use of green space in the context of increasing urban capacity, *Planning Practice and Research*, 15(3): 193–210.

Openshaw, S. (1984) *The Modifiable Areal Unit Problem*, Geo Books, Norwich.

Pelizaro, C. (2005) A Spatial Decision Support System for the Provision and Monitoring of Urban Green spaces, *Published doctoral thesis*, Eindhoven University of Technology, Eindhoven.

Ponjé, M.M.W. and Timmermans, H.J.P. (2003) Development and Test of a Polytomous Conjoint Choice Model and Experiment to Predict Trips to Urban Parks, (Tech. Rep.), Urban Planning Group, Faculty of Architecture, Eindhoven University of Technology, Eindhoven.

Ponjé, M.M.W., Timmermans, H.J.P. and Waerden, P. (2005) Factors Influencing Awareness Levels of Urban Green spaces: A Partially Exploded Logit Analysis, (Tec. Rep.) Urban Planning Group, Faculty of Architecture, Eindhoven University of Technology, Eindhoven.

Pozsgay, M.A. and Bhat C.H. (2001, January) Destination choice modeling for home-based recreational trips; analysis and implications for land-use, transportation, and air quality planning, Paper presented at the 80th Annual Meeting of the Transportation Research Board, Washington, D.C.

Randall, T.A., Churchill C.J. and Baetz, B.W. (2004) AGIS-based decision support system for neighbourhood greening, *Environment and Planning B*, 30: 541–563.

Sprague, R.H.Jr. (1980) A framework for the development of decision support systems, *Management Information Systems Quarterly*, 4(4): 1–26.

Talen, E. and Anselin, L. (1998) Assessing spatial equity: an evaluation of measures of accessibility to pubic playgrounds, *Environment and Planning A*, 30: 595–613.

Voogd, H. (1983) *Multicriteria Evaluation for Urban and Regional Planning*, Pion, London.

Wachs, M. and Kumagai, T.G. (1973) Physical accessibility as a social indicator, *Socio-Economic Planning Science*, 7: 437–456.

Wegener M (2001) New spatial planning models, *International Journal of Applied Earth Observation and Geoinformation*, 3(3): 224–237.

Xiang, W-N. and Clarke, K.C. (2003) The use of scenarios in land-use planning, *Environment and Planning B*, 30(6): 885–909.
Yates, P.M. and Bishop, I.D. (1998) The integration of existing GIS and modelling systems: with urban applications, *Computers, Environment and Urban Systems*, 22(1): 71–80.
Yeh, A.G. and Qiao, J.J. (2005) Model objects: a model management component for the development of planning support systems, *Computers, Environment and Urban Systems*, 29(2): 133–157.

Additional Reading

Hopkins, L. D. (1999) Structure of a planning support system for urban development, *Environment and Planning B*, 26: 333–343.
Nicol, C. and Blake R. (2000) Classification and use of green space in the context of increasing urban capacity, *Planning Practice and Research*, 15(3): 193–210.
Randall, T. A., Churchill C. J. and Baetz, B. W. (2004) AGIS-based decision support system for neighbourhood greening, *Environment and Planning B*, 30: 541–563.
Wesseling C. G., Karssenberg, D. J., Burrough, P.A. and Van Deursen, W. P. A. (1996) Integrated dynamic environmental models in GIS: the development of a dynamic modelling language, *Transactions in GIS*, 1: 40–48.
Xiang, W-N. and Clarke, K. C. (2003) The use of scenarios in land-use planning, *Environment and Planning B*, 30(6): 885–909.

Part III
New Methods for Planning Support

Chapter 11
Agent-Based Modelling: A Dynamic Scenario Planning Approach to Tourism PSS

Peter A. Johnson and Renee Sieber

11.1 Introduction

Creating simplified models of a system to better understand its workings has long been an effective method of solving complex problems (Bradley and Schaefer 1998; Gilbert and Troitzsch 2005). Models help to formalize theories and can be used to test hypotheses before committing to implementation (Holling 1978). Agent-based Modelling (ABM) is one branch of computerized simulation modelling that shows particular promise as a tool for planning support. Recent work has explored the application of ABM to study volatile gasoline market dynamics (Heppenstall *et al.* 2006), urban sprawl (Benenson and Torrens 2004), the conversion of Amazonian forest into farmland (Deadman *et al.* 2004) and economic development in response to climate change in the remote Canadian north (Berman *et al.* 2004). These research examples push the use of ABM from simplified theoretical models towards more detailed representations that incorporate real-world data (Alessa *et al.* 2006). These next generation examples of ABM have the potential to fill a role as a planning support system (PSS) within variety of planning and policy development areas.

Tourism is one area that could benefit from the integration of ABM and PSS. Planning for tourism development is a complex undertaking, as it affects processes involving multiple individuals interacting over time and space. Traditional approaches to tourism planning have focused on *ad hoc* decision making, and past performance evaluation of individual impacts, rather than examining system changes and effects. If tourism planners incorporated a system-wide perspective into their planning activities, it is argued that better decisions could result (Hall 2000; Jamal *et al.* 2004). The visioning of future scenarios and evaluation of a range of actions, all aimed to increase the benefits derived from tourism development,

P. A. Johnson (✉)
Department of Geography
McGill University
805 Sherbrooke Street West
Montreal, Quebec, Canada H3A 2K6
E-mail: peter.johnson2@mail.mcgill.ca

S. Geertman, J. Stillwell (eds.), *Planning Support Systems Best Practice and New Methods*, 211
DOI 10.1007/978-1-4020-8952-7_11, © Springer Science+Business Media B.V. 2009

are the main functions of tourism planning (Yeoman *et al.* 2005). These types of goal-directed plans are often developed in a community consultative environment, where a set of agreed future states are broken down into functional steps required for their achievement (Hall 2000). Scenarios can be used to anticipate these various 'what if?' questions and are often used as a tool to generate discussion amongst stakeholders (An *et al.* 2005), for example, on whether to focus on supply (e.g. creating tax incentive for bed and breakfast establishments) or demand (e.g. increasing marketing efforts to attract a specific type of tourist). It is this iterative experimentation to which the ABM method shows particular promise to address specific planning challenges, integrating various stakeholder viewpoints, evaluating a course of action and creating impetus to effect change (Walker *et al.* 1999; Zellner 2007).

This chapter presents research on the design and application of an ABM-based PSS to study tourism in Nova Scotia. Nova Scotia is Canada's easternmost mainland province, with a formerly resource-based economy that is becoming increasingly invested in tourism development. We begin by reviewing select literature on ABM and tourism planning. The reader is then introduced to *TourSim*, an ABM-based PSS that we developed for tourism planning. We demonstrate the use of *TourSim* to generate a planning scenario set in Nova Scotia that examines the effect of port of entry on tourist visitation levels. Reflections on the development and application of *TourSim*, drawn from interviews with tourism planning practitioners, inform a broader discussion of the use and adoption of an ABM-based tourism PSS.

11.2 Using Agent-Based Modelling to Provide Planning Support

ABM is a computer-based simulation approach that uses heterogeneous 'agents' to represent real-world decision makers, possessing characteristics of autonomy, communication, reactivity and proactivity (Wooldridge and Jennings 1995). These agents tend to be represented as graphic points that move and interact within a 2D world that represents reality. Here we choose to run our agents (i.e. simulated tourists) against a geo-referenced base, although agents can interact in social and other networks. Using an ABM, one can set initial parameters and grow that system forwards in time, comparing the results of separate model runs (Bonabeau 2002; Grimm and Railsback 2005). Agents are assigned behavioural rules that model socio-economic characteristics and simulate preferences and needs. Agents act out these rules on a varied and competitive landscape, generating system-level outcomes from individual-level interactions. These behaviours are often based on heuristic pathways or bounded rationality, rather than an optimizing equation (Manson 2006).

ABM is used in a variety of capacities, from simple models focused on theoretical exploration, to empirically-based representations of real-world systems. Creating a model that represents a real-world system is an approach that has potential for conducting policy-relevant research (Parker *et al.* 2003). Recent studies have focused on the application of ABM to planning problems. Bolte *et al.* (2006) advocate the use of ABM in developing alternate futures of landscape change in

Oregon. They point to ABM as an integrative tool that acknowledges the coupled nature of human/natural systems and the variety of spatial and temporal scales that characterize their systems of study.

A tourism-specific example of ABM implementation comes from Berman *et al.* (2004) who use ABM as a visioning tool, showing economic outcomes in a remote community in the Canadian north, specifically the impacts of road construction, re-training and climate change. This type of policy application is evaluated by Ligmann-Zielinska and Jankowski (2007), who use ABM to generate alternate scenarios of urbanization effects in a small community in Washington State. The authors point to several advantages of an ABM-based PSS, as opposed to traditional equation-based models. Key amongst these benefits include a better representation of individual decision making through the use of heterogeneous agents, making ABM a more intuitive construct for community consultation. Additionally, the creation of multiple futures, with an emphasis on iterative experimentation, underlines the role that an ABM can play in a collaborative planning environment, particularly one marked by uncertainty and multiple viewpoints (Lempert 2002; Ligmann-Zielinska and Jankowski 2007). It is these tasks of future visioning, scenario development, community consultation and integrating multiple sources of input that provide the strongest application areas for ABM within planning.

Despite these entry points, integrating ABM within PSS presents several barriers. At the development stage, when deciding how to represent a complex real-world system as an ABM, tension frequently exists between the desire for simplicity and a level of detail required to accomplish planning goals. A model will never perfectly reflect reality, and decisions to increase both the realism and complexity of the model often inhibit analysis (Bankes *et al.* 2002). This friction between simplicity and realism is manifested in concerns of model validation (Parker *et al.* 2003) and the use of rich data sources to parameterize models (Robinson *et al.* 2007). Historically, simplicity in model creation was promoted, using theoretical landscapes and simplified assumptions to create models that developed theory rather than policy (Bonabeau 2002). When considering the use of ABM in planning, a model that represents reality is essential. Validating an ABM, or ensuring that it adequately represents the system under study, is a process that can require large data sets for statistical or pattern-oriented comparison of model output to real-world data (Grimm and Railsback 2005). As ABM moves into the planning arena, model builders must aim for a higher level of validity, yet ensure that models do not become so complex as to preclude analysis by non-modellers (Bolte *et al.* 2006). The use of realistic landscapes (especially those informed by geospatial data) and validation with robust datasets mediates these needs of realism and simplicity that are required for the use of ABM within planning.

11.3 *TourSim:* An ABM-Based Approach to Tourism PSS

TourSim is an ABM-based PSS whose purpose is to simulate the interactions between tourists and destinations. It is designed to help policy makers visualize and evaluate impacts of external processes as well as proposed planning decisions. Building the *TourSim* model followed an iterative process, aiming to create the simplest model that still provided a realistic view of tourism dynamics (Gilbert and Troitzsch 2005; Grimm and Railsback 2005). It is conceived to be an open system, so that complexity and user-specific detail can later be added.

11.3.1 TourSim Model Structure

Creating a tourism PSS such as *TourSim* requires the developer to follow several steps. Grimm and Railsback (2005) present a six stage model development cycle: formulate the purpose; assemble hypotheses; choose model structure; implement the model; analyse the model; and communicate the model results. We rely on tourism literature to hypothesize that a supply-demand relationship characterizes much of the tourism system. Tourists provide a demand according to their preferences, and destinations contain a supply of product to satisfy this demand. At model structure stage, agent behaviours are characterized by preferences for accommodation and activities. Accommodation choices include types of hotels, campgrounds and resorts. Activities include visits to national parks, historic sites and a variety of outdoor pursuits. Each tourist is heterogeneous, possessing multiple preferences, potentially selecting any location that meets at least one accommodation and activity preference. The landscape on which these agents interact consists of the supply of accommodation and activity at each destination.

Implementation comprises the choice of software development environment and data and the actual programming of the model. For *TourSim,* the implementation environment of choice is *AnyLogic* (http://www.xjtek.com). Compared to other environments, *AnyLogic* was chosen due to a user-friendly interface, visualization options and ability to develop web-based versions for further communication of model results, all factors that are important for use in planning. Tourist agents are based on data from the Canadian Travel Survey (CTS), collected from 2000 to 2004, and the International Travel Survey (ITS), collected from 2001 to 2004. These are both products of Statistics Canada. Tourist agent preferences were parameterized using a method developed by Berger and Schreinmachers (2006), where survey data are randomly assigned to agents according a sample distribution. This method results in the larger group of agents replicating the sample distribution. Using the thousands of records of unique preference data from the CTS and ITS allowed us to represent agent heterogeneity. The use of large, multi-year surveys, such as the CTS and ITS, gives us a high level of confidence that the sample is representative (Robinson *et al.* 2007). Destination (supply) data was sourced from the Nova Scotia Department of Tourism, Culture, and Heritage tourism portal (www.novascotia.com), which contains a database of accommodation and activities. From this database, 35 destinations were selected for use in *TourSim.* These destinations were plotted on a base map of Nova Scotia, drawn from the Digital Atlas of Canada (www.atlas.gov.ca). This map is used for visualization purposes, placing *TourSim* components in the appropriate geo-referenced context. In addition to the main supply-demand dynamics, *TourSim* includes a daily straight line travelling distance and length of stay preference, each informed by the CTS and ITS distributions. This selection of variables and interactions is considered the base version of *TourSim*, to which specific planning scenarios and further dynamics can be added.

Validation ensures that the model sufficiently resembles the real system and provides confidence in the results. For *TourSim,* this involves matching the modelled tourist visitation data to the total provincial visitation numbers published by the Nova Scotia Government. Whereas these numbers are suitably close, this is only a preliminary measure of model validity. A more robust method is under development that will validate the model against tourist visitation in each destination, attempting to replicate monthly variations in tourist distribution. The final step in the model development cycle is communicating model results. *TourSim* allows users to run the model themselves through a web-based user interface. This feature can be used to create an ongoing dialogue of collaborative development with the user, allowing the developer to quickly add features, and distribute new versions of *TourSim* in response to feedback on the utility and validity of the model, as well as incorporate feature and scenario requests.

Rather than focus on 'launching' a finished product for site-specific users to then either accept or reject, we have distributed the current version of *TourSim* as the first step in an ongoing model development process. This prototype is available at http://toursim.wordpress.com. *TourSim* was introduced to the Canadian tourism research community at the Travel and Tourism Research Association of Canada (TTRA-Canada) annual conference in October 2007. A web-based version of *TourSim* was then distributed to a select group of tourism planning experts to gather feedback on the usability and applicability of the model. These experts work throughout Canada in tourism planning at both the provincial and local scale. Experts were contacted with a follow-up telephone interview after viewing a presentation that introduced the *TourSim* scenario detailed in this chapter. This type of development cycle blurs the distinction between 'user' and 'developer', as feedback from users is constantly incorporated into the model. An iterative, collaborative model development precludes the production of a finished product, as indeed all versions of *TourSim* are experiments that aim to continually improve on the state of the tool. Reflections from this panel of experts are used to inform the discussion in Section 11.4.

11.3.2 TourSim Model Use

Figures 11.1 and 11.2 show the user interface of *TourSim*. In the first pane (Fig. 11.1), users are introduced to the model and can set initial system conditions. The user then runs *TourSim* through a Java-based map loaded into a web browser (Fig. 11.2). A tool bar at the top of the browser window allows control of the model execution. A date counter is visible in the lower left corner. The user views agent routes from location to location on a map of Nova Scotia, as well as a bar chart of total tourist arrivals in the selected destination areas. The simulation unfolds as tourist agents access Nova Scotia at one of seven ports of entry. Agents then randomly evaluate destinations to satisfy their accommodation and activity preferences. If the first destination does not match, then an agent evaluates another destination until a successful match on both accommodation and activity is found. Next the tourist agent checks whether the destination is within a particular straight-line distance threshold, and if so, moves to that location, causing a route to symbolically widen, as more trips take place along it. If the destination is outside of the tourist agent's distance threshold, the agent returns to the start of the decision-making loop. In the default setting, *TourSim* runs for one calendar year, which is compressed into a few minutes of actual run time. The time setting can be modified to suit the time scale of the particular scenario. When the model finishes its run, parameter settings and bar graph results are available for import to spreadsheet software for further analysis.

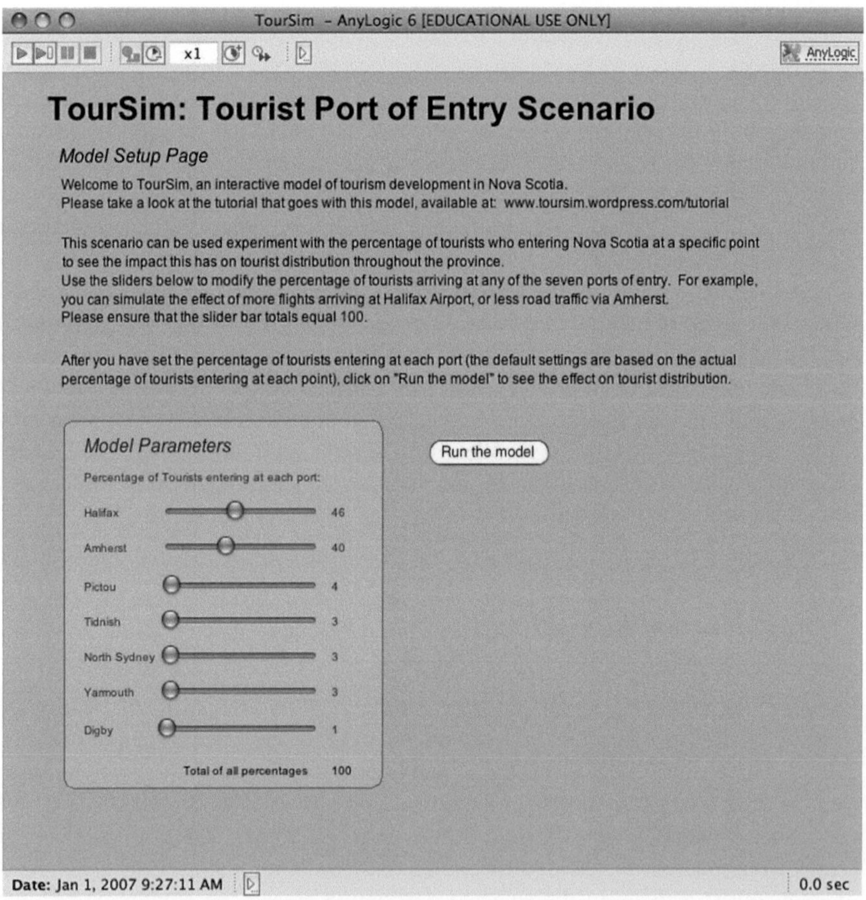

Fig. 11.1 *TourSim* model setup page showing slider bars used to manipulate model parameters

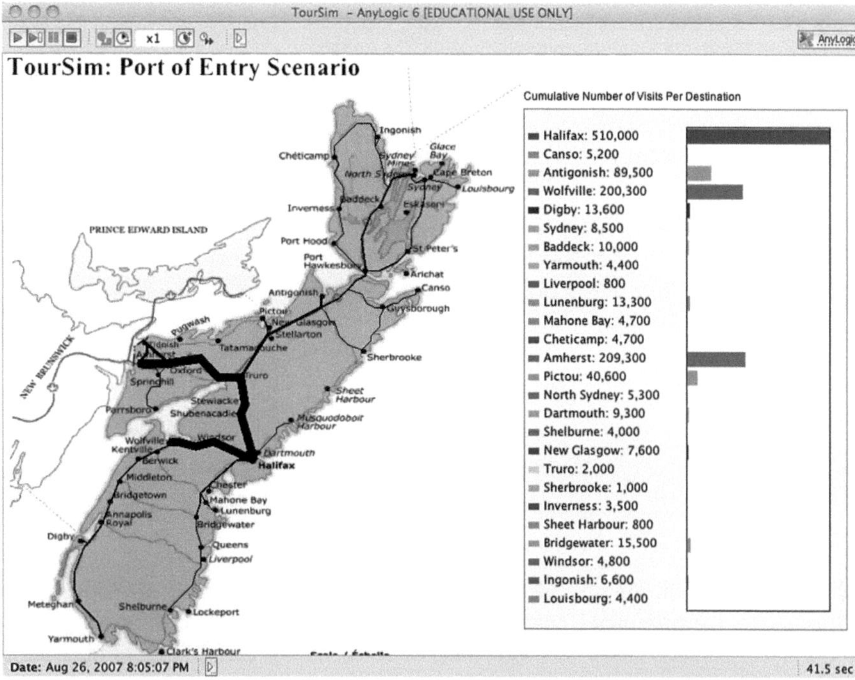

Fig. 11.2 Main *TourSim* model page, including map of Nova Scotia; the width of each route represents the number of tourist trips with bar chart showing cumulative visitation (See also Plate 23 in the Colour Plate Section)

11.3.3 Using TourSim to Experiment with Tourism Planning Scenarios

We designed *TourSim* to allow experimentation with a host of tourism-related scenarios, such as the effects of developing new tourism infrastructure on tourist distribution (e.g. establishing a local music festival), 'what if' questions concerning global tourism trends (e.g. an upturn in the value of the Canadian dollar relative to the American dollar) and changes to tourism demand as a result of geopolitical or economic events (e.g. a greater focus on travelling domestically).

We illustrate *TourSim* with a scenario that examines changes to the tourist port of entry on total tourism distribution. These entry points include four ferry (Caribou, North Sydney, Yarmouth, and Digby), two road (Amherst, Tidnish) and one airport access point (Halifax International Airport). Ferry linkages to Nova Scotia have received increased attention, as financial difficulties of ferry operators have caused service disruptions and threatened the closing of two ferry ports. Varying the percentage of the total tourist pool that enters at different ports of entry can be used to develop scenarios, such as quantifying the benefits of adding extra flights at Halifax Airport or marketing approaches to attract more driving tourists from adjacent provinces. We looked at three alternatives: (i) a 'business as usual' or base scenario,

Table 11.1 Share of tourists entering at each port, under three scenarios, and resulting change in visitation levels

	1 Base scenario	2 Increase ferry scenario		3 Decrease ferry scenario	
Port of entry	Share of tourists (%)	Share of tourists (%)	Tourist visitation change (%)	Share of tourists (%)	Tourist visitation change (%)
Amherst	40	38	26	41	7
Halifax Airport	46	44	2	47	0
Yarmouth	3	6	74	1.5	7
Digby	1	2	6	0.5	–5
North Sydney	3	3	58	3	0
Tidnish	3	3	1	3	–13
Pictou	4	4	–38	4	15

where the shares of tourists entering at each port of entry match observed averages collected by the provincial government; (ii) increased emphasis on ferry connections, introducing a larger share of tourists at Yarmouth and Digby, compared to the base scenario; and (iii) a reduced emphasis on ferry connections, with a smaller share of tourists entering through the above ports of entry. The adjustments for each scenario are made on the model setup page (Fig. 11.1). This allows users to modify the share of tourists entering at each port. Note that when these parameters are adjusted, the total of all share percentages must still equal 100 per cent. Table 11.1 lists each port of entry, the percentage share of tourists arriving under each scenario, and model results in the form of the percentage change in total tourist visitation levels compared to the visitation levels observed under the base share scenario. Common parameters throughout each scenario include tourist preferences, destination supply and number of tourists entering the model per simulated day.

Table 11.1 shows that by varying the share of tourists entering at selected ports of entry, system level effects are generated. Under scenarios 2 and 3 (the increased ferry, and decreased ferry emphasis, respectively), not only do the destinations show changes in total visitation, but often other ports of entry show changes in total visitation as well. Whereas some of the relationships between port of entry and a rise or fall in visitation seem intuitive, such as a rise in entry at Yarmouth and Digby resulting in a positive increase in total tourist visits (although at very different degrees of effect), not all destinations where port of entry was altered show a similar degree of response. This type of non-intuitive finding can be used as a step towards better understanding the effects of change at one location on an entire system. For example, even though the initial share of tourists entering at Caribou, Tidnish and North Sydney was held constant, they experienced large fluctuations in visitation as a result of changes at other entry points. These results suggest that changes to ports should be discussed at a provincial scale, instead of remaining the problem of the impacted community. *TourSim* benefits local and regional tourism planners by allowing them to see how local initiatives could fare as a response to system conditions, and in turn, what system effects they themselves contribute to.

11.3.4 TourSim Areas of Application

The port of entry scenario indicates how *TourSim* can be used to study tourism system dynamics. These results show that changes in tourist port of entry can lead to a wide variety of changes in tourist dispersion. For tourism planners, this tool can be used within a community consultation process to focus discussions and development of new ideas, or decide between competing options. *TourSim* can be expanded to address many other tourism planning concerns such as evaluating the effects of competition between destinations, economic impact of tourist visitation, effects of targeted marketing on specific tourist generating areas, and diversifying the mix of accommodation and activity choices within Nova Scotia. Additionally, our development and use of *TourSim* shows that tourism dynamics can be formalized in an ABM, creating an environment with which to explore theories of tourism development and change.

11.4 Discussion

Notwithstanding these initial findings, further development of *TourSim* into a tool for planning support requires consideration of a number of factors, including how well *TourSim* represents the tourism system, how easy both the development and user environments are to use, and how *TourSim* can be applied within real world planning situations. These factors mirror the study of constraints, or 'bottlenecks' that have challenged the widespread adoption of PSS as well as the adoption trajectory followed by GIS (Budic 1994; Nedovic-Budic 1998). Vonk *et al.* (2005) outline a number of adoption constraints, including: (i) awareness of and experience with PSS; (ii) technical issues such as software user interface and data availability; and (iii) the overall applicability of a PSS towards planning tasks. We consider each in turn as a way to evaluate and refine *TourSim*.

11.4.1 Awareness of and Experience With an ABM-Based PSS

ABM and similar approaches have found application in business logistics, manufacturing, and telecom infrastructure planning, yet this has not translated into general use of ABM. Applications of ABM-based PSS are few, either in municipal planning more generally or tourism planning more specifically. This low level of penetration within municipal planning departments contrasts with the more rapid uptake of GIS in 1990s (Budic 1994). The early commercialization of GIS software provided a professional-level product, with consistent resources for its development, and correspondingly raised awareness of GIS, further driving the development of the technology within a base of users. Comparably, there is no dominant commercially available ABM software program and there is a lack of consensus on a common approach to ABM development.

Current applications of ABM within planning practice use a variety of software programs, ranging from in-house custom development (Bolte *et al.* 2006; Deadman and Gimblett 1994) to those based on open-source platforms (Berman *et al.* 2004; Brown 2001), as well as proprietary, closed-source platforms developed as commercial products, with copyright protection to inhibit alternate uses and user customization (Ligmann-Zielinska and Jankowski 2007). We continue this latter tradition in our choice of *AnyLogic*, a commercial, copyright-protected development environment that finds most of its application within manufacturing and business logistics. Selected due to its relatively advanced graphic development interface (Fig. 11.3), *AnyLogic* does not yet possess the critical mass of users who can share experiences and push development. The lack of a common, accessible and user-friendly environment on which to build successful ABM implementations continues to retard adoption in planning practice. Until a clearly dominant ABM environment emerges that can provide both a user-friendly interface and attract a community of users, ABM development will continue to split ABM-related resources and expertise into diffuse nodes.

Awareness of and experience with ABM was highlighted by the panel of tourism experts as an important consideration in the adoption of *TourSim*. Comments highlighted the importance of who was promoting and developing *TourSim*, and the credibility (or lack thereof) that this would engender. It was noted that consultant or private enterprise driven development and 'release' of a product like *TourSim* would be met with a large amount of scepticism by tourism planning practitioners. The panel instead proposed targeted development of *TourSim* in collaboration with a specific jurisdiction, who would then act as 'champions' for the model. These comments underline the conflict between the need for both a common ABM development environment and the simultaneous requirement to develop specific models and applications in a site-specific, collaborative manner.

11.4.2 Technical Issues

Expert interviews with tourism planners reinforced the idea of *TourSim* as a user-friendly way to manipulate large datasets (in this case, the CTS and ITS), generating better value from those products. Additionally, feedback from planners has pointed to the use of such dynamic representations of tourism as an effective way to advocate specific tourism issues to decision makers. Figures 11.1 and 11.2 show how end users interact with *TourSim*. Graphic sliders and activation buttons should facilitate use by those with a general level of computer skill. We feel that, with the help of a tutorial and documentation, the *TourSim* scenarios could be used by non-technical users with some previous experience using web-based applications. This simplified user front-end is built using the *AnyLogic* development environment (Fig. 11.3). *AnyLogic* provides graphic tools to assist users with basic structure but for more complex functions, programming is required. From a developer perspective, a programming requirement is not unusual for this type of customized ABM. However, from a planners' perspective, the level of programming knowledge required to expand or alter the structure of the model, add new data sources or change visualization options is high and may slow adoption of *TourSim*. Due to these greater technical requirements, using *TourSim* within planning practice would be likely to require in-house or contracted expertise to build simulations for less technical users to then operate. This mirrors some of the early GIS analyses in planning but contrasts with current GIS, where insights from human-computer interaction

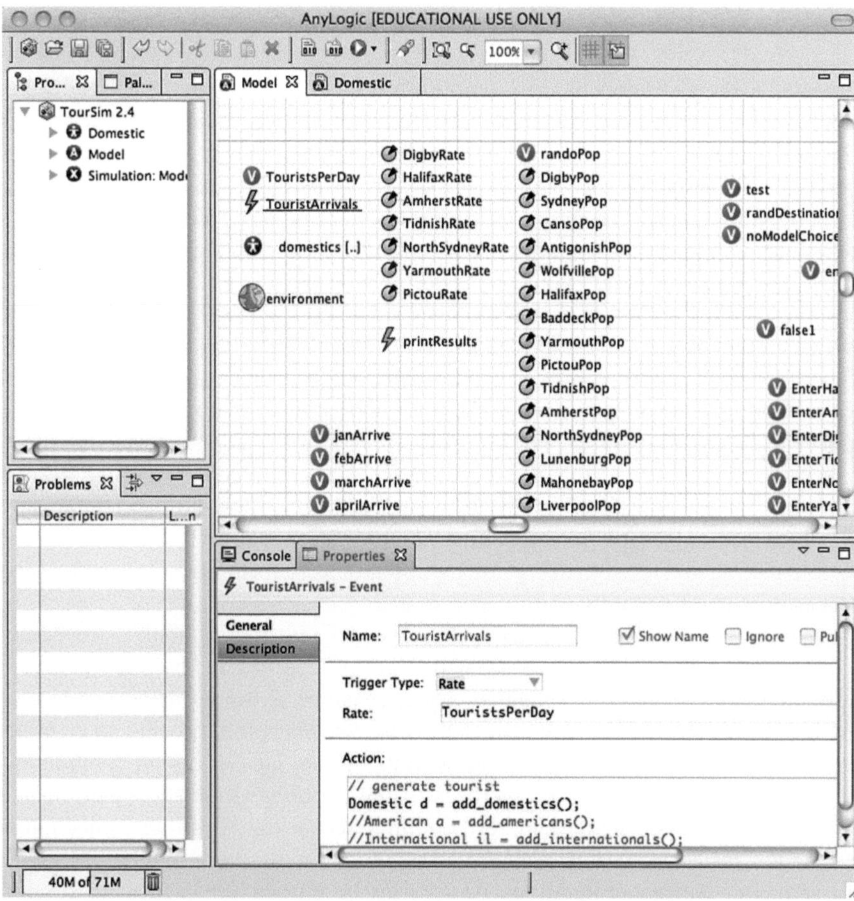

Fig. 11.3 *AnyLogic* developer environment. Based on Eclipse IDE, this interface provides for organization of model variables, Java programming and debugging

research have made GIS more broadly accessible and responsive to the needs of non-technical users (Haklay and Tobon 2003; Talen 2000). Much research on ABM involves models built, tested, used and evaluated largely by those individuals who developed them, rather than being developed and evaluated by potential end users of the product (McIntosh *et al.* 2007). Recent advancements in computer interface design has increased the accessibility of many types of command line based software. By replacing current environments that require pure computer programming code with clickable menus, reusable graphic components, and a 'what you see is what you get' style interface, ABM development will require less specialized knowledge, opening it up to a wider range of users. In the near term, it is still likely that end users will rely on a model developer, or 'chauffeur', to carry much of the technical programming weight when implement revisions to the model.

11.4.3 Overall Application to Tourism Planning Tasks

Essential to the development and promotion of any PSS is an assessment of how that technology integrates with planning tasks and how it adds value to existing methods. Advocates of the use of ABM within planning point to it as a method that integrates a variety of data, allows for evaluation of comparable options, and visualizes complex interactions and future outcomes (Bonabeau 2002; Zellner 2007). Consultations with tourism planning experts have identified several planning situations where the *TourSim* model would be useful. These include its use as a method to develop thinking and expectations on investment opportunities, such as marketing, the development of regional attractions, or expanding transportation infrastructure. Other identified improvements of the *TourSim* model include the ability to model the economic value of each tourist trip, and to observe the spatial characteristics of this expenditure. In order to better tie *TourSim* to tourism planner needs, the panel of experts recommended that the model move away from showing simple tourist numbers, and towards showing dollars spent, particularly how it differs depending on where the tourist trip originated. This addition was seen to be beneficial as an aid to determining where advertising dollars might be best spent in order to attract the highest yield type of tourist.

Tourism planning in Nova Scotia has evolved into a consultative procedure, with government agencies working with local groups and industry to guide the development of tourism in a manner that is sensitive to the needs and desires of the host communities. To support this process, *TourSim* could provide an environment to capture the ideas of a tourism working group. Ideally, a joint community-industry-government consultation process using *TourSim* would allow participants to identify areas of concern or interest, outlining scenarios to be programmed into a simulation, tested and evaluated by the group. Accordingly, these simulations would run through a number of iterations. As *TourSim* becomes a tool to focus discussion and ideas, it can allow group members to test and visualize various options, particularly risky ones, without investing actual resources.

11.5 Conclusions and Future Directions

Tourism planning is a process marked by complexity, multiple stakeholders and outcome uncertainty. Compared to urban or municipal planning which aims to provide services to taxpayers, tourism planning has the additional difficulty of developing a product that must compete effectively within a global marketplace. In a location such as Nova Scotia, where tourism represents a vital part of the economy, the desire to develop new attractions and increase tourism benefits must be tempered by the risk inherent in these plans. Tools that can visualize and evaluate scenarios serve as a starting point for community consultation on developing and evaluating a range of tourism policies. This is a strength of *TourSim*, as it can be used to experiment with possible outcomes.

In its current state, *TourSim* simplifies tourism processes in Nova Scotia. Various data and interactions have been excluded and, while this makes *TourSim* easier to operate and analyse, results may miss a sense of validity. To increase the validity of *TourSim*, incorporating aspects of destination popularity, accommodation capacity, and event and seasonal-driven inputs of tourists, such as for summer festivals could help. Adding these dynamics will enable a wider range of scenarios and analysis options, providing a more compelling tool for tourism planning. As *TourSim* is further introduced into tourism planning in Nova Scotia, we will be soliciting ongoing feedback from numerous planners about the possible integration of *TourSim* into their site-specific tasks. Additional steps include a user needs analysis of planners operating at various scales, to provide an indication of where and how *TourSim* could be modified to better provide value to tourism planners. This type of collaborative PSS development will continue to help identify bottlenecks to adoption, build awareness of ABM-based PSS, and provide for a test bed for future implementations of *TourSim*.

As an emerging technology, the use of ABM as a PSS in tourism planning holds potential, but is accompanied by significant hurdles. Adoption constraints include a lack of awareness and a lack of experience using ABM, technical issues, and the fit of ABM with planning tasks. *TourSim* is affected by all of these restrictions, particularly when compared to more conventional GIS-based PSS. The development of *TourSim* is one step in translating the benefits of an ABM approach into applied planning practice, building on the notable work of others (Berman *et al.* 2004; Heppenstall *et al.* 2006; Ligmann-Zielinska and Jankowski 2007) to promote this technology. Clearly there is still much work to be done, particularly in raising awareness of the technology, and mitigating a host of constraints to adoption.

The use of empirical datasets to inform model parameters, and the integration of ABM functionality into desktop GIS are important areas of research that will shape the further development of ABM-based PSS. Recent advances in techniques to translate large datasets into model parameters (Berger and Schreinemachers 2006; Robinson *et al.* 2007) represents a significant step towards improving the quality and realism of agents represented in an ABM. This movement of ABM towards more realistic models increases the potential for ABM application as a PSS. The integration of ABM functionality within commercial desktop GIS is another area of future development that could also improve adoption and awareness aspects of ABM-based planning tools (Brown *et al.* 2005). By coupling ABM with an established base of users and computer skills, ABM-based PSS can be implemented as an additional tool within the GIS tool box. Integrating ABM with GIS may provide the tools for creating dynamic models but it will not necessarily increase awareness of the ABM's potential. Continued academic research and applied projects are needed to evaluate the use and impacts of ABM-based PSS in specific planning situations, building a wider base of knowledge.

References

Alessa, L.N., Laituri, M. and Barton, M. (2006) An "all hands" call to the social science community: establishing a community framework for complexity modeling using agent based models and cyberinfrastructure, *Journal of Artificial Societies and Social Simulation*, 9(4): 6. http://jasss.soc.surrey.ac.uk/9/4/6.html.

An, L., Linderman, M., Qi, J., Shortridge, A. and Liu, J. (2005) Exploring complexity in a human-environment system: an agent-based spatial model for multidisciplinary and multiscale integration, *Annals of the Association of American Geographers*, 95(1): 54–79.

Bankes, S., Lempert, R. and Popper, S. (2002) Making computational social science effective: epistemology, methodology and technology, *Social Science Computer Review*, 20(4): 377–388.

Batty, M. (2004) Dissecting the streams of planning history: technology versus policy through models, *Environment and Planning B*, 31: 326–330.

Benenson, I. and Torrens, P.M. (2004) Geosimulation: object-based modeling of urban phenomena. *Computers, Environment and Urban Systems*, 28(1): 1–8.

Berger, T. and Schreinemachers, P. (2006) Creating agents and landscapes for multiagent systems from random samples. *Ecology and Society*, 11(2): 19. http://www.ecologyandsociety.org/vol11/iss2/art19/.

Berman, M., Nicholson, C., Kofinas, G., Tetlichi, J. and Martin, S. (2004) Adaptation and sustainability in a small Arctic community: results of an agent-based simulation model, *Arctic*, 57(4): 401–414.

Bolte, J.P., Hulse, D.W., Gregory, S.V. and Smith, C. (2006) Modeling biocomplexity – actors, landscapes and alternative futures, *Environmental Modelling and Software*, 22: 570–579.

Bonabeau, E. (2002) Agent-based modeling: methods and techniques for simulating human systems, *Proceedings of the National Academy of Sciences USA*, 99(3): 7280–7287.

Bradley, W.J. and Schaefer, K.C. (1998) *The Uses and Misuses of Data and Models: The Mathematizations of the Human Sciences*, SAGE Publications, Thousand Oaks, California.

Brown, D.G. (2001) Project SLUCE, http://www.cscs.umich.edu/research/projects/sluce/.

Brown, D.G., Riolo, R., Robinson, D.T., North, M. and Rand, W. (2005) Spatial process and data models: toward integration of agent-based models and GIS, *Journal of Geographical Systems,* 7: 25–45.

Budic, Z.D. (1994) Effectiveness of geographic information systems in local planning, *Journal of the American Planning Association*, 60(2): 244–263.

Deadman, P. and Gimblett, H.R. (1994) The role of goal-oriented autonomous agents in modeling people-environment interactions in forest recreation, *Mathematical and Computer Modelling*, 20(8): 121–133.

Deadman, P., Robinson, D., Moran, E. and Brondizio, E. (2004) Colonist household decisionmaking and land-use change in the Amazon rainforest: an agent-based simulation, *Environment and Planning B*, 31: 693–709.

Gilbert, N. and Troitzsch, K.G. (2005) *Simulation for the Social Scientist* (Second edition), Open University Press, Maidenhead.

Grimm, V. and Railsback, S.F. (2005) *Individual-based Modeling and Ecology,* Princeton University Press, Princeton, New Jersey.
Haklay, M. and Tobon, C. (2003) Usability evaluation and PPGIS: towards a user-centered design approach, *International Journal of Geographical Information Science*, 17(6): 577–592.
Hall, C.M. (2000) *Tourism Planning: Policies, Processes and Relationships,* Prentice Hall, Harlow.
Heppenstall, A., Evans, A. and Birkin, M. (2006) Using hybrid agent-based systems to model spatially-influenced retail markets, *Journal of Artificial Societies and Social Simulation*, 9(3): 2. http://jasss.soc.surrey.ac.uk/9/3/2.html.
Holling, C.S. (1978) *Adaptive Environmental Assessment and Management*, John Wiley and Sons, New York.
Jamal, T., Borges, M. and Figueiredo, R. (2004) Systems-based modeling for participatory tourism planning and destination management, *Tourism Analysis*, 9(1–2): 77–89.
Lempert, R. (2002) Agent-based modeling as organizational and public policy simulators, *Proceedings of the National Academy of Sciences USA*, 99(3): 7195–7196.
Ligmann-Zielinska, A. and Jankowski, P. (2007) Agent-based models as laboratories for spatially explicit planning policies, *Environment and Planning B*, 34: 316–335.
Manson, S.M. (2006) Bounded rationality in agent-based models: experiments with evolutionary programs, *International Journal of Geographical Information Science*, 20(9): 991–1012.
McIntosh, B., Seaton, R.A F. and Jeffrey, P. (2007) Tools to think with? Towards understanding the use of computer-based support tools in policy relevant research, *Environmental Modelling and Software*, 22: 640–648.
Nedovic-Budic, Z. (1998) The impact of GIS technology, *Environment and Planning B*, 25(5): 681–692.
Parker, D.C., Manson, S.M., Janssen, M.A., Hoffmann, M.J. and Deadman, P. (2003) Multi-agent systems for the simulation of land-use and land-cover change: a review, *Annals of the Association of American Geographers*, 93(2): 314–337.
Robinson, D.T., Brown, D.G., Parker, D.C., Schreinemachers, P., Janssen, M.A., Huigen, M. *et al.* (2007) Comparison of empirical methods for building agent-based models in land use science, *Journal of Land Use Science,* 2(1): 31–55.
Talen, E. (2000) Bottom-up GIS: a new tool for individual and group expression in participatory planning, *Journal of the American Planning Association*, 66(3): 279–285.
Vonk, G., Geertman, S. and Schot, P. (2005) Bottlenecks blocking widespread usage of planning support systems, *Environment and Planning A*, 37: 909–924.
Walker, P.A., Greiner, R., McDonald, D. and Lyne, V. (1999) The tourism futures simulator: a systems thinking approach, *Environmental Modelling and Software*, 14: 59–67.
Wooldridge, M. and Jennings, N.R. (1995) Intelligent agents: theory and practice, *The Knowledge Engineering Review*, 10(2): 115–152.
Yeoman, I., Galt, M. and McMahon-Beattie, U. (2005) A case study of how VisitScotland prepared for war, *Journal of Travel Research*, 44(1): 6–20.
Zellner, M.L. (2007) Generating policies for sustainable water use in complex scenarios: an integrated land-use and water-use model of Monroe County, Michigan, *Environment and Planning B*, 34: 664–686.

Additional Reading

Axelrod, R. and Tesfatsion, L. (2006) On-line guide for newcomers to agent-based modeling in the social science. http://www.econ.iastate.edu/tesfatsi/abmread.htm.
Bankes, S. (2002). Agent-based modeling: a revolution? *Proceedings of the National Academy of Sciences USA*, 99(3): 7199–7200.

Epstein, J.M. and Axtell, R. (1996) *Growing Artificial Societies: Social Science from the Bottom Up,* The Brookings Institution, Washington, D.C.
Gimblett, H.R. (2002) *Integrating Geographic Information Systems and Agent-Based Modeling Techniques for Simulating Social and Ecological Processes,* Oxford University Press, New York.
Milne, S. and Ateljevic, I. (2001) Tourism, economic development and the global-local nexus: theory embracing complexity, *Tourism Geographies*, 3(4): 369–393.
Reed, M.G. (1999). Collaborative tourism planning as adaptive experiments in emergent tourism settings, *Journal of Sustainable Tourism*, 7(3–4): 331–356.

Chapter 12
Google Maps Mashups for Local Public Health Service Planning

Maurizio Gibin, Pablo Mateos, Jakob Petersen and Phil Atkinson

12.1 Introduction

One of the key challenges in public health service planning is to reduce health inequalities at the local level. This often entails understanding the detailed profile of a local area's population and ensuring equal access to health services. Accessibility to health services, personal behaviour and lifestyles, community influences, living and working conditions, educational attainment and health literacy can all impact upon an individual's health, and their aggregated effect is clearly manifested at neighbourhood level.

Monitoring such population's characteristics and health outcomes, and targeting health initiatives on the groups with most need has proved increasingly difficult for primary care health authorities. This challenge is especially critical in London's increasingly multicultural society, characterised by a highly transient population with ties to all over the world. Traditional planning support tools and data sources have failed to keep up with these challenges due to the rapid geographic and demographic changes occurring in London's inner boroughs at small area level since the last census of population in 2001.

This chapter describes a successful geographic visualisation tool for supporting public health service planning developed through a Knowledge Transfer Partnership (KTP) between University College London (UCL) and Southwark and Camden Primary Care Trusts in London. This tool is based on a simple implementation of *Google Maps* application programming interface (API) as a framework for geographical visualisation of the population characteristics at small area level (geographical units comprised of 285 people on average). Innovative population datasets have been developed by UCL through several collaborative projects, based on new and frequently updated information sources available at the individual level such

M. Gibin (✉)
Lecturer in Geographic Information Science
Birkbeck College
School of Geography, Room 706, University of London, Malet Street
London WC1E 7HX, UK
E-mail: m.gibin@bbk.ac.uk

S. Geertman, J. Stillwell (eds.), *Planning Support Systems Best Practice and New Methods*,
DOI 10.1007/978-1-4020-8952-7_12,

as patient registers, hospital admissions, births and deaths registrations, electoral roll registrations and neighbourhood geodemographic classifications. One of these derived datasets, termed *Onomap* and which is based on the origin of people's names, estimates population counts by very fine ethnic groups at small area level, providing a rich multicultural atlas of London's neighbourhoods.

This chapter is structured in three sections, which address the content developed to meet the challenges described above, as well as technical details of the visualisation system presented here. Section 12.2 summarises the characteristics of the population datasets developed by UCL that form the core content of the visualisation tool presented here: the 'Multicultural Atlas of London', geodemographic profiling and hospital admissions. Section 12.3 describes the technical aspects of building mapping mashups and the specific solution developed for this case, the *London Profiler*. Finally, Section 12.4 offers some conclusions and proposes future avenues for research and development in mapping mashups.

12.2 Monitoring Population to Improve Primary Care Service Provision

Given the challenges in public health service planning presented in the previous section, it is clear that there is a strong need for planning support tools that allow public health planners a more effective way to understand the changing nature of the population in rapidly changing cities like London. Two general aspects are required

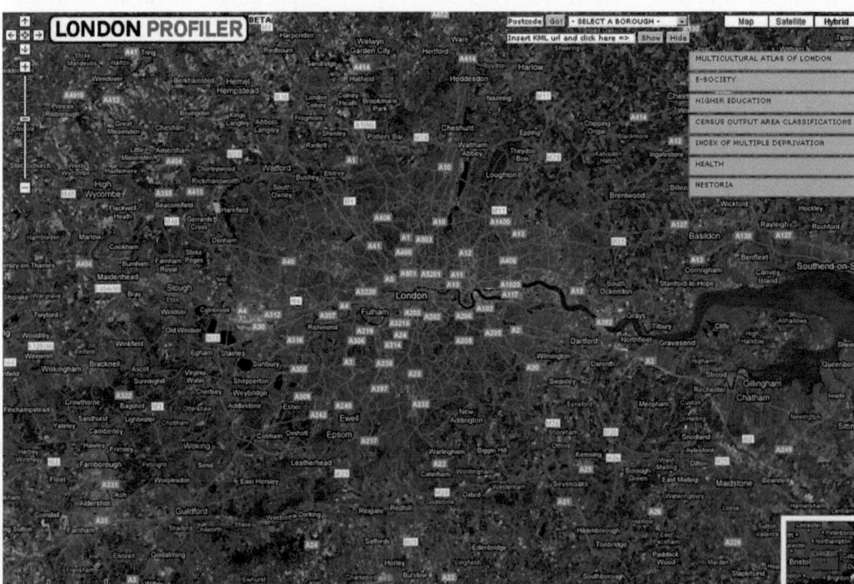

Fig. 12.1 *London Profiler*: full London view, hybrid layer (See also Plate 24 in the Colour Plate Section)

to put together such planning support tools; content and technology. This section describes the content developed for the geographic visualisation tool known as the *London Profiler* (www.londonprofiler.org) (Fig. 12.1).

This tool is comprised of a set of innovative datasets about London's population that can be frequently updated and are available at a geographically disaggregated level as to monitor changes at the neighbourhood scale. There are seven different types of population datasets included in the *London Profiler* web tool. Three of them will be introduced in detail through the next three subsections, while the four others will be summarised in Section.12.2.4.

12.2.1 Onomap: A Classification of Ethnicity Based on Names and the Multicultural Atlas of London

UCL Department of Geography has developed a new methodology to classify populations and neighbourhoods into groups of common cultural ethnic and linguistic origin using surnames and forenames (Mateos *et al.* 2007). This methodology, which is now known as *Onomap*, is a response to a growing set of pressures on national and local government to understand and identify the detailed composition of ethnic groups in today's increasingly multicultural societies. Ethnicity classifications are often hotly contested, but still greater problems arise from the quality and availability of official classifications, with knock-on consequences for our ability to meaningfully subdivide populations (Aspinall 2002). Name analysis and classification has been proposed as one efficient method of achieving such sub-divisions in the absence of ethnicity data (Nanchahal 2001), and may be especially pertinent to public health and demographic applications. However, previous approaches to name analysis have been designed to identify one or a small number of ethnic minorities, and not complete populations (Mateos 2007).

UCL's 'Multicultural Atlas of London' intends to celebrate the diversity of cultures represented in many of the neighbourhoods of contemporary London. It represents the results of an innovative initiative through which the UK Electoral Register has been classified using *Onomap* names classification. *Onomap*'s methodology is based on an alternative ontology of ethnicity that combines some of its multidimensional facets: language, religion, geographical region and culture, as encapsulated in the origin of people's forenames and surnames, used as a proxy for their probable ethnicity. It is a methodology developed using data collected at very fine temporal and spatial scales, and made available, subject to safeguards, at the level of the individual. Such individuals are classified into 185 independently assigned categories of cultural, ethnic and linguistic groups, based on the probable origins of their names (Mateos *et al.* 2007).

The UK Electoral Register is a public register that contains names and addresses of all adults that are entitled to vote, comprising over five million people in London in 2001. The version used here is from 2001, prior to a change in legislation that allowed the option to 'opt out' from the publicly available version. Using *Onomap* classification applied to the Electoral Register, the geographical distribution of 18 of the most symbolic ethnic groups in London have been mapped at output area level: Bangladeshi, Chinese, English, Greek, Indian, Irish, Italian, Jewish, Nigerian and Ghanaian, Other Muslim, Pakistani, Polish, Portuguese, Russian, Sikh, Sri Lankan, Turkish, and Vietnamese. Many of these cultural, ethnic and linguistic groups are not collected by the census of population or official surveys, hence these datasets represent an innovative approach to measuring cultural diversity in London. Furthermore, making them easily accessible to public health analysts through the visualisation techniques presented here has been a major breakthrough for local population monitoring initiatives.

12.2.2 Geodemographic Profiling of Local Population

Geodemographics is defined as "*the study of population types and their dynamics as they vary by geographical area*" (Birkin and Clarke 1998, p. 88). One of the few premises of geodemographics is the observation that we, on average, tend to have a lot in common with those living around us when we look at age, family structure, income, occupation, interests and patterns of consumption. The commercial sector has exploited this question for a number of years used neighbourhood or geodemographic classifications for direct mailing and market analysis. The characteristics of different neighbourhood types discerned from geodemographic classifications are sometimes summed up in the concept of 'lifestyle'. The public sector has recently started to embrace geodemographics and lifestyle management, since it is deemed to bring a better understanding of the local neighbourhoods in 'tailoring' services and highlighting areas of particular interest or concern (Longley 2005). The field pioneering this adoption has been public health, where geodemographics has supported actions to reduce inequality of health outcomes and access to services, such as, for example, targeting patients with information about healthier choices and the location of local facilities.

The Office for National Statistics (ONS) has developed geodemographic classifications of neighbourhoods from both the 1991 and 2001 UK censuses, the latter termed the Output Area Classification or OAC (Vickers and Rees 2007). OAC comprises a comprehensive and accessible neighbourhood classification that allows users to perform geodemographic analyses free of charge through a robust and transparent methodology. A single national system like the OAC enables a level of comparison across the country, so that, for example, large surveys can be projected to most areas. Yet in our experience, for the planning health services at a local level in the inner-city boroughs of London, OAC appears overly vague, with swathes of central London ascribed to a single dominant 'Multicultural' category. This suggests that London on the whole is very different to the rest of the country and that a national system comes with a substantial loss of local variability. In order to tie in OAC with our work with public service data, we decided to modify the OAC to produce the London OAC or LOAC (Petersen *et al.* 2007). This is based on the

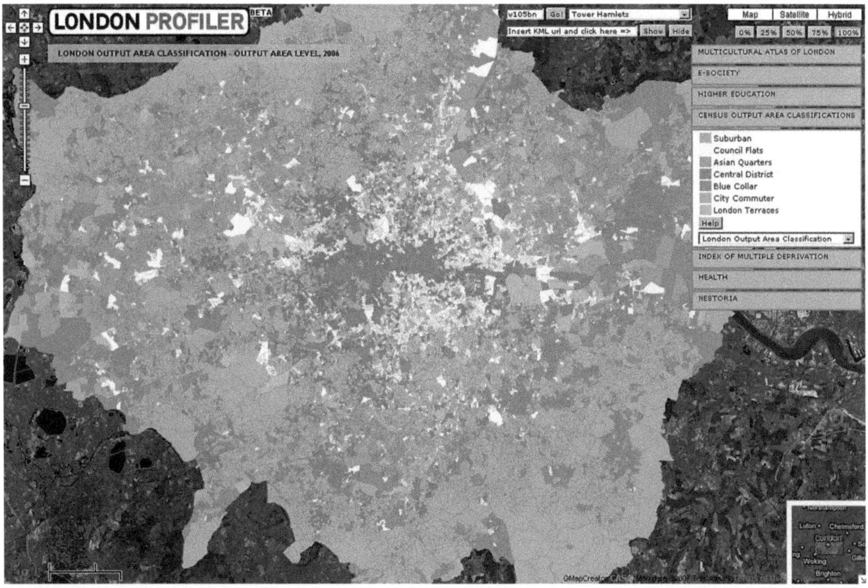

Fig. 12.2 *London Profiler*: full London view, London Output Area Classification layer, 100 per cent transparency, satellite layer (See also Plate 25 in the Colour Plate Section)

same methodology as OAC with regards to input variables and clustering algorithm, but focuses just on London for data and standardisation. LOAC is offered as an alternative classification of neighbourhoods, and analyses suggests that it presents a more 'natural' way of dividing up the city for a number of social, economic and demographic variables (Petersen *et al.* 2007).

The map of LOAC (Fig. 12.2) shows London with a core of regenerated and privately owned flats ('Central District') and rings with post-war social housing in apartment blocks ('Council Flats') interspersed with older, now part publicly and part privately owned, terraced housing ('London Terraces'). Further from the centre, we find traditionally are more affluent and contained a higher proportion of privately owned housing ('City Commuter') and areas characterised by British Asian communities ('Asian Quarters'). The outer ring consists of satellite settlements ('Suburban') and areas traditionally inhabited by manual labourers in the industry areas to the east of the city ('Blue Collar'). This map of the LOAC, as well as a separate one of the OAC, have been included in the final geographical visualisation tool presented in this chapter.

12.2.3 Hospital Care Needs and Public Health Campaigns

In the UK, long-term health conditions account for 80 per cent of General Practitioner (GP) consultations, 60 per cent of hospital bed days, and 67 per cent of medical emergencies. The 10 per cent of patients with the most severe health care needs account for 55 per cent of inpatient days and the top 5 per cent of patients account for 40 per cent. In all, 17 million people in the UK are estimated to suffer from a long-term disease like arthritis, asthma, back pain, chronic airways obstruction, diabetes or diseases of the nervous system (Department of Health 2004). The problems of long-term diseases are growing alongside the ageing of the population, but the effects of lifestyle changes with richer food and less exercise is already adding to the burden of some long-term diseases, such as heart diseases and diabetes (Type 2). Reaching those population groups and individuals with the highest needs, and providing them the most appropriate level of health care, is a substantial task for health authorities. Mapping long-term needs has a significant potential not least to planners, but also for the users. For example, they can look forward to receiving treatment closer to home in community clinics, GP surgeries or indeed at home (Department of Health 2006).

Hospital admission data in the UK provide a very detailed picture of local health care needs and in this work we would like to illustrate how admission maps can be used to support public health campaigns. One of the first health themes in the *London Profiler* is diabetes risk presented as an admission ratio; i.e. the number of admissions for an area divided by the expected number were the age and sex specific rates to be the same as for the whole of London. An admission ratio of 100 is the London average; 50 is half and 200 the double. Profiling the diabetes admissions with a geodemographic system, for example the LOAC system, highlights neighbourhood types where admissions are more or less common (Fig. 12.3). Here

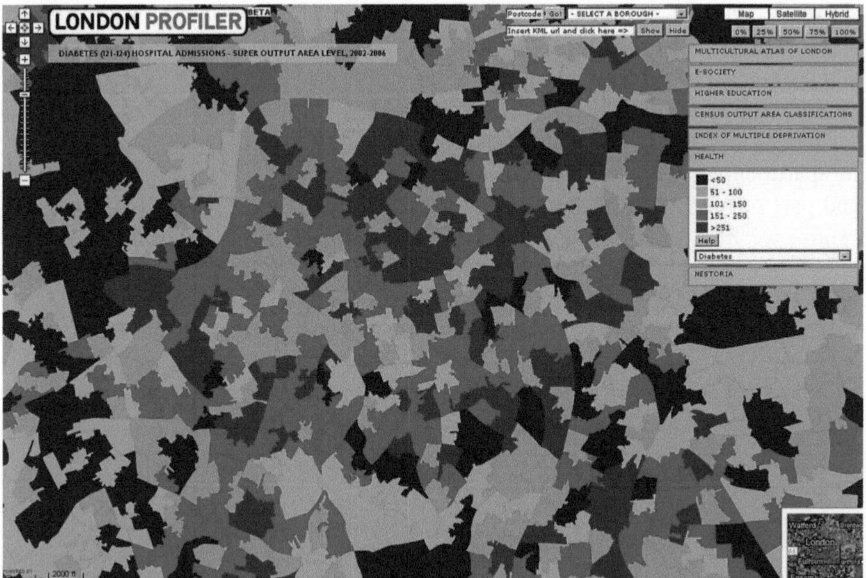

Fig. 12.3 *London Profiler*: health tab, HALT dataset, diabetes hospital admission ratio, 100 per cent transparency (See also Plate 26 in the Colour Plate Section)

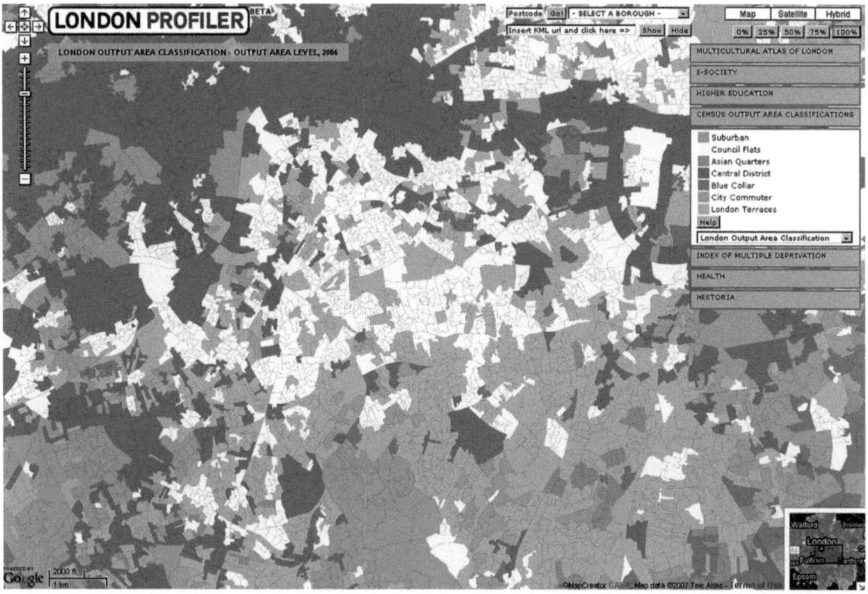

Fig. 12.4 *London Profiler*: London Output Area Classification layer, 100 per cent transparency (See also Plate 27 in the Colour Plate Section)

we see 'Council Flats' in particular, but also 'Blue Collar' and 'London Terraces' as having a higher risk than average (Fig. 12.4). Used in this way, geodemographics can be a reasonable way to perform exploratory data analysis of disease patterns and potentially geodemographics can also be useful providing the broader social context for diseases with a 'lifestyle' component.

12.2.4 Other Population Datasets

In addition to the three UCL research projects so far presented in this section, four other datasets are included in the *London Profiler* tool:

- *E-society*: a geodemographic classification developed as part of a project at UCL that presents a detailed classification of neighbourhoods based on information about levels of awareness of information and communications technologies (ICTs), usage patterns and attitudes to their effects upon quality of life. The classification provides a valuable and accessible means of studying the 'e-society' and people's engagement with new information and communications technologies (Longley *et al.* 2006).
- *Higher education*: The 'Participation Of LOcal AReas' (POLAR) is a classification of neighbourhoods according to rates of higher education young

participation (18–19 years old). It indicates higher education participation rates at ward level shown as quintile bands. The classification was developed by the Higher Education Funding Council for England (HEFCE) and is used in the allocation of widening participation funding to HEFCE-funded institutions. In addition to the main classification which relates to absolute participation rates, HEFCE have provided a series of supplementary data at ward level which relate to participants in higher education, all of which are available on the *London Profiler* website.

- *Index of Multiple Deprivation (IMD)*: The IMD was created in 2004 by the Department for Communities and the Local Government (DCLG) as a method of identifyingdeprived areas across UK. The IMD is presented as a series of rank and scores that classify every area according to its deprivation, measured on seven different domains (Fig. 12.5).
- *Residential property data*: These are coming from a commercial property search engine called Nestoria (www.nestoria.com). This website allows the user to filter through a property database and visualise the results in *Google Maps*. Data can also be downloaded as a Keyhole Markup Language (KML) file and opened in *Google Earth*. The Nestoria tab on *London Profiler* allows replicating the process of filtering the property database and loading the KML file without the need to access the Nestoria website. Data are directly streamed from the website as a typical Web 2.0 point mash up dataset.

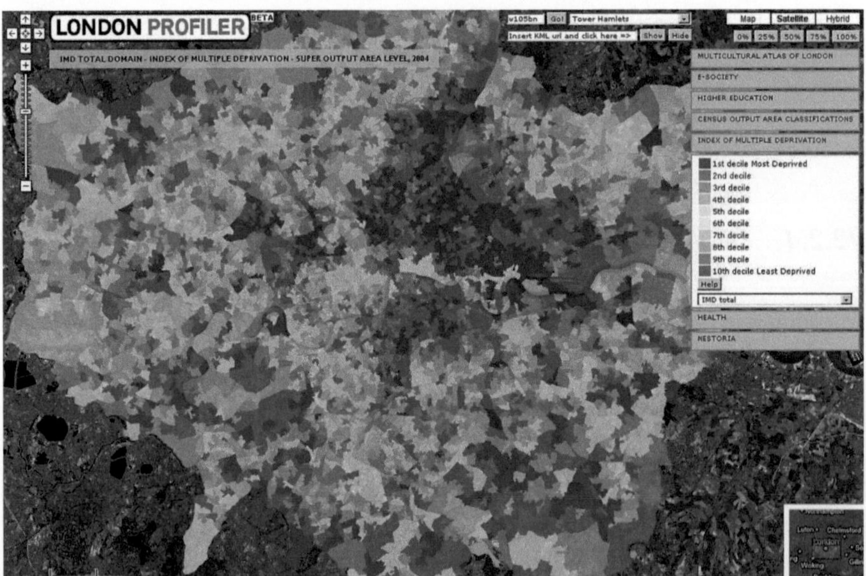

Fig. 12.5 *London Profiler*: full London view, Index of Multiple Deprivation tab, IMD total layer, 75 per cent transparency, satellite layer (See also Plate 28 in the Colour Plate Section)

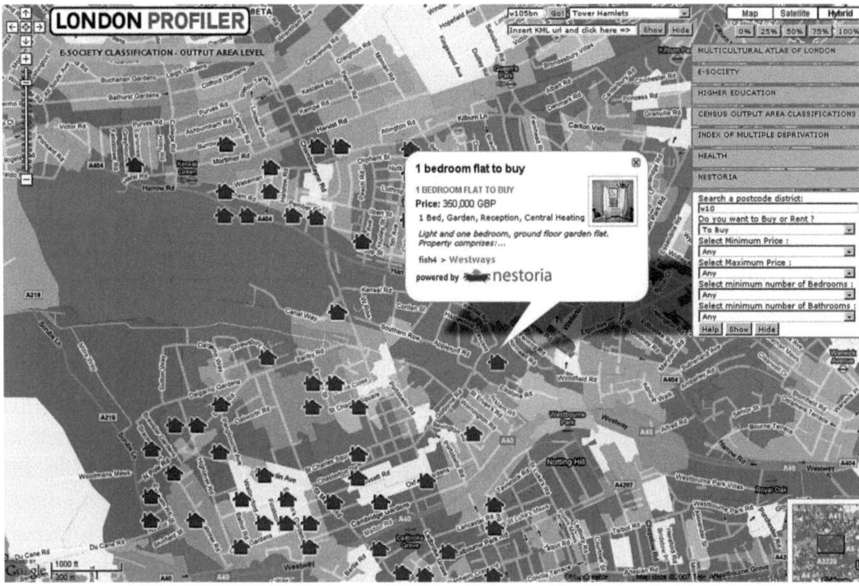

Fig. 12.6 *London Profiler*: postcode level, e-society classification tab and layer with a KML file overlay from Nestoria data, 100 per cent transparency, hybrid layer (See also Plate 29 in the Colour Plate Section)

12.3 Geographic Visualisation Through *Google Maps*: Mapping Mashups and The *London Profiler*

12.3.1 Mapping Mashups

The term geographic visualisation or geovisualisation (GVis) refers to spatial data and *"can be applied to all the stages of problem-solving in geographical analysis, from development of initial hypotheses, through knowledge discovery, analysis, presentation and evaluation*" (Buckley *et al.* 2000). There is increasing realisation of the potential for 'geography' to provide the primary basis for innovative visualisation and knowledge exploration (Dodge *et al.* 2006).

GVis applications on the internet are typically offered through a type of technology developed in the 1990s commonly known as WebGIS. A WebGIS combines the features of a common geographical information system (GIS) in an internet environment using an ordinary web page as the front end. Different examples of WebGIS available on the internet cover a wide range of GIS applications, from environmental to social sciences. These websites provide users with several tools to visualise, query and sometimes edit the datasets. Nonetheless, the diffusion of WebGIS has been very limited, because of technical as well as human factors, both at the institutional level supplying the service and at the user level. Some of these limitations will be mentioned here.

First of all, the design of a successful WebGIS website requires developers with very specialised programming and database skills, as well as a great deal of work paid to the design of the end user interface. The end user must be able to access the different tools available in the WebGIS in an accessible and easy-to-use way. Second, WebGIS solutions are very expensive to develop, both in terms of software platforms (although there are also open source solutions available) and developing their content, since they usually require very expensive proprietary data, sorting out cumbersome copyright issues with geographic data for every single area of concern, and developing customised tools that are only meant for a specific vertical market.

Because of this last problem of the high cost of developing content, most of WebGIS websites offer little contextual geographic information (GI). Such contextual GI typically allows the user to intuitively relate the specialised data provided by the WebGIS (e.g. road accidents, or weather forecasts) with the geographical context in which these actually takes place. For example, in the application presented in this chapter, the specialised data to be visualised are population datasets by administrative areas in London and the GI context would be the road and public transport network, the names of the neighbourhoods and aerial photography.

The diffusion of *Google Maps* and *Google Earth* since 2005 has increased the use of GI among internet users and fuelled new ways of deploying GI in an effective and easy way that is rapidly and intuitively understood by any sort of public. Mapping mashup is a term recently introduced to refer "*to hybrid web applications that combine data or software from two or more sources*" (Monmonier 2007, p. 373). The most widely used mapping mashup technology is based on *Google Maps*, a free web service that has been deemed to "*dramatically raised users' expectations [with] its fluid movements, intuitive user experience and competent cartography*" (Fairhurst 2005, p. 57). Although *Google Maps* is not a complete GIS tool, the availability of a free *Google Maps* API stimulates people with basic programming skills to build their own applications using *Google Maps* as a visualisation interface 'mashed-up' with their own geographic data. In *Google Maps* mashups, the user's specialised data are given instant geographical context through detailed satellite and aerial imagery, place names, administrative boundaries, road and street networks and point of interest data (such as underground or rail stations), seamlessly available throughout the world.

The simple cartographic design of the *Google Maps* base layers, which are integrated into all mashups based on *Google Maps* API, make them the ideal background for thematic map overlays. *Google Maps* enable users to visualise thematic map layers made up from boundaries of unfamiliar size and location, and provided by third parties, within the context of local and scalable geographical features. Compounding essential geographic information on the 'where' with the 'what' needs careful visual design (Tufte, 1990). The colour scheme choice for the thematic overlay must avoid tones (Harrower and Brewer 2003) that can be confounded with the underlying *Google Maps*. A solution employed in the example presented in this chapter uses a combination of thematic overlay transparency and the Google 'hybrid map' layer, as described in the following section (Fig. 12.7).

Fig. 12.7 *London Profiler*: postcode level, Multicultural Atlas of London tab, Bangladeshi population layer, 50 per cent transparency, hybrid layer

12.3.2 The London Profiler Website

The Centre for Advanced Spatial Analysis (CASA) at UCL released in 2006 a freeware application to simplify thematic mapping in *Google Maps*, termed *GMap Creator* developed by Richard Milton as part of the GeoVUE Project (a node of the ESRC National Centre for e-Social Science). This simple application takes native GIS files (typically in *shape file* format) and creates a *Google Maps* mashup in a very straightforward fashion. Users have to provide a projected *shape file* containing the data to be visualised, choosing a variable from the attribute data and a colour scheme for the final thematic map to be created. The thematic layer is transformed in a series of image tiles of equal size (256 × 256 pixels), the number of tiles being determined by the *Google Maps* zoom level selected. Finally, an *html* file is created that places the numerous tiles on top of the *Google Maps* interface using the previously mentioned API. *GMap Creator* output is a webpage that contains the mapping application in a standard format and layout.

The *London Profiler* project (www.londonprofiler.org) has harnessed *GMap Creator* to visualise population data alongside industry standard web interfaces to develop comprehensive profiles of the public that can be used by the public to understand local geographies of public service delivery. Its principal deliverable is a web tool to visualise population administrative data at small area level (census output areas), to be made accessible to a mass audience of specialist users as well as the general public via an easy to use website. The *London Profiler* project has been conceived as an initial prototype of larger future planning support systems based on mapping mashups. All the thematic maps are created through *GMap Creator* and then managed through *Google Maps* API embedded in the front-end user webpage.

London Profiler enables users to build up a picture of the geodemographics of Greater London from data on population attributes such as the distribution of cultural and ethnic groups, the level of deprivation, the extent of e-literacy, level of higher education, and health-related problems together with the free geodemographic classifications available in the UK (OAC and LOAC). Some of the datasets displayed are the result of different research projects carried out at UCL, such as the Multicultural Atlas of London, the OAC standardised to London (LOAC), Hospital Admissions for Long-Term diseases (HALT), and the e-society classification. Other datasets, such as the IMD, POLAR and residential property data, have been taken from publicly available sources. These seven domains in *London Profiler* have been fully described in Section 12.2, and duplication is avoided here.

London Profiler users can select to map one attribute within those seven domains (each of which offer different attributes to map), and focus on an area within London searching by postcode or one of the 33 London Boroughs. They can also decide to add contextual GI to the theme map being displayed, such as an area's physical environment—through the aerial photographs—and its place names and streets—through the road network and points of interest layer. Furthermore, they can change the level of transparency of the map layer being visualised in order to combine it with such contextual geographical information in more meaningful ways, depending on the zoom level at which the map is being displayed. Finally, the user might choose to add any other customised map layers through the standard Google KML file format by simply adding its internet address location to the web page (for example from a bespoke map prepared in 'Google My Maps'). Overall, the web layout of *London Profiler* is simple and very easy to use since it is targeted to the general user, and not only analysts which are used to GIS tools. Some examples of the visualisations that can be created using *London Profiler* are shown in Fig. 12.1 to Fig. 12.7.

12.4 Conclusions and Further Research

The mapping mashup visualisation tool presented in this chapter, termed the *London Profiler*, has allowed Southwark and Camden Primary Care Trusts to create a picture of their populations at the small area level, typically the census output area. This is crucial in areas such as inner London where identifying so called 'hard to reach' groups is key to reach the public health targets specified by the government. Using *London Profiler*, a number of public service planning functions can be easily supported in an inexpensive way. In this chapter we have shown one example of its application to public health service planning, in which it can support the objective of tackling health inequalities through the targeting of initiatives to the specific local population at risk or in need. This has been achieved through a very simple to use web based mapping interface, and thus accessible to any person, in this case working in a Public Health department.

The 'mashing up' of different population data with base geographic information has provided planners with an efficient tool to improve primary health care service provision and monitor General Practices' performance. The health domain in *London Profiler*, and in particular the Hospital Admissions for Long Term illnesses (HALT) dataset, can alert local health authorities and GPs about admission hotspots in their catchment areas. Such catchments areas' of GPs, which for example were created for Southwark PCT using individual level patients registration geocoded data (Gibin *et al.* 2007), can be easily overlaid on top of all the thematic layers in the *London Profiler* enabling exploratory spatial analysis between a GP's location, the catchment area of the GP's registered population, and that area's socioeconomic and demographic characteristics. The e-society classification domain also revealed to be very useful in pioneering the use of new technologies for health care service provision. By overlaying their catchment areas on top the e-society classification in *London Profiler*, Southwark Primary care Trust's GPs, were able to assess the engagement in new information and communication technologies (ICTs) of their patients, and therefore promote the use of internet booking systems for GP appointments and repeated prescriptions, as well as identifying population groups that would not be likely to use this channel.

Through the examples presented in this chapter, we have intended to present a new and innovative approach to GVis termed 'mapping mashups'. The technological platform adopted in this example is based on the freely available *Google Maps* API, but other similar technologies have recently emerged (e.g. Microsoft Local Live or Yahoo Maps).

'Mapping mashups' offer a typically free solution for online exploratory cartographic visualisation, and as such they can prove an inexpensive and effective platform for visual communication of population data such as in the public health example presented here. The *Google Maps'* graphic user interface (GUI) is very intuitive and requires hardly any guidance to the end user, be it either external or internal to an organisation. *Google Maps* mashups can be easily and inexpensively developed, offering public service institutions great time and cost saving benefits when compared to much more complex mainstream WebGIS technologies. The example of mapping mashups presented in this chapter can be applied to other service planning scenarios. Its flexible and inexpensive features clearly make it the platform of choice for the development of future planning support systems where geographical visualisation plays a central role.

Further research in this area will necessarily have to address three main challenge areas: cartography, technology and obsolescence. From a cartographic perspective, overlaying thematic data on top of a rich and detailed reference map involves different visualisation problems. The thematic map classes need to be clearly identifiable as well as the reference data underneath. A typical way of intermediate in this problem is using layer transparency to alter each layers weight and allow the visualisation of a proportion of either the thematic map or the base maps layers. The level of transparency must be specifically tuned to allow the user to distinguish between the different thematic classes displayed, in conjunction with the choice of an appropriate classification algorithm, number of classes to be used, and the colour ramps assigned to them. Another interesting cartographic issue is how to accommodate changes of geographical scale in the way data is displayed, in particular the simplification of thematic area boundaries.

Technology is another key driver in further developments of mapping mashups such as the *London Profiler*. One reason of the great success of websites based on *Google Maps*, versus those based on traditional WebGIS solutions, is the fast speed at which the map is served to the user. Most of WebGIS systems display ‘on the fly’ geographic information stored in the server according to a set of parameters supplied by the user. This approach is very flexible but levies the bulk of the workload on the server hosting the website, while in *Google Maps* and similar websites the workload is shifted to the internet browser. By doing so, maps are not created on the fly but hosted as pre-rendered images that are served very quickly, minimising the server workload. The major drawback of this approach is the lack of flexibility, since all the maps must be pre-rendered and stored on the server. Further research on mapping mashups, and indeed future enhancements of *London Profiler*, will have to attempt to integrate the two approaches described above. Some commercial GIS software vendors are actually moving in this direction. For example the new version of ESRI’s ArcGIS Server includes a feature to create a ‘cached’ image version of a map for different scales so that when a user zooms to a certain scale the map tiles are served faster.

The last but most important issue of all is to prevent obsolescence and ensure an adequate level of website update in order to keep user interest. The pre-rendered map approach, as in the *London Profiler* example described in this chapter, requires an intensive amount of work to update the content of the thematic layers or to add new ones. Website longevity could also be extended by allowing users to add more content to these mapping mashups, through facilities such as the KML file overlay in London Profiler to add customised map layers, or the Map Tube project at UCL (www.maptube.org).

Mapping mashups is a recent innovation in the visualisation of geographic information whose future looks even brighter than that already enjoyed in its recent short and successful history since they appeared in 2005. This chapter has presented one example of mapping mashups applied to a local public health service planning. As discussed through the chapter, this approach offers an inexpensive, flexible, scalable and intuitive method of communicating geographical information that will most likely eventually replace traditional ‘top-down’ or ‘turn-key’ WebGIS solutions for most basic geographical exploration functions of planning support systems.

Acknowledgements We are grateful to Camden and Southwark Primary Care Trusts to have provided part of the data and support to develop sections of the work described in this chapter. The work presented here has been partially funded by Knowledge Transfer Partnerships KTP programmes 37 and 666, ESRC PTA 026-27-1521, and ESRC RES-348-25-0015.

References

Aspinall, P.J. (2002) Collective terminology to describe the minority ethnic population: the persistence of confusion and ambiguity in usage, *Sociology* 36(4): 803–816.

Birkin, M. and Clarke, G. (1998) GIS, Geo-demographics, and spatial modeling in the UK financial service industry, *Journal of Housing Research,* 9(1): 87–111.

Buckley, A. (2004) Geographic visualization, In Mcmaster, R.B. and Usery, E.L.A. (eds.) *Research Agenda for Geographic Information Science*, Crc Press, Boca Raton, Fl, pp. 270–296.

Department of Health (2004) *Improving Chronic Disease Management*, Department of Health, London.

Department of Health. (2006) *Our Health, Our Care, Our Say: A New Direction for Community Services*, Department of Health, London.

Dodge, M., McDerby M. and Turner, M. (2006) *Geographic Visualization across the Social Sciences*, NCeSS seminar. University of Manchester, 12–13 June. Available at: www.ncess.ac.uk/events/ASW/visualisation/ (Accessed 10 August 2007).

Fairhurst, R. (2005) Next-generation web mapping, *Society of Cartographers Bulletin,* 39: 57–61.

Harrower, M.A. and Brewer, C. (2003) ColorBrewer.org: an online tool for selecting color schemes for maps, *The Cartographic Journal*, 40(1): 27–37.

Gibin, M., Longley, P. and Atkinson, P. (2007) Kernel density estimation and percent volume contours in general practice catchment area analysis in urban areas, Paper presented at Geographical Information Science Research UK (GISRUK) 2007, National University of Ireland, Maynooth.

Longley, P.A. (2005) A renaissance of geo-demographics for public service delivery, *Progress in Human Geography,* 29: 57–63.

Longley, P.A., Chao Li and Webber, R. (2006) The UK Geography of the E-Society: A National Classification, *CASA Working Paper* 111 http://www.casa.ucl.ac.uk/publications/workingPaperDetail.asp?ID=111 (Accessed 20 January 2008).

Maguire, D.J. and Longley, P.A. (2005) The emergence of geoportals and their role in spatial data infrastructures, *Computers, Environment and Urban Systems,* 29(1): 3–14.

Mateos, P. (2007) A review of name-based ethnicity classification methods and their potential in population studies, *Population Space and Place*, 13(4): 243–263.

Mateos, P., Webber, R. and Longley, P.A. (2007) The cultural, ethnic and linguistic classification of populations and neighbourhoods using personal names, CASA *Working Paper 116,* Rep. ISSN 1467–1298, Centre for Advanced Spatial Analysis, University College London, London.

Monmonier, M. (2007) Cartography: the multidisciplinary pluralism of cartographic art, geospatial technology, and empirical scholarship, *Progress in Human Geography,* 31(3): 371–379.

Nanchahal, K., Mangtani, P., Alston, M. and dos Santos Silva, I. (2001) Development and validation of a computerized South Asian names and group recognition algorithm (SANGRA) for use in British Health-related studies, *Journal of Public Health Medicine,* 23(4): 278–285.

Petersen, J., Ashby, D. I. and Atkinson, P. (2007) Health applications for open geo-demographics, *Paper presented at Geographical Information Science Research UK 2007,* National University of Ireland, Maynooth.

Vickers, D. and Rees, P. (2007) Creating the UK National Statistics 2001 output area classification, *Journal of the Royal Statistical Society: Series A Statistics in Society* 170(2): 379–403.

Tufte, E. (1990) *Envisioning Information*, Graphic Press, Cheshire.

The Websites

http://www.londonprofiler.org
http://www.onomap.org
http://www.maptube.org
http://www.nestoria.co.uk
http://www.google.com/apis/maps/
http://www.google.com/apis/maps/documentation/reference.html

Additional Reading

Gibin, M., Singleton, A. and Longley, P. (2007) Exploratory cartographic visualisation of health and higher education through Google Maps API, Paper presented at Geocomputation 2007, National University of Ireland, Maynooth.

Hudson-Smith, A., Milton, R., Batty, M., Gibin, M., Longley, P. and Singleton, P. (2007) Public Domain GIS, Mapping & Imaging Using Web-based Services, *Centre for Advanced Spatial Analysis Working Paper 120*, University College London, London. Available at: http://www.casa.ucl.ac.uk/publications/workingPaperDetail.asp?ID=120 (Accessed 31 December 2007).

Chapter 13
ArcGIS Processing Models for Regional Environmental Planning in Bavaria

Jörg Schaller, Thomas Gehrke and Nathan Strout

13.1 Introduction

Over the past five years, the Geographical Information Systems (GIS) laboratory headed by Jörg Schaller at the University of Technology Munich (TUM) has worked on the development of new GIS software processing tools and models for environmental planning and management. The research cooperation between the University of Technology in Munich, ESRI Germany and ESRI Inc. in Redlands and the University of California in Redlands involved the development of new applications of ESRI's *ModelBuilder* software for planning support, and is based on the existing experience of the four institutions. The cooperation focused on the development of GIS environmental modelling technology to provide new applications in the field of regional environmental planning and assessment, and the applied model and software tools were tested using existing regional databases from Bavarian and Californian sources, and applying them in actual regional planning projects.

The chapter will therefore describe the methodology of the related Geo-Database generation, the *ModelBuilder* technology and the geoprocessing applications for regional planning. The result of the successful research cooperation between the four institutions has been the development of general GIS processing model structures for the regional Landscape Development Concept (LDC) in the Munich planning region and an urban growth model for the Munich and Redlands regions. The graphical interface that is used for creating models allows an entire processing package to be run over and over again with the click of a mouse, which means that a GIS analyst can change a parameter quickly and easily. This optical interface structure makes GIS planning processes simple to understand, and allows more efficient teamwork, even via the Internet.

The Environmental Model applications themselves were carried out to deliver results for political planning decisions. The presentations of the results to regional governmental bodies and to the involved public were very successful, since the

J. Schaller (✉)
ESRI Geoinformatik GmbH
Ringstrasse 7, 85402 Kranzberg, Germany
E-mail: j.schaller@esri-germany.de

S. Geertman, J. Stillwell (eds.), *Planning Support Systems Best Practice and New Methods*,
DOI 10.1007/978-1-4020-8952-7_13, © Springer Science+Business Media B.V. 2009

workflow diagrams were excellent for easy explanation of the complex planning and assessment steps that had been carried out. The new GIS-based planning and modelling technology therefore strongly supports the acceptance of modelling and planning results by decision makers and the public. The two selected planning examples will illustrate the advantages for optimising and speeding up planning procedures from data collection to support for decision making, and for the new possibilities for documenting planning and decision procedures for the public.

13.2 Background

In order to support actual regional planning projects, the ESRI *ModelBuilder* technology was used to support complex planning issues. During 'normal' GIS work in a planning process, GIS analysts use one tool after another to complete their task(s) and must always be aware which working step they are completing at any one time. If the data changes (e.g., new protection areas have been established), most of the work has to be repeated. It is also not really feasible to work with a team on a project unless detailed data documentation has been written in a final form. In contrast, the *ModelBuilder* technology opens up a workspace in which the user can 'plan' the GIS jobs to be carried out, and ultimately run them. Parameters and input data can be changed again at any time, and the model can be run again with these new settings with a click of the mouse. Therefore, it is not necessary to open the whole GIS data processing model before running the model. Since all tasks are documented graphically, the work can therefore be planned by a team, since all working steps are traceable.

Actual regional planning activities requirements necessitate intensive geo-data processing in which many existing geo-datasets must be processed to create output datasets using geoprocessing tools, scripts or models. Accordingly, the *ModelBuilder* innovation allows complex actions to be displayed in the form of flux diagrams in a graphical user interface. Thanks to the diagram, every step of the whole process can be traced and documented on the fly. Planners themselves are able to run the necessary operations because they are very simple and intuitive, since datasets or tools can be selected by drag and drop into the model window and connected graphically with arrows according to the planning and processing workflow. Together with the model, other planners, decision makers, and, not least, the public, can follow the planning process. The result can be stored and used as tool for future applications, or for providing the application to other planners for similar planning requirements with different Geo-Databases, e.g., as a template for necessary modifications (McCoy 2003).

13.2.1 ModelBuilder Window

The *ModelBuilder* window (Fig. 13.1) consists of a display window in which users build a diagram of the processing model, a main menu and a toolbar that they can

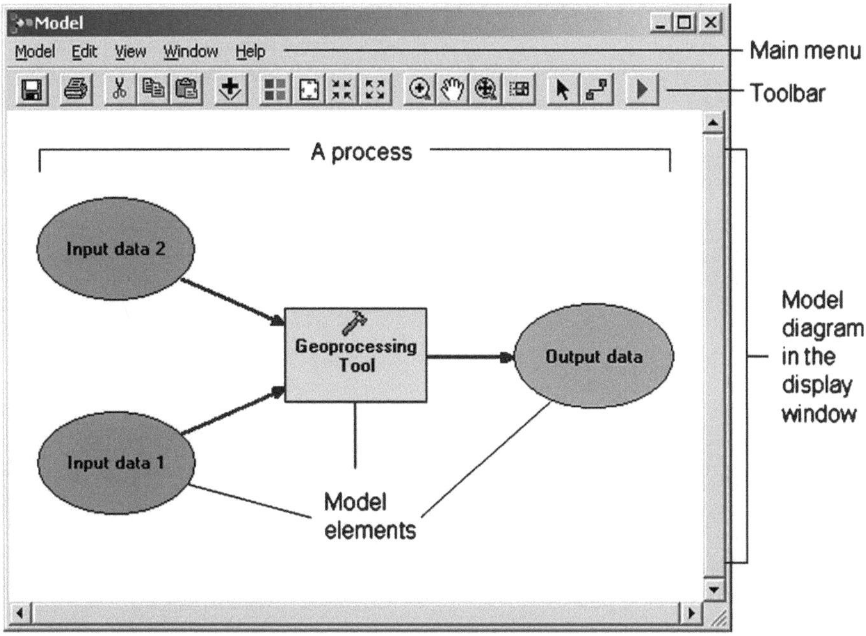

Fig. 13.1 ModelBuilder window

use to interact with elements in the model diagram. They can run a model from within the *ModelBuilder* window or from its dialogue box.

One of the easiest ways to author and automate a workflow and keep track of a geoprocessing task is to create a model. A model consists of one *process*, or, more commonly, multiple processes strung together. A process consists of a *tool* – a system tool or a custom tool – and its *parameter* values. Examples of parameter values include input and output data, a cluster tolerance and a re-classification table. The model presented here allows the user to perform a workflow, modify it and repeat it over and over again with a single click of the mouse. ArcGIS 9.x presents a comprehensive set of geoprocessing tools that work with all the supported data formats, including Geo-Database features. It also offers a completely new framework for working with these tools that enables the user to open them individually, combine them in a visual modelling environment, write scripts in standard scripting languages, and run the tools as commands in a command line window.

13.2.2 Personal Geo-Database and ArcToolbox

The new ArcGIS Personal Geo-Database format (Fig. 13.2) has been used to store all geographical information. Unlike the older shapefile and grid data formats, in which all data are stored in separate files, the Geo-Database stores all relevant information in one single database. As based on the MS Access format, some information (tables) can also be read but not written with Access. As shown in Fig. 13.2, a

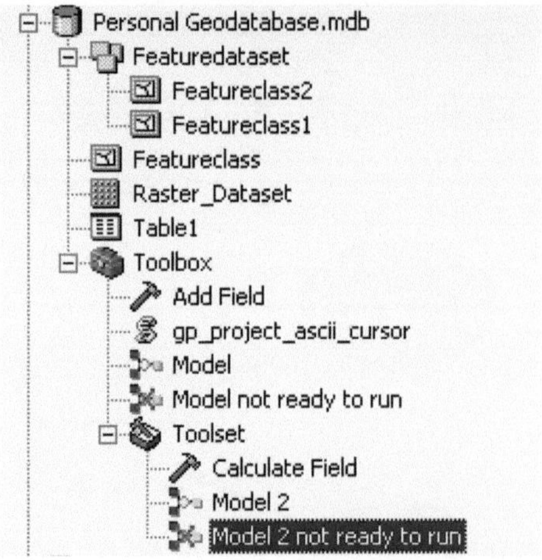

Fig. 13.2 Structure of a personal Geo-Database

Geo-Database can store geographical information directly as feature classes (vectors), raster datasets and tables. Information with the same spatial extent and geographical features can also be combined into feature datasets.

Additionally, toolboxes can be integrated into the Geo-Database. A toolbox can contain tools, scripts, toolsets and models. Normally, only a handful of tools are used within a model. For quicker accessibility, it is advisable to copy these tools into one central toolbox. This is even more convenient if self-compiled scripts are being used in one model. In ArcGIS, the Toolbox is a dockable window that is integrated into all the ArcGIS Desktop applications. For example, when users run tools from the ArcToolbox window in ArcMap, they can use the layers of the current map as inputs, and the outputs can be directly added to the map as new layers. In ArcGIS *ModelBuilder*, these tools can be dragged and dropped into the model and set up. If teamwork is required, only one database has to be spread within the team, and all data, tools and models are contained within it. Models that are not ready to run are crossed out (e.g., if some data or a tool is missing).

13.2.3 Model Elements

As described in the previous section, an ArcGIS model consists of model elements. These can be divided into three categories: input; processes; and output.The model elements are represented in the user interface in form of graphical symbols such as ovals and boxes in different colour.

13.2.3.1 Input Data

Input data are represented by a dark blue oval. They can be raster and vector data, as well as single input values such as numbers. By default, the oval shows the name of the input data, but can be renamed afterwards. Input data can also be specified as a parameter (oval shows the letter P next to it). Input data specified as a parameter appears in the model parameter dialogue box, and can be changed/replaced there before running the model.

13.2.3.2 Input Expression

Instead of input data, a SQL expression can be used. Such an expression is represented by a light blue oval. A dialogue box can be used to generate the expression. If a tool property is set as a parameter, the P symbol does not appear next to the tool icon, but instead a new input parameter oval will be visible in the model.

13.2.3.3 Tools

The input data is normally processed by a tool. The GIS software offers a huge amount of predefined tools that can be dragged and dropped into the model and connected to the input data. The tools settings can be altered in the properties dialogue.

13.2.3.4 Scripts

A script can be used as a tool. The scripting language can be Python, VBScript, Jscript or Pearl, whereby Python is used for most of the predefined scripts. The settings for a script can be altered the same way as for a tool, while the source code can be modified directly from within the model.

13.2.3.5 Embedded Models

Sometimes, a calculation needs several processes to be completed (e.g., a specialised tool is missing and must therefore be done by two or more tools). In order to keep a model clearly arranged, it is wise to put these processes into a separate model and to embed this into the main model. This is extremely useful if the calculation has to be executed several times in a model. The whole model appears as only one tool, with a small model symbol on top. For modifications, the embedded model can be opened and modified within the main model. The selected tools, scripts or embedded models are represented as yellow boxes in the user interface.

13.2.3.6 Output Data

Output data are represented by a green oval. The type of output data is dependent on the tool producing this output. Only the name and path of the output can be changed. Sometimes, a tool produces several outputs (e.g., the distance tool that derives a distance grid and a direction grid). If this remains unused, it is symbolised by a white oval.

13.2.3.7 Model Parameter Dialogue Box

Before running a model, the dialogue box with all model parameters opens (Fig. 13.3). All properties set as a parameter can be changed here (e.g., data path, re-classification table, SQL expression).

13.2.4 Sub-models

In order to keep a model clearly arranged, it is most important to keep it 'readable' for other persons (e.g., clients, planning partners in the team). A model for urban growth, for example, cannot be done with a handful of processes. It is therefore useful to arrange the processes into sub-models with specific topics. The following application examples will show the issue-based selection of sub-model topics and their integration to final planning results (Schaller *et al.* 2006).

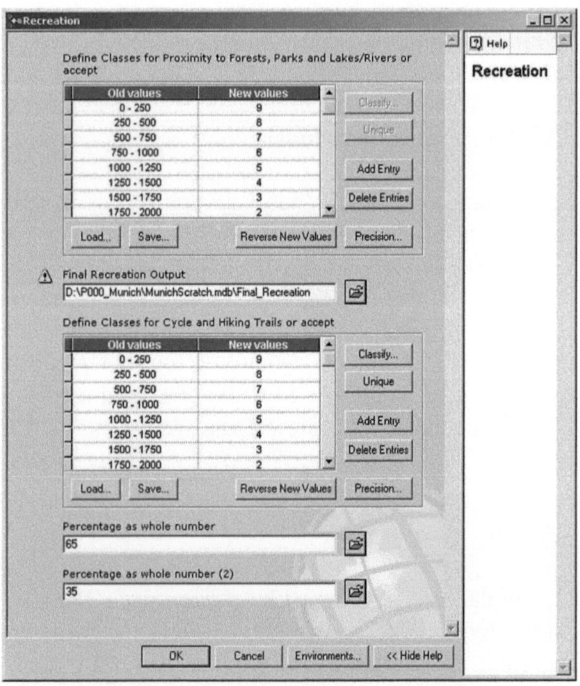

Fig. 13.3 Model parameter dialogue box

13.3 Application Examples

13.3.1 Munich Planning Region

The planning region of Munich, the capital city of Bavaria in southern Germany, is an economically strong and dynamically growing region. The Munich region has a population of 2.5 million distributed over 5,504 km^2 in 199 municipalities (*Gemeinden*) contained within nine districts (*Landkreise*). Munich city has a population of 1.2 million spread over 310 km^2. The continuing growth requires additional land development for infrastructure, commercial and housing requirements.

13.3.2 The Landscape Development Concept (LDC)

The dynamic growth of the region is having an increasing impact on natural resources such as soils and water and air quality, and is associated with land development, intensification of land use and increasing pressures on nature and the landscape. The result is a decline of living quality and loss of natural landscape quality. Sustainable development of this regional growth is not yet guaranteed, while the inhabitants of the region place high demands on maintaining their existing outstanding quality of rural landscapes and the open spaces of the city and region for recreation. The regional government, therefore, decided to create a planning concept for sustainable development based on GIS and environmental planning methods and advanced modelling technologies for further application, with permanent updating to follow up the impacts of growth and the results of planning measures for sustainable development. The LDC therefore makes an important contribution to the sustainable protection and development of the natural resources of the region. It also supports decisions for the

Fig. 13.4 Land use map and digital terrain model (DTM) of the Munich planning region (Schaller and Schober 2007) (See also Plate 30 in the Colour Plate Section)

regional plan in the field of environmental protection, as well as for the maintenance of nature reserves and visual landscape quality (Schaller and Schober 2004).

13.3.2.1 The Goals of LDP

The main general goals of this regional planning concept are:

- maintenance, protection and development of nature and landscape as compensation for intensively used urban areas;
- protection and sustainable development, and utilisation of the natural resources;
- development of open-space and green-belt concepts for the city and region as a foundation for the maintenance of living quality in the region; and
- proposals for sustainable land use and land use development, especially for commercial, housing, transportation and water management, and also agriculture, forestry and recreation demands.

To support this LDC, complex spatial issues need to be solved for the planning workflow and the display of results for effective decision support: What quality and sensitivity do the natural resources have in the region? What are the actual environmental impacts on population, nature and visual landscape values? What are the actual planning and development scenarios for the next 20 years? What qualitative and quantitative changes are caused by these developments, and how can these be compensated by planning measures? How can these complex inter-relationships be communicated to decision-makers and the public? (Schaller and Schober 2004).

13.3.2.2 The LDC Geo-Database

The first step was the creation of an issue-based Geo-Database structure containing all necessary spatial information to solve these planning issues, such as abiotic resources, biotic resources, land use changes and planning data (Arthur and Zeiter 2004; Booth *et al.* 2004). These data include the following: topographical data – digital terrain model (DTM); nature protection and biotic resources; geology, soils and soil quality; surface water and groundwater resources; local climate, air quality and ventilation; noise, eutrophication and other impacts; administrational boundaries; statistical data related to municipalities; actual land use data from aerial photographs; and municipality and infrastructure master plans. The data structure is related to the main issues of the project, documented in the personal Geo-Database and can be visualised with VISIO. Figure 13.5 depicts an example of this documentation for the soil database and related thematical maps.

13.3.2.3 Assessment of Natural Resources and Land Use Impacts

The natural resources of the planning region are assessed on the basis of function, quality and development potential. Actual and planned land use types are assessed

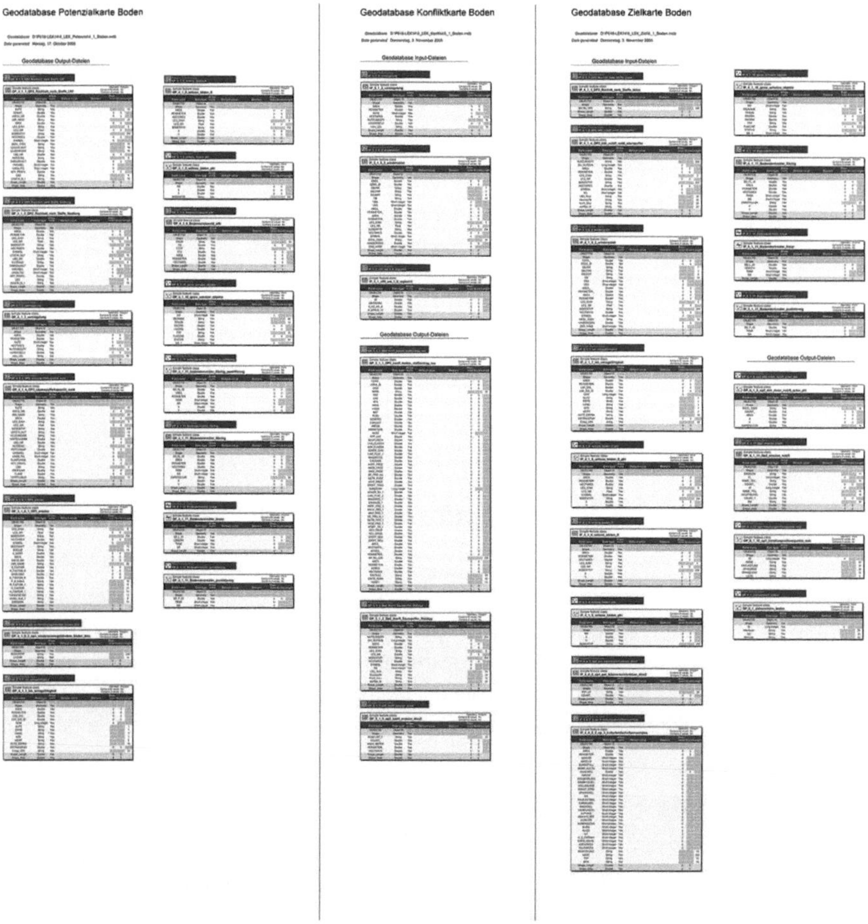

Fig. 13.5 Example of a VISIO documentation of part of the LDC Geo-Database

by intensity and impacts on natural resources. The derived conflicts were modelled and general and specialist goals for sustainable land use and development formulated from the results of the conflict analysis for each abiotic or biotic resource, as well as impacts on human requirements. Typical examples of abiotic resource functions or environmental risks that can be either reduced or exacerbated by land use impacts are described in Table 13.1.

13.3.2.4 Creation of Resource Assessment Maps

The first step of the model application was therefore the creation of resource assessment maps that provided the basic information for the conflict and impact analysis. As an example, Fig. 13.6 depicts a sectional map of the soil potential and yield classes for agricultural land use (Schaller 2004).

Table 13.1 Examples of abiotic functions or risks impacted by land use

Soils	• Buffer capacity for absorbable substances • Erosion hazard
Water	• Groundwater recharge • Floodplain functions (retention) • Flooding risk
Air quality, local climate, noise	• Energy exchange functions • Ventilation functions • Noise impact
Biotic resources	• Wildlife preservation functions • Species diversity and biotope network functions
Recreation	• Recreational functions of the landscape

13.3.3 Processing Model Applications for LDC

In order to estimate possible impacts of actual land use or land use changes, very simple relationships between resource potentials and sensitivity of impacts were initially described in conceptual relationship models. Figure 13.7 depicts examples of these conceptual models and a table of the assessment criteria used, in this case in relation to soil resource functions. For each conceptual base model, a detailed

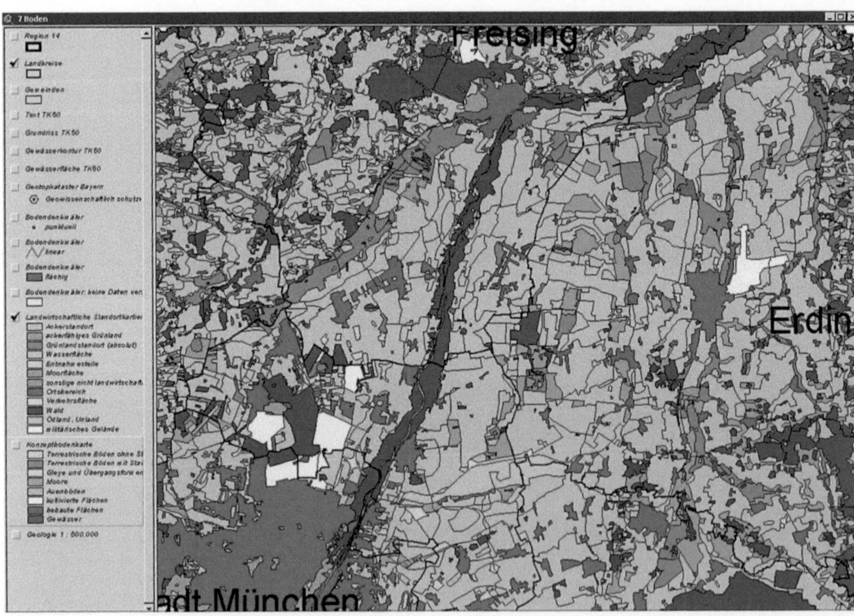

Fig. 13.6 Soil potential and yield classes for agricultural land use (See also Plate 31 in the Colour Plate Section)

model was developed in a second design phase with a direct relationship to the existing Geo-Database and scientific assessment criteria from literature and expert ratings. In some cases, based on the issues to be solved and the additional input of expert know-how, the primary Geo-Database had to be expanded. Figure 13.7 depicts the related detailed model structure for calculation purposes, e.g., the soil erosion hazard using the Geo-Database.

13.3.4 Sustainable Development Goal Maps

In order to develop general and detailed goals for sustainable land use and resource protection, the resource function maps were overlaid with land use maps in a processing model to illustrate the actual conflicts between land use and resource functions. Based on the intensity of conflicts in relation to the protection and sustainable development goals, general goal maps have been calculated and displayed as the model output in the form of thematic maps. Figure 13.8 depicts such an example of a result map for the protection and development goals for groundwater and surface water resources in the region.

13.3.5 Urban Growth Model

In order to estimate possible impacts on natural resources and visual landscape quality in the future, an urban growth model framework was developed. The model for urban growth in regions is divided into three parts: six sub-models, each handling one specific topic; the main model called 'settlement suitability'; and a model adding the time dimension.

Although the input data for the models are available as vector data, most of the calculations are done in raster format. For performance reasons and for storage space efficiency, a raster resolution of 100 × 100 m has been chosen that can easily be rendered more precisely at a higher spatial resolution if required. The raster format allows some mathematical operations to be carried out more quickly and easily, e.g., a weighted overlay or distance calculations. Suitability classes used in the following models comprise ten classes with a range from 0 to 9: 0 is unsuitable, 1 very low suitability, 5 medium suitability, and 9 is very high suitability (Schaller *et al.* 2006).

13.3.5.1 Sub-models

In order to keep the growth model clearly arranged, the following sub-models were defined: exclusion of 'taboo' areas, restrictions for urban development, landscape suitability, land use, proximity to existing infrastructures, and recreation potentials. These sub-models each deliver a final result for the specific topic. They are

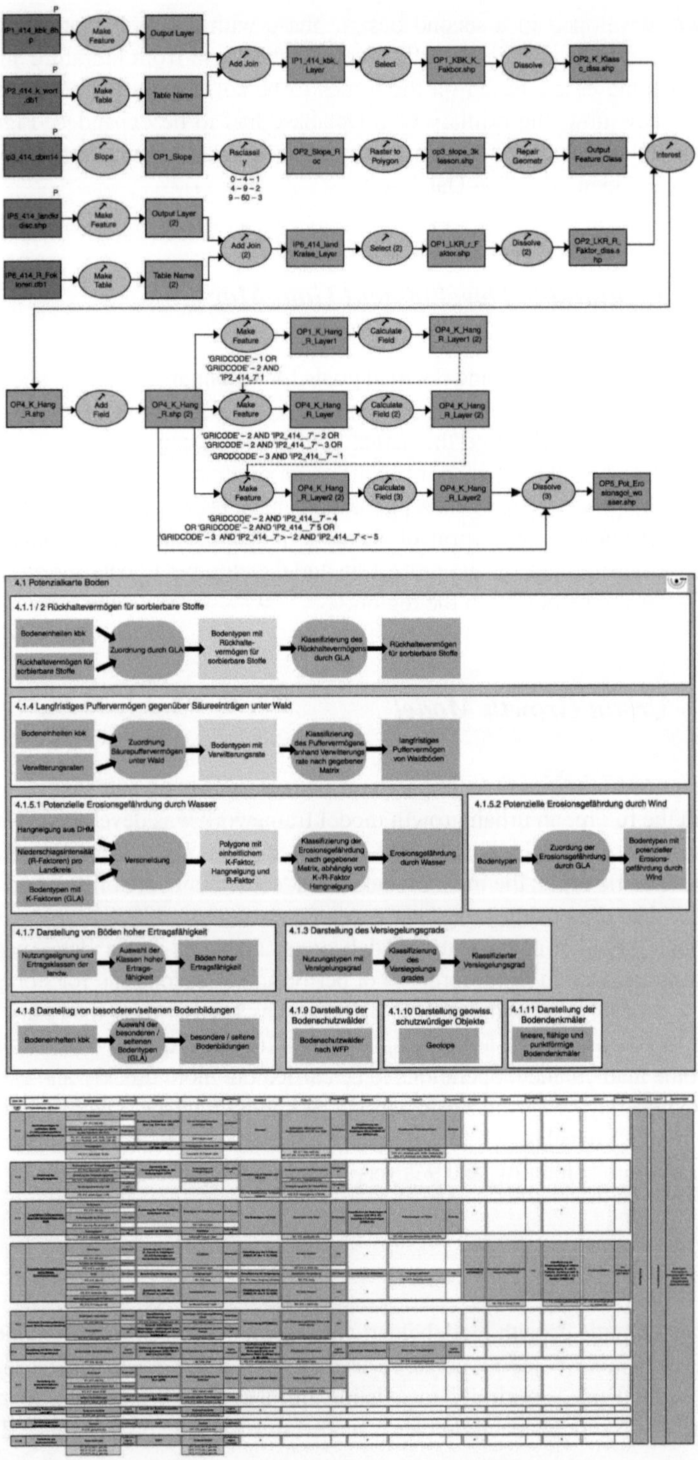

Fig. 13.7 Conceptual and detailed processing models for resource assessment

Fig. 13.8 Protection and sustainable development goals of groundwater and surface water resources in the region (Schaller and Schober 2007)

implemented into the main model as embedded models. Areas that are unsuitable for settlement are handled in the ‘exclusions sub-model’ and include: water surfaces, water protection areas, flood areas, biotopes, flora fauna habitat (FFH) areas, special protected areas (SPA) for birds, Ramsar bird protection habitats, and nature protection areas (NSGs). These are treated as taboo areas, and are excluded from any further examination.

13.3.5.2 Restrictions for Urban Development

The restriction model is more complex than the exclusion model. All restrictions are given a suitability value in the range from 0 to 9 that the planner can change before running the model. By assigning a value of ‘0’, it is also possible to regard a category as exclusion. In the second step, the planner can weight the restrictions against one another. A tool checks whether the entered values add up to 100 per cent, or otherwise reports an error. This mechanism has also been used in some of the following models. In the restriction model, the following categories are examined: forest reservations, greenbelts, dividing green corridors, landscape retention

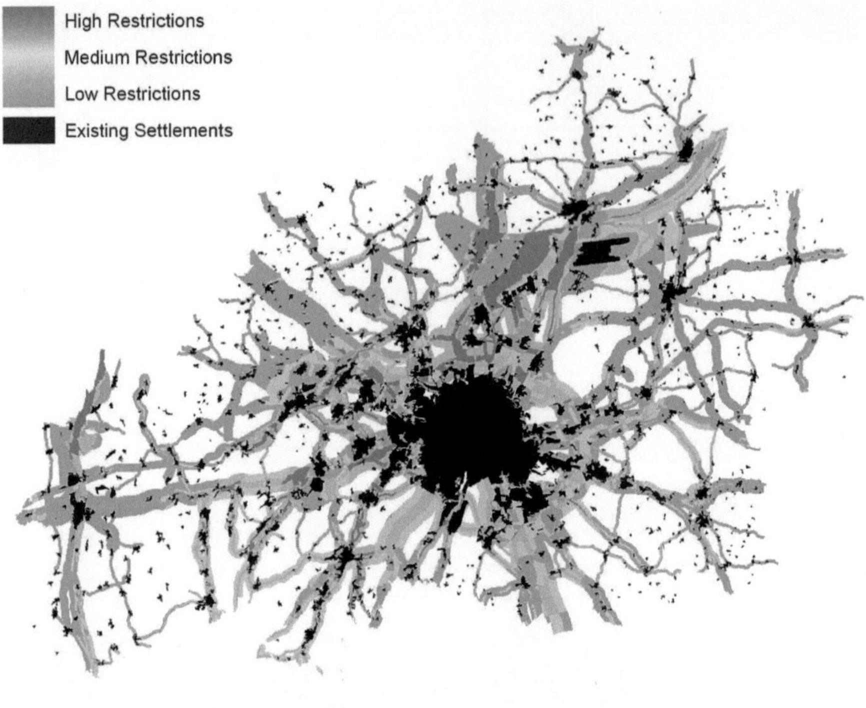

Fig. 13.9 The final restrictions raster (white spaces = no restriction) (See also Plate 32 in the Colour Plate Section)

areas, landscape protection areas (LSGs), and noise impacts from transportation and traffic. After the model has verified that the weighting adds up to a hundred per cent, the suitability is calculated for all six categories. In the following steps, these are added up to one geometry showing the settlement suitability for all restriction categories. The result map of the restriction model (Fig. 13.9) shows the number of restrictions per raster cell, as all six categories have been weighted almost equally. In this example, low represents only one restriction, whereas high stands for up to five restrictions.

13.3.5.3 Landscape Suitability

This sub-model covers two topics that are very much related to one another. Firstly, it evaluates the suitability of the soil potential where new settlements should not be built, for example: gley and mixed forms of groundwater-influenced soils, cultivated soils with high yield classes for agriculture, existing settlements, water bodies, wet terrestrial soils, very dry terrestrial soils, and wetlands and swamps. Secondly, it analyses the slope of the relief and reclassifies it to the ten suitability classes. With the weighting factors, it is possible to balance slope and soil suitability.

13.3.5.4 Land Use

The land use sub-model is a simple model that assigns a suitability value to each of the land use categories. Thus, it is more likely that agricultural areas are used for new settlements than, for example, forests. The land use data has been derived from aerial photograph mapping and generalised to the following 11 categories: unused areas; forests; agricultural; military; recreation; mines; transportation, traffic; settlements; industrial, commercial; lakes, rivers, and other land uses.

13.3.5.5 Proximity to Existing Infrastructure

The idea behind this sub-model is that people indeed want to live in 'green surroundings'. Because, however, most people have to work in the city or a bigger town, and the children have to go to school, the proximity to existing settlements and public transportation as well as a good road connection are important factors for future settlements. Therefore, the proximity model deals with the following categories: proximity to Munich, proximity to other settlements, proximity to railroads, proximity to roads, and proximity to the airport. Cities and towns offer many jobs to the population, which means it is not very advisable to live very far away from these places. Big cities normally offer far more jobs than smaller towns do, and so they are handled separately, and the weighting can be set independently.

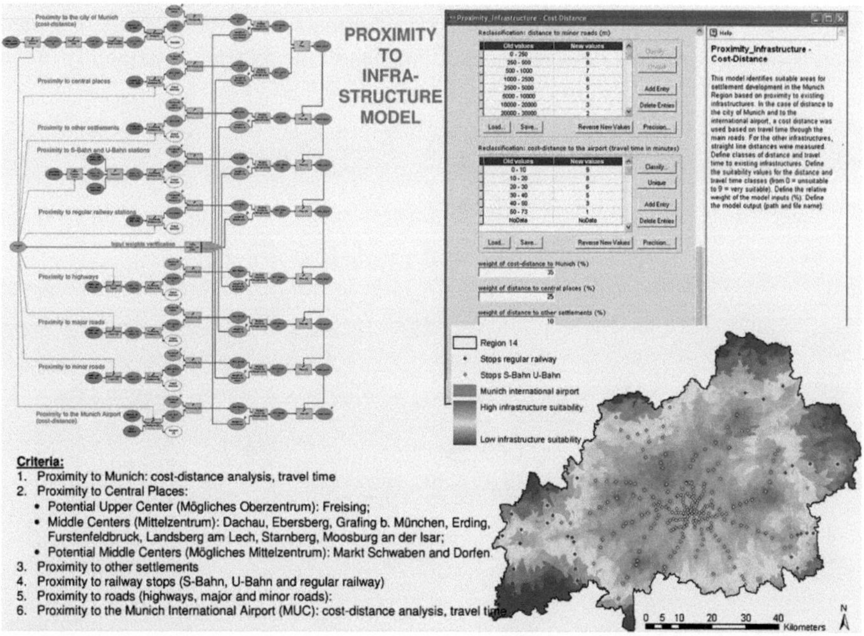

Fig. 13.10 The final proximity model and map result (Mattos 2007) (See also Plate 33 in the Colour Plate Section)

For all people in the region, transportation is a very important factor, as hardly anyone lives within walking distance from their workplace. The proximity to roads for individual traffic and to railways for public transportation are handled separately in this model. Finally, the distance from an international airport and its connections is important for the population, and for business. The model calculates the Euclidean distance to the five categories separately, and reclassifies the distances into the ten suitability classes. Finally, the partial results are added to the final result. In the proximity model, the suitability classes can be defined individually, and a weighting among the five categories can be set. Figure 13.10 depicts the proximity model and its result. The resulting raster dataset clearly shows the massive influence of the city of Munich and the relatively minor influence of the airport owing to the weighting values chosen. However, the medium-weighted categories such as roads also are clearly visible.

13.3.5.6 Recreation

The idea behind the recreation sub-model is that people prefer to live close to places that are suitable for recreation. This can be a forest, park, river or lake, or also cycle or hiking trails that are used for sport. In this sub-model, the categories cycle trails, hiking trails, lakes, forests, and parks are appended and the Euclidean distance is calculated and reclassified into nine classes for suitability. Figure 13.11 depicts the structure of the recreation sub-model.

In the model properties, the user can modify the class ranges and suitability values, and also alter the weighting of nature-oriented over more sports-oriented recreation. Figure 13.12 depicts the resulting map of the recreation suitability of the

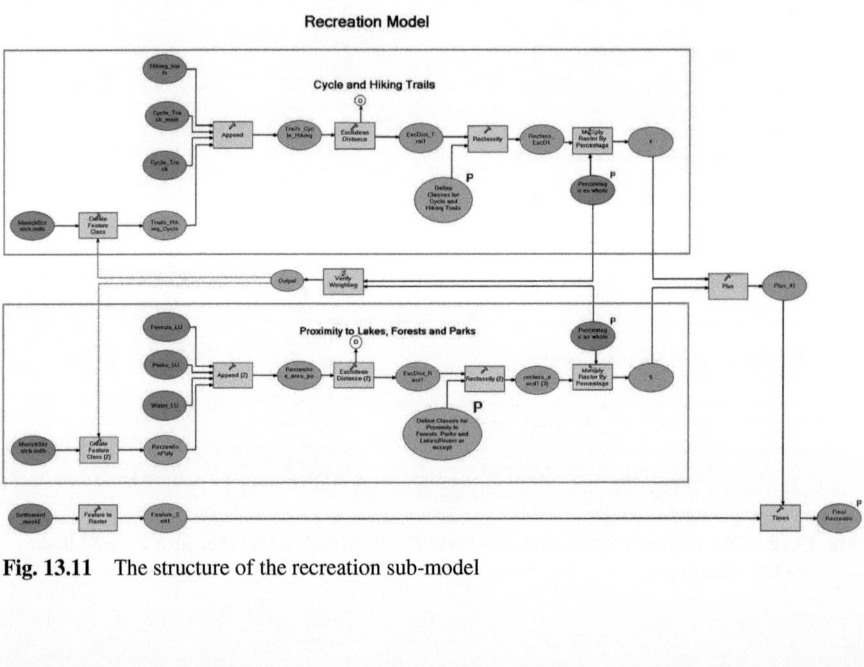

Fig. 13.11 The structure of the recreation sub-model

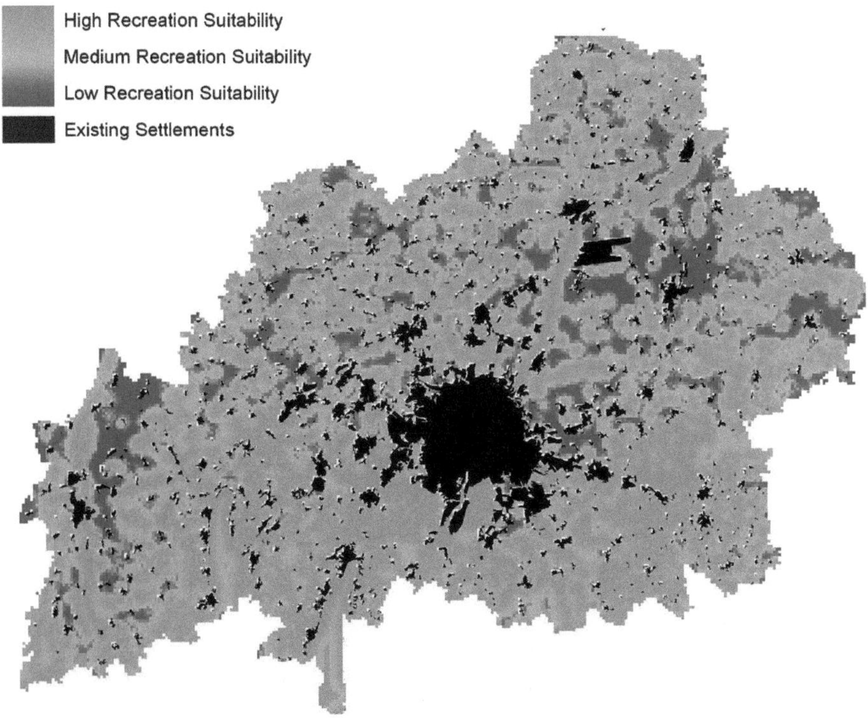

Fig. 13.12 The final recreation raster map (See also Plate 34 in the Colour Plate Section)

region. The resulting raster dataset gives an impression of where people would most like to live (green) from a recreation point of view.

13.3.5.7 The Main Model

The main model is called 'Settlement Suitability', because it finally calculates the suitability for new settlements (the higher the value, the more likely it is that the urban area will grow here first). In the model dialogue box, the user can again specify a weighting of all the sub-models (apart from the exclusions that are taboo zones) that are implemented here. The model first verifies whether or not the values add up to 100 per cent. The model processes the sub-models itself. It contains, or, if already executed, takes their results and weights them as specified in the dialogue box. Apart from the exclusions, all other models can be weighted. The final result of the main model is a raster dataset showing values from 0 (white = not suitable for settlements/excluded areas) up to 9 (red = very suitable for settlements) (Fig. 13.13). The higher the number, the more likely it is that the city will 'grow' into these areas first during the next decades. This growth rate will be analysed in the following model.

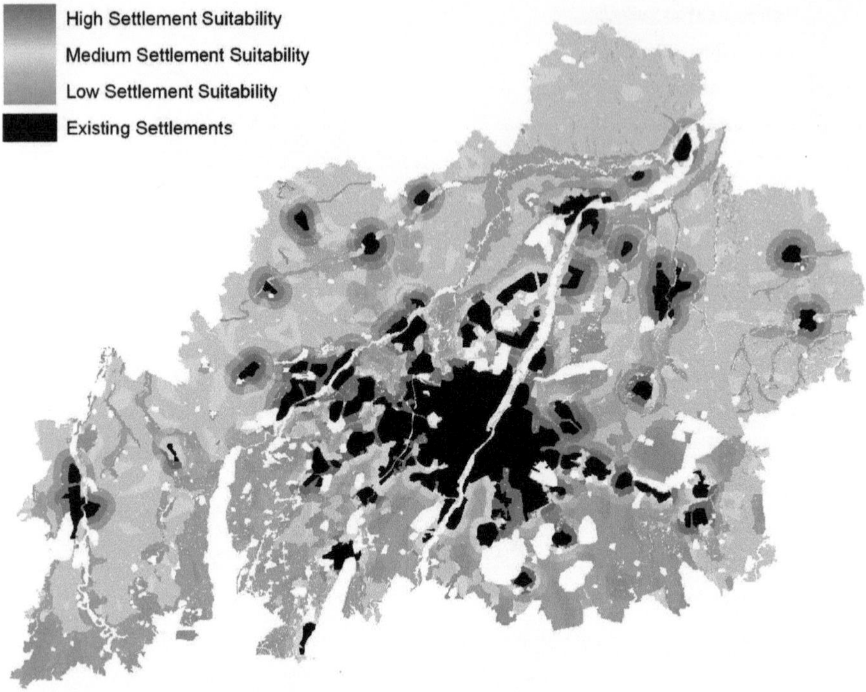

Fig. 13.13 The resulting map for settlement suitability (white spaces = exclusions) (See also Plate 35 in the Colour Plate Section)

13.3.5.8 Time Scale

In the preceding model, the general feasibility for urban growth into the surrounding region has been determined, whereby not only the growth of the city of Munich, but also that of the surrounding towns and villages have been taken into account. With this model, a time dimension will be added (Fig. 13.14). Therefore, the currently existing settlement area will be excluded from the examination.

In the model parameter dialogue box, the user can specify the number of years for which the urban growth will be calculated. The idea of the model is to 'fill up' the region year by year, taking the most likely places for new settlements first and the least likely ones last. The assumed growth rate has been determined from historical maps of the city of Munich. The area of these 'footprints' from the years

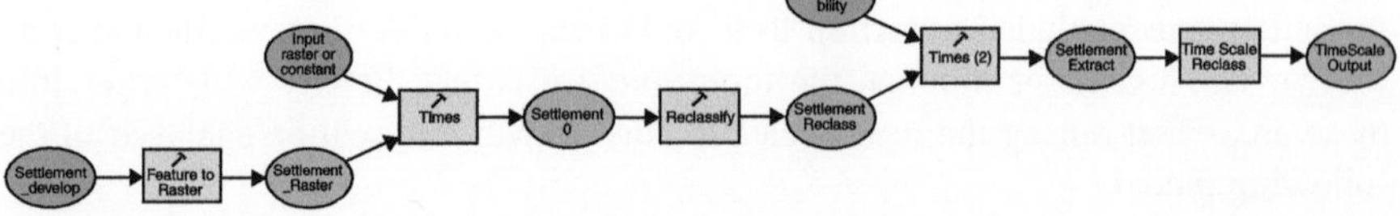

Fig. 13.14 Time model

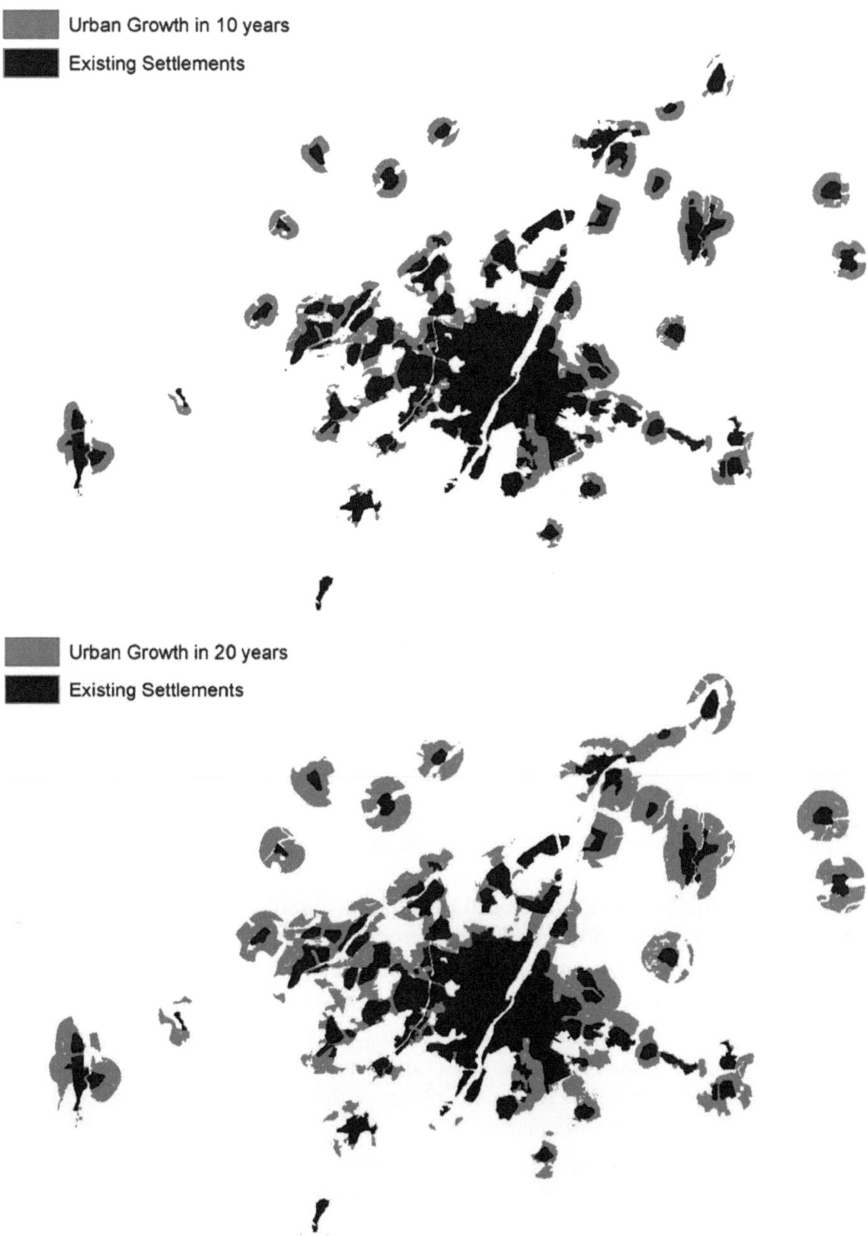

Fig. 13.15 Assumed growth in the Munich region in the next 10 and 20 years

1958–2000 has therefore been calculated and statistically analysed. In the time scale model, the assumed growth for the next decades can be calculated. The resulting maps (Fig. 13.15) show the assumed extent of settlements after 10 and 20 years (Schaller *et al.* 2007).

13.4 Conclusions and Future Research

On the whole, the application of the GIS *ModelBuilder* geoprocessing technology in a actual and large regional environmental planning project for the region of Munich has proved very successful. Nevertheless, the model application for the planning support showed some limitations: *ModelBuilder* processing models can be created only for processes that can be fully automated in relation to the content of the existing Geo-Database; there are certain planning processes that cannot be fully automated, such as data collection and selection, expert rating or expert selections with different data aggregation and weighting types, *et cetera*; and some calculations can be carried out with models, but must be finalised by manual interactions such as the creation of spatial units of visual landscape quality.

On the other hand, it has been shown that the application of this technology has many advantages in the form of an effective and convincing planning support system: better data and process documentation; easy modifications of workflows, updates, GIS processes; simple re-calculation after data and model parameter updates; simple modification of assessment criteria or changes in methodology; better documentation of workflows; models and GIS tools and scripts can be integrated in the personal Geo-Database; the method provides an easily understandable common language between the planners involved; and there is high acceptance by decision-makers and the public, since all planning steps and decisions are highly transparent.

The region of Munich, the districts and municipalities, the public and non-profit organisations, and also commercial and industrial decision makers can use the complex information for their assessments and decisions for the future development of the region. In future, the Internet will be the discussion panel for the different regional players.

Therefore, the following steps need to be taken. Firstly, the delivery of the Geo-Database and the developed models to the regional government, city and regional planners, and also to the municipalities for decision support and own model applications. Secondly, the implementation of the Geo-Database and the models on the Internet for public access. Thirdly, the creation of a monitoring procedure for land use changes and related database updates. Fourthly, the implementation of an Internet-based discussion forum for planners, municipalities and the public to improve the models and the scenario applications. Finally, the provision of models for public use via ‘model on the web’ technology.

The developed model applications can be improved by future teaching and research activities for planning support systems in the following fields: development of a general Geo-Database design instrument to support the generation of standard planning data descriptions according to INSPIRE and the local planning legislation; development of a 3D GIS model design for real-time simulation of a region and virtual 3D; visualisation of time-related model output or display of planning variants or decisions; development of standard schemes and structures of models and sub-models to be applied to the standard Geo-Databases 2D and 3D visualisation as planning support tools; integration of dynamic modelling technologies for more sophisticated modelling of developments over time (replacement of the time model); and development of a training model package for regional environmental and urban planning designed for planner workshops and university-level curricula.

References

Arthur, D. and Zeiler, M. (2004) *Designing Geo-Databases Case Studies in GIS Data Modeling*, ESRI Press, Redlands, CA.

Booth, B., Shaner, J., MacDonald, A., Sanchez, P. and Pfaff, R. (2004) *ArcGIS9 Geo-Database Workbook*, ESRI Press, Redlands, CA.

Mattos, C.O. (2007) *GIS-Based Modeling of Settlement Development Scenarios in the Munich Region*, Master of Science thesis submitted to the School of Forest Science and Resource Management, Sustainable Resource Management Program, University of Technology (TUM) Munich-Weihenstephan.

McCoy, J. (2003) *ArcGIS 9—Geoprocessing in ArcGIS*, ESRI Press, Redlands, CA.

Schaller, J. (2004) *GIS for Resource Management and Environmental Planning in the Region of Munich,* Redlands Institute, ESRI Colloquium Series, University of Redlands, Redlands, CA.

Schaller, J. and Schober, H.M. (2004) Das Landschaftsentwicklungskonzept (LEK) für die Region München, In Zehlius-Eckert, W., Gnädinger, J. and Tobias, K. (eds.) Landschaftsökologie in Forschung, Planung und Anwendung. Friedrich Duhme zum Gedenken, Freising, Weihenstephan.

Schaller, J. and Schober, H.M. (2007) *Landschaftsentwicklungskonzept Region München (LEK14),* Schlussbericht im Auftrag der Regierung von Oberbayern, unpublished final report, Freising, Kranzberg.

Schaller, J., Gehrke, T., Kemp, K. and Strout, N. (2006) *GIS-Processing Model Development for Environmental Planning and Management*, Research Report, Bavaria California Technology Center (BaCaTec), Erlangen.

Powerpoint Presentations of the Topic at Scientific Conferences

Schaller, J. (2006) *ArcGIS processing models for regional environmental planning in Bavaria*, Presentation at the 21st ESRI EUC, Athens, Greece, 6–8 November.

Schaller, J. (2006) *GIS Processing models for regional urban environmental planning and management*, Presentation at the 44th Annual Conference of the Urban and Regional Information Systems Association, Vancouver, British Columbia, 26–29 September.

Schaller, J. (2007) *ArcGIS—ModelBuilder Applications for Regional and Development Planning in the Region of Munich (Bavaria)*, Presentation at the 16th Czech ESRI GIS User Conference 7–8 November 2007 Praha, Czech Republic.

Schaller, J. (2007) *Sustainable development of the region of Munich using advanced GIS Geo-Database and environmental modelling technologies*, Presentation at the 7th International Workshop on Geographical Information System (IWGIS'07), Peking, 14–15 September.

Additional Readings

Ahern, J. (1999) Spatial concepts, planning strategies and future scenarios: a framework method for integrating landscape ecology and landscape planning, In Klopatek, J.M. and Gardner, R.H. (ed.) *Landscape Ecological Analysis: Issues and Applications,* Springer-Verlag, New York. pp. 175–201.

Batty, M. and Xie, Y. (2005) Urban growth using cellular automata models, In Maguire, D., Batty, M. and Goodchild M.F. (eds.) *GIS, Spatial Analysis and Modeling,* ESRI Press, Redlands, CA, pp. 151–172.

Burrough, P.A. (1986) *Principles of Geographical Information Systems for Land Resources Assessment,* Clarendon, Oxford.

By, R.A. de, Knippers, R.A., Weir, M.J.C., Georgiadou, Y., Kraak, M-J., Westen, C.J. van and Sun, Y. (2001) *Principles of Geographical Information Systems: An Introductory Textbook,* Institute for Geo-Information Science and Earth Observation, Entschede.

ESRI (2000) *ModelBuilder* for ArcView Spatial Analyst 2, ESRI, Redlands, CA. (ESRI White Paper – May 2000). http://downloads.esri.com/support/whitepapers/other_/model_bldravsa2.pdf.#.

ESRI (2003) *ArcGIS 9: Geoprocessing in ArcGIS,* ESRI, Redlands, CA.

ESRI (2005) *ArcGIS 9: Getting Started with ArcGIS,* ESRI, Redlands, CA.

ESRI (2006) *ArcGIS 9: Building Geodatabases Tutorial,* ESRI, Redlands, CA.

ESRI (2007) *Spatial Analyst – Data Analysis Tutorial,* ESRI, Redlands, CA.

Fabos, J.G. (1985) *Land-use Planning: From Global to Local Challenges,* Chapman & Hall, New York.

Goodchild, M.F. (2005) GIS, spatial analysis and modeling overview, In Maguire, D., Batty, M. and Goodchild, M.F. (eds.) *GIS, Spatial Analysis and Modeling,* ESRI Press, Redlands, CA, pp. 1–18.

Krivoruchko, K. and Crawford, C.A.G. (2005) Assessing the uncertainty resulting from geoprocessing operations, In Maguire, D., Batty, M. and Goodchild, M.F. (eds.) *GIS, Spatial Analysis and Modeling,* ESRI Press, Redlands, CA, pp. 67–92.

Maguire, D.J. (2005) Towards a GIS platform for spatial analysis and modelling, In Maguire, D., Batty, M. and Goodchild, M.F. (eds.) *GIS, Spatial Analysis and Modeling,* ESRI Press, Redlands, CA, pp. 19–40.

Maguire, D.J., Batty, M. and Goodchild, M.F. (eds.) (2005) *GIS, Spatial Analysis and Modeling,* ESRI Press, Redlands, CA.

Maxwell, T. and Voinov, A. (2005) Dynamic, geospatial landscape modeling and simulation, In Maguire, D., Batty, M. and Goodchild, M.F. (eds.) *GIS, Spatial Analysis and Modeling,* ESRI Press, Redlands, CA.

Parker, D.C. (2005) Integration of geographic information systems and agent-based models of land use: prospects and challenges, In Maguire, D., Batty, M. and Goodchild, M.F. (eds.) *GIS, Spatial Analysis and Modeling,* ESRI Press, Redlands, CA, pp. 403–422.

Steinitz, C. and Rogers, P. (1970) *A Systems Analysis Model of Urban Change: An Experiment in Interdisciplinary Education,* MIT Report 20, Massachusetts Institute of Education Press, Cambridge.

Chapter 14
Geo-Information Technologies in Support of Environmental and Landscape Planning

Mariolina Besio and Nadia Quadrelli

14.1 Introduction

Knowledge bases have been built to support three environment and landscape planning processes to assess and decide where and how to:

- locate environmental systems (waste disposal and processing, water supply and processing) in areas included within the EU's Objective 2 programmes for the Liguria Region;
- design strategies, zoning and rules for the plan of the Cinque Terre National Park, which prioritise high value landscape preservation; and
- establish a new layout for strategic railway and motorway infrastructures for the metropolitan area of Genoa, tackling the question of compatibility with the affected areas

The knowledge bases, built for the three planning support systems (PSS), require: the investigation of multiple phenomena to organise data, the analysis of data relationships to elaborate meaningful information and the formulation of synthetic models to represent the environmental systems involved in the planning process. Geographical information systems (GIS) projects, characterised by similar methodological procedures, were designed to process the complex knowledge, developed on the basis of a three-levels interpretation and cognitive procedure.

At the first level, official data available from public administrations and local institutions were processed on the basis of general concepts, categories of phenomena, feature classes and values consistent with the models, defined for each environmental and/or landscape planning process. At the second level, the geographical data and thematic maps were analysed. The data were processed, on the basis of morphological and topological approaches, to generate information about the geographical distribution and the simple relations between the phenomena considered

M. Besio (✉) · N. Quadrelli
Università di Genova
Dipartimento DIPARC
Genova, Italy
E-mail: besio@leonardo.arch.unige.it

S. Geertman, J. Stillwell (eds.), *Planning Support Systems Best Practice and New Methods*,
DOI 10.1007/978-1-4020-8952-7_14,

by the models. At the third level, manifold relations between data and information were considered and analysed. The data and information, implemented at the first and the second level, were processed on the basis of the framework of the relations of the models to produce geographical representations that synthetise in an iconic way the environment and/or landscape knowledge.

The innovative aspect of this method lies in the transfer of conceptual models, used to interpret the environmental systems and to represent the landscape profiles, into the methodological procedures adopted to organise the data and to query the database. For this reason, the aspects common to all three experiences were illustrated, which are significant not only in terms of the technological instrument adopted, but above all for the way in which they were used.

In the following sections, the general outlines of the shared methodology are introduced, i.e. the theories and methods followed on the three-levels of the cognitive procedure, in order to reproduce the environmental systems on which planning operates. Subsequently, an explanation is provided of how the general methodological approach was applied in the three cases of construction of the complex knowledge supporting the environmental and landscape planning systems. The chapter concludes with some suggestions for further methodological and research development in the area.

14.2 Conceptual and Methodological Background

The knowledge building projects, to support the three planning systems, are all based upon general approaches, sharing a conceptual model to describe what the environmental system is and which cognitive procedures are required to represent it via GIS technologies. Environmental aspects are crucial in the production of knowledge; a conceptual model regarding the ‘environmental system’ was our previous concern.

Individuals and their communities carry out activities that establish relationships with and modify the environmental system in which they live. The new environmental systems, the landscape transformations and the new projects for railways and motorways not only represent additions, but also interrelate with and modify existing environmental systems and/or landscape profiles. In order to decide which projects and transformations should be acceptable, we should select those phenomena, elements, structures and relationships that are significant for the objectives and the assessment of the environmental and landscape planning process.

At the theoretical level of ‘environmental system’ we considered the environmental system accordingly as a unit of various phenomena with interrelated structures. We assumed the phenomena are composed of many elements; the elements are interrelated in structures; the elements and structures are organically connected into a complex, holistic and hierarchic environmental system: complex as it is made up of many elements, which are related to each other; holistic, as each of their elements is implicit in all the others; hierarchic, as we can represent the environment at different levels from the more general to the more detailed of defination and territorial dimensions. Each more general level implicitly contains those more determined, which, in turn, inherit the concepts and the categories of the more general (Frixione 1994; Marchetti 2002). In operational applications of the conceptual model, we could not examine the whole environmental phenomena, elements and relations (Alexander 2005; Salingaros 2000). In moving from the theoretical considerations to the operational implementation into the planning process, we should activate only those relations relevant to their purpose (Bertalanffy von 2004; Le Moigne 1995; Salingaros 2007; Waldrop 1996).

We envisaged the following environmental system: in the first case, an environmental system consisting of buildings, houses and inhabitants, which determine different types of settlement; in the second, an environmental system consisting of natural and antropic phenomena, which determine different rural organisation and landscape profiles; and in the third, environmental systems of the urban, rural and natural areas, in relation to the existing planning systems and transport network.

Before illustrating the three study cases, it is useful to summarise also the conceptual model we used to categorise the objects of reality. To data representing the objects we attributed implicit concepts and explicit properties, in order to improve the environmental systems and/or the landscape profiles. Concepts are mental representations of reality, which codify experience through categories; categorizing reality is a mental process, which considers objects as parts of a whole. The categories are made up of a series of objects whose properties are not visible (implicit or intrinsic). Because they depend on the concepts (mental pictures) they represent as well as on reasoning. Concepts are found at different hierarchical levels with different degrees of generality and abstraction (e.g. building-house-terraced house). The superordinate concepts create fewer logical inferences than the subordinate ones. A class is the series of objects with explicit properties (characteristics) attributed to them (e.g. three-storey houses, green-coloured houses). Through concepts, categories and feature classes, we associated geographical objects (data) with the phenomena constituting environmental systems. The codification of the cognitive experience through concepts enabled us not only to structure simple data, but also to carry out analyses, which produced derived information and directed reasoning which, in turn, yielded complex and synthetic knowledge.

We developed the cognitive procedure to build up knowledge about the 'environmental systems' at three levels, where the phenomena, elements and relations have been filtered through the concepts, categories and feature classes, relating to the models of environmental systems (Brachman and Levesque 1985; Retake 1995; Smith 1995; Zilli 1999). At the first level (Fig. 14.1), we associated the environmental system with concepts, categories of phenomena, classes of objects/elements and their attributes; moreover we selected belongings and hierarchical relations between the attributes. We broke down analytically the environmental system into sub-concepts: the *natural environment* subsystem, the *man-made environment* subsystem, and the *civil environment* subsystem. We considered only the physical phenomena

Fig. 14.1 Data-base chart

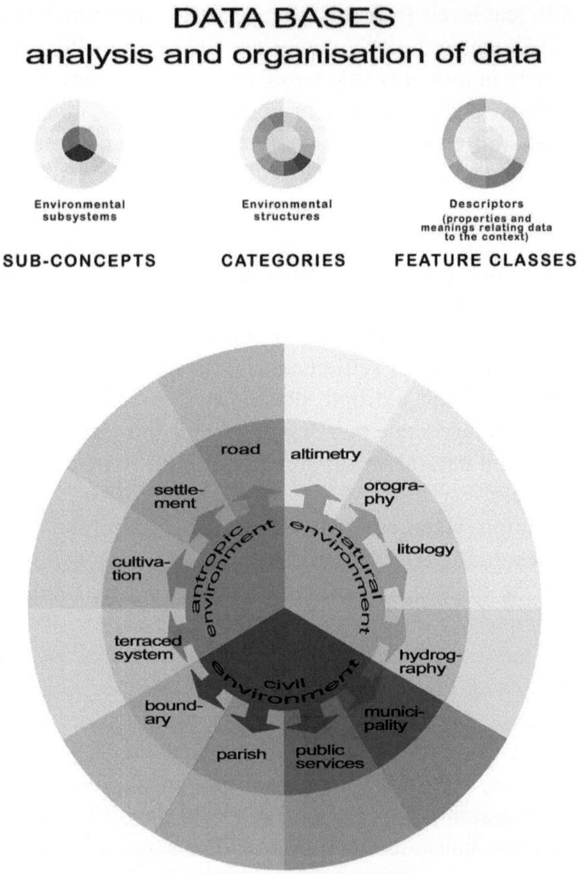

and ascribed to each one of the three sub-concepts. To the natural environment categories we associated the physical phenomena which characterise the natural features of a specific geographical context such as air, water, soil, spontaneous shrub and/or vegetation. To the man-made environment categories we associated the phenomena which have modified the natural asset of a geographical contexts such as settlements, roads and connecting infrastructures, cropping and farming. To the civil environment categories we associated the phenomena relating to the intangible and symbolic structures established by man in order to organise the society, the economy, the institutions, the culture: religious, civil and administrative poles, administrative and cultural borders, property limits, cadastre layout, symbolic buildings, for example.

We analysed the available data: ascribing each of them to one of the three subsystems (natural, anthropic and civil), assigning each to different categories and classifying each by different properties and attributes. We catalogued, selected, conceptualised, categorised and classified the data, implemented the database and drafted thematic maps (Laurini 1992; Laurini and Milleret-Raffort 1989; Oosteron

1993). The term 'descriptors' is used for the data representing the objects with properties and attributes, relating to the planning systems issues.

The same conceptual model produced different representations for each of the three PSS, not only because we considered different categories of phenomena, but also because of different meanings (concepts and attributes) of the same phenomena and elements. We used the same data in a different planning context with different categories, properties and meanings according to different issues. For instance, we used the same data from the Liguria Region technical map-land cover and settlement-albeit with different meanings for the Park plan and evaluation of the rail and motorway projects. In the first case, we analysed data with meanings that represent urban concentration and functions, whilst in the second case, we analysed the same data with meanings that represent landscape forms and profiles.

At the second level (Fig. 14.2), we analysed and reconnected the simple relationships between the elements (descriptors) in an attempt to reconstruct — the environment system. We described physical and spatial phenomena and their simple relations in the light of the decisions of the planning systems and analysed the morphological configurations of the geographic data and their topological relations. We carried out quantification operations in order to identify size, weight and recurrence relations and qualification operations in order to identify form, polarisation,

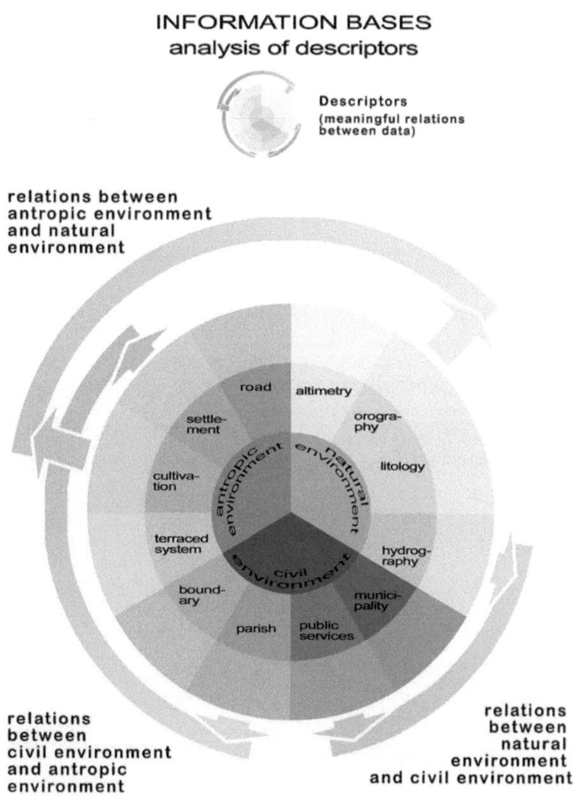

Fig. 14.2 Information-base chart

structures and reciprocal correspondence relations. We implemented the information base, representing a second level of data obtained through basic geographical data processing. We drafted maps, generated by the basic thematic maps, representing the morphological structure of the phenomena and their correspondence (Berry 1995; Bailey and Gatrell 1995; Longley and Batty 1996). 'Descriptors' represent not only original data, but also the data resulting from their interrelations.

At the third level (Fig. 14.3), we reconstructed the environmental system according to the conceptual models adopted. We produced a synthesis of data and information, generating models of environmental and/or landscape systems. In order to represent the simultaneous and structured occurrence of multiple phenomena, we processed data synthetically rather than analytically and adopted the selective and synthesis-oriented tree-structured logic so as to identify meaningful relationships for the context of each planning process. We subjected the data and information to multiple querying, in order to define territorial entities indicating unitary environmental systems and corresponding landscape profiles. We implemented the knowledge base, representing a third level of data, which was obtained via the synthetic, iconic and semantic processing of data and information bases. At the end, we drafted maps representing synthetically the environmental systems and the corresponding

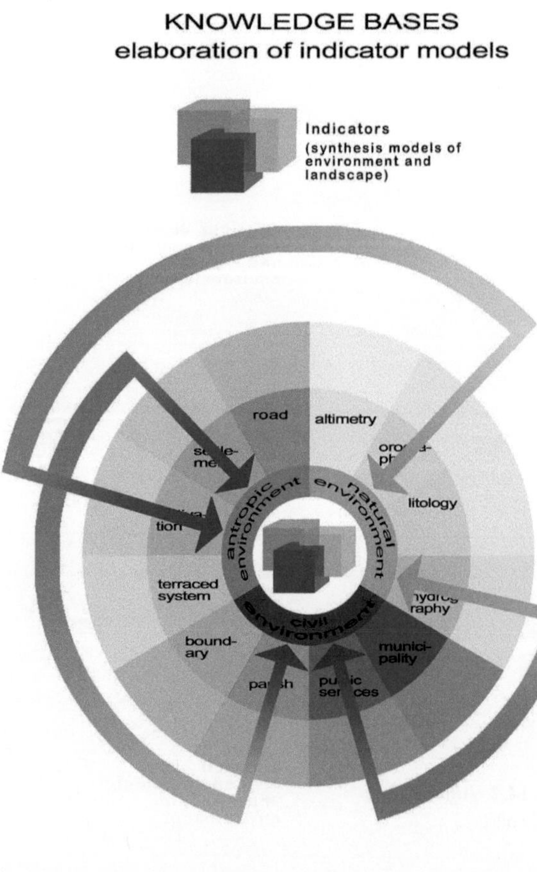

Fig. 14.3 Knowledge-base chart

landscape profile (Bessie 1999; Goodchild 1993; Morin 1993). 'Indicators' are the synthetic data corresponding to different environmental systems and/or landscape profiles.

14.3 Three Experiences of Methodology Application

14.3.1 The Population and Housing Database Organised on the Basis of Environmental Processes of Urban Extension

This project was realised on behalf of the Liguria Region's Environment and Territory Department. It was implemented in those areas of the Liguria Region included within the EU's Objective 2 programmes that represent these most urbanised of the Region. Information and knowledge produced should enrich SITAR's data, information and communication technology (ICT) instruments and geographical information systems (GIS) projects, providing support for the Liguria Region's environmental planning design and control. For the areas already urbanised, it supports the decisions about the planning of environmental systems such as urban waste treatment, water supply and control of pollutant emissions. For the new urbanisation, it supports the procedures evaluating the environmental impact of new residential development, the definition of zones and rules for the plans of constituent of the municipalities.

In the operational application of the conceptual model to different cases of the planning process, we consider only those phenomena, elements, categories, properties, *et cetera,* that can be linked to the tasks and objectives of the planning process and to the human activities involved in it. In this case, the human activities that should be considered are connected to producing, collecting and to processing the domestic waste and to consuming water and electric power for domestic uses. The environmental system, directly linked to these activities, is the environment of the urban and regional settlement, and the most significant phenomena that interact into the system are the dwellings and the communities that people live in.

The purpose of the project is to process the numerical and quantitative data, produced by the Italian Statistics Institute's (ISTAT) 1991 Census, so that they can generate environmental and qualitative information and knowledge. We had to resolve the general cognitive problem to process the numeric available data, concerning the buildings and the demography, so as they produce not only quantitative knowledge, but also qualitative knowledge about the geographical distribution (morphology) of the forms and the structures of the settlements and of the way of dwelling of the communities and inhabitants.

The first operational problem was to select the data and to assign them the meaning useful to represent the form of the settlement and type of dwelling; in other words to identify the 'descriptors' of the phenomena considered. The second problem was to analyse the geographical distribution of the phenomena characterising the settlement form and type of dwelling; in other words to analyse the geographical distribution and consistency of the ‘descriptors’ and to classify them. The third problem was to identify the manifold relations between the ‘descriptors’ that allow one to indicate the territorial entities where the settlement environmental systems are organised and structured; in other words to produce the ‘indicators’ of the geographical distribution of different settlement environmental systems. The data were processed according to a three-level cognitive process (Fig. 14.4).

At the first level, the database was built by analysing, organising and geo-referencing numeric data. The original data, taken directly from the household census sheets were catalogued, aggregated and geo-referenced. Conceptualisation of the settlement environmental system refers to two sub-concepts of built-up and civil environmental sub-systems. The first is represented by the data on houses and buildings and the second by the data on the inhabitants and the families. To these concepts, categories were associated that can express useful meaning to represent the settlement form the type of dwelling and the way of living in the houses. In the case of the house concept, the categories were associated with the house structure, the use and the comfort of the property. For the building concept, the categories were associated with the structure, use and age of the building. The categories associated with the inhabitant concept were identity, cultural level and job. Finally, for the family concept, the categories were associated with structure, property of house and social condition. Then to each category values were given to the feature classes, as illustrated in Fig. 14.4. The value of each numeric datum, extracted from the household census sheets, was associated with the geographic coordinates of the centre of the corresponding census section. They represent the first level ‘descriptors’.

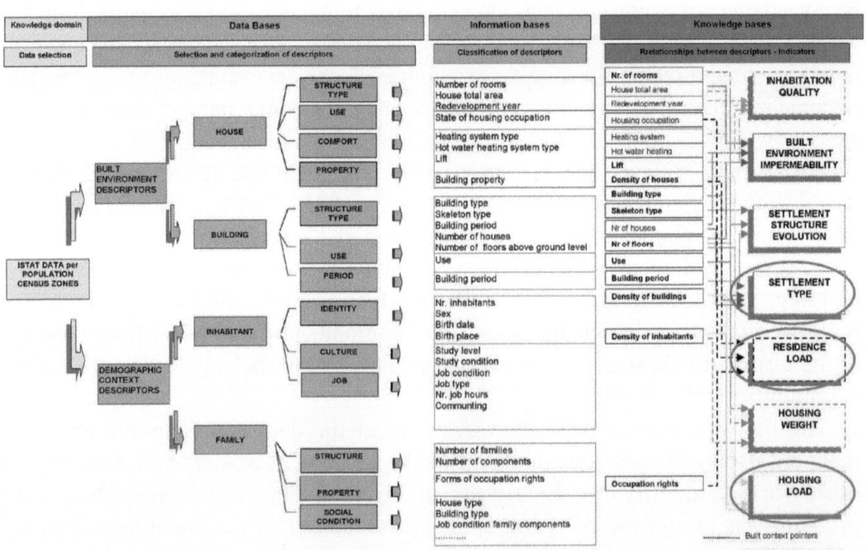

Fig. 14.4 The cognitive procedure: a methodological outline of building up knowledge

At the second level, the information base was established by the geographic processing of the first level data. Each census section is assigned a dominant value for each class of data, which represents conventionally all the present values. The 'descriptors' were assigned value classes, created *ad hoc* by calculating the parameters of the values of each class on the basis of analyses carried out on the all the values present in the Objective 2 areas. These data describe the geographical distribution of the 'descriptor' values on the census sections. Thematic maps illustrate how the descriptors are distributed in the territory of the Objective 2 areas (i.e. their morphology). At the third level, the knowledge base was constructed by means of multiple geographic processing of the first and second level data. The data were subjected to multiple querying, formulated by following a conceptual model that organised the data in such a way that it was possible to identify different environmental systems relating to the settlement, housing and dwelling. These data describe the geographical distribution of the 'indicator' values on the census sections. A breakdown was obtained of municipality territories into environmental units characterised by different types of settlement, types of dwelling and ways of living. Three such indicators were produced: 'settlement type', 'housing load', and 'residence load'.

The indicator 'Settlement type' represents the geographical distribution of various forms of settlement, distinguished from one another in terms of different densities, different aggregations of buildings and houses and different uses. It was constructed on the basis of relationships amongst the descriptors of the structure

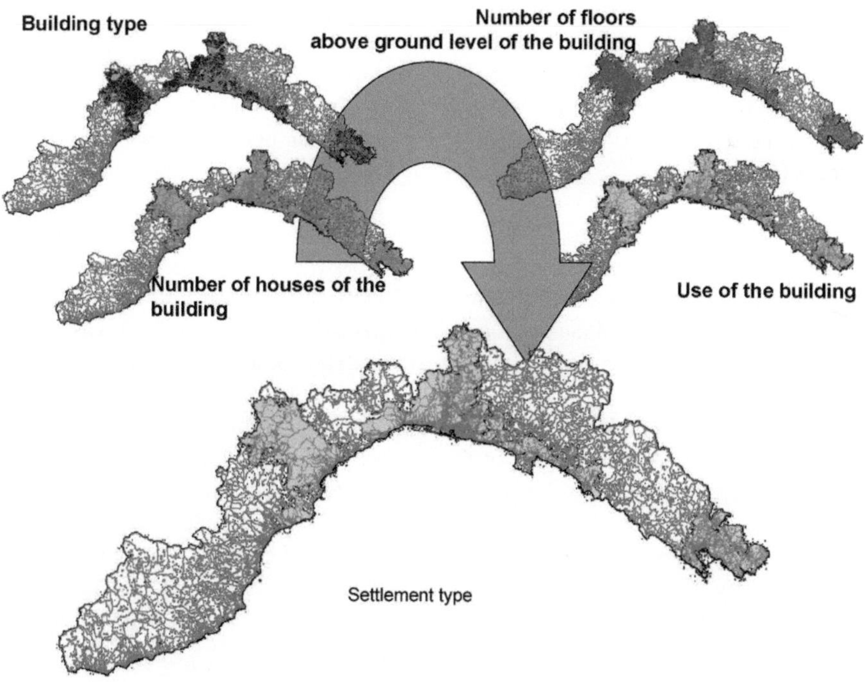

Fig. 14.5 The building up of the 'settlement type' map

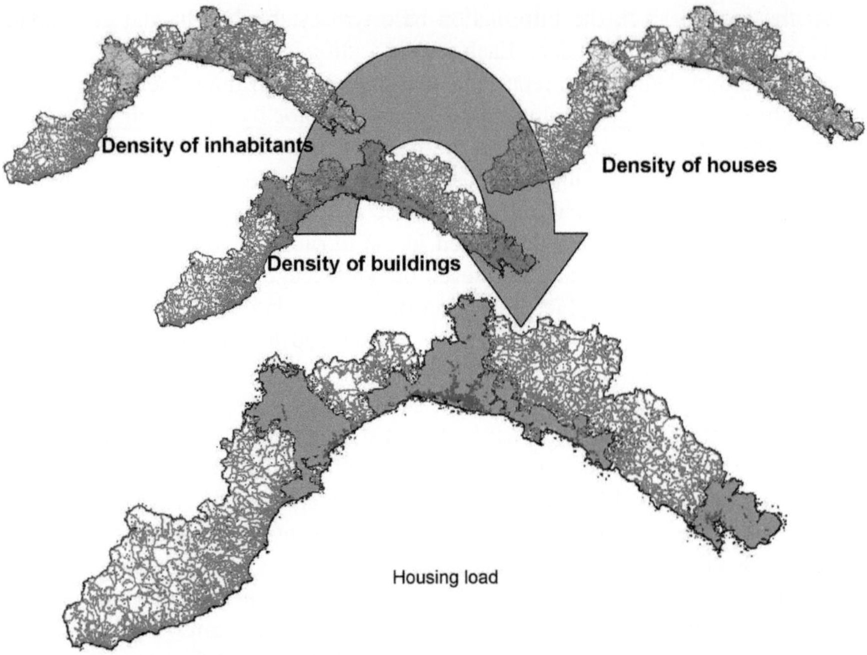

Fig. 14.6 The building up of the 'housing load' map

of buildings, of the number of houses per building, of the number of floors above ground level in the buildings and of the uses of the buildings. The indicator was then assigned values representing the relationship between the settled and unsettled spaces: 'closed' if the ratio was in favour of the first; 'semi-open' if the ratio was equal; 'open' if the ratio was in favour of the second; 'empty' if there were no settled spaces (Fig. 14.5).

The indicator 'housing load' represents geographical distribution of various types of dwelling, relating to the characteristics of the houses and of the inhabitants. It was constructed on the basis of relationships between the descriptors of the density of buildings per square kilometre, the density of houses per square kilometre and the density of inhabitants per square kilometre. The housing load was calculated exclusively with regards to settled spaces. The values are: 'high', 'medium' and 'low' (Fig. 14.6).

The indicator 'residence load' represents the geographical distribution of the various ways the inhabitants live in the houses, relating to the temporary or permanent duration of stay, to the property and to the house characteristics. It was constructed on the basis of relationships between the descriptors of the state of housing occupation, of the forms of occupation rights (i.e. ownership, tenancy) and of the density of homes per square kilometre. The residence load was calculated exclusively with regards to settled spaces. The values are: 'high', 'medium' and 'low' (Fig. 14.7).

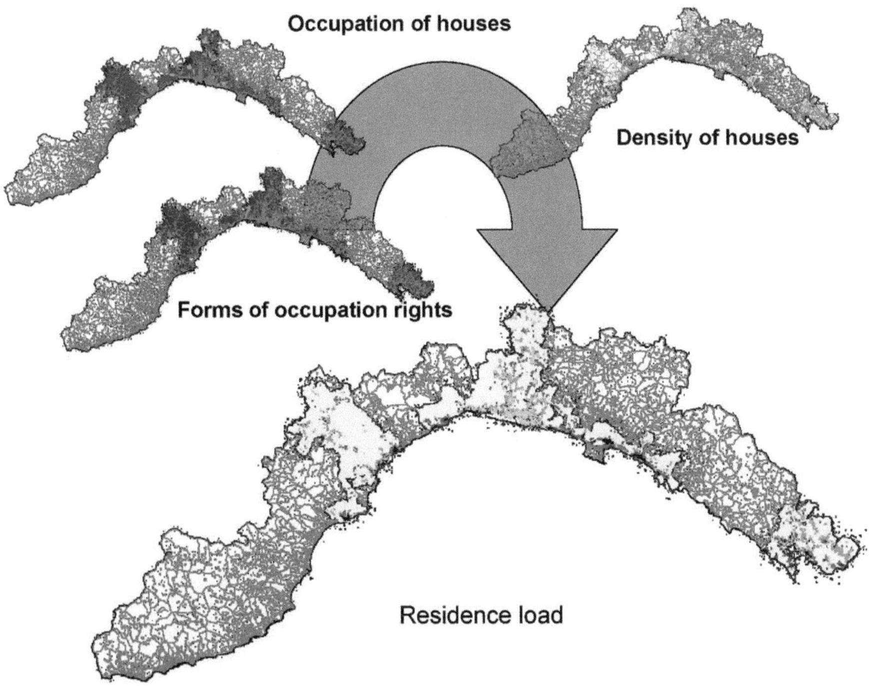

Fig. 14.7 The building up of the 'residence load' map

14.3.2 *Study of Foundation Description (Descrizione fondativa) Supporting the Cinque Terre National Park Plan*

This project was carried out on behalf of the Cinque Terre National Park as part of drawing up the Park Plan, requiring the official documentation of a 'Foundation Description'. The document represents the knowledge which logically and clearly justifies the choices to design the zoning maps and to establish the rules and the norms regulating the transformation of different zones of the plan. The first divides the territory into zones on the basis of a weight of natural values, whilst the second establishes those actions permitted in each zone.

The territory of the Cinque Terre was established as a National Park as a result of its natural value. However, the area's UNESCO heritage site status is also due to its man-made terraced landscape. The Park Plan, therefore, has to protect at the same time both the Cinque Terre's natural environmental system and its rural terraced landscape. Italian law establishes that the plans for the parks should divide the territory into four zones, where the possibilities of transformation increase as the value of the natural environment decreases: zone A involves all over preservation; zone B is orientated preservation; zone C is conservation areas; and zone D involves urbanised areas and coastal centres.

The Foundation Description should produce knowledge that recognises the Cinque Terre's natural environment values together with the values of its terraced landscape. In the planning process, both natural and man-built physical phenomena, elements, categories and properties were considered that could be appreciated as components of an environmental system produced as a project of the territorial architecture. The local communities have realised the project in a collective way during the past centuries, to secure their own survival and that of the coming generations. The natural conditions of the climate, soil, water, original vegetation, *et cetera,* were the resources for developing the project. The civil conditions of competence, knowledge, economy, politics, have been the structural means to develop the project. The man-made works, their spatial organisation, morphology and typology, are the visible results of the project. The natural condition, the man-made works, the civil conditions and their interrelations should be represented and evaluated by the 'Foundation Description'.

The environmental system was considered as a territorial eco-system consciously realised by human beings. The organism hosted in the eco-system is the local community and the natural and man-made phenomena interrelated is the sheltering environment. The human activities involved in the planning process are those that can modify the eco-system equilibrium such as the natural conditions of the soil, water and vegetation, and the man-made works of the rural settlement such as terraces, crops and roads. Furthermore, the institutional activities relating to the existing planning system must be considered. Conceptualisation of the environmental system considers the natural and man-made environment as a unique eco-system (interaction between man and nature) and necessarily takes into account for its protection the civil environmental system (Besio and Quadrelli 2003).

The purpose of the decision support system project was to process geographical and qualitative data to locate different territorial units, characterised by organic and unitary relations between the natural conditions and the man-made works, to evaluate their integrity and to recognise different evolving processes. Only geographic data were processed to design different zones including paper maps, raster and vector maps, technical and regional thematic maps, each produced by the Liguria Region's SITAR, thematic maps prepared by specialists, land-based and aerial photographs, and on-site surveys of terrace structures.

The first problem was to select, organise and to make uniform the data to represent the natural, man-made work and civil phenomena; in other words to build the 'first descriptors' of the phenomena. The second problem was to select the simple relations between the data, representing the meaningful interactions between nature and man; in other words to build the 'second level descriptors' corresponding to phenomena that represent typical relationships between natural and man-made phenomena. The third problem was to identify the manifold relations between the 'descriptors' that allow one to locate the ecosystems corresponding with the four zones of the plan. Moreover, also different evolving processes in the ecosystems were recognised to evaluate their integrity and risk, equilibrium or breakdown, corresponding with sub-zones of the first zones. In other words, the 'indicators' of the ecosystems and of their status were produced,

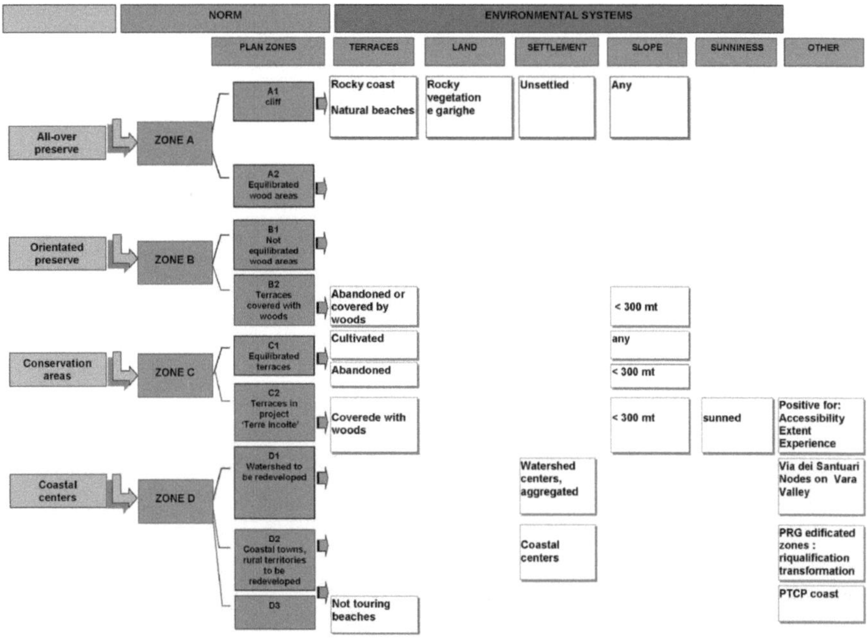

Fig. 14.8 A methodological outline of building up the zoning of the plan

corresponding also to different landscape profiles, which should support the decisions about the zoning of the plan (Fig. 14.8). Here, too, data processing followed a knowledge construction methodology developed over three progressive levels.

At the first level, selecting, standardizing, integrating, organising and implementing data were carried out in constructing the database. The concept of ecosystem (human) was split into the sub-concepts of natural environment, man-made environment and civil environment. For the natural environment, the categories that express the availability of natural resources to support human settlement were associated, such as relief, acclivity, exposure, geomorphology, lithology, natural vegetation. For the man-made environment, the categories of the rural settlement project were associated, such as rural buildings, access and mobility, soil and water arrangement (dry wall terraces and drainage channels), crops and monuments. The man-made environment was then divided into two sub-conceptual groups: the rural environment and the urban environment. To the former were assigned the categories of crop, rural settlement, inter-farm access and mobility, land and water-use structures. To the latter were assigned the categories of urban settlement, polarity and services, accessibility with the external territories, *et cetera*. To the civil environment the categories of ownership or tenancy, administrative divisions, planning regulations currently in force, *et cetera,* were assigned. The concept of instability was also introduced and broken down into two main types, natural and man-made. The first includes the categories of susceptibility to landslides, erosion, flooding, fire. The second includes the categories such as abandoned buildings, unmaintained roads, collapsed terraces. Then, to each category, feature classes values relating to the landscape profiles were assigned. Thematic maps, representing the first level 'descriptors' were produced.

At the second level, the information base was established. Using a morphological approach, the different data and their relationships were analysed, thereby generating information relating to the form and structure of the phenomena as well as to the status of equilibrium and/or conservation. For instance, the terraced areas with woodland vegetation or inaccessible rural settlements were classified as areas susceptible to risk or of instability as defined above. Derived maps were drawn which highlight the morphological configuration of each data set, the relations between different data sets and the evaluation of the status of equilibrium of the elements represented. Maps, representing a second level of 'descriptors', were produced. At the third level, a knowledge base was implemented by multiple processing of the first and second level data. They were subjected to multiple querying which produced synthetic geographical representations of environmental ecosystems and their associated landscape outlines. They represent 'indicators' of the different effectiveness of the zoning of the Park Plan (Fig. 14.9).

At the strategy and policy level of zoning, the indicators in support of the planning decisions are the natural ecosystem, the rural ecosystem and the urban ecosystem

B zones — Zoning and verify criteria

		B zones	B1	B2
zoning	NATURAL SYSTEM	geology		
		geomorphology		Collapses areas
		sunniness		adverse
		slope		elevated
		Natural vegetation	equilibrated wood areas	not equilibrated wood areas or bushes
	ANTHROPIC SYSTEM	agricolture		
		terraces	sporadic	Abandoned or ruined
		mobility	local	local
		settlement	Unsettled or rural episodico	Unsettled or rural episodico
verify	PLANNING SYSTEM	PRG/PUC	Agricultural zones (E)	Agricultural zones (E)
		PTCP	ANIMA	ANIMA, ISMA, ISCE
		SIC	SITI NATURA 2000	SITI NATURA 2000
		Risk zones		R3, R4 (D.L.180/'98)

Fig. 14.9 The query framework producing the zoning of the plan corresponding to the environmental system-landscape outlines

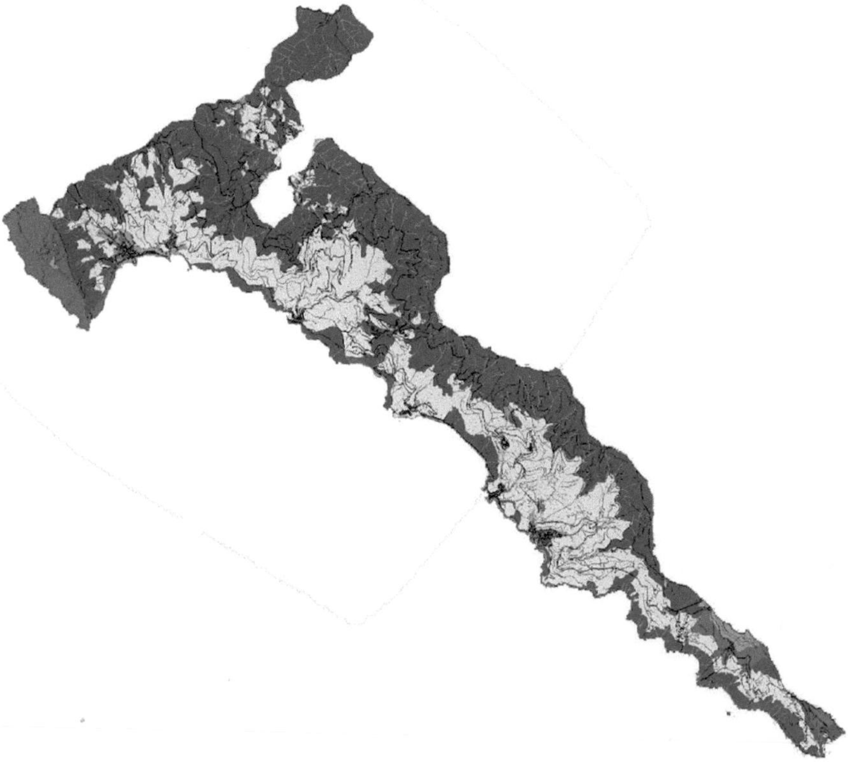

Fig. 14.10 Zoning map of the Cinque Terre Park Plan at the strategic level

and their landscape profiles (Fig. 14.10). At the legislative level of zoning, more detailed 'indicators' in support of the planning decisions are the processes of transformation in progress into the ecosystems as a result of the actions of inhabitants and tourists (Fig. 14.11).

A distinction is made between the zones with the abandonment of farming and renaturalisation of agricultural vegetation and the zones subject to pressure from tourism and urbanisation. This distinction corresponds to zones in which different norms and rules control the transformations. At the level of projects aiming for 'recovery, landscape and environmental renewal, and sustainable development', the rural environment system is represented in greater detail by the 'rural settlement ecosystem', which highlights different areas in terms of organisation and image (not considered in this chapter). The final definition of the Park Plan's zoning was established after the 'indicators' representing the ecosystems were compared with the zoning of planning tools currently in force and presented for consultation with local government representatives and local residents.

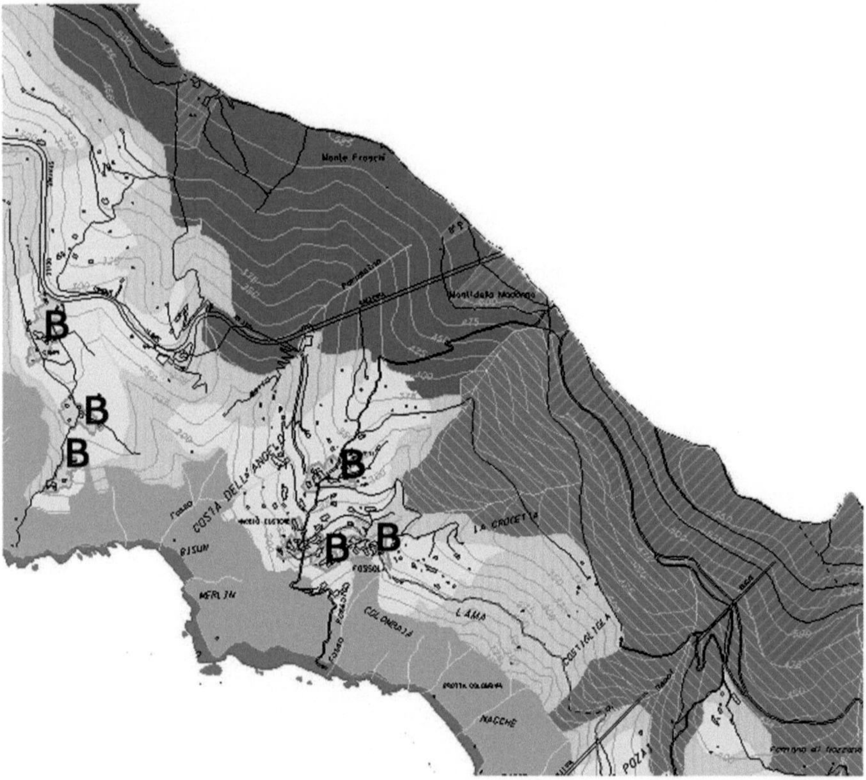

Fig. 14.11 Extract from the zoning map of the Cinque Terre Park Plan at the legislative level (See also Plate 36 in the Colour Plate Section)

14.3.3 Methods and Instruments for the Planning of Sustainable Transport Systems, Integrated with the Management of Large-Scale Areas

This project was carried out on behalf of Genoa City Council. Support was provided for the evaluation, control and realisation of new strategic rail and motorway projects for the port and city of Genoa within the framework of the EU's transport network ('Corridor 5'). It was implemented only for those territories of the Greater Genoa City Council area involved in the projects. Although many institutional bodies, from national to local level, are involved in the projects, their design is the responsibility of the companies providing rail and motorway infrastructures nationally. The local impacts of the projects should be evaluated and the sustainability of new transport links in terms of the environmental compatibility in the areas affected, the consistency with the planning system, the integration with pre-existing transport infrastructures and with urban and port area nodes should be assessed. Moreover, the project set out also to strengthen the GIS capabilities of the Genoa City Council's Territorial

Information System (SIT). An evaluation process was constructed and implemented into the GIS technologies using scenario methodologies. Partial scenarios were produced to evaluate separately the rail and motorway projects with regard to the areas affected, the planning system in force and the pre-existing transport infrastructures. The partial scenarios were integrated together in a single procedure to perform a comprehensive and a punctual evaluation of the projects (Besio *et al.* in press).

The new strategic rail and motorway links impact with the different urban and territorial areas affected: the port area connected to national and international trade; the urban area connected to the ordinary life of the neighbourhoods; the rural areas; the natural areas. The first scenario considers the environmental system corresponding to the concept of an 'urban region' environmental system, composed of the port, urban, rural and natural areas. The new, strategic rail and motorway projects were decided at national level outside of the local and regional decisional planning process. It was necessary therefore to assess their impact on the planning activities of the public administrations of the Municipality of Genoa, of the Province of Genoa and of the Liguria Region. Consequently, the second scenario addresses the planning system that indicates different zoning regulations for urban and rural and natural zones: zones in which no transformations are allowed; zones in which any transformations planned require particular attention; and zones in which new projects are forecast.

Although the new strategic rail and motorway projects interconnect with the pre-existing network of rail and urban roads through only a few interconnection nodes, the effect on urban transport and mobility will be significant. The third scenario represents the urban network of transportation infrastructures and of traffic attraction poles, as a structure that operates at hierarchical levels: neighbourhood, urban unit and metropolitan area.

A considerable amount of heterogeneous data in terms of format (paper, raster, vector), content (generic technical maps, land-use maps, vegetation maps, woodland and forest maps, risk evaluation maps, maps illustrating the zoning of current landscape instruments, maps illustrating rail and motorway projects and major urban development projects) and origin (maps of the Genoa City Council, the Province of Genoa, the Liguria Region, the Genoa Port Authority, other maps produced directly) was used.

The most difficult problem was to integrate into a sole environmental system the 'environmental systems' represented by the scenarios. Conceptually, this was nearly impossible because they refer to different knowledge domains. The problem was solved by processing data with a procedure that had previously evaluated the rail and motorway layout, with each single scenario and then jointly evaluating the single evaluations. Moreover, the evaluation was done both for the stretches of land between the nodes of the rail and motorway network and for each single point of the layout. The information and knowledge generated from the scenarios was organised in a hierarchical way; from more general to more detailed. Standardizing and organising the data in a hierarchical structure such as that adopted in the Corine land-cover database successfully resolved the problem. Data processing followed a knowledge construction methodology developed over three progressive cognitive levels (Fig. 14.12).

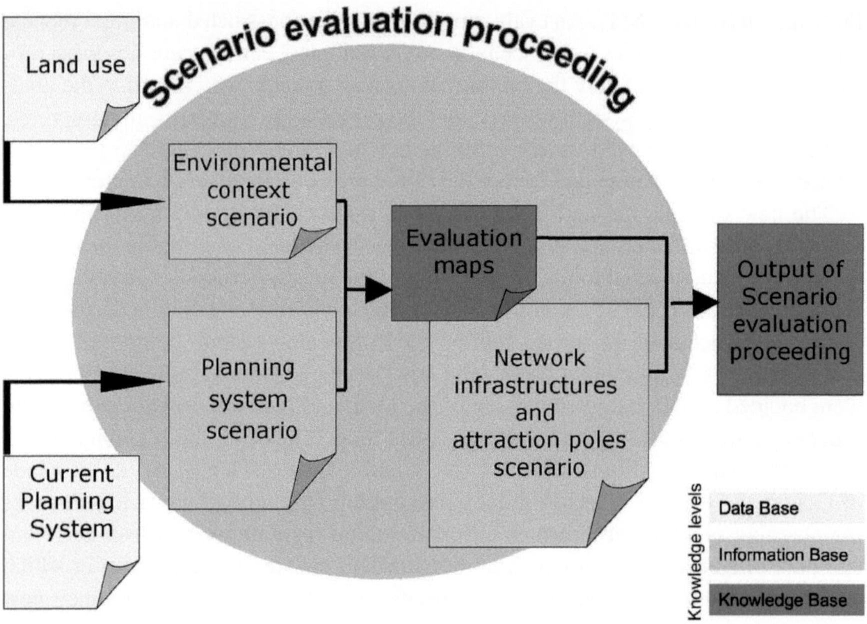

Fig. 14.12 The cognitive procedure: a methodological outline of scenario evaluation

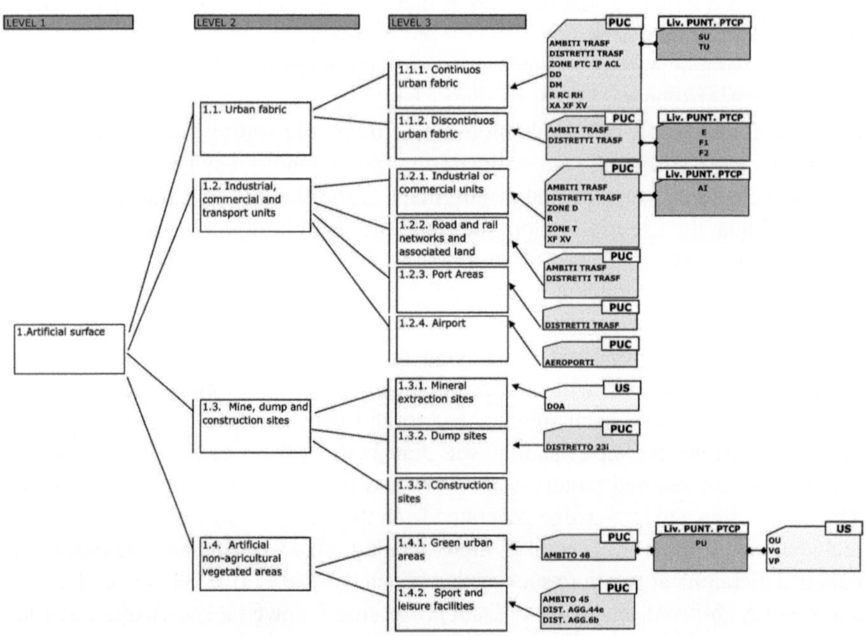

Fig. 14.13 Land-use data classification by Corine land-cover model

At the first level, the database was built up by selecting, standardising integrating and organising available data: the land-use data was used in order to produce the scenario of 'urban region environmental system'; the planning system in force data was used in order to produce the planning system scenario; and the existing rail, motorway, roads network and the new rail and motorway links were used in order to produce the network infrastructures and attraction poles scenario. All data were organised hierarchically adopting the Corine land-cover model, which allows for evaluation both at the large-scale area and in the single points of the layout.

In regard to the first scenario, as can be seen in Fig. 14.13, at the first hierarchical level the concept of 'urban region environmental system' was split into the sub-concepts: port system, urban system, rural system and natural system. At the second hierarchical level, each system was assigned categories of data representing the different elements within; i.e. urban tissues, vegetation, functional use. At the third hierarchical level, each category of data was assigned different classes of values.

With regard to the second scenario, as can be seen in Fig. 14.14, the concept of 'regulation of the territory, environment and landscape transformations' was split into the sub-concepts of conservation of the status quo, transformation with particular attention and the likelihood of transformation projects. Moreover, there is

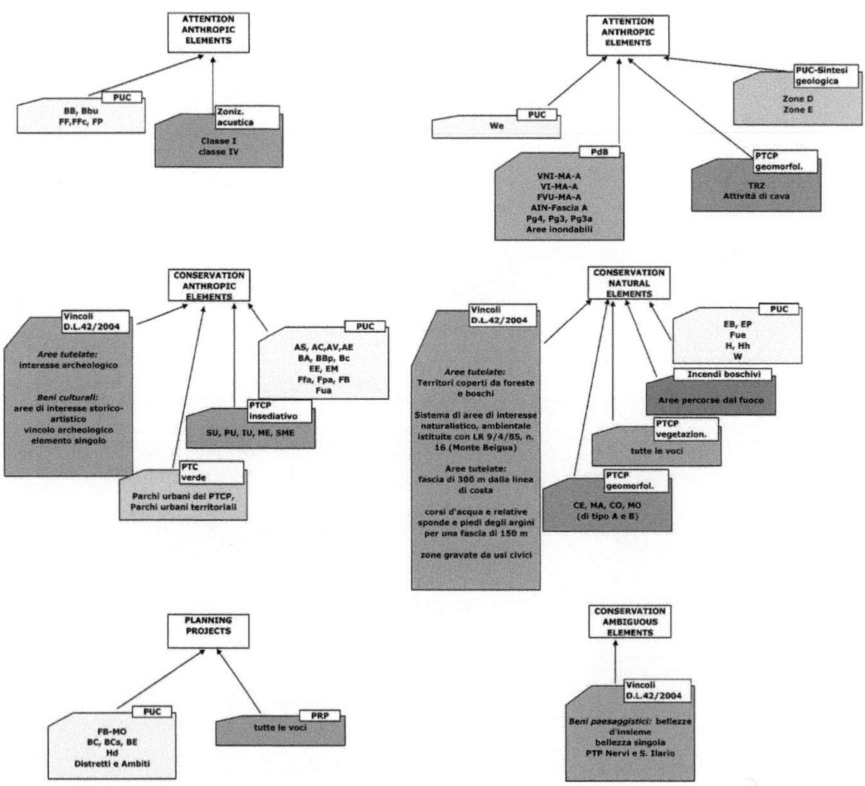

Fig. 14.14 Planning system data classification

a distinction between rules applied to natural or to anthropic elements. The sub-concepts have been applied to different categories of planning tools: landscape and hydro-geological constraints, the municipality of Genoa urban plan, the landscape regional plan, the basin provincial plan. The classes of values of zoning of each category of planning tools were assigned to the regulation concepts. In this way, the zoning and the rules of different planning tools were standardised by the same three sub-concepts: conservation, attention and project.

In the third scenario, as shown in Fig. 14.15, the different elements of the transport and mobility infrastructure network (links and nodes), pre-existing and projected, were categorised and classified hierarchically on the basis of urban functionality of neighbourhood, urban unit and metropolitan area. This first level of data corresponds to the 'descriptors' of the previous experiences of DSS. Thematic maps were produced for all the phenomena analysed.

At the second level, the procedure to evaluate the rail and motorway projects in relation to each single scenario was constructed. This level corresponds to the information base constructed in the previous experiences of DSS construction. The first scenario, evaluating compatibility of the new rail and motorway layout with the area affected, allows one to: offer an oversight of projects relating to the logistics and economic functioning of the port; define actions that limit the impact on urban

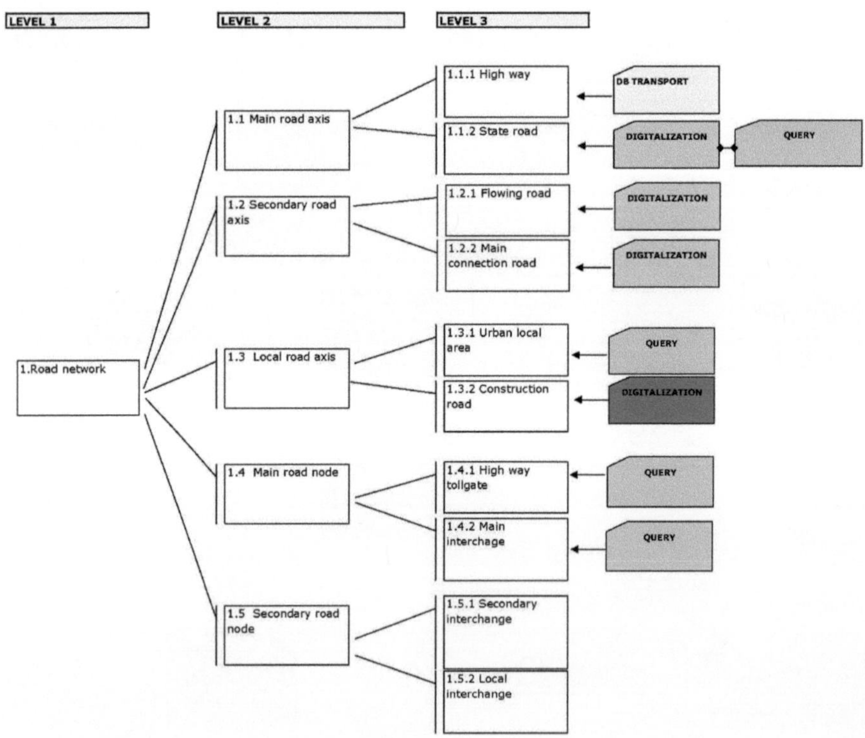

Fig. 14.15 Road network data classification

Fig. 14.16 Urban region environmental system scenario (See also Plate 37 in the Colour Plate Section)

environments; and protect the landscape and prevent hydro-geological risk on the rural and natural environment (Fig. 14.16).

The second scenario aims to compare project outlines of the new rail and motorway with the planning system in force, evaluating the congruence with its zoning,

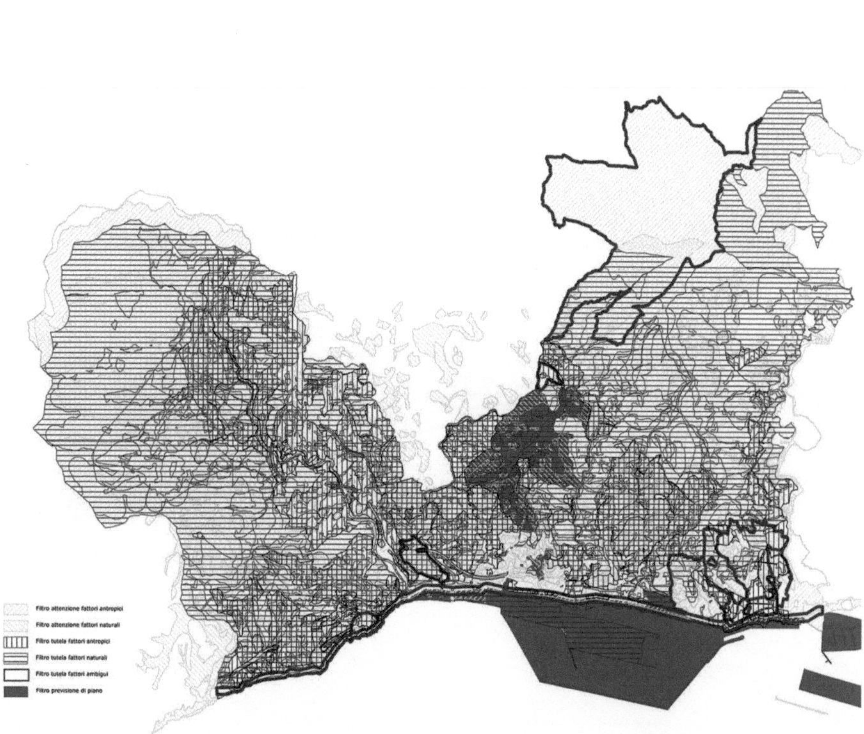

Fig. 14.17 Planning system scenario (See also Plate 38 in the Colour Plate Section)

Fig. 14.18 Network infrastructure scenario (See also Plate 39 in the Colour Plate Section)

rules and restrictions for the regulation of territorial, environmental and landscape transformations (Fig. 14.17).

The third scenario, evaluating the integration of new rail and motorway projects with existing transport networks, seeks to obtain at the same time a detailed and a global view of transport routes and nodes so as to evaluate improvements made to transport networks at national level and assess possible improvements also at local level (Fig. 14.18).

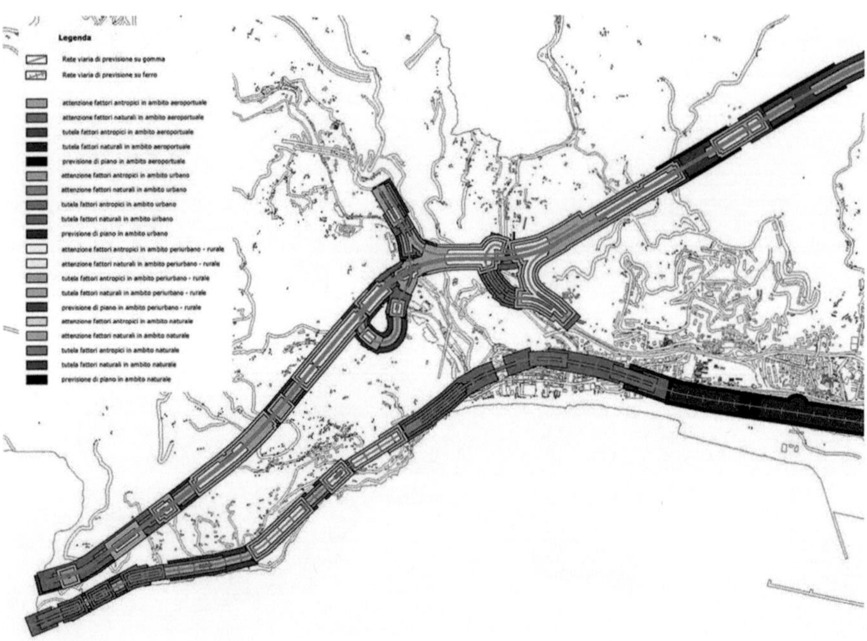

Fig. 14.19 Final scenario evaluation at the general level

Each scenario is analysed in terms of morphological configuration and of hierarchical relationships between different geographical scales. Derived maps were prepared to highlight the impact of railway and motorway projects on the scenarios. At the third level, the whole evaluation procedure was constructed to assess the rail and motorway projects with regard to the three partial scenarios as a whole. The scenarios were subjected to multiple querying, codified in the procedure that evaluated contemporaneously: compatibility with the areas affected; consistency with the planning system; and integration with the urban transportation network of the new railway and motorway projects.

The rail and motorway projects were compared to the three scenarios so as to highlight the contextual features along with an evaluation of criticality and opportunity. It is possible to visualise immediately the implications of positioning a stretch of motorway or railway on a specific urban region area, recovering the values of the three scenarios. The assessment is carried out on different scales: at the global scale, a synthetic visualisation is possible of the compatibility with the urban region crossed, of the suitability with the planning system regulation and of the integration with the pre-existing transport stretches in different affected areas (Fig. 14.19); at the local scale, it is possible to have the detailed sheets of values of indicators and descriptors (Fig. 14.20). The procedure of assessment, based on scenario methodology, corresponds to the level of the 'indicators' of the previous experiences.

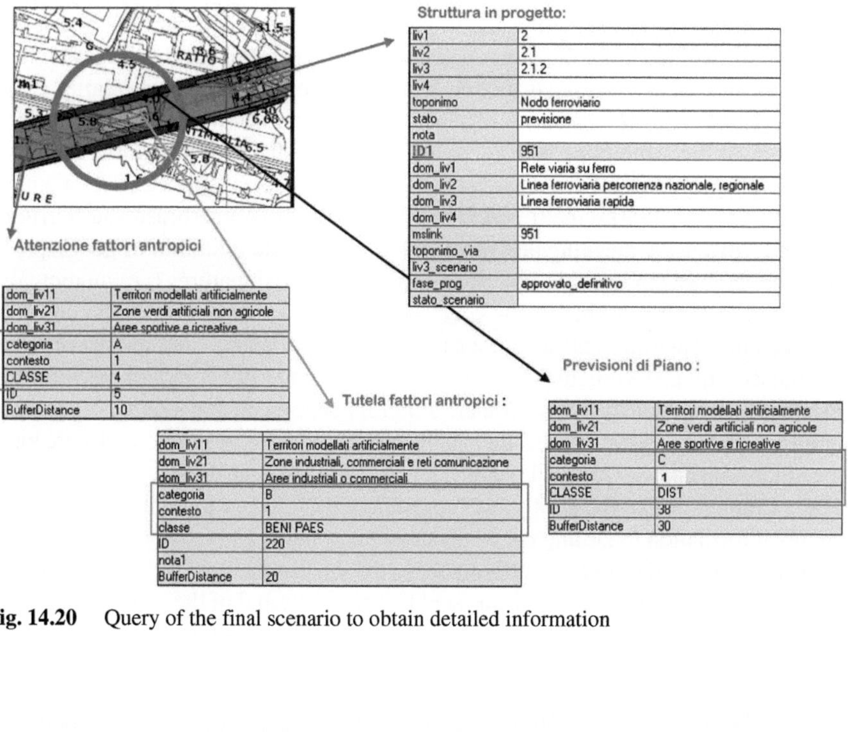

Fig. 14.20 Query of the final scenario to obtain detailed information

14.4 Conclusions and Future Directions

In conclusion, it is useful to reflect how each experience contributed to highlight a particular aspect of the knowledge construction methodology and, on the basis of the experiments, make some more general observations. In the first experiment, our aim was to identify environmental units characterised by different settlement forms and ways of dwelling. Generally, ISTAT numerical data relating to buildings and population are used to obtain quantitative statistical values which are overlaid and applied homogeneously over all municipal territories. In this case multiple numeric data sets had to be treated in combination in order to provide qualitative knowledge. In each municipality different settlement forms and types of dwelling were attributed blurred and qualitative values, produced by the interrelationships between several numeric data values. With this objective in mind, we analysed the meaning that the group of different numeric data might have with regards to settlement morphology and dwelling use. The meaning was linked not only to the choice of data, but also to the way of processing them. It was necessary to consider the logical process from the definition of the conceptual model, as far as the choice of data classes and values. We processed the data according to an ordered sequence of three levels of analysis and interpolation. Firstly, the settlement environmental system was broken down into their simple elements: houses, buildings, inhabitants, and families. Then the single relationships connecting them in a significant way regarding the settlement form and the way of dwelling were selected. Finally, the multiple relations, that locate different settlement environmental systems that represent different settlement form and the way of dwelling, were reconstructed (Besio 2001).

In the second experiment, we wanted to distinguish territorial units (zones), which correspond at the same time to environmental systems and landscape profiles. We investigated the meaning of 'environmental system' and associated 'landscape profile' in a conceptual model. We considered that only one reality exists. Consequently, if different terms of territory, environment and landscape are used to designate the same portion of land, it is not the reality that changes: each of the three terms denotes a different aspect of the reality and these three aspects are interrelated (Assunto 1980; Ghetti and Degli Alberi 1977). We assumed the 'territory' within an extensive meaning and represented it by the delimitation of geographical areas on the basis of specific functions and geographical features (i.e. municipality territories at the beginning of the process, the zoning of the plan at the end). We assumed the 'environment' within a structural, morphological and relational meaning (Giacomini and Romani 2005; McHarg 1969; Tiezzi 1992) and represented it by the manifold relationships between meaningful phenomena that assume organic and unitary structure. We assumed the 'landscape' within an aesthetic and perceptive meaning (Assunto 1975; Bell 1999; Dewey 1934; Merlau-Ponty 1945; Pareyson 1991; Council of Europe 2000) and represented it by the cartographic images conveying information regarding the environmental systems and territorial organisation forms. The territorial level is subordinate to the environmental level. The landscape level is superordinate to the environmental level. At the end of the cognitive process, the landscape level entailed the cartographic images, which define the boundaries of environmental systems and underlying territorial organisations. The theoretical-conceptual model of the 'rural settlement ecosystem', in which man-made and natural phenomena interact, has been defined. A morphological approach was adopted in order to analyse the phenomena, to interpret their meanings through the choice of the elements and their feature classes and to build up the model of the environmental system and the associated landscape profiles through the interpolation of their spatial structures (Besio and Quadrelli 2003).

In the third experiment, the produced knowledge should be useful both on an urban region scale, utilising evaluation criteria related to port logistics and urban planning policies, and at a local level, taking into account the evaluation criteria of the quality of life, of the landscape and of the hydro-geological risk of the areas affected. In order to satisfy the evaluation requisite, a scenario technique was used. In order to satisfy the inter-scale evaluation, the hierarchical structure of data was adopted so as to produce structured inter-scalar knowledge of the phenomena examined. A conceptual framework was developed which ranges from general to detailed concepts and concentrates on defining the data available. The advantage of this is to obtain models of environmental systems, albeit of different definitions, also when data is scarce and/or when it comes from different sources with different meanings and different levels of approximation.

Decisions relating to the three planning processes concern the projects of the future of territories and people. Knowledge can only be useful for environmental and landscape future projects if data and phenomena are represented synthetically. Knowledge can only be useful if it is understood by the people who are affected by it (local government authorities and residents) and therefore should be communicated in a clear and simple way. The processing of knowledge required the adoption of complex procedures, which demanded high technical skills and expertise. How ever, while developing the cognitive procedures in three different levels we tried to simulate commonsense reasoning and convey the results through simple graphics that can be easily understood by a non-expert audience (Frixione 2002; Vaccari 2002; Magnaghi 2005).

On the basis of the experiments, we can make some more general observations. GIS technologies were useful in two ways. At an operational level, as processing together manifold geographical data that should be manually impossible; as processing procedures, which are formalised and demonstrable and make planning processes more transparent. At an epistemological level, as the application of conceptual models and cognitive procedures helped to clarify the environment and landscape conceptual model and paradigms for implementing and processing the GIS projects (Besio *et al.* in press).

In all three experiments, we experienced both technical and cognitive problems yet not resolved, which could require further analyses and studies. With regards to problems of a technical nature, some elaborations of data came from outside the GIS technological process. They are essentially the operative phases of: structuring of data coming from different sources according to the same concept, categories and classes; morphological analyses and classification of the geographic data;

interpolation of various different phenomena within a single territorial unit; and of representation of cartographic results according to expressive graphic language. With regard to problems of a cognitive nature, in order to make the cognitive process seamless and more sequential, further work is required in formal ontologies necessary to acquire non-homogenous data in manner which is coherent and homogenous with the projects (Guarino 1999; Simons 1987; Teller, Lee and Roussey 2007; Varzi 2005) in spatial analyses of geographic data so as to identify morphological and semantic entities; and in topological and mereological analyses so as to mark off geographic areas in which significant interactions amongst different phenomena take place (Casati and Varzi 1996, 1999; Hillier 1999). The integration of geo-technologies with image elaboration technologies so as to produce widely understandable expressive graphic results is a typical technological problem.

The current power of geo-information technologies, together with a theoretically infinite availability of data, makes it possible to process an enormous quantity of information and knowledge. At the same time, such a capacity risks producing an information overload which can be just as problematic as an information shortage. Recent experimentation has extended the use of GIS technologies from generic multi-use geographic databases to knowledge bases structured in order to provide decision support in the planning process.

Nevertheless as to more general problems, we came across insufficient skills and competence in the area of cognitive modelling and the cognitive sciences in general. Knowledge production required the use of suitable conceptual models to interpret the phenomena subject to planning systems, as well as appropriate methodological procedures to transfer the models into processing data. Theoretical and conceptual models in the initial conceptualisation, categorisation and classification stages define the ontology of the geographic objects represented by the data. The significance of this extended role lies not in data and instruments, but in cognitive processes. This certainly represents a major drawback to extending the application of geo-information technologies to the more complex and flexible levels necessary to provide knowledge support to the planning process. To overcome this problem, closer interdisciplinary ties between experts working in the geo-technologies field and planning are needed.

References

Alexander, Ch. (2005) *The Nature of Order: An Essay on the Art of Building and the Nature of the Universe, Books 1–4,* The Centre of Environmental Structures, Berkeley, CA.

Assunto, R. (1975) *Il Paesaggio e l'Estetica*, Giannini, Napoli.

Assunto, R. (1980) Paesaggio, ambiente, territorio: un tentativo di precisazione concettuale, In *Rassegna di Architettura e Urbanistica*, 47/48, pp. 27–35.

Bailey, T. and Gatrell, A. (1995) *Interactive Spatial Data Analysis*, Longman Scientific, London.

Bell, S. (1999) *Landscape: Pattern, Perception and Process*, Spon, New York.

Berry, J. (1995) *Spatial Reasoning for Effective GIS*, Geoinformation, Fort Collins.

Bertalanffy von, L. (2004) *Teoria Generale dei Sistemi: Ffondamenti, Ssviluppo, Applicazioni*, Mondadori, Milano.

Besio, M. (1996) Verso un'integrazione di conoscenze simboliche e di conoscenze metaforiche dell'ambiente costruito, In Maciocco, G. (ed.) *La Città in Ombra: Pianificazione Urbana e Interdisciplinarità*, Milano, Angeli, pp. 221–255.

Besio, M. (1999) The application of cognitive models to environmental planning: the role of GIS technologies, In *Conference Proceedings of CUPUM'99, Computers in Urban Planning & Urban Management*, Venezia, September.

Besio, M. (2001) Costruire conoscenza qualitativa a partire da dati numerici quantitativi, proceedings 1rd conferenza nazionale input 2001, *Informatica e Pianificazione Urbana e Territoriale: Democrazia e Tecnologia*, Isole Tremiti, September.

Besio, M. and Quadrelli, N. (2003) Strategia di conoscenza & strategia di piano, proceedings 3rd Conferenza nazionale Input 2003, *Informatica e Pianificazione Urbana e Territoriale; Costruzione e Gestione Della Conoscenza*, Pisa, July.

Besio, M., Bisio, L. and Quadrelli, N. (in press) Scenari per valutare i progetti delle nuove infrastrutture in area Genovese, In Magnaghi, A. (ed.) *La Costruzione di Scenari Strategici per la Pianificazione del Territorio: Metodi e Tecniche*, Alinea, Firenze.

Brachman, R. and Levesque, H. (ed.) (1985) *Readings in Knowledge Representations*, Morgan Kaufman, Los Altos, CA.

Casati, R. and Varzi, A. (1996) *Buchi e Altre Superficialità. I Buchi Sono Oggetti Reali? Illusioni Cognitive? Entità Metafisiche?* Garzanti, Milano.

Casati, R. and Varzi, A. (1999) *Parts and Places: The Structures of Spatial Representations*, MIT Press, Cambridge, Massachusetts.

Cosgrove, D. (1984) *Social Formation and Symbolic Landscape*, Croom Helm, London.

Council of Europe (2000) *European Convention of the Landscape*, Congress of Local and Regional Authorities of Europe, Firenze.

Dewey, J. (1934) *Art as Experience*, Balch, New York.

Dretske, I. (1995) Meaningful perception, In Osherson, D.N. (ed.) *An Invitation to Cognitive Science, Visual Cognition Vol 2*, MIT Press, Cambridge, Massachusetts, pp. 331–352.

Frixione, M. (1994) *Logica, Significato, Intelligenza Artificiale*, Angeli, Milano.

Frixione, M. (2002) Il senso comune, In Besio, M. (ed.) *Il Vino del Mare. Il Piano del Paesaggio tra i Tempi della Tradizione e i Tempi della Conoscenza*, Marsilio, Venezia, pp. 105–117.

Giacomini, V. and Romani, V. (2005) *Uomini e Parchi*, Angeli, Milano.

Ghetti, P.F. and Degli Alberi, R. (1977) *Proposta per una Lettura Integrata del Territorio*, ed. Istituto di Ecologia, Parma, I.

Goodchild, M. (1993) *Environmental Modelling with GIS*, Oxford University Press, Oxford.

Guarino, N. (ed.) (1999) *Formal Ontology on Information Systems*, IOS Press, Amsterdam.

Laurini, R. and Milleret-Raffort, F. (1989) *Les Bases Dedonnéès en Geomatique*, Hermes, Paris.

Hillier, B. (1999) The hidden geometry of deformed grid, *Environment and Planning B*, 6: 169–191.

Laurini, R. (1992) *Fundamentals of Spatial Information Systems*, Academic Press, London.

Le Moigne, J.L. (1995) *La modellisation des Systemes Complexes*, Dunod, Paris.

Longley, P. and Batty, M. (1996) *Spatial Analysis: Modelling in GIS Environment*, Bell & Bain, Glasgow.

Magnaghi, A. (2005) Il ritorno dei luoghi nel progetto, In Magnaghi, A. (ed.) *La Rappresentazione Identitaria del Territorio. Atlante, Codici, Figure del Progetto Locale*, Alinea, Firenze, pp. 7–19.

Marchetti, M. (2002) Metodologie per una cartografia di uso del suolo multilivello e multiscala: analisi e sperimentazione applicative, *Documenti del Territorio*, 49: 33–51.

McHarg, I. (1969) *Design with Nature*, Doubleday & Company, Inc. Garden City, New York.

Merlau-Ponty, M. (1945) *Phenomenologie de la Perception*, Galimard, Paris.

Morin, E. (1993) *La Conoscenza della Conoscenza*, Feltrinelli, Milano.

Oosteron, P. (1993) *Reactive Data Structures for Geographic Information Systems*, Oxford University Press, Oxord.

Pareyson, L. (1991) *Estetica: Teoria della Formatività*, Bompiani, Milano.

Salingaros, N. (2000) Structure of pattern language, *Architectural Research*, 4: 149–61.

Salingaros N.A. (2007) Complessità e coerenza urbana, *Planum, the European Journal of Planning, topics/themes on line*, 1–32, September, from http://planum.net.

Simons, P. (1987) *Parts a Study in Ontology*, Mouton de Gruitel, Berlin.

Smith, E. (1995) Concepts and categorisation, In Osherson, D. (ed.) *An Invitation to Cognitive Science, Thinking Vol 3*, MIT Press, Cambridge Mass, pp. 3–35.

Teller, J., Lee, J. and Roussey, C. (eds.) (2007) *Ontologies for Urban Development*, Springer, Berlin.

Tiezzi, E. (1992) *Il Capitombolo di Ulisse: Nuova Scienza Estetica della Natura, Sviluppo Sostenibile*, Feltrinelli, Milano.

Turri, E. (1990) *Semiologia del Paesaggio Italiano*, Longanesi, Milano.

Vaccari, E. (2002) Il territorio degli uomini ed i sistemi dinamici complessi, In Besio, M. (ed.) *Il Vino del Mare. Il Piano del Paesaggio tra i Tempi della Tradizione e i Tempi della Conoscenza*, Marsilio, Venezia, pp. 147–159).

Varzi, A. (2005) *Ontologia*, Laterza, Bari.

Waldrop, M. (1996) *Complessità*, Instar, Torino.

Zilli, M. (1999) *Rappresentazioni ed operatività del piano: tra linguaggio disciplinare e linguaggio del senso comune*, Tesi di laurea, Facoltà Architettura, Università di Genova.

Part IV
Participation and Collaboration in PSS

Chapter 15
Enhancing Comprehensive Planning with Public Engagement and Planning Support Integration

Scott N. Lieske, Steve Mullen and Jeffrey D. Hamerlinck

15.1 Introduction

Standing at the forefront of proven methods for integrating planning support with land planning processes is a set of techniques for developing comprehensive plans with consistently high levels of community support. This public engagement based planning process has been used successfully for developing several comprehensive plans in the Rocky Mountain Region of the western United States. In this process, planning support instruments (PSI) are used to collect information and a planning support system (PSS) is used to integrate public values in the development of a concept plan which becomes the basis of the comprehensive plan.

A strong public engagement process, integrating public values and preferences with geographic data, and public feedback on preliminary products results in a quality plan with a high likelihood of being adopted and effectively implemented. The objective of this chapter is to overview this planning process and PSS methods by describing a collaboration between Foresee Consulting, Inc., the University of Wyoming, Winston Associates, Inc. and the Albany County, Wyoming Planning Office. This overview extends the literature addressing public engagement in the planning process, provides insight into the debate about public values leading to the development of a quality plan, and allows case based generalizations which serve to offer further insight into enhancing the use of planning support.

S. N. Lieske (✉)
Wyoming Geographic Information Science Center, Department 4008
1000 E. University Avenue, University of Wyoming
Laramie, Wyoming 82071, USA
E-mail: lieske@uwyo.edu

S. Geertman, J. Stillwell (eds.), *Planning Support Systems Best Practice and New Methods*, DOI 10.1007/978-1-4020-8952-7_15, © Springer Science+Business Media B.V. 2009

15.2 Public Engagement, Quality Plans and Planning Support

15.2.1 Public Engagement for Improved Planning

The ongoing trend in planning toward greater public engagement is well documented in the literature. Dating back to the 1980s (e.g. Alexander and Faludi 1989), public engagement discussions provide critical context for many of the issues surrounding planning in the last decade of the 20th century and the first decade of the 21st century. In the literature, public engagement is referred to in various ways: "*communicative action and interactive practice*" (Innes 1995) "*consensus building*" (Innes 1996), "*communicative planning*" (Innes 1998), "*public participation*" (Harris 1999) "*citizen participation*" (Bleiker and Bleiker 2000) "*planning as reasoning together*" (Klosterman 2000), "*participatory planning*" (Geertman 2002), "*collaborative planning*" (Healey 2003) and here as "*public engagement in the planning process*". The concept is well and succinctly explained as "*...making decisions with stakeholders, rather than making decisions for stakeholders*" by Pettit and Pullar (2004, p. 1).

There are numerous benefits from public engagement in the planning process. Public engagement brings information, knowledge and expertise to planning issues (Brody 2003; Burby 2003). Public engagement facilitates group learning, which can result in more widely shared views and reduced value conflicts (Brody 2003; Harris 1999; Innes 1996). Public engagement processes can bring about a shift from the defence of competitive interests to collaboration and consensus building (Geertman 2006; Harris 1999), give stakeholders a sense of ownership of the process outcomes (Brody *et al.* 2003; Burby 2003) and bring potentially contentious issues to the surface earlier rather than later in the process, thereby helping to avoid disruptions toward the end of planning efforts (Burby 2003; Kelly and Becker 2000).

What is particularly interesting about the trend toward public engagement in planning is that public engagement can lead to higher quality plans and plans that are more likely to be adopted and implemented. The literature, although not conclusive, provides several arguments supporting this assertion. Berke and French (1994) positively correlate public engagement and quality plans. They summarize the components of a quality plan as: (i) establishing existing conditions based on facts;(ii) clear goals based on shared local values; (iii) frequent community-wide exchanges of information; and (iv) specific and action oriented policies (Berke and French 1994). Wide exchanges of information and developing plans based on shared local values are fundamental to today's planning processes, including the example presented in this chapter.

The literature also associates public engagement with successful plan adoption and implementation. Innes (1996), while arguing for the legitimacy of comprehensive planning, correlates consensus building and agreement with plan adoption. Korfmacher (2001) references Forino (1991) in stating that when citizens have contributed to and been educated in a decision process they are more likely to support outcomes and facilitate implementation. Burby correlates planning process generated consensus with political support for a policy proposal. In looking at planning processes in the states of Florida and Washington, Burby found broad stakeholder participation which achieved understanding and agreement resulted in strong plans and increased the likelihood of plan implementation (Burby 2003).

The key element in public support leading to adoption and implementation is achieving a significant degree of shared agreement (a major step along the path toward consensus). Shared agreement results in political support. And, as implementation and adoption are political processes (see also Healey 2003), political support should enable both adoption and implementation. Burby (2003) supports this argument explicitly. He references Bardach (1977) who first observed "*...the politics of adoption should carry over to implementation*" (Burby 2003, p. 34). Citing Innes and others (Innes *et al.* 1994; Innes 1996; Innes and Booher 1999) Burby observes that stakeholder involvement helps to create political capital which in turn helps in getting legislation adopted and implemented. Burby references Healey (1994) and Creighton (1992) respectively in saying stronger attempts at consensus building will be more likely to affect government decisions and that stakeholder involvement can create support for plan implementation. Healey (1994) argues that stronger attempts at consensus building, with integration and consistency of plan elements, result in a plan with more influence over subsequent events (i.e. more implementable). Burby focuses in on why this works in discussing the idea that public engagement can bring about a sense of ownership for participants which in turn can lead to adoption (Burby 2003).

This sense of ownership is the political capital for plans to be adopted and implemented (Brody 2003; Burby 2003; Godschalk and Burby 2003). Brody (2003) generates a similar argument by looking at plan updates in order to gauge their effectiveness in mitigating problems of natural hazards. One of the conclusions of Brody's work is that citizen participation can enhance plan quality and may result in more successful plan implementation. When participation is based on collaboration and/or develops agreement, plans are stronger and more likely to be adopted and implemented. Mastop and Faludi (1997) suggest, as in the PSS process detailed below, the public has a significant role in assessing whether a plan is reasonable and relevant. They advance an interesting idea that public engagement may bring about a situation where a plan can work before it exists because people do not wait for plans to be adopted before acting on shared understanding (Mastop and Faludi 1997). Finally, Talen (1997) offers support for the argument that public engagement leads to desirable planning outcomes from the opposite perspective, by looking at what makes plans fail: lack of consensus, low community support, uncertainty and lack of data and knowledge.

15.3 Planning Support—Instruments and Systems

Coinciding with the growing recognition and employment of public engagement in the planning process has been the emergence of information technology based strategies and tools for community decision making or planning support.

Geertman (2006) defines planning support as *"dedicated information, knowledge, and instruments that people actively involved within formal [planning] practices can receive to enlighten... their planning tasks and activities"* (p. 864). Geertman suggests the way to bring about planning support is through PSI, defined as computer based tools dedicated to the support of spatial planning tasks (Geertman 2006). Others have more broadly defined these instruments to refer to a wide range of planning tools fostering civic engagement and public decision making. Some but not all of these instruments support cooperative or collaborative work. Some are also not digital technologies, and not all of the digital tools incorporate geospatial data and information. Recognized categories include: (i) information resources; (ii) community process tools; (iii) visualization tools; (iv) impact analysis tools; (v) geographic information systems; (vi) predictive modelling techniques; and (vii) regional resource centres (Alliance for Regional Stewardship 2001; Boyd and Chan 2002; Snyder 2003).

Within the literature, the term PSS has evolved as a subset of PSI representing, *"geo-information-technology-based [planning support] instruments that incorporate a suite of components (theories, data, information, knowledge, methods, tools) which collectively support all or some part of a unique professional planning task"* (Geertman 2006 p. 864). More specifically, PSS, in an application oriented definition, *"...bring together the functionalities of geographic information systems (GIS), models, and visualization, to gather, structure, analyze, and communicate information in planning"* (Vonk *et al.* 2005, p. 910). PSS extend GIS capabilities in analysis and problem solving and add design, decision making and communication capabilities (Nedović-Budić 2000).

There are two dominant trends in the literature surrounding PSS: the utility of planning support tools (Couclelis 2005; Nedović-Budić 2000; Snyder 2003) and the under-utilization of these tools (Geertman 2002, 2006; Geertman and Stillwell 2003a,b; Nedović-Buditć 1998, 2000; Vonk *et al.* 2005). The utility of PSS are broadly supported in the literature. Couclelis (2005) lists the development of scientifically based insights, communication, the testing of alternative planning strategies ('what may be'), assistance with visioning ('what should be') and assistance with storytelling ('what could be') as some of the potential benefits of PSS. Snyder (2003) indicates PSS could transform planning decision making in two ways: first, by shifting planning from the currently typical regulatory approach to a more forward looking pro-active performance based approach; and second, by enabling public involvement in planning and decision making processes. Nedović-Budić (2000) lists what PSS are expected to do: help with data management, analysis, problem solving, design, decision making and communication as well as facilitate group understanding in collaborative planning processes. Specific to integrating preferences and values from public involvement, PSS can incorporate decision analysis techniques such as multi-criteria analysis (MCA) which help in assimilating values and moving past values disagreements in the planning process (Malczewski 1999).

In spite of the potential benefits of PSS, the literature makes clear usage of PSS is not on par with utility. As a general summary, Geertman (2006) puts forth that planners have not embraced tools available to them and goes on to address the mismatch between supply, demand and applications of PSS. Vonk *et al.* (2005) summarize existing research and conclude PSS usage is not widespread, listing many general (e.g. institutional) and specific (e.g. too complex) reasons for under-utilization. Little has been published on the effectiveness of PSS or other spatial decisions support systems in specific planning applications (Aggett and McColl 2006). However, ways to more effectively implement and expand use of these tools are actively being explored (see Couclelis 2005; Klosterman and Pettit 2005; Vonk *et al.* 2005). The PSS literature has evolved to the point where there are well informed recommendations for enhanced utilization and broader implementation. Some of the most important are a shared commitment to well-defined methods and the ability of PSS to provide needed outputs for a substantial user community (Klosterman and Pettit 2005). In order to reach this desirable state, Vonk *et al.* (2005) recommend disseminating information and knowledge about PSS through real world example projects.

Based on the broad trend of greater public engagement in planning processes and the theoretical and researched arguments that public engagement can lead to higher quality plans, increased plan adoption and enhanced implementation, there exists an opportunity to define methods for planning support (PSI and PSS) use in support of public engagement planning processes. The planning process described in this chapter is based on a relatively new method which integrates public values with geographic data through a public engagement process. The process results in high levels of community support which lead to the development of a quality plan and enhanced probabilities for plan adoption and implementation.

15.4 Applying Planning Support Technologies in a Public Engagement Process

The methods described below were applied in a county comprehensive plan update in Albany County, Wyoming, USA. Albany County is located in southeastern Wyoming, in an arid, high plains portion of the Central Rocky Mountains. It is a large county (4,307 sq. miles or 11,114 sq. km) and sparsely populated. The county government and 25,688 of the county's 30,360 people are located in the city of Laramie. The other incorporated town is Rock River, population 202 (Wyoming Division of Economic Analysis 2007). In terms of land use, approximately one third of the county is state or federally owned public range and forest land, with an additional 62 per cent of the land in the county classified as agricultural. Only 4 per cent is residential, while commercial and industrial lands comprise less than 1 per cent of the total land area. The Denver metropolitan area begins approximately 100 miles to the south along the fast growing Front Range region of Colorado.

With its roots dating back to the Standard City Planning Enabling Act of 1928, comprehensive planning can be defined in the USA as an effort to address the entire range of interrelated land-use, transportation, and growth issues facing a community (Jurgensmeyer and Roberts 2003). Comprehensive planning is characterized by inclusivity in three dimensions: (i) geographic coverage, i.e. a plan for an entire planning service area, such as a city or county; (ii) subject matter, where a land-use based physical planning focus is integrated with transportation, housing, environmental and social concerns; and (iii) time horizon, addressing both near- and long-term (i.e. 20–50 years) goals (Kelly and Becker 2000).

Development of the Albany County Comprehensive Plan began in October of 2006. The plan was necessary to address issues of utility infrastructure, service provision, environmental protection and community character brought about by steady population increases, exurban development around Laramie, and concerns with protecting a valuable ground water resource located within and adjacent to Laramie. The effort was envisioned as year-long process, with county planning department staff supported by consulting services from the University of Wyoming and a private sector planning consultant team.

Figure 15.1 outlines the three major phases of activity for the Albany County comprehensive planning process. Phase I, Project Organization, involved attaining buy-in and support from county commissioners and members of the county planning and zoning board, as well as other elected officials and key decision makers in local government and the business community. Phase II, Public Engagement, included a five stage process which resulted in a draft preferred alternative, or concept plan, for a county-wide general land-use plan. Finally, Phase III, Plan Development and Adoption, included development of policies and strategies for implementation, culminating in the plan's adoption by the Board of County Commissioners.

An important task under Phase I was the establishment of a Citizens Advisory Committee (CAC) for the plan's development. Selected from a pool of volunteer citizens submitting background and opinion questionnaires, the group consisted of 19 men and women representing different interests and geographic areas of the county. Members participated in monthly meetings, served as liaisons with various sectors of the county population, assisted with public input activities, evaluated technical reports, and reviewed interim and final planning documents. The CAC was supported by a smaller Technical Advisory Committee (TAC) comprised of local experts with professional backgrounds ranging from law and economics to hydrology and wildlife biology. While portrayed in Fig. 15.1 as a strictly linear

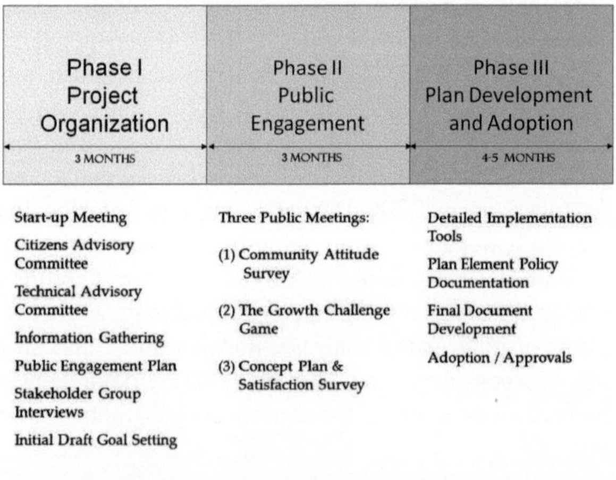

Fig. 15.1 Albany County comprehensive planning process

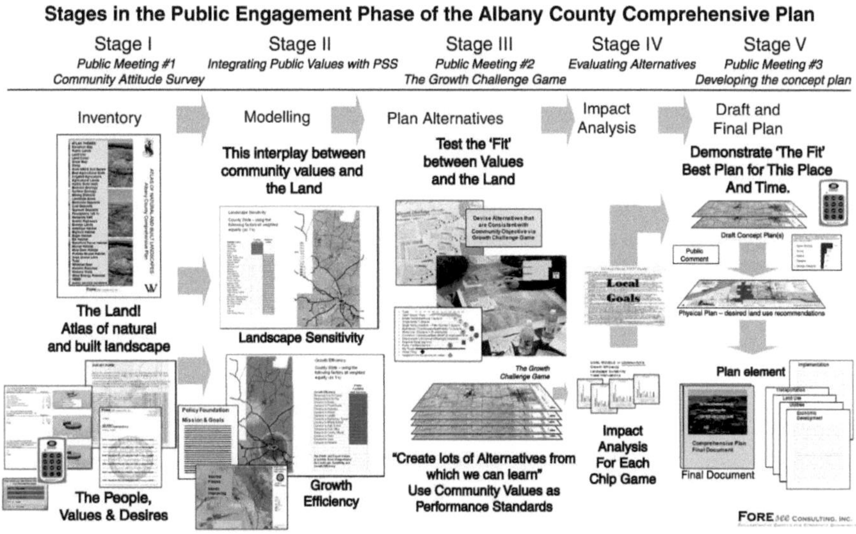

Fig. 15.2 Public Engagement phase of the Albany County comprehensive planning process

process, the CAC and TAC framework established in Phase I was actively engaged throughout the entire planning process and critical to eventual plan adoption.

Phase II of the county's overall comprehensive planning effort was designed around a five-stage public engagement process including three PSI-enabled public meetings (Fig. 15.2) and two major inter-meeting stages where behind the scenes use of PSS supported the public engagement process. The remainder of Section 15.4 details these activities.

15.4.1 Stage I: The Community Attitude Survey

The purpose of Stage I, Public Meeting Number 1, was to let people know they have a role in the planning process, provide necessary and shared background, and to survey community attitudes and values. The first meeting followed the Bleiker technique for citizen participation where a problem is identified, the legitimacy of the agency to solve the problem is established and the legitimacy of the process is established (Bleiker and Bleiker 2000). In comprehensive planning, the problem usually focuses on accommodating growth while at the same time maintaining quality of life and levels of service provision. The meeting began with an overview of this comprehensive planning process, the three meeting public engagement process, a discussion of the importance and benefits of planning, and a description of how a comprehensive plan fits and works within the legal framework of planning in Wyoming. Part of the background material was a wall-mounted atlas consisting of large paper maps of geographic features relevant to planning including elements of the built environment, resource lands (e.g. prime agricultural lands) and problems and hazards (e.g. steep slopes). Discussion of the comprehensive planning process, including past successes, served to establish the legitimacy of the process. The discussion of the legal framework of comprehensive planning established the Albany County Planning Department as the legitimate agency to proceed with development of the comprehensive plan.

A critical element in gaining support from all process participants was integrating their values and direction in all stages of plan development. Two PSI were used to gather input at the first public meeting: keypad polling and a paper survey form. Keypad polling continues to prove itself as an effective tool for generating high quality public input. The anonymous nature of feedback creates a level playing field in a public setting so no individual or issue dominates discussion and everyone's opinion is heard (Snyder 2006). Keypad polling also serves to educate people on majority opinions. For example, a citizen may enter a meeting with the impression fellow citizens do not value wetlands. The citizen may learn, in fact, the majority of the community does value wetlands. Another benefit is with near instantaneous processing and presentation of results, people enjoy the polling process. Anecdotal evidence indicates keypad polling participants take their responses seriously. One could argue that if they are voluntarily present, then they are committed to valid participation, regardless of their level of engagement. Further, responses to questions about the value of keypad technology posed in actual planning processes by both the authors and other PSS researchers indicate that the strong majority of planners and citizen participants find the technique to be 'very helpful' and/or 'engaging' as opposed to merely 'fun and different' or 'of little value' (Field and Snyder 2006; Hamerlinck 2007).

Keypad polling typically begins with a lighthearted question used to familiarize people with the technology. Next are a series of demographic questions that are both easy to answer and provide a sense of who is participating. This introduction is followed by the 'community attitude survey'. The community attitude survey covered a range of topics relevant to planning and development in the county: adequacy of health care, law enforcement, fire protection, roads, schools, libraries, affordable housing, shopping, aesthetics, growth and growth management concerns (with choices among traffic, character, housing affordability, *et cetera*), building codes, nuisance standards, agriculture and government fiscal solvency. Developing best practices suggest when asking citizens to respond with keypad polling to technical issues it is incumbent upon keypad polling surveyors to provide enough information so the public can respond in an intelligent manner. Background slides for educational purposes may proceed questions where appropriate.

Another important step in developing the comprehensive plan was establishing community goals and objectives. Goals, based on the previous (1997) comprehensive plan were presented at the first meeting as being still relevant to the new comprehensive plan. Keypad polling and the community attitude survey were used to gather public input and verify the relevance of those goals. The community attitude survey also resulted in several, sometimes overlapping, objectives for the plan. Examples include: new development should locate adjacent to existing development;

agricultural land should be preserved; and the Albany County legacy (identity and character) should be amplified (Albany County 2007, p. 3; Appendix A).

Plan goals and the entire list of objectives served as a basis for establishing 'measures of a quality plan'. These measures were the protection and/or maintenance of: (i) historic legacy; (ii) agricultural resources; (iii) growth efficiency; (iv) resource protection; (v) water availability; (vi) frontage evaluation; and (vii) access evaluation. The measures of a quality plan are used later in the process as the basis for evaluating citizen generated development alternatives.

The paper survey at the first meeting asked the public to rank factors that affect development in two broad categories: sensitive lands (e.g. floodplains, hazards, wetlands *et cetera*) and factors important to growth efficiency (e.g. distance to elementary schools, proximity to paved roads and quick response time for emergency services). Respondents could chose from 1 (least important) through 5 (most important). The components of the sensitive lands survey were identical to the resource lands and problems and hazards presented in the atlas. After the first meeting, a PSS was used to create weighted maps of sensitive lands and factors important to growth efficiency that incorporated these public values. These two maps begin to define 'where not to grow' and 'where to grow' in presenting both a spatial expression of community values and the interplay between community values and land.

The PSS used in landscape sensitivity modelling, growth efficiency modelling and the overall method is the Scenario 360 module of *CommunityViz®*. Developed by the Orton Family Foundation (Rutland, VT), *CommunityViz* is a modular system built on the *ArcGIS* platform (ESRI, Inc.). Scenario 360 provides functionality for creating and assessing the potential impacts of specific, proposed land-use actions by facilitating scenario building and monitoring change in a series of associated indicators. It extends the quantitative capabilities of *ArcGIS* by allowing spreadsheet like calculations to be performed on geographic data layers and associated tables (Kwartler and Bernard 2001).

15.4.2 Stage II: Modelling Landscape Sensitivity and Growth Efficiency

15.4.2.1 Landscape Sensitivity Modelling

Landscape sensitivity modelling uses multiple criteria analysis (MCA) techniques rooted in McHarg's (1969) concept of overlay mapping. The landscape sensitivity model was built of data elements consisting of resource lands (e.g. wetlands, riparian areas and prime agricultural soils) and problems and hazards (e.g. flood plains, steep slopes, areas of potential landslide). In order to transform values of the disparate set of thematic layers to comparable units (or overlay), digital versions of the resource layers and problems and hazards layers were presence/absence coded and unioned to form a single composite data layer with component parts as separate attribute columns.

Average surveyed values for each data element were integrated into the PSS as assumptions—a *CommunityViz®* term for a changeable input value (Orton Family Foundation 2006). Assumptions for importance range from 1 to 5. The unioned layer of sensitive lands was then made dynamic, another *CommunityViz* term, which enabled the quantitative functionality of Scenario 360. In this context, dynamic functionality allowed values indicating feature presence to be weighted by public input survey data and summing of these weighted values to create a total sensitivity score for each polygon in the study area. The larger the score, the more sensitive land is to future development. Mapped output (Fig. 15.3) demonstrates a range of public opinion based on importance to sensitive lands with low, lightly coloured values being the least sensitive lands and higher, darker values representing more sensitive lands. By weighing and combining resource lands and problems and hazards lands the public determines which lands in the county are sensitive to development and which lands are not. Results indicated the public was most concerned with groundwater recharge areas, other hydrological features including creeks, rivers and lakes; and a several components of wildlife habitat (Albany County 2007).

15.4.2.2 Growth Efficiency Modelling

Growth efficiency modelling is a PSS-enabled effort to evaluate efficient locations for future growth based on the physical locations of service origins (e.g. the locations of fire stations, elementary schools *et cetera*), population distribution, and public survey data. It is a two part process. The first portion of the model can be looked at as a simple production constrained spatial interaction model (Fotheringham and O'Kelly 1989). Spatial interaction is based on relationships between service origins, destinations (population centres), a measure of destination attractiveness (population) and the direct distance separation between origin and destination. The attributes used to represent service origins are a custom made layer of civic facilities: airports, county offices, elementary schools, fire stations, high schools, hospitals, landfills, middle schools, parks, police stations, post offices, roads, trails and a university. Spatial interaction is modelled in the PSS using the dynamic attribute functionality of *CommunityViz®* to multiply the population of each census block in the county by the distance to each civic facility. The spatial interaction model results in a set of flows representing cost of service provision which are used to establish performance measures (indicators) for community norms for growth efficiency.

The second major component of the growth efficiency model is a public value weighted MCA model similar to the landscape sensitivity model. The geographic basis of the growth efficiency model calculations and outputs was a grid layer with cells of one square mile. For model construction, the grid had to first be attributed with the distance from each cell to each service origin. The next calculation was level of service for each cell for each factor important to growth efficiency. The level of service is a Boolean value where one indicates the cell is less than or equal to the community norm specified distance and zero indicates the cell falls outside the community norm specified distance. In a similar fashion to the landscape sensitivity

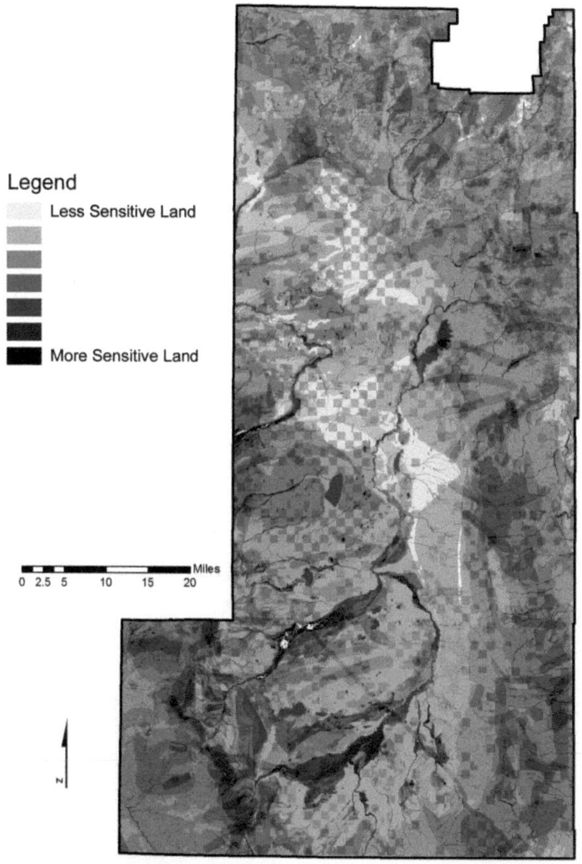

Fig. 15.3 Value weighted landscape sensitivity (See also Plate 40 in the Colour Plate Section)

model, the weighting component of the growth efficiency MCA model is based on results from polling at the first public meeting incorporated into the PSS as a set of assumptions of importance for each factor important to growth efficiency. In the PSS, ones and zeros used to depict cells above or below existing service limits are multiplied by the appropriate weighting values then summed for each cell to create an overall level of service for each cell. The map output (Fig. 15.4) combines all of the factors important to growth efficiency into a summary representation showing the range of level of service provision throughout the study area. On the map, darker areas represent higher levels of service provision and lighter areas represent lower levels of service provision. In effect, the growth efficiency summary map is a citizen created spatial expression about efficient locations for future growth.

The contrasting use of vector outputs in landscape sensitivity modelling (Fig. 15.3) and raster outputs in growth efficiency modelling (Fig. 15.4) is intentional. The strengths of the raster data model for processing large amounts of data are well known. These strengths are drawn upon in growth efficiency modelling where people feel comfortable with the general nature of the presentation of data in the raster environment. On the other hand, experience has shown with a raster

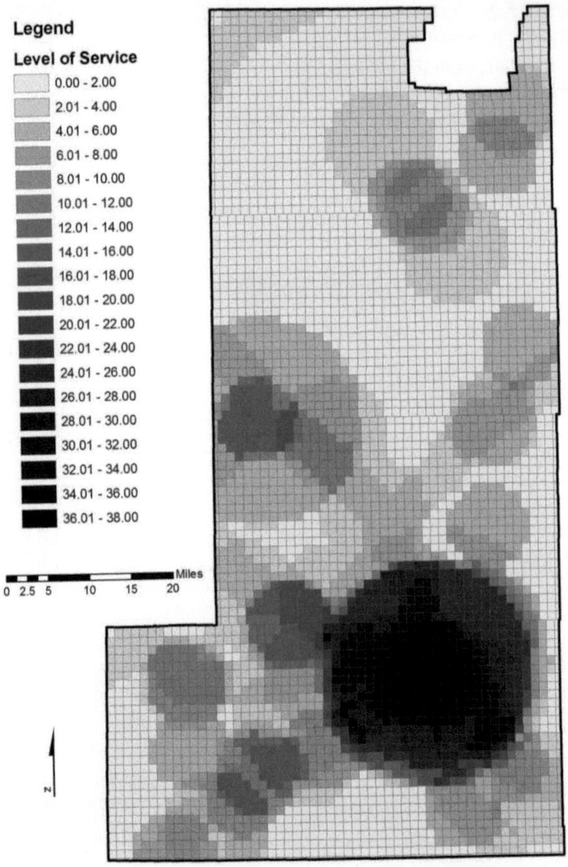

Fig. 15.4 Value weighted growth efficiency (See also Plate 41 in the Colour Plate Section)

portrayal of landscape sensitivity people start to distrust the generalizations visible in the outputs of the model. In a landscape sensitivity model utilizing vector data people recognize individual components of the model in the output and feel comfortable with those outputs. The acceptance of vector based landscape sensitivity model outputs are more than worth the trade off with the increased demand in computational resources.

15.4.3 Stage III: The Growth Challenge Game

Public Meeting Number 2 was the citizens' opportunity to express their personal notion of a land-use future through collaborative game playing. The 'Growth Challenge Game' (often referred to as the 'chip game') is a PSI used by citizens to create (and by the consultants to collect and analyse) a range of possible future development patterns. From a planner's perspective, the game is an opportunity to generate discussion in the community on planning issues, develop a shared form of reference for addressing those issues, and create a large number of growth scenarios which can be evaluated as a range of possible futures.

The growth challenge game consists of rules, a set of land-use chips (Fig. 15.5) and a game board. The amount of chips provided to players is related to the planning horizon of the current planning effort. The distribution of chip types is often a reflection of existing land-use densities, so participants can answer the question: If we keep doing what we have been doing, are we going to end up where we want to be? In addition to playing chips, each table is given a set of bank chips. Players are encouraged to swap chip types with the bank to create what they believe is the best land-use for the future. The game board is composed of a map of sensitive lands (Fig. 15.3) overlain with a transparent grid. Citizens are able to see how sensitive an area of land is prior to recommending a land-use change. When placed on the game board, the landscape sensitivity map has been proven successful at influencing where citizens place future development.

The meeting began by grouping participants around tables in groups of six to eight along with a facilitator for each table. Background information and context were provided including growth projections, evaluative criteria based on plan goals, objectives derived from the first meeting, and the measures of a quality plan. The challenge of the game was to determine where growth should go by placing a large number (in this case based on a 40-year growth projection) of residences, jobs and public services around the county. Participants were allotted three hours to play the game. The key rule was there had to be agreement from all members of a group before a chip was placed on the board. Consistently among the different groups, the first half of the allocated time was spent discussing community issues surrounding growth and development. At roughly the halfway point, facilitators began to nudge the game players along, encouraging them to place chips. Hittle summarizes a typical game: "*When citizens are asked to determine where growth should go, they are faced with the tradeoffs required in determining density, land use, transportation, proximity to services, and the myriad considerations that go into planning for growth... The process of considering alternatives and creating scenarios forces participants to learn about challenges and strategies in planning, but also to work with neighbors who may have very different opinions about how and where the community should grow*" (Hittle 2007).

15.4.4 Stage IV: Evaluating Growth Alternatives

Growth challenge game generated alternatives were brought into a PSS analysis framework and evaluated by *CommunityViz®* based impact assessment models that quantitatively contrasted the performance of the alternative with sensitive lands, growth efficiency and the measures of a quality plan. The growth challenge game presents an application example addressing some of the long standing problems of integrating PSS into planning put forth by Harris (1999). With the game,

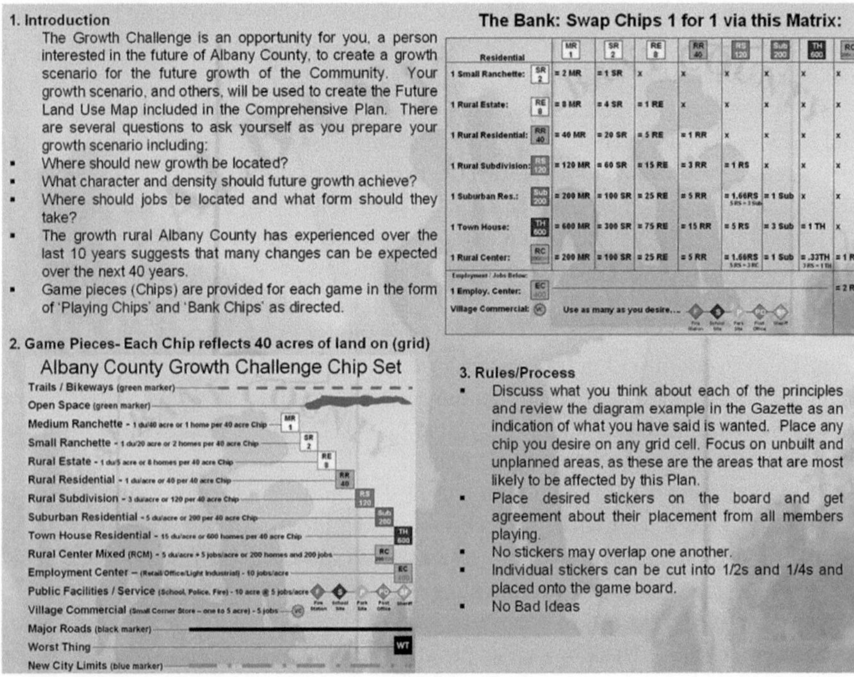

Fig. 15.5 Rules of the growth challenge game

developing alternative plans is quick, easy and informative. Evaluating alternatives with PSS, especially in the context of agreed upon goals, is time consuming but also straightforward. Digitizing and analyzing the results of the game were accomplished after the second meeting and took about a month to complete. The chip game with PSS evaluation is very similar to what Harris described as sketch planning, the rapid generation of alternatives with state of the art evaluation of the implications of the alternatives (Harris 1999).

15.4.5 Stage V: Developing the Concept Plan

The third and final public meeting centred on presenting the results of the growth challenge game, presenting a preliminary concept plan and gathering citizen evaluation of the concept plan. Game performance was presented for each alternative generated as a series of indicator charts. After presenting the results of the chip game, the initial pieces of a concept plan were presented. Keypads were used to collect reactions to the elements of the concept plan, address the relevance of the measures of a quality plan as evaluation criteria, provide clear direction toward a final plan that achieves community objectives and generate feedback on the public engagement portion of the planning process.

As a representative example of the presentation of the components of the concept plan, the presentation of 'development tiers' is over-viewed here. Development tiers were defined as areas where growth is encouraged. They are areas that are easy to service (high growth efficiency) where development would bring about minimal impact on natural resources and sensitive lands. Outside these tiers agriculture will be encouraged and other development discouraged. In the meeting, the concept of the tiers was explained and maps of preliminary tier locations were presented. Meeting participants were asked: Do you agree development tiers are an appropriate method to enable good development while assuring efficient service delivery? Countywide, 82.9 per cent of respondents agreed or strongly agreed. Presentation of the other components of concept plan proceeded in a similar fashion with similar polling results.

The outcome from the third meeting was a refined concept plan and a set of desired land-use recommendations that became the basis of the components (land-use, transportation, economic development, *et cetera*) of the comprehensive plan. Public input on the concept plan *"...strongly influenced the final Comprehensive Plan"* (Albany County 2007, p. 4; Appendix). Because attendees had been active since the first meeting in shaping community goals, addressing issues, indicating opportunities and constraints, developing alternative futures, and shaping the preferred plan they feel a high degree of ownership over the final products.

The three meetings of the public engagement process are looked upon as a minimum. If, at the third public meeting, there is controversy or the group can't get beyond two to three alternatives to agree on a concept plan it may be necessary to do a fourth meeting. At the fourth meeting the consultants will go farther into potential design elements in order to communicate the planning implications of alternatives and build shared agreement. For example, visualization software may be used to illustrate examples of development densities, the implications of transfers of development rights or other issues at hand.

Beyond the public participation phases of the process is the development of a set of politically palatable tools which allow the plan to be implemented. The goal is to devise implementation tools to enable community agreed upon 'good growth' to occur in a logical sequence. Once tools for plan implementation are decided upon, the plan is ready for adoption. Typically in the plan adoption phase the consultants step aside and let the local planning agency take the lead. The adoption phase usually takes more time than the organization and public engagement phases of plan development and requires local knowledge of the political landscape. Time, budgets and the realization that local agencies have a better knowledge of the political situation than consultants usually preclude adoption as part of the scope of work of the planning consultants. Once the plan is adopted, the same set of impact criteria and PSS models can be used to measure the performance of proposed projects for compliance with the comprehensive plan. These models become a valid implementation tool to assist communities with achieving their vision and demonstrate the plan fulfils community desires.

15.4.6 Discussion

The third public meeting typically includes a keypad polling question to gauge citizen's overall satisfaction with the planning process and products. In Albany County the question was: Do you feel the emerging products are representative of values expressed by participants? Results for Albany County are shown in Table 15.1.

Other communities where these methods have been implemented have experienced similar results. In Brush, Colorado 83 per cent of respondents supported or strongly supported the plan. In Buena Vista, Colorado, again, 83 per cent of respondents indicated their product is representative of the citizens who participated. In Fort Lupton, Colorado 77 per cent of respondents agree or strongly agree that the product outcomes of the process represent citizen values. In Berthoud, Colorado 74 per cent of respondents supported or strongly supported the draft land-use plan. Although the questions vary slightly, the results summarized in Table 15.2 indicate strong support for this PSS supported comprehensive planning process and its outcomes. The similar satisfaction survey results from Albany County and four communities in Colorado indicate methods result in consistent outcomes.

In describing the process, the consultants state using these methods and tools to effectively engage a community, realize a common vision, and create ownership in the final product results in a quality plan with a high degree of community support. Moreover, a quality plan with political support enjoys a high likelihood of being adopted and implemented. Clearly, the process generates significant public support which can be seen in both polling results from the evaluation of the preliminary concept plan (demonstrated for development tiers) and in the process satisfaction survey question results above. A public engagement process based on community goals and objectives supported with tools such as key pad polling, gaming and PSS dramatically improves a planner's ability to demonstrate to participating citizens they are directly influencing planning outcomes. The approach continually demonstrates to participants that their values are dictating the outcomes, thus heightened levels of support for the final plan. The process results are nicely summarized in the Albany County plan: "*Most importantly, the plan is the reflection of the community's values and a statement of the community's united vision. The roots of this comprehensive plan are based in public input. It identifies what the community wants by defining goals and identifying ways to achieve those goals through land uses, policies and actions*" (Albany County 2007, p. 13).

Table 15.1 Albany County process summary question results

	Count	%
Strongly Agree	27	33.8
Agree	44	55.0
Disagree	8	10.0
Strongly Disagree	1	1.2
Total	80	100.0

Table 15.2 PSS supported comprehensive planning satisfaction survey results (per cent)

	Agree or strongly agree	Unsure or neutral	Dislike or strongly dislike
Albany County, Wyoming	88.8	0.0	11.2
Berthoud, Colorado	74.0	15.0	11.0
Brush, Colorado	83.0	17.0	0.0
Buena Vista, Colorado	83.0	14.0	3.0
Fort Lupton, Colorado	77.0	15.0	9.0

15.5 Conclusions and Future Directions

The methods presented in this chapter and the various threads from the literature highlighted herein suggest several additional avenues for future research. These include public engagement leading to higher quality plans, greater plan adoption and more effective plan implementation; technical aspects of PSS model development and further recommendations for enhanced planning support adoption.

This research cannot definitively prove the significant public support developed in this process in turn leads to a high likelihood of plan adoption and/or plan implementation. The opportunity exists for further research addressing if these methods in fact lead to these desired outcomes. Within a short while it will be clear whether the communities listed in Table 15.2 have or have not adopted the plans resulting from this process. However, the ultimate goal of a planning process is plan implementation (Kelly and Becker 2000). Given more time it will be possible, although it is significantly more challenging than addressing adoption, to evaluate plan implementation resulting from this process in these communities (Brody *et al.* 2006).

On the technical side, validation of landscape sensitivity and growth efficiency modelling will help legitimize and standardize both methods. As both landscape sensitivity and growth efficiency modelling are based in part on MCA modelling an initial framework for this research are the MCA best practices outlined in Malczewski (1999, 2000). For growth efficiency modelling, although in terms of logic and outputs the technique seems intuitively sensible, it would be a worthwhile exercise to test the model by contrasting results and verifying methods through comparison with the other spatial models addressing service provision and associated costs, such as fiscal impact models (Burchell *et al.* 1989).

In terms of PSS adoption, this project overview makes an important point relevant to the thread in the PSS literature evaluating supply directed (software developers) versus demand directed (planners) PSS development (Geertman 2002, 2006; Geertman and Stillwell 2003b). This project makes clear PSS can function successfully when used by contracted consultants in a supporting role for local government. In this example, PSS enjoy a favourable benefit cost analysis for all parties involved. In discussing the demand side (planner driven) of the PSS equation the literature has yet to acknowledge the first question asked by planners and elected officials when contemplating a PSS project. The question is, invariably: How much does it cost? A discussion of planning support is not complete without mentioning costs. While hardware and software prices continue to drop, the expertise to both work with planners and run a PSS, essentially a salary for one or more people, is a significant consideration. On the supply side, it would be wise for PSS developers and those who use PSS to begin developing the business case, a detailed and convincing cost benefit analysis, for PSS use. This could include PSS developers integrating cost benefit analysis into their product with new features supporting project time and cost tracking.

Another common theme in the PSS literature is the requirement for ease of use (e.g. Geertman 2002; Geertman and Stillwell 2003b; Klosterman 2000; Klosterman *et al.* 2003). The methods described in this project example rely upon behind-the-scenes expertise, suggesting that PSS ease of use may not always be critical for effective utilization. Here, the role of the expert is similar to that of the intermediary as technology operator, specified in classic decision support system configurations (Densham 1991; Sprague 1980). In contrast, easy to use planning support implements (in this case, keypad polling and the growth challenge game) are a direct part of the public engagement process. We suggest the future trend in planning support implementation may be a shift toward easy to use PSI for public engagement, in turn de-emphasizing the necessity for more accessible PSS. This will require further integration of PSI and PSS where PSI function as an interface for non-novice driven PSS.

Part of what makes the methods described in this chapter successful is the production of needed outputs (a comprehensive plan with significant public support) for a substantial user community (local governments that need comprehensive plans or plan updates). As PSS further prove capable of generating a variety of needed outputs at a variety of scales we anticipate seeing a shift in the planning support literature from lack of utilization to new ideas for implementation based on these successes.

The methods put forth in this project example will continue to be used and refined in comprehensive planning in the Rocky Mountain Region of the western United States. The work holds promise as the beginning of a shared commitment to well-defined methods for using PSS in comprehensive planning. There is an opportunity for future research to evaluate and refine the models and methods behind this process to create a greater awareness of successful and repeatable PSS use. And, it is the authors' hope that Klosterman and Pettit's (2005) recommendation for government support and Nedović-Budić's (2000) recommendation for support of local planning by major U.S. and international funding agencies help lead to greater adoption of planning support, better planning and a better built environment.

References

Aggett, G. and McColl, C. (2006) Evaluating decision support systems for PPGIS applications, *Cartography and Geographic Information Science*, 33(1): 77–92.

Albany County (2007) *Albany County Comprehensive Plan August 22, 2007 Draft*, Albany County Planning Department, Laramie.
Alexander, E. and Faludi, A. (1989) Planning and plan implementation: notes on evaluation critieria, *Environment and Planning B*, 16: 127–140.
Alliance for Regional Stewardship (2001) *Empowering Regions: Strategies and Tools for Community Decision Making,* Alliance for Regional Stewardship, Palo Alto.
Bardach, E. (1977) *Implementation: What Happens to a Bill After it is Adopted?,* MIT Press, Cambridge, Massachusetts.
Berke, P. and French, S. (1994) The influence of state planning mandates on local plan quality, *Journal of Planning Education and Research*, 13 (4): 237–250.
Bleiker, H. and Bleiker, A. (2000) *Citizen Participation Handbook for Public Officials and Other Professionals Serving the Public.* 13th Edition. Institute for Participatory Management & Planning, Monterey.
Boyd, S. and Chan, R. (2002) *Placemaking Tools for Community Action,* Concern, Inc. Environmental Simulation Center, U.S. Department of Energy, Denver.
Brody, S.D. (2003) Are we learning to make better plans? A longitudinal analysis of plan quality associated with natural hazards, *Journal of Planning Education and Research*, 23(2): 191–201.
Brody, S.D., Godschalk, D.R. and Burby, R.J. (2003) Mandating citizen participation in plan making: six strategic planning choices, *Journal of the American Planning Association*, 69: 245–65.
Brody, S.D., Highfield, W.E. and Thornton, S. (2006) Planning at the urban fringe: an examination of the factors influencing nonconforming development patterns in southern Florida, *Environment and Planning B*, 33: 75–96.
Burby, R.J. (2003) Making plans that matter: citizen involvement and government action, *Journal of the American Planning Assocation*, 69(1): 33–49.
Burchell, R.W., Listokin, D. and Dolphin, W.R. (1989) *The New Practitioner's Guide to Fiscal Impact Analysis*, Center for Urban Policy Research, New Brunswick.
Couclelis, H. (2005) Where has the future gone? Rethinking the role of integrated land-use models in spatial planning, *Environment and Planning A*, 37(8): 1353–1371.
Creighton, J. (1992) *Involving Citizens in Community Decision Making: A Guidebook*, Program for Community Problem Solving, Washington, DC.
Densham, P.J. (1991) Spatial decision support systems, In Maguire, D.J., Goodchild, M.F. and Rhind, D.W. (eds.) *Geographical Information Systems: Principles and Applications*, John Wiley and Sons, New York, pp. 403–412.
Field, P. and Snyder, K. (2006) Introduction of keypad polling and initial baseline questions, Presentation at the Interactive Wind Summit, Manchester, VT, 11 February, (Results of the survey component of an unpublished PowerPoint presentation). www.town.manchester.vt.us/windforum/questions/Keypad%20Polling%20Results.ppt.
Fiorino, D.J. (1990) Citizen participation and environmental risk: a survey of institutional mechanisms. *Science, Technology and Human Values* 15(2): 226–243.
Fotheringham, A.S. and O'Kelly, M. (1989) *Spatial Interaction Models: Formulations and Applications*, Kluwer, Dordrecht.
Geertman, S. (2002) Participatory planning and GIS: a PSS to bridge the gap, *Environment and Planning B*, 29: 21–35.
Geertman, S. (2006) Potentials for planning support: a planning-conceptual approach, *Environment and Planning B*, 33: 863–880.
Geertman, S. and Stillwell, J. (2003a) Planning support systems: an introduction, In Geertman, S. and Stillwell, J. (eds.) *Planning Support Systems in Practice,* Springer, Heidelberg, pp. 3–22.
Geertman, S. and Stillwell, J. (2003b) Interactive support systems for participatory planning, In Geertman, S. and Stillwell, J. (eds.) *Planning Support Systems in Practice*, Springer, Heidelberg, pp. 25–44.
Hamerlinck, J.D. (2007) Planning Support Systems: Exploring the Role of Information Technology in Local Government Planning, presented at the Annual Meeting of the Western Planner

Association, Dickinson, North Dakota, USA, August 9, 2007 (results of the survey component of an unpublished PowerPoint presentation). http://www.planitwyoming.org/conferences.html
Harris, B. (1999) Computing in planning: professional and institutional requirements, *Environment and Planning B*, 26: 321–331.
Healey, P. (1994) Development plans: new approaches to making frameworks for land-use regulation, *European Planning Studies*, 2(1): 39–55.
Healey, P. (2003) Collaborative planning in perspective, *Planning Theory*, 2(2): 101–123.
Hittle, J. (2007) Tech Toolbox: The Chip Game, *Scenarios E-Journal*, The Orton Family Foundation. Retrieved July 2, 2007, from http://www.orton.org/index.cfm?fuseaction=page.viewPage&pageID=657&nodeID=1.
Innes, J.E. (1995) Planning theory's emerging paradigm: communicative action and interactive practice, *Journal of Planning Education and Research*, 14: 183–189.
Innes, J.E. (1996) Planning through consensus building, *Journal of the American Planning Association*, 62(4): 460–473.
Innes, J.E. (1998) Information on communicative planning, *Journal of the American Planning Association*, 64: 52–63.
Innes, J.E. and Booher, D. (1999) Consensus building and complex adaptive systems, *Journal of the American Planning Association*, 65: 412–423.
Innes, J.E., Gruber, J., Neuman, M. and Thompson, R. (1994) Coordinating growth and environmental management through consensus building, *C P S Briefs*, 6(4): 1–8.
Jurgensmeyer, J.C. and Roberts, T.E. (2003) *Land-use Planning and Development Regulation Law*, West Group, St. Paul, Minnesota.
Kelly, E.D. and Becker, B. (2000) *Community Planning: An Introduction to the Comprehensive Plan*, Island Press, Washington, DC.
Klosterman, R.E. (2000) Planning in the information age, In Hoch, C.D. (ed.) *The Practice of Local Government Planning,* International City/County Management Association, Washington DC, pp. 41–57.
Klosterman, R.E. and Pettit, C. (2005) An update on planning support systems, *Environment and Planning B*, 32: 477–484.
Klosterman, R.E., Siebert, L., Hoque, M.A. Kim, J.W. and Parveen, A. (2003) Using an operational planning support system to evaluate farmland preservation policies, In Geertman, S. and Stillwell, J. (eds.) *Planning Support Systems in Practice,* Springer, Heidelberg, pp. 391–407.
Korfmacher, K.S. (2001) The Politics of Participation in Watershed Modeling. *Environmental Management*, 27 (2): 161–176.
Kwartler, M. and Bernard, R.N. (2001) CommunityViz: an integrated planning support system, In Brail, R.K. and Klosterman, R.E. (eds.) *Planning Support Systems: Integrating Geographic Information Systems, Models and Visualization Tools*, ESRI Press, Redlands, pp. 285–308.
Malczewski, J. (1999) *GIS and Multicrtieria Decision Analysis*, John Wiley & Sons, Inc., New York.
Malczewski, J. (2000) On the use of weighted linear combination method in GIS: common and best practice approaches, *Transactions in GIS*, 4(1): 5–22.
Mastop, H. and Faludi, A. (1997) Evaluation of strategic plans: the performance principle, *Environment and Planning B*, 24: 815–832.
McHarg, I. (1969) *Design with Nature*, Natural History Press, Garden City, New York.
Nedović-Budić, Z. (1998) The impact of GIS technology, *Environment and Planning B*, 25: 681–692.
Nedović-Budić, Z. (2000) Geographic information science implications for urban and regional planning, *URISA Journal*, 12(2): 81–93.
Orton Family Foundation (2006) *Scenario 360™ v. 3.2 Quick Reference*, Orton Family Foundation and Placeways, LLC.
Pettit, C. and Pullar, D. (2004) Collaborative decision-making processes for maintaining biodiversity: two Australian case studies, International Environmental Modelling and Software Society Conference Proceedings, Osnabrück.
Snyder, K. (2003) Tools for community design and decision-making, In Geertman, S. and Stillwell, J. (eds.) *Planning Support Systems in Practice,* Springer, Heidelberg, pp. 99–120.

Snyder, K. (2006) Putting democracy front and center: technology for citizen participation, *Planning Magazine*, 72(7): 24–29.

Sprague, R.H., Jr. (1980) A framework for the development of decision support systems, *MIS Quarterly*, 4(4): 1–26.

Talen, E. (1997) Success, failure and conformance: an alternative approach to planning evaluation. *Environment and Planning B*, 24: 573–587.

Vonk, G.A., Geertman, S. and Schot, P. (2005) Bottlenecks blocking widespread usage of planning support systems. *Environment and Planning A*, 37: 909–924.

Wyoming Division of Economic Analysis (2007) *Wyoming Incorporated Place Population Estimates: April 1, 2000 to July 1, 2006,* Retrieved August 8, 2007, from Wyoming Division of Economic Analysis: http://eadiv.state.wy.us/pop/SUB-06EST.htm

Additional Reading

Balram, S. and Dragicevic, S. (eds.) (2006) *Collaborative Geographic Information Systems*, Idea Group Publishing, Hershey, Pennsylvania.

Craig, W.J., Harris, T.M. and Weiner, D. (eds.) (2002) *Community Participation and Geographic Information Science,* Taylor and Francis, London.

Mullen, S. (2006) Community visualization: the key to democratic urban planning, *Geoworld*, November 2006: 24–27.

Rugg, R. (2003) A framework for the use of geographic information in participatory community planning and development, *URISA Journal*, 15: 75–80.

Snyder, K. (2005) Tech toolbox: keypad polling, *Scenarios E-Journal*, The Orton Family Foundation. http://www.orton.org/

Chapter 16
Integrating Analytical and Participatory Techniques for Planning the Sustainable Use of Land Resources and Landscapes

David Miller, Nathan Vogt, Maria Nijnik, Eduardo Brondizio and Stefano Fiorini

16.1 Introduction

Planning the sustainable use of land resources and landscapes must be a process in which stakeholders and the public work together to establish common guidelines for understanding the options of, and the implications for, future land uses. Key aspects of this process are the establishment of meaningful knowledge bases and tools, and methodologies based on the enhanced involvement of stakeholders in making decisions, and their subsequent implementation. The gradation of power or control in public participation proposed by Arnstein (1969), extending from 'citizen control' to 'manipulation', provides a conceptual basis for considering the evolution in political thinking about participation in areas such landscape planning.

This chapter discusses the integration of analytical and participatory techniques for planning the sustainable use of land resources and landscapes using two examples, one from South America and the second from Europe. The first example considers land use in the Amazon, and the second, the socio-economic, ecological and visual aspects of land-use changes in a European landscape. Each example involved active participation of stakeholders and the public in the process of decision making.

A framework is presented for the Amazonian example, which comprises methodologies and survey instruments for multi-level, integrated assessments of land-use and land-cover change. The framework was developed in collaboration between Indiana University and several Brazilian institutions. It adopts an historical ecological approach (Brondizio 2006) and applies a range of tools from the social, ecological and geographic sciences in fieldwork and laboratory analysis (Moran and Ostrom 2005). The framework and methodologies are being used by scientists in the Amazon Initiative (AI)—Land Degradation Assessment (LDA) thematic network as a tool for responding to land degradation problems occurring at farm to regional

S. Fiorini (✉)
The Macaulay Land Use Research Institute (MLURI)
Aberdeen, Scotland, UK
E-mail: s.fiorini@macaulay.ac.uk

S. Geertman, J. Stillwell (eds.), *Planning Support Systems Best Practice and New Methods*, 317
DOI 10.1007/ 978-1-4020-8952-7_16, © Springer Science+Business Media B.V. 2009

scales across the Amazon basin. A specific example was chosen of the application of the approach to agro forestry in Brazil which, in principle, can be adjusted and applied elsewhere.

In Europe, an example is presented of the development and testing of approaches to aid the assessment and implementation of options for sustainable landscape management. Visualisation tools, spatial models and scenarios of proposed changes in land use have been tested and evaluated for facilitating participation in managing landscape change (Miller *et al.* 2005). The focus was on the contribution of factors to change in the visual landscape, and on the development of proper institutions and capabilities for stakeholders' involvement, to aid in the assessment and implementation of options for sustainable landscape management. The involvement of stakeholders and the public has provided a means of developing their capability for landscape planning and management.

The outcomes from the experiences of these two actual planning activities, applied under the very different conditions of the Amazon and Europe, are used to illustrate stakeholder and community involvement in assessing environmental problems and potential solutions. These experiences suggest that wider stakeholder involvement in decision making has had a high level of participant satisfaction, and an increased understanding of the issues associated with landscape change. Comparisons of the similarities and differences between the studies provide a basis for discussion of common, and locally distinctive, guidelines and good practices in landscape use and planning.

16.2 Background

The Brundtland Commission (World Commission on Environment and Development 1987) identified the importance of information and participation in issues relating to the management and planning of change. Subsequent policies for sustainable development (Scottish Executive 2006; United Nations 2005) emphasise needs for education and capacity building to increase levels of public and professional engagement in environmental decision making. Such public policies set the context for political and social aims for the use of land, and the multiple functions it fulfils.

The challenge of developing methodologies and procedures to integrate analysis, for the study of rural development and land-use change, has been the subject of research at the Anthropological Center for Training and Research on Global Environmental Change (ACT), and the Center for the Study of Institutions, Population, and Environmental Change (CIPEC), in collaboration with colleagues from several Brazilian institutions. ACT-CIPEC research projects are focused on questions integrating socio-demographic, institutional and environmental aspects of land-use and agrarian change. Integrative methodologies, and strategies for research design, aim at linking data collected at the level of the household and farm lot to larger regions. Research instruments include ethnographic techniques, survey protocols, ecological assessment methods and procedures to use multi-temporal remote sensing data and geographic information technologies.

In Europe, the importance of the sustainable use of natural resources through linking socio-economic objectives for land use with a need to maintain or enhance its environmental quality and cultural functions has become of increased policy importance (Council of Europe 2003; European Commission 2005). The European Landscape Convention (Council of Europe 2003) recognizes the role of landscape in delivering on this mix of functions, as well as a resource favourable to economic activity. However, it is the effects of change on the visual landscape which are amongst the most widely perceived impacts of changes in land use (Lange and Bishop 2005). Visually representing the real world, and potential alterations, is essential for landscape planners to communicate their thinking to the wider public (Ball *et al.* 2008; Bishop 2007). The Scottish Executive (2007), in reforming the planning process, also noted that 3D modelling has potential for *"engaging communities and assisting planners and Councillors to visualise and assess the visual impact of development proposals"*. One tool being used to support such engagement is a 'Virtual Reality Theatre', with associated software tools, for use in public venues.

In order to support a sound stewardship of rural areas, the relationship between visual qualities and other functions of landscape, such as biodiversity, cultural heritage, amenity and sustainable production, was the subject of a recently completed project on European Commission (EC) funded project entitled 'Visualization tools for public participation in the management of landscape changes' (VisuLands). The project focused attention on the contribution of the location, dimension and geographic context of land-cover types and specific landscape elements to the visual landscape. The project consisted of partners in six countries across the European Union (as of 2003), with an international end-user group which assessed outputs for their relevance for operational use or strategic planning. Each country also had defined local end-users, the nature and role of which was designed to suit the management scenarios of that country.

The following sections include descriptions of how these two interdisciplinary projects use participatory techniques for planning and management of land resources. In each case, the process of stakeholder and public participation required as much consideration as the development of new technologies and tools.

16.3 Methodology for Assessing Processes of Land-use/Cover Change

A framework is presented for training and collaboration used amongst researchers from the different disciplines involved in the AI, a project supported by the United States Agency for International Development (USAID). The first phase of the project was implementation of collaborative activities between the International Center for Tropical Agriculture (CIAT), institutions in the Land Degradation Assessment thematic network of the AI Consortium, and the ACT. ACT students and scientists have developed and applied tools for capturing the values, objectives, and perspectives of local stakeholders on the feasibility and sustainability of their use of land resources in diverse social and ecological settings across the Amazon basin. Findings on the sustainability of resource use by traditional populations in Amazonia have at times run counter to conventional wisdom. One reason for this is that land-use change assessments and planning often occur at regional or national levels, with assumptions about the role of local factors. A multi-level approach, which aggregates fine scale social and ecological processes to explain change observed at broader scales, has shed light on these assumptions and brings the perspective of the local actor to the assessment (Boucek and Moran 2004; Brondizio and Siquiera 1997; Moran *et al.* 2005).

Cernea (2005), writing on the role of social sciences within the Consultative Group on International Agricultural Research (CIGAR), argues that such institutions should consider how to put 'culture back in agriculture'. He describes a number of factors contributing to distancing social science researchers from mainstream agricultural development programmes. We argue that integrated socio-cultural and environmental methodologies can contribute to bridging this gap. Studying multi-faceted, multi-level phenomena of land use and agrarian change requires integrating processes taking place at the level of the farm lot as well as larger regions and commodity chains. Agricultural decision making at the level of a farm lot reflects factors working at the international level (e.g. commodity prices), national and regional (e.g. credit policies) as well as the local level (e.g. farmer's experience, available technology, soil fertility). Although these decisions are local (e.g. abandoning a pasture, deforesting land), they aggregate to form regional patterns of

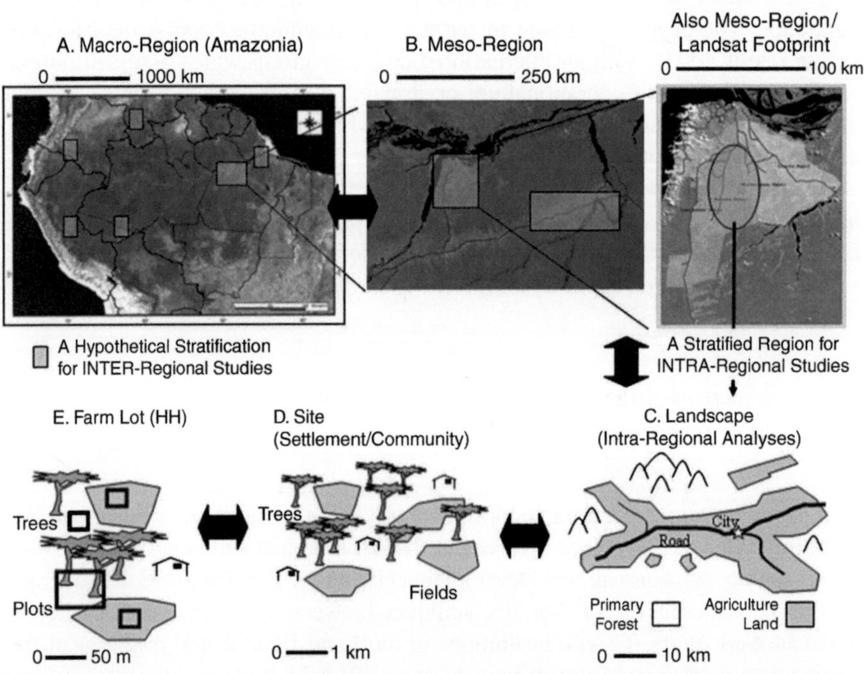

Fig. 16.1 Level of analysis and suggested terminology for bridging social and biophysical units of analysis (adapted from Green *et al.* 2005; ACT 2005_2008)

land-cover change. Understanding how these factors and levels interact requires expertise from both the social and physical sciences, and the sharing of methodologies, including a common language and terminology, and the development of shared hypotheses whilst drawing on disciplinary-based research approaches.

Such understanding also takes involvement of the actors, decision takers and end users. They are a source of technical information and policy, and aid in shaping the direction of investigation, influencing the problematic concepts, logical frameworks, analytical strategies, methods and scales of analysis. A framework to understand land use requires attention to both the scale and unit of analysis, and units of observation. Green *et al.* (2005) suggest a terminology intended to bring together heuristic notions of aggregation levels, and their spatial extent, from social and ecological perspectives. In this methodological framework, the terms chosen to depict different levels of spatial extent are 'macro-region', 'meso-region' (e.g. administrative units), 'landscape/micro-basins', 'settlement and community', and 'household and farm lot' (Fig. 16.1), where the examples are taken from the Brazilian Amazon.

16.3.1 Analytical Strategies for Assessing and Planning Land-use/Cover Change

Two common analytical strategies for land-use studies are the so-called 'driving forces' and 'process-pattern' analysis (Brondizio 2005). In both cases, there is an attempt to link changes in regional land-cover patterns, observable in a time series of satellite images, to the human and ecological factors that underlie them. Driving forces include the underlying, or fundamental, forces shaping land-use decisions, and proximate forces, which are the direct activities manipulating the biophysical environment, and mediating factors between them (Brondizio 2005; Geist and Lambin 2001; Millennium Ecosystem Assessment 2005). The challenge in land-use studies has been to characterize the pathways and feedback mechanisms between proximate and underlying driving forces, given the factors that mediate them, which can operate at multiple levels and spatial extent (Brondizio 2005; Geist and Lambin 2001). Process-pattern analysis, in comparison, aims at integrating datasets collected at the farm or community levels, and their contextual peculiarities. This supports greater insight into regional patterns of land cover observed in the satellite images. In doing so, this strategy seeks regional contextualization of variability of social and ecological processes observed in community and local level studies.

A hypothesis-driven study of land-use change processes often focuses on trying to understand the role of a particular factor in shaping the observable patterns of land-cover, or land-management problems at a particular level. A multi-level framework is necessary to account for the other factors that affect the process under investigation, of which the sampling strategy illustrated in Fig. 16.2 is an example. In order to understand the factors affecting deforestation, the region was divided into areas characterized by common history of occupation, characteristics of social

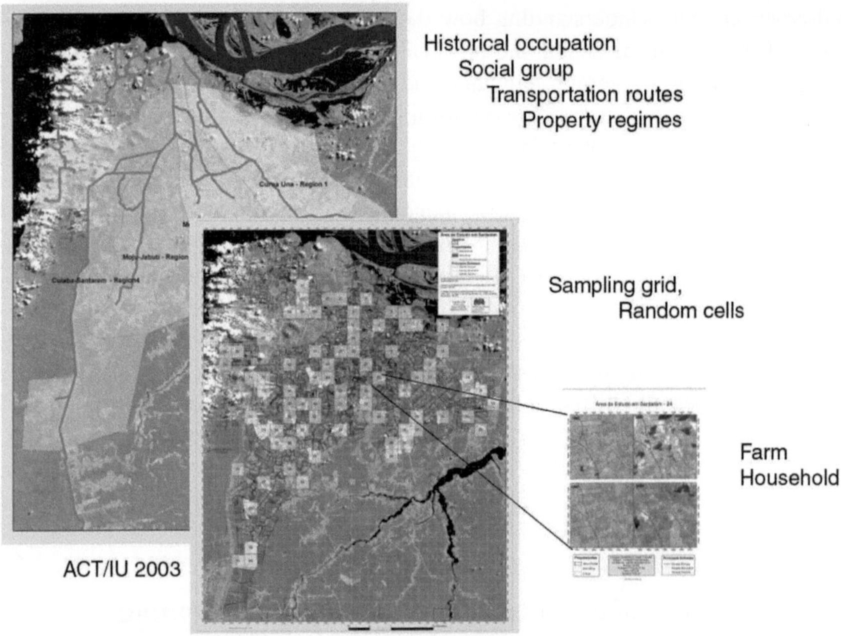

Fig. 16.2 Stratifying levels of analysis; images of the ACT study areas, Santarem-Belterra Region in the Brazilian state of Para (See also Plate 42 in the Colour Plate Section)

groups, transportation routs and property regimes. Smaller areas were then randomly selected, and interviews conducted in households randomly selected within grids.

The direct involvement of stakeholders was in the approach taken, and could not be substituted by literature-based derivations of specific processes driving land-use change between several locations. This was because, as discussed by McConnell and Keys (2005: 349), *"the directionality of the relationships between causal variables is multiple: high market demand can encourage investment in land capital, or it can spur the degradation of geographically dispersed resources"*. Full understanding of processes and relationships can only be achieved through the involvement of those directly affected or shaping these relationships and the directionality of change.

16.3.2 The Methodological Framework for Integrated Assessment of Change

The choice of social and ecological survey instruments, or protocols, applied within the integrative framework for assessing land-use change depends on the specific questions and hypotheses identified in the process of designing the collaborative study. The protocols, and questions or issues, illustrated at each level of analysis derive from those developed by, and of interest to, scientists. These protocols and questions are intended to provide examples of research design which allow comparable sampling design and data integration between social and ecological studies of land-use change. The framework presented in Table 16.1 summarizes different levels of analysis (Table 16.2–16.7), aimed at illustrating a research design for elucidating key social and environmental factors influencing land-cover change. For each level, social and ecological processes relating to land use change are presented, together with an integrated assessment of change in the patterns of land cover.

The investigations of social and ecological processes address issues such as: (i) examples of influencing factors; (ii) examples of stratification strategies; (iii) information on sampling design; (iv) examples of data sources; and (v) examples of survey instruments. The Integrated Assessments of change in land-cover patterns contain notes of research questions, regarding specific social-ecological processes, relevant for addressing land-use issues. A study may be conducted at any level or at multiple levels but the framework provides a heuristic tool to create awareness of factors influencing the extent, rate, and direction of change at other levels that need to be considered, or controlled for, in an integrated assessment, such as that of agroforestry in the Amazon.

16.3.3 Participatory Perspective to Land-use Change: Agro-Forestry in Marajo Island, Para, Brazil

Proposals for future land uses can be made by researchers and extension agencies without consideration of how local land-use systems actually work and connect with external market chains. However, the land use of regions and communities can be shaped by employing robust research approaches which are able to take account of stakeholders' perspectives of the issues being investigated. The example which follows shows the evaluation of classified remotely sensed imagery with the involvement of stakeholders.

Population growth in urban Amazônia has created markets for regionally preferred food sources such as the *açaí* palm fruit (*Euterpe oleracea* Mart.), which is a regional staple food consumed by rural and urban populations alike. Since the late 1970s, the Amazon region has seen intensification of the production system following an increase in market demand. Today, *açaí* fruit is the most important source of income for the majority of riverine households, and the main economic activity of most municipalities, of the Amazon estuary. This production system builds upon existing knowledge and technology, such as the management of floodplain forests and the planting of *açaí* palm agro-forestry in multi-cropping systems. Despite its high economic productivity and the level of agro-forestry manipulation, such areas are often viewed as being only for extraction, with the work of producers and their management knowledge treated as extractivism (Brondizio 2008). Unlike a system based upon extractivism, the management and planting of *açaí* agro-forestry require

Table 16.1 Methodological framework for multi-level, integrated assessments of change in land-cover pattern (resource condition)

Levels of analysis (heuristic units defined according to research questions)	Investigating social processes related to land-use change	Integrated assessments of change in land-cover patterns	Investigating ecological processes related to land-use change
	Elucidating social-economic-cultural factors underlying patterns of land-cover change. Stratify levels of analysis based on social, political, and cultural organization of the population. Inform sampling design. Define key data sources and survey instruments.	Define research questions and hypothesize key social and ecological variables and processes explaining change in resource use and management decisions (and in turn observable land-cover patterns or land management problems). Develop a research design and sampling framework controlling for social and environmental variability at each level of analysis.	Elucidating environmental factors influencing patterns of land use and land-cover change. Stratify levels of analysis based on biophysical characteristics of the land (topography, soil, rainfall patterns, watershed, *et cetera*). Inform sampling design. Define key data sources and assessment instruments.

clear input of specialized agricultural and forestry labour in order to maintain and increase the productivity of crop stands.

The rigid boundary drawn between different food production systems has led to the characterization of forested areas, as in agro-forestry systems, as unproductive, or at best, under the category of agro-extractivism. Consequently, it is common to see land-cover classifications of the estuary which disregard *açaí* agro-forestry as a land-use class, despite its importance as a land-use system in the region (Fig. 16.3). Brondízio (2008) and Brondízio and Siqueira (1997) argue for a re-interpretation of local agro-forestry land-use systems as being intensive, and a change in the economic identity of local producers from extractivists to forest farmers. While *açaí* agro-forestry represent the most significant land-use system in the estuarine area, a generalized classification system of land cover (Fig. 16.3) in the area would, or could, ignore this production system, treating it as any other forest cover without regard for its important economic and social roles. Therefore, the most important regional land-use systems could be omitted from maps and models, as would a whole social group managing these forest and agro-forestry areas.

The coarse spectral, radiometric, and spatial resolution of satellite images does not always distinguish important land-use classes in forest biomes. However, equally important is the way researchers conceptualize land-use systems and how they should be represented in the landscape. These decisions and analyses carry

Table 16.2 Macro-regional level of analysis

Levels of analysis (heuristic units defined according to research questions)	Investigating social processes related to land-use change	Integrated assessments of change in land-cover patterns	Investigating ecological processes related to land-use change
Macro-regional	Broad descriptions of factors at this level are adequate. Examples of factors to focus on include major and planned transportation routes to key national and international markets, long term historical occupation of broader region, population distributions, economic development programs, national forestry or agriculture policies, etc… Data sources: historical census data (population and economy), baseline demographic and political maps, archival resources, policy documents, etc…	Computer model depicting the temporal interaction of few broadly defined variables underlying land cover change: economic policy (credit programs), infrastructure expansion (roads), total population change, and commodity market prices. Characterize interaction among demographic flows, economic and production areas, development programs, and political-cultural organization and changes. Macro-level land cover models may be integrated to climate, topography, and soil data. Describe historical trajectories (fluctuation in rates) of land use and cover change, areas of expansion and rapid population turn-over. Describe broad historical depth of regional occupation and resource use: Develop timelines and historical cartography (occupational history of region, change in regional access, etc…).	Examples of ecological factors to focus on at this level may be broad patterns of climate, relief, soil types and vegetation. Data sources: historical maps/databases of rainfall and temperature distributions; soil types, relief, and coarse resolution remote sensing data (e.g., MODIS or AVHRR) and thematic land cover maps. That is, data sets used in broad vegetation classification schemes. Regional sampling of major land cover and land use types (composition, seasonality, structural characteristics, etc…).

Table 16.3 Meso-regional level of analysis

Levels of analysis (heuristic units defined according to research questions)	Investigating social processes related to land-use change	Integrated assessments of change in land-cover patterns	Investigating ecological processes related to land-use change
Meso-Regional	Examples of social factors to focus on include trajectories of occupation, migration and emigration, the resource use and social and cultural organization of different groups, areas of conflicts and disputes, key markets for forest resources and cash crops, resource policy regimes and overall land tenure history. Articulate administrative units and develop timelines and his torical cartography (occupational history of region, change in regional access, *et cetera*). Stratify region by political and administrative units, census tracts, ethnic divisions, or land tenure and ownership categories. Survey instruments and data sources for characterizing broad, social processes per above strata include exploratory fieldwork (rapid participatory assessment of key processes with community representatives and key informants), archival research, census analysis, image land-use/cover pattern.	Pay attention to broad relationship between occupational histories, routes and forms of access, resource ownership, national or regional policies influencing land-use, demographic and economic history, and decadal change in land-cover. Measure and characterize historical rates of land cover change, areas of expansion and fast deforestation and re-growth dynamics. Describe historical depth of tional history of region, change in regional access, *et cetera.* Characterize interaction among demographic flows, economic and production areas, development programs, and political-cultural organization and changes. Describe cash crops produced and degree of commodity change development. Also at this level the quantitative focus is typically on "modelling" of contemporary, social and ecological forces on land-cover change. Qualitative or descriptive relationships may be conducted, also based on the multiple layers of spatially explicit, ancillary datasets. analysis. Appropriate key informants per strata include agriculture and forestry officials/extension workers, customary or formal leader at broader administrative levels, ministry of natural resource officials, *et cetera* who are familiar with State or Department level policies, planning, and histories related to natural resources for the study area.	Examples of ecological factors to focus on include broad soil types, landforms, differences in seasonal changes in weather and change in land-cover patterns. Stratify biophysical environment of meso/micro-region, geographically analyze historical change in land-cover per strata. Instruments and data sources include time-series assemblage, processing, and analyses of remote sensing data. Field assessment of soil andvegetation sampled over environmental gradients and/or compartments. Develop multi-level land cover classification systems and characterize major land-use/cover pathways. Utilize broader-scale classifications at this level.

Table 16.4 Landscape/intra-regional level of analysis

Levels of analysis (heuristic units defined according to research questions)	Investigating social processes related to land-use change	Integrated assessments of change in land-cover patterns	Investigating ecological processes related to land-use change
Landscape/ intra-regional	Examples of social factors to focus on include spatial and temporal patterns of change in resource allocation, property regimes for different resources, economic systems and land uses. Also identify interventions (by NGOs, *et cetera*), commodity chain of major production systems, structure of local government and its relation to resource use, location specific resource policies. Stratify and sample by social groups and institutional arrangements, community distributions, observable land-cover dynamics in a time-series of satellite images, patterns of economic and land use activities, *et cetera*. Survey instruments and data sources for characterizing social factors.	Focus on comparative analyses within a region and variability among different social groups. Important processes include spatial and temporal patterns in social groups settlement, resource allocation, historical rates of land cover change, areas of expansion and rapid population turnover, areas of fast deforestation and re-growth dynamics. Also important is the temporal and spatial shifts in major production systems, development of commodity chains for the major production systems and interventions. Characterize history of location specific policies or laws influencing resource use (e.g. forest and land management programs) in local government and problems or successes of them.	Examples of ecological factors to focus on include types of vegetation cover and their spatial patterns, soils types, topography, monthly distributions of temperature and rainfall. Stratify biophysical environment geographically (finer scale soil types, landforms, *et cetera*). Instruments and data sources include time-series assemblage (Landsat products or higher resolution, remotely sensed products), processing, and analyses of remote sensing data; field assessment of soil and vegetation sampled over

Table 16.4 (continued)

Levels of analysis (heuristic units defined according to research questions)	Investigating social processes related to land-use change	Integrated assessments of change in land-cover patterns	Investigating ecological processes related to land-use change
	Include participatory assessments of land-cover with community representatives and key informants; rapid assessment tools applied at the community/group level; multi-temporal remote sensing data, census data at the level of municipality and/or tracks/sectors. Key informants as above and also cooperative leaders organized around agriculture or natural resources, local government leaders, natural resource extension agents of local government, *et cetera* who are familiar with local administrative level policies, planning, and histories related to natural resources for the study area. Regional image print out for exploratory fieldwork.	Describe the structure of local government, who participates and how are they chosen, what are their values (and level of resource allocation) toward agricultural development and also resource conservation programs. Conceptual analysis may be conducted of the interaction among landforms, roads and rivers (market access routes), demographic dynamics, property regimes and land-cover dynamics upon patterns of land use and settlement.	different environmental compartments and vegetation types and/or areas characterized by distinct land cover spatial pattern. Utilize multi-level land-cover classification schemes (Utilize broader-scale classifications at this level as above level. Again, classes chosen will depend on issues to be addressed).

Table 16.5 Settlement/community level of analysis

Levels of analysis (heuristic units defined according to research questions)	Investigating social processes related to land-use change	Integrated assessments of change in land-cover patterns	Investigating ecological processes related to land-use change
Settlement/ Community	Examples of social factors to focus on include forms of social organization, community conflicts, socio-economic factors, migration patterns, demographic trends, cultural norms shared by community (values), type and number of businesses and philanthropic organizations, degree of market access, knowledge of agro-forestry technologies and ecological services provided to community by each land-use type in it. Stratify by social groups and/or community (e.g., by age, location, size, *et cetera*), resource user groups, patterns of land-cover trajectories, patterns of informal types of land ownership, *et cetera.* Survey instruments and data sources primarily used for characterizing important social factors at community level include participatory assessments as above regional level or use of household surveys that can be aggregated to represent a community or settlement. Appropriate informants at this level include long-term occupants/elderly of, administrators of community and heads of community organizations, resource user groups, *et cetera.*	Focus on community/settlement level processes, such as understanding forms of social organization, history of occupation, economic systems, migration patterns, collective activities, *et cetera.* Characterize rates of demographic and economic changes resulting from changes in production system and forms of resource use. Characterize any locally coordinated activities to produce, manage, or sell a resource in the community, attributes of customary institutional arrangements that regulate those collective activities, and how these arrangements emerged and are maintained. Characterize historical land-cover patterns, change in structure and composition of vegetation in key, community land-cover types (i.e. structure and composition of trees in agro-forestry systems, home-gardens, and forest patches), land-use pathways, *et cetera.* Conceptual analysis of interaction among settlement pattern, landforms, roads and access, formal and informal property regimes and relationship to land use systems. Capture history of land and economic conflicts and their impact on settlement and land use.	Examples of ecological factors to focus on include land-cover patterns and their change, structure and composition of land-cover types, land-use pathways. Stratify biophysical environment geographically (finer scale soil types, landforms, vegetation types, *et cetera*). Instruments and data sources include time-series assemblage, processing, and analyses of remote sensing data; field assessment of soil and vegetation sampled over environmental gradients and/or compartments, image interviews, GPS, *et cetera.* Utilize multi-level land-cover classification schemes (Utilize finer-scale classifications at this level. Again, classes chosen will depend on issues to be addressed). Instruments to produce finer-scale classifications for remotely sensed images include training sample collection at this level.

Table 16.6 Farm/household level of analysis

Levels of analysis (heuristic units defined according to research questions)	Investigating social processes related to land-use change	Integrated assessments of change in land-cover patterns	Investigating ecological processes related to land-use change
Farm/Household (HH)	Variety of ethnographic methodologies Key social factors to focus on include demographic and economic history of the household, family labour organization, HH composition, access to financial resources and technology; social organization (and position in them), kinship and social networks, economic and land-use decisions. Survey Instruments and data sources include participatory assessments at the family and farm level, lot mapping, structured and semi-structured socio-demographic and socio-economic surveys.	Important processes include demographic history of household, social organization and kinship networks overt-time, and change in economic and land-use decisions. Conceptual and statistical models depicting the interaction between household life cycle and land use and cover change. Comparative analysis of households: economic and land use portfolio, resource management and conservation strategies. HH position or involvement in collective activities that regulate resources important to the HH and community, rates of abandonment and expansion of activities, *et cetera.* Analysis of the inter-linkages between market signals (e.g. demand and price), household decision-making, and land cover change.	Key ecological factors include agricultural plots, resource use/condition (soil, water, pasture, forest, secondary forest, *et cetera* within the holding). Instruments and data sources for assessing resource use and condition on the farm include field inventories, image interviews, GPS, soil sampling, vegetation inventory.

Table 16.7 Individual level of analysis

Levels of analysis (heuristic units defined according to research questions)	Investigating social processes related to land-use change	Integrated assessments of change in land-cover patterns	Investigating ecological processes related to land-use change
Individual	Key factors to focus on include life history, knowledge system, social and political participation and social networks.	Characterize knowledge of resource use and management. Build narratives of social-environmental histories representative of the larger population. Describe perception and value of environmental resources, views of environmental and resource degradation, value of participating in collective activities to govern/use/conserve common-pool resources of community/settlement, and any solutions to environmental and land-use issues.	Largely as above. Other key factors to focus on include individual plants/trees of personal or economic importance to individual. Sample individual plant species (and other resources) based on transects with key informants, herbarium collection building.

significant political and economic implications. The integration of ethnographic and participatory assessments can help to bring greater detail to the analysis of land-use/cover patterns in remotely-sensed images and to help to better represent land users and their management of forest landscapes. The results indicate that the combination of diverse approaches and disciplinary (and often inter-disciplinary) perspectives, bringing together the diversity of stakeholders' attitudes, needs and impacts is fundamental for supporting decisions over policies and actions affecting changes in land-use practices.

16.4 Participatory and Visualisation Techniques for Sustainable Use of Land Resources and Landscapes

The improvement of understanding of the outcomes of landscape planning decisions is a political priority, which builds on the aspirations of Brundtland (WCED 1987), reinforced by international and national policies over subsequent years (Table 16.8). Progressively, these policies have led to investment in improving access to environmental information (as per European Union 1998b), increasing public engagement, and an evolution of processes leading to greater citizen control (as termed in the framework of Arnstein 1969).

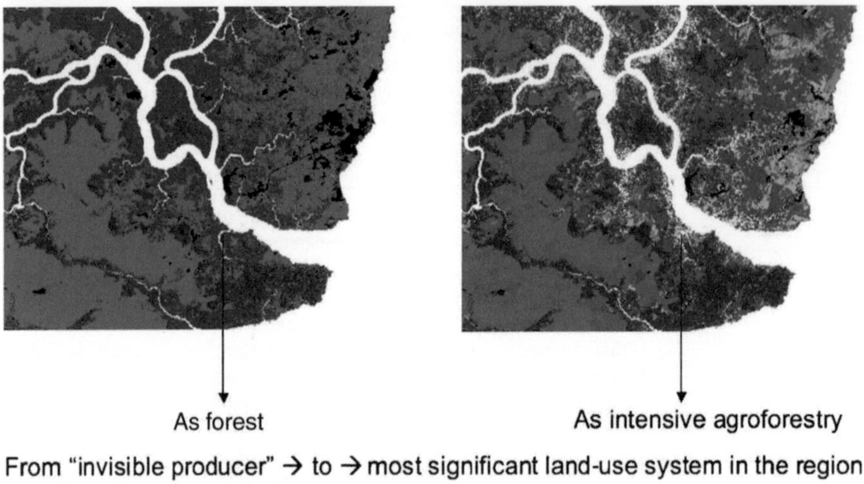

Fig. 16.3 Outputs from an ethnographic (participatory assessment) in which local input improved the classification process with the identification of areas of intensive agro-forestry (right-hand image) (See also Plate 43 in the Colour Plate Section)

Recognition of the interactions and balance between economic, environmental and social pillars of sustainability was reflected in a priority area of the EU (2001) Sixth Environmental Action Programme, in relation to protection of the wider countryside. The Programme recognises the importance of *"policy-making based on participation and sound knowledge"*, and that *"involvement of stakeholders will be central to its successful implementation."* To deliver on this priority there is a prerequisite for raising public awareness, and for providing people with opportunities to influence this change toward an improved quality of life in rural areas. A principal aim is therefore to enable the direct involvement of stakeholders in the development of landscape quality, and use public participation in the identification of objectives and in the evaluation of future landscapes with respect to their multiple functions.

The charging of governance with obligations relating to public participation has gradually permeated to national policies and regulations, resulting in guidelines for planning systems which acknowledge *"community involvement and dialogue, and negotiation"* as part of a process *"that respects the rights of the individual while acting in the interest of the wider community"* (Scottish Executive 2002). The inclusion of actors with multiple objectives in the planning process improves its potential to accommodate both socio-economic and environmental requirements (Scottish Executive 2007).

This section describes the involvement of stakeholders in the identification of the aims and functions demanded of the landscape (Landscape level in Table 16.4), and the use of virtual reality tools to facilitate information dissemination and consultation (Settlement/community level in Table 16.5).

Table 16.8 Milestones in progress towards planning of landscape change

Event	Outcome
The Brundtland Report: (1987)	Our Common Future: Detailed analysis of sustainable development
Rio Earth Summit (1992)	Agenda 21 – A Programme of Action for Global Action in all Areas of Sustainable Development. The Rio Declaration on Environment and Development
Aarhus Convention (1998)	Promoting public participation in decision-making, environmental awareness and access to related information
Doha Round of World Trade Talks (2001)	Removal of production-based support for agriculture, to be replaced by environmentally friendly land management
Johannesburg (UN 2002)	Affirmation of importance of broad-based participation in policy formulation, decision making and implementation at all levels
European Landscape Convention (Council of Europe, 2003)	Promotion of 'people centred landscapes', participatory planning and increased public understanding of change
United Nations (2005)	Increasing education and awareness of sustainable development

Source: Adapted from Miller *et al.* (2005)

16.4.1 Stakeholder Preferences and Attitudes Towards Landscapes and Change

Stakeholders were consulted to obtain their perspectives of multi-functional purposes of landscapes. In particular, those who directly interact with land-use systems, at a strategic or operational level, to identify appropriate courses for future management and planning. The 'people included' principle that identifies a creative management between the integrity of ecosystems and the livelihoods of people, living and working in the environment was employed in this research. Further details of the development and testing of the planning tools applied to a range of different case studies in Miller *et al.* (2005).

The details which follow are for a case study in north-east Scotland (Clashindarroch Forest). The design plan for the Clashindarroch Forest area was due for review, thus providing a real case for testing how visualisations might contribute to the stakeholder participatory decision-making process. The flowchart in Fig. 16.4 presents the interactions between user needs and data capture or analysis, and the stapes in which visualisation tools are used, and assessments of their fitness for purpose.

To quantitatively identify and analyse stakeholder opinions a Q-methodology was used, followed by a discourse analysis to explain the results obtained using statistical methods. Q-methodology is a quantitative means for examining human values and beliefs which enabled the identification and assessment of subjective structures, attitudes and perspectives of the public, from the standpoint of the people being observed. This approach provided insights into respondents' preferences, identified criteria of particular importance.

The sample included end-users with responsibilities for land management and planning, as well as the public. These were then analysed with respect to respondents' socio-economic backgrounds to reveal stakeholder and public attitudes, and of the elements of which they are composed (see Miller *et al.* 2005; Nijnik and Mather 2007, 2008; Nijnik *et al.* 2008).

Figure 16.5 shows that four alternative options have been identified concerning landscapes and land-use management decisions. The first impartial scenario is, broadly, an equal distribution of peoples' preconception towards financial investment (+3); environmental pillar (+3); social pillar (+2); economic pillar: farming (+4), and industrial/urban development (+2). The second environmental option reveals a quite strong environmental preference: with the environmental pillar (+5); social pillar (+2) and greater financial investment (+3). The second policy option rejects the economic pillar, with (–4) for farming activity and (–3) for industrial/urban development. The third essential economic alternative considers the development of farming activity (+5) in combination with industrial/urban development (+4) in rural landscape to meet the requirements of the social pillar (+4), whereas the importance of the environmental pillar (–3) and of financial investments (–3) is underestimated. The fourth fair economic option is similar in preference to the third option but with less pronounced inflection towards the other pillars: farming activity (+5); social pillar and industrial/urban development (+4); environmental pillar (–2) and the financial investment (–1). For more information on the method and on the results of its application, see Miller *et al.* (2005), and Nijnik and Mather (2008).

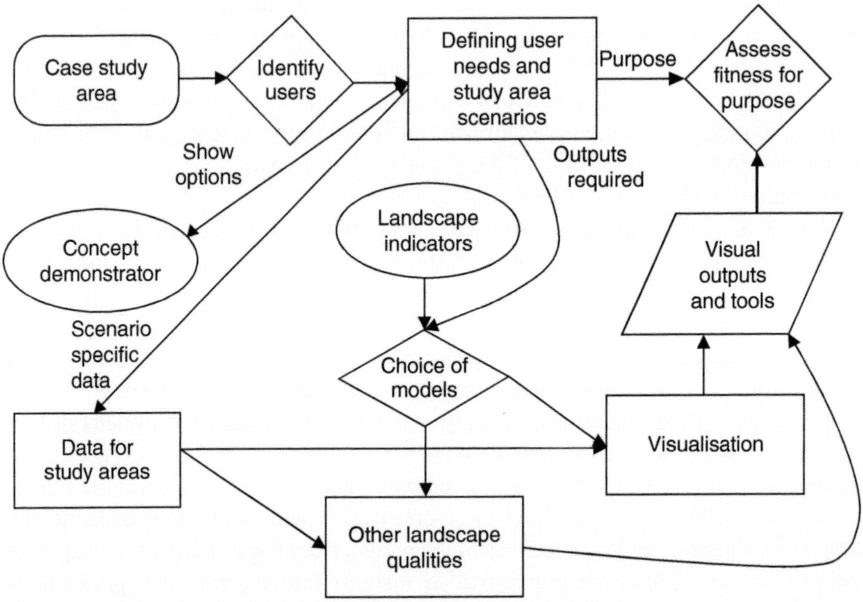

Fig. 16.4 Summary flow chart for the VisuLands research

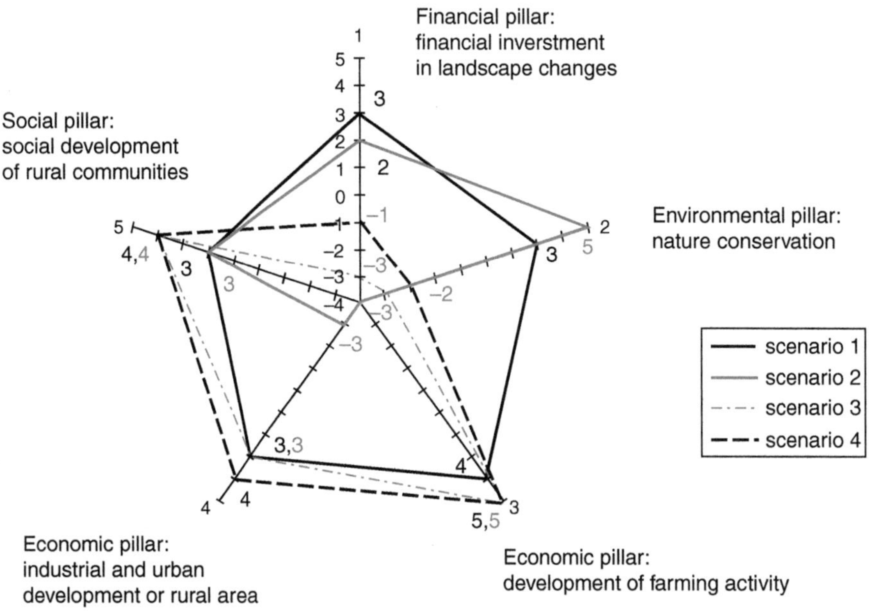

Fig. 16.5 Analysis of preference surveys, showing attitudinal diversity of stakeholders towards landscape changes

16.4.2 Awareness Raising and Participation in Landscape Planning and Management

Drawing on the results of the stakeholder and public attitudes and preferences for landscape change, the designs and plans for future management of the site were developed further with respect to layout and distribution of woodland species. This information was then used to develop representations of scenarios of proposed changes in land use, specifically in relation to the introduction of native woodlands in areas of pasture and moorland. Visualisation tools were then used to test public preferences for different scenarios of future landscape change.

The scenarios developed for the Clashindarroch area were: (i) maximising on the proportion of native woodland species (i.e. biodiversity), (ii) maximize on timber woodland (i.e. economic return), and (iii) diverse lands cover of moorland, forestry and agriculture (i.e. no change). These scenarios were presented in a mobile Virtual Reality Theatre, designed to support the sharing of views by audiences (Fig. 16.6), with electronic voting whilst navigating through landscape models. This was followed by a phase of knowledge transfer and raising of public awareness of the issues associated with each scenario. The output was an analysis of the preferences for the scenarios of change and an evaluation of the effectiveness of the associated programme of awareness raising.

The Theatre enabled scenarios of change to be presented to several audience groups, using a 'drive-through' of the area. As part of the presentations software functions were used to switch on and off groups of features (e.g. new woodlands), or movement of model features to audience selected locations in the landscape. Theses functions supported tests on audience preferences for landscapes under the scenario of change.

16.4.3 Survey of Preferences for Landscape

The participants completed a landscape preference survey, using images at eight set viewpoints in the model and at two points in time: the present and 100 years into the future. Participants scored the landscape view using voting handsets (Fig. 16.7), with the change in woodlands as the focus of the interest. The approach taken for the sessions was: (i) a short introduction to provide a context for the landscape planning process; (ii) a pre-set route through the study area enabling participants to 'experience' moving through the landscape in real-time; (iii) stops at eight viewpoints (Fig. 16.8) at which participants were invited to record their opinion of the landscape view and keywords that they would use to describe the view, using electronic voting handsets.

Table 16.9 shows an example of the central sections of the views at two viewpoints, under current conditions and the scenario of change. The principal keywords used by participants are included beside each image. The results for viewpoint 1 were not recorded so as to familiarise the audience with the facility, the views and the operation of the voting system.

In total, 139 responses were received for the test and survey. Only 20 per cent of respondents were employed in an activity directly related to land management or in a related advisory capacity. However, 35 per cent of respondents considered themselves to live in the countryside and 17 per cent in a village. Table 16.10 summarizes a finding that the scenes which are least preferred are those described as bleak, barren and bare, compared to the preferences for diverse, varied and scenic views. However, the discussion periods which followed the voting revealed groups whose preferences were for landscapes characterized as barren and bleak, which provide different types of experiences than those in the latter category. This range of views is reflected in the overall mean of scenes described with words such as barren or open being 3.2 (with a standard deviation of 0.61).

Experience is of considerable importance in people's preferences for landscapes, and in this regard the virtual environment can only convey an impression of the landscape and the changes proposed. Survey replies included comments that implied consideration of the view on the left, centre or right of the screen, with reference to the view on the 'left' and 'right'. This could be interpreted as suggesting that the Theatre did convey a sense of 'immersion' in the landscape. Further details of the surveys can be found in Miller *et al.* (2005).

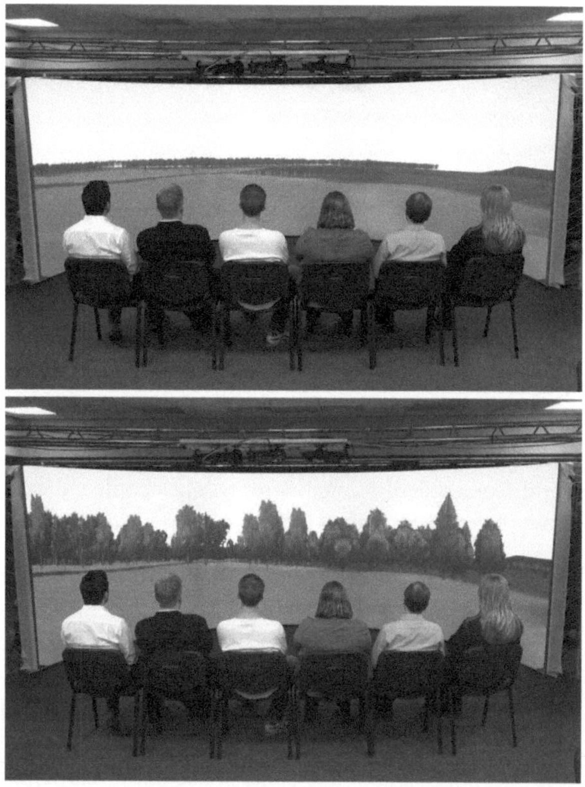

Fig. 16.6 Visualisation of 'before' and 'after' option for a change in woodland cover

Fig. 16.7 Voting on landscape preferences using electronic handsets in the Virtual Landscape Theatre

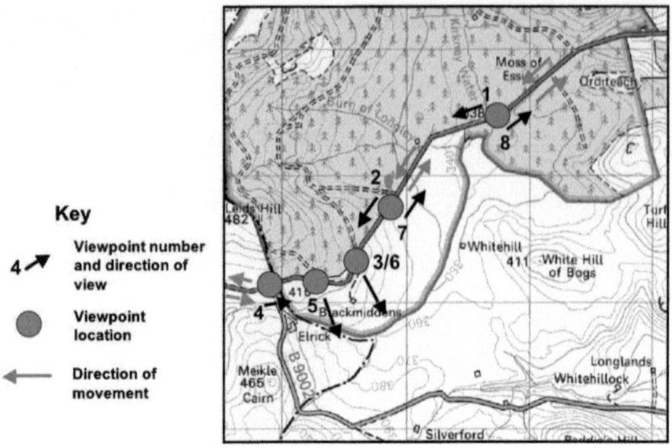

Fig. 16.8 Route of walk-through of landscape using the Virtual Landscape Theatre (Map reproduced from Ordnance Survey, Crown Copyright, under license to Macaulay Institute GD27237′2002) (See also Plate 44 in the Colour Plate Section)

In general, feedback from the participants was good, albeit with lessons to be learnt from specific remarks regarding people's experience in the Theatre, with a range of comments, such as:

- 'slightly sick-making'
- 'a much better impression than a computer screen'
- 'I would like to have sat back to see everything'.

The voting method via hand-held electronic devices seems to have been an effective way of collating quantitative data quickly and efficiently. However, through observing participants, it became clear that not all were voting in time (hence their vote was not counted) and some chose not to participate in the voting at all. In the latter instance, the use of a more discursive approach through the virtual journey proved very effective, perhaps more so than the voting/word association method,

Table 16.9 Extract of the centre of the view of the landscape as projected in the Virtual Landscape Theatre, with associated descriptive phrases for two viewpoints

Viewpoint Number	Current Situation	Keywords	Prospective Landscape	Keywords
2		Bare, stark, bland, spacious, uninteresting		Sheltered, improved, varied, interesting, pleasant
5		Barren, dull, open, boring, empty, unrestricted		Diverse, more interesting, better view

Table 16.10 Summary of ranking of landscape preferences and the associated keywords

	Least preferred	SD	Most preferred	SD
Mean ranking	1.55	0.46	4.23	0.34
Example keywords/phrase	Boring, barren, dull, bleak, bare		Diverse, varied, interesting, scenic	

in that it provided greater freedom to expand on answers and gave more time to consider each landscape. The virtual landscape tour appeared effective in engaging the public, providing a means of communicating environmental information and potential change in a comprehendible manner and thus enabling them to become involved in the decision-making process (Orland *et al.* 2001). Some aspects of the scenarios were pre-set in the model (e.g. density of woodland), thus limiting the nature of participation. However, the role of the facility has stimulated further use in participatory processes, opening it up to the public in the very early stages of the decision-making process and increasing transparency.

16.5 Conclusions and Further Research

The logical framework developed by ACT-CIPEC researchers (Brondizio 2005, 2006) shows how social and ecological scientists integrate disciplinary methodologies to assess processes of land use, and aid in land-use planning, from very local to regional scales. The framework outlined identifies examples of factors influencing change, stratification strategies, information on sampling design, examples of data sources, and examples of survey instruments for land-use change assessment and planning at different levels of analysis. Research instruments include ethnographic techniques, survey protocols, ecological assessment methods, and procedures to use multi-temporal remote sensing data and geographic information technologies during fieldwork and laboratory analysis. The actors, the decision takers, are at the centre of this approach by both providing information and shaping the direction the investigation take, an approach to explicitly capture the perspectives and values of the local stakeholder.

There are, however, challenges to successfully following through with this framework as there is a need for commitment to more interdisciplinary research. This is challenging, as it requires a significant investment of time to develop a common language, integrating those of the relevant disciplines, for a particular research topic or planning goal. Further challenges include the need for familiarization with the methodologies of different disciplines, and the development of an understanding of any trade-offs required between them. Underlying successful implementation of the framework is the development of specific research questions or planning goals for guidance, and the participation of actors and stakeholders at each needed level of analysis, study and planning phases.

The results of the VisuLands project show the potential for techniques and visualisation tools for public participation in decision making to improve the quality of planning for land-use change. The responses of professional and public stakeholders across the partner countries and case study areas suggest that the recognition of the European Landscape Convention (2000) of a public wishing to "... *play an active part in the development of landscapes*" is borne out. As with the work in the Amazon area, the decision-making process relies upon human and technical factors. Human factors include: the attitude towards participatory decision making of those who design and facilitate the planning process; adequate resourcing of the process to meet participatory objectives; and the perception of the role of visualization tools as a participatory tool. Technical factors include: the incorporation of necessary technological features in visualization tools for the particular stage, and context, of its use in the planning process; and the inclusion of appropriate levels of information content (including data on non-visual information) in visualisation tools to communicate information to those involved in the process.

Technological advances in landscape visualization offer approaches to representing landscapes of the past, present or future. However, they do not necessarily inform as to the information required for interpretation of the consequences of environmental change, nor how it should be tailored to different types of audiences. In addition, given the diversity of tools available, there are no transparent assessments of their effectiveness. This offers a challenge to further improvement of tools so that they are relevant, accessible and offer meaningful information for aiding decision making or an understanding of the consequences of environmental change (Fig. 16.9).

Areas of further research include the adaptation of interactive tools to enable different options for users to switch between scenarios, including changing the period of time over which the landscape changes may take place. Due of the range of requirements for linking technology, methodology and design of imagery, the research will gain from interdisciplinary co-operation between landscape-related

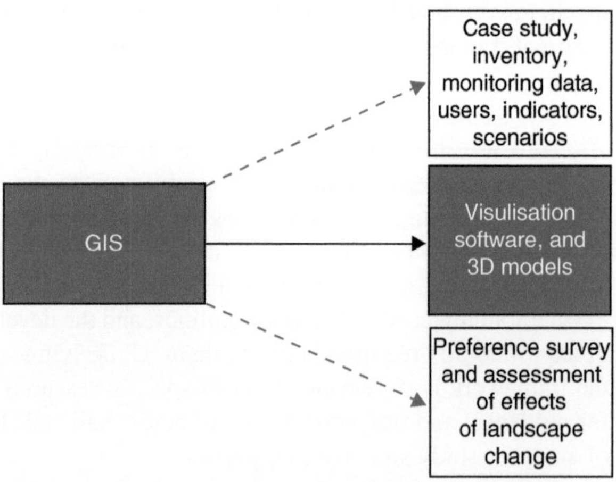

Fig. 16.9 Categorisation of participatory methods and techniques

sciences, with disciplines drawn from computer visualization, planning, design, social sciences as well as prospective end-users or their disciplines.

Widening the understanding of processes of change in landscapes amongst different stake holding groups should form the basis of planning their future (Council of Europe 2003; European Union 2001). The evolution of such processes is subject to largely social considerations at each of the levels identified in Table 16.1, but the acceptability and effectiveness may be greater in one area compared with another.

A categorization of participatory methods is shown in Fig. 16.10, in which the representation has been adapted from the concepts proposed by van Asselt *et al.* (2001) and Ball (2002), with a focus on divergence mapped on one axis and aspirations of participation on the other. For example, the top-left quadrant equates diversity and democratisation and contains methods developed, and processes initiated, by stakeholders. This represents initial, informative and exploratory stages in the decision-making process reflecting the plurality and diversity of values in society, and can reflect spontaneous civic movements which emerge to address special concerns.

The search for methods and tools for decision support in land planning and environmental policy areas has spurred scientists in Europe, and other parts of the world, into working in an interdisciplinary manner with concepts arising from areas of environmental assessments to improve public policy (e.g. Brondizio 2005; Mansvelt 1997; Munier *et al.* 2004; Potschin and Haines-Young 2003). Whether

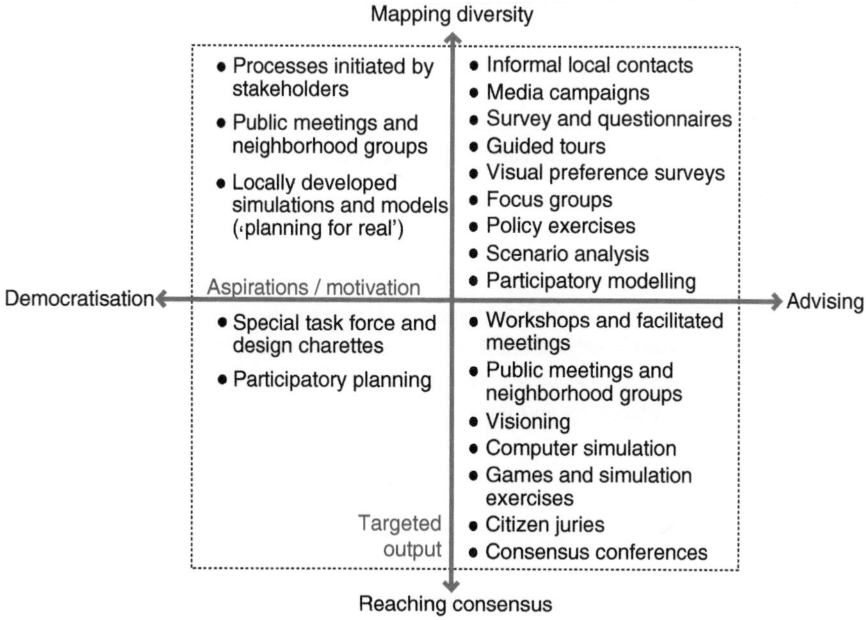

Fig. 16.10 Geographic information systems (GIS) provide inputs to different types of tools used in the assessment of landscape changes

the aim is awareness raising and information dissemination, or facilitating citizen actions, a framework which involves wider stakeholder participation requires to be adequately supported, not only in relation to the provision of information but also with the quality of awareness-raising. The impact of such participation may equate not to the level but rather the quality of involvement. This aspect of Arnstein's ladder of participation, and the related discussion by Sheppard (2005), will require greater consideration, as the risk of failing to deliver on expectations raised amongst participants could be the greatest challenge to the aims of public policy.

In the examples presented, stakeholder and public perspectives of landscape and land-use change have been combined with participatory techniques to enable stakeholder values, objectives and preferences to be incorporated into an analysis of options for future land uses. The process of research, scientific networking and communications with end users has led to identification of design features, and criteria for the development and use of techniques, methods and tools. However, more research is required on the how, and to what extent, stakeholder involvement affected decision making. Among the issues which arise are how stakeholder perspectives are best incorporated into new policies and programs that affect land-use and landscape change, and whether the increased social capital created through participation in research and consultation translates into more effective implementation of policies.

The approaches have built on research and participatory methodologies widely tested in the social sciences, and introduced new approaches that make use of advanced technologies. Further research should address the effectiveness and impact of technological applications in fostering stakeholder participation. The geographic, cultural, institutional or demographic context might impair the application of certain approaches, or distract from the content and the objectives of the participation. So, a question remains as to how the methodologies used affect the opportunity and ability of the stakeholder to freely and effectively contribute to the planning process?

The regions presented in this chapter have their own regulatory frameworks and cultural considerations which govern or influence the level of, and opportunities for, participation in land management and planning. In each, there is a pressure towards more participatory, interdisciplinary and holistic approaches for landscape planning and management to link sustainable development goals with local level priorities and practices. However, the effectiveness of the participation will depend upon the legal framework, socio-economic characteristics and geography of an area. Although the case studies discussed have operated at different scales and in different forms, the integration of technical, analytical and participatory techniques for planning the sustainable use of land resources and landscapes appeared to be of importance in both.

References

Arnstein, S. (1969) A ladder of citizen participation, *Journal of American Institute of Planners*, 35(4): 216–224.

Ball J. (2002) Towards a methodology for mapping regions for sustainability using PPGIS, *Progress in Planning*, 58: 81-140.

Ball, J., Capanni, N. and Watt, S. (2008) Virtual reality for mutual understanding in landscape planning, *International Journal of Social Sciences*, 2(2). http://www.waset.org/ijss/v2/v2-2-12.pdf.

Bishop I.D. (2007) Developments in public participation and collaborative environmental decision-making, In Lovett, A. and Appleton, K. (eds.) *GIS for Environmental Decision-Making, Innovations in GIS*, CRC Press, pp. 181–191.

Brondizio, E.S. (2005) Intraregional analysis of land-use change in the Amazon, In Moran, E.F. and Ostrom, E. (eds.) *Seeing the Forest and the Trees: Human-Environment Interactions in Forest Ecosystems*, MIT Press, Cambridge, Massachusetts.

Brondizio, E.S. (2006) Landscapes of the past, footprints of the future: historical ecology and the analysis of land use change in the Amazon, In Balée, W. and Erikson, C. (eds.) Time and Complexity in *Historical Ecology: Studies in the Neotropical Lowlands*, Columbia University Press, NY, pp. 365–405.

Brondizio, E.S. (2008) *The Amazonian Caboclo and the Açaí palm: Forest Farmers in the Global Market*. New York: The New York Botanical Garden Press. 403 pages.

Brondizio, E.S. and Siquiera, A.D. (1997) From extractivists to forest farmers: changing concepts of Caboclo agroforestry in the Amazon estuary, *Research in Economic Anthropology*, 18: 234–279.

Boucek, B. and Moran, E.F. (2004) Inferring the behavior of households from remotely sensed changes in land cover: current methods and future directions, In Goodchild, M.F. and Janelle, D.G. (eds.) *Spatially Integrated Social Science*, Oxford University Press, Oxford.

Cernea, M. (2005) Studying the culture of agriculture: the uphill battle for social research in CGIAR, *Culture and Agriculture*, 27(2): 74–87.

Cohen, W. and Levinthal, M. (1990) Absorptive capacity: a new perspective on learning and innovation, *Administrative Science Quarterly*, 35: 128–152.

Council of Europe (2003) *European Landscape Convention, Committee of Ministers of the Council of Europe*, Florence, 2000.

European Commission (2005) European Commission's review of progress towards the delivery of the Sixth Environmental Action Programme, http://ec.europa.eu/environment/ newprg/.

European Union (1998a) *Communication from the Commission to the Council and European Parliament on a Forestry Strategy for European Union, COM(1998)649*, 03/11/1998, Brussels.

European Union (1998b) Convention on Access to Information, Public Participation in Decision-making and Access to Justice on Environmental Matters, Aarhus, 25 June.

European Union (2001) *Sixth Environmental Action Programme, COM (2001)31.*

Geist H.J. and Lambin E.F. (2001) What drives tropical deforestation? A meta-analysis of proximate and underlying causes of deforestation based on sub-national case study evidence, *LUCC Report Series, no. 4*, LUCC International Project Office, International Geosphere-Biosphere Programme, Louvain-la-Neuve. http://www.geo.ucl.ac.be/LUCC/pdf/LUCC%20Report%20%20Screen.pdf.

Green, G.M., Schweik, C.M. and Randolph, J.C. (2005) Linking disciplines across space and time: useful concepts and approaches for land-cover change studies, In Moran, E.F. and Ostrom, E. (eds.) *Seeing the Forest and the Trees: Human-Environment Interactions in Forest Ecosystems*, MIT Press, Cambridge, Massachusetts.

Lu, D., Mausel, P. and Moran, E.F. (2002) Linking Amazonian secondary succession forest growth to soil properties, *Land Degradation and Development*, 13: 331–343.

Macaulay Institute (2007) *Putting People in the Planning Picture*. www.macaulay.ac.uk/videos/vlt/ index.php.

McConnell, W.J. and Keys, E. (2005) Meta-analysis of agricultural change, In Moran, E.F. and Ostrom, E. (eds.) *Seeing the Forest and the Trees: Human-Environment Interactions in Forest Ecosystems*, MIT Press, Cambridge, Massachusetts.

Mansvelt, J.D. (1997) An interdisciplinary approach to integrate a range of agro-landscape values as proposed by representatives of various disciplines, *Agricultural Ecosystem Environment*, 63: 233–250.

Millennium Ecosystem Assessment (2005) *Ecosystems and Human Well-being a Framework for Assessment*, Island Press.

Miller, D., Morrice, J., Anderson, L., Durozard, E., Fidalgo, B., Fry, G., Gaspar, J., Gibon, A., Hassan, R., Hislop, M., Horne, P., Huet, B., Ladet, S., Lange, E., Leandro, N., Messager, P., Mottet, A., Nijnik, M., Ode, A., Pascoa, F., Pinto, L., Quine, C., Schroth, O., Schwarz, G., Shepherd, N., Tveit, M., Vitry, A., Watts, K. and Wissen, U. (2005) *Visualization Tools for Public Participation in the Management of Landscape Change,* Final Report for the European Commission, VisuLands, Macaulay Institute, U.K.

Moran, E. and Ostrom E. (2005) *Seeing the Forest and the Trees: Human-Environment Interactions in Forest Ecosystems*, The MIT Press, Cambridge, Massachusetts.

Moran, E.F., Brondízio, E.S. and Van Wey L.K. (2005) Population and environment in Amazônia: landscape and household dynamics, In Entwisle B. and Stern, P.C. (eds.) *Population, Land Use and the Environment*, National Academies Press, Washington, DC.

Munier, B., Birr-Pedersen, K. and Schou, J.S. (2004) Combined ecological and economic modelling in agricultural land use scenarios, *Ecological Modelling*, 174: 5–18.

Nijnik, M., Mather, A. and Hill, G. (2005) From multifunctional forestry toward post-productivism? The Multifunctional Role of Forests: Policies, Methods and Case Studies, University of Padova, 1–12.

Nijnik, M. and Mather, A. (2007) Analysing institutions and public perspectives to identify the future of British forests, In Reynolds, K.M. (ed.) *Sustainable Forestry: From Monitoring and Modelling to Knowledge Management and Policy Science*, CABI publishers, pp.171–188.

Nijnik, M. and Mather, A. (2008) Analysing public preferences for woodland development in rural landscapes in Scotland. *Landscape and Urban Planning*, 86: 267–275.

Nijnik, M., Zahvoyska, L., Nijnik, A. and Ode, A. (2008) Public evaluation of landscape content and change. *Land Use Policy,* 26: 77–86.

Orland, B., Budthimedhee, K. and Uusitalo, J. (2001) Considering virtual worlds as representations of landscape realities and as tools for landscape planning, *Landscape and Urban Planning*, 54(14): 139–148.

Potschin, M.B. and Haines-Young, R.H. (2003) Improving the quality of environmental assessments using the concept of natural capital: a case study from Southern Germany, *Landscape and Urban Planning*, 63: 93–108.

Skole, D. and Tucker, C.J. (1993) Tropical deforestation and habitat fragmentation in the Amazon: satellite data from 1978 to 1988, *Science*, 260: 1905–1910.

Scottish Executive (2002) *Scottish Planning Policy: The Planning System.* http:// www.scotland.gov.uk/library/planning/spp1.pdf.

Scottish Executive (2006) *Learning for Our Future: Scotland's First Action Plan for the UN Decade of Education for Sustainable Development*, Scottish Executive.

Scottish Executive (2007) *Planning Advice Note 81: Community Engagement "Planning with People"*, Scottish Executive.

Sheppard, S.R.J. (2005) Participatory decision support for sustainable forest management: a framework for planning with local communities at the landscape level, *Canadian Journal of Forest Research*, 33(7): 1515–1526.

United Nations (1992) International Institutional Arrangements, In *Agenda 21 - A Programme of Action for Global Action in all Areas of Sustainable Development, The Rio Declaration on Environment and Development*, United Nations Conference on Environment & Development. http://www.habitat.ing.org.

United Nations (2002) *World Summit on Sustainable Development: Plan of Implementation*, United Nations, Johannesburg. http://www.johannesburg summit.org/ html/documents/summitdocs/ 2309_planfinal.htm.

United Nations (2005) *United Nations Decade of Education for Sustainable Development*. http:// portal.unesco.org/education/.

Van Asselt, M.B.A., Mellors, J., Rijkens-Klomp, N., Greeuw, S.C.H., Molendijk, K.G.P., Beers, P.J. and van Notten, P. (2001) Building Blocks for Participation in Integrated Assessment: A Review of Participatory Methods, International Centre for Integrated Studies (ICIS), Maastricht.

Vogt, N., Bahati, J., Unruh, J., Green, G., Banana, A., Gombya-Ssembajjwe, W. and Sweeney, S. (2006a) Integrating remote sensing data and rapid appraisals for land-cover change analyses in Uganda, *Land Degradation and Development*, 17: 31–43.

Vogt, N., Banana, A., Gombya-Ssembajjwe, W. and Batí, J. (2006b) Understanding the stability of forest reserve boundaries in the West Mengo Region of Uganda, *Ecology and Society*, 11(1): 38. [online] URL: http://www.ecologyandsociety.org/vol11/iss1/art38.

World Commission on Environment and Development (WCED) (1987) *Our Common Future*, (Brundtland Report), Report of United Nations Commission.

World Trade Organisation (2001) Ministerial Declaration Doha 2001, WT/MIN(01)/DEC/1, World Trade Organisation.

Additional Reading

Hunsberger, C. and Kenyon, W. (2008) Action planning to improve issues of effectiveness, representation and scale in public participation, *Journal of Public Deliberation, 4(1), Article No 1.*

Lovett, A., Appleton, K and Jones, A (in press) GIS-Based Landscape Visualization: The State of the Art, In Mount, N., Aplin, P., Priestnall, G. and Harvey, G. (eds.) *Innovations in GIS 13*, Taylor and Francis.

Moran, E. and Ostrom, E. (2005) *Seeing the Forest and the Trees: Human-Environment Interactions in Forest Ecosystems*, The MIT Press, Cambridge, Massachusetts.

Miller, D., Morris, J., Coleby, A. (in press) The provision of visualization tools for engaging public and professional audiences, In Bishop, I. and Lange, E. (ed.) *Visualization in Landscape and Environmental Planning,* Taylor and Francis, *Horne.*

Nijnik, M., Zahvoyska, L. and Ode, A. (in press) Public evaluation of landscape content and change: several examples from Europe, *Land Use Policy.*

Richards, C., Sherlock, K. and Carter, C. (2004) Practical approaches to participation, *SERP Policy Brief No.1.,* Macaulay Institute, Aberdeen.

Chapter 17
New Ways of Supporting Decision Making: Linking Qualitative Storylines with Quantitative Modelling

Hedwig van Delden and Alex Hagen-Zanker

17.1 Introduction

To explore how people will live and work in Europe, what the landscape will look like and what the environmental consequences will be in some 35 years from now, the PRELUDE project (EEA 2007) of the European Environment Agency developed five different land-use scenarios for Europe. The project was carried out according to a Story And Simulation (SAS) approach in which, iteratively, storylines developed in participatory sessions are underpinned by land-use models. *Storylines* in this context are defined as narratives about future developments in Europe. They provide qualitative information on a broad range of issues in an integrated context.

The project, set up and coordinated by the European Environment Agency, engaged about 25 stakeholders representing a range of European and international agricultural, industrial, governmental and environmental organisations, alongside facilitators of the stakeholder process and quantitative modellers working at European or regional scales. In three workshops, stakeholders developed scenarios and storylines. In between workshops, modellers prepared model simulations on the basis of the qualitative information from these storylines. During the workshops, the stakeholders and modellers discussed how to improve the storylines and model applications to arrive at creative, plausible and consistent scenarios.

This chapter focuses on elements of the project that were part of the regional interpretation and modelling of the five European storylines and scenarios and discusses the following questions:

- Are planning support systems (PSS), currently used for what-if analysis in a more traditional planning practice, capable of handling extreme drivers and shocks?
- How can qualitative storylines be translated into quantitative model inputs?

H. van Delden (✉)
Research Institute for Knowledge Systems (RIKS)
PO Box 463, 6200 AL Maastricht, The Netherlands
E-mail: hvdelden@riks.nl

S. Geertman, J. Stillwell (eds.), *Planning Support Systems Best Practice and New Methods*, DOI 10.1007/978-1-4020-8952-7_17,

- How can the use of a PSS contribute to the development of scenarios in a participatory process?

17.2 Background

In this chapter, a scenario is defined as a story – that can be told in both words and numbers – offering an internally consistent and plausible explanation of how events unfold over time (Gallopín *et al.* 1997; Kok and Van Delden 2007). Participatory development of qualitative storylines has been used in various disciplines and has, over the past years, entered the world of planning practice. Meanwhile, integrated land-use models – often developed as part of a PSS – are starting to provide support for this practice as well. Since both approaches have benefits and drawbacks, integrating quantitative models and more qualitative participatory methods to develop scenarios is currently being advocated as a promising way forward (EEA 2001; Kok *et al.* 2007). This chapter illustrates with a case study in what ways both approaches can complement each other, what problems still have to be solved and what possible future research can take this integration to the next level.

The PSS we use in this project is *Metronamica* (White and Engelen 1997; Engelen *et al.* 2003) and has been selected because it is:

- dynamic: simulation of yearly land-use and indicator maps, enabling the inclusion of feedback loops which are crucial to support consistency in the storylines;
- spatial: global, regional and local scales are used; spatial interaction and detail at local level are crucial since the interest of the project is in land-use developments and their environmental consequences;
- visual: output is presented in the form of dynamic maps, time charts and Excel tables; and
- flexible: the system is able to capture a wide range of drivers (external factors, policy options, autonomous behaviour) in ranges that exceed traditional policy exercises.

17.3 Method

The storylines are the starting point for the regional modelling. Where possible, meaningful, qualitative information from the storylines is quantified. Furthermore, results from the modelling at European scale, as well as historical data for each of the three regions have been instrumental in quantifying the storylines. Quantification of the five scenarios is largely carried out using the methodology of the European research projects VISIONS (White *et al.* 2004) and MedAction (Van Delden *et al.* 2007; Kok and Van Delden 2007). The methodology applied has been enhanced with a view to improve the consistency, completeness and robustness of the storyline interpretation and translation. The approach is generic, even though details of the implementation are specific to the *Metronamica* PSS. The methodology follows six steps:

step 1: setting the boundaries and application of the PSS;
step 2: calibration of the regional applications;
step 3: regional interpretation of European scenarios;
step 4: quantification of narrative storylines;
step 5: model runs and analysis of results; and
step 6: feedback to the stakeholder group.

This process is iterative and interactive so that the input for the model(s) can be fine-tuned in a number of rounds. As the models are generally fast (a simulation covering 30 years will run in the order of minutes), and as the outputs are very visible and tangible (time series of land-use and indicator maps presented as movies), this iterative loop is run through in the presence of stakeholders and experts. Experience has shown that this enables a quick fine-tuning of the model and greatly facilitates communication, thus enabling a quick refinement and validation of the stakeholder intentions and the derived storylines.

The following sections detail the six steps. Each of the steps will start with a general description of the step followed by an example of the PRELUDE project to illustrate and explain the process.

17.3.1 Step 1: Setting the Boundaries and Application of the PSS

In the first step the study areas are selected, the model specifics such as scale, resolution and boundaries are defined and the PSS is applied to the selected regions. For comparing the spatial impact on different countries and regions, three case regions of similar size, but with reasonably different characteristics were chosen: Estonia, the Netherlands and northern Italy. Selection criteria included: diversity in the location of the regions (i.e. lowland western Europe, southern Europe and central and eastern Europe); the range of contrasting circumstances and pressures (i.e. coastal areas, urbanised lowlands and mountainous rural areas); and the difference in the period they have been EU member states and hence been exposed to EU spatial policies.

17.3.2 Step 2: Calibration of the Regional Applications

Metronamica integrates a number of sub-models, most notably a constrained cellular automata (CCA) land-use model and a regional spatial interaction model. Each requires calibration of a considerable number of parameters. The calibration process is a persisting scientific and technical challenge. In a typical calibration procedure, the model is iteratively run for a period in the past. The parameters are then adjusted to optimise agreement between simulation and reality. The availability and quality of spatial data are crucial to the calibration.

For the case study areas, new applications have been set up with CORINE land-cover data (Haynes-Young and Weber 2006). For the Netherlands, additional data from national agencies are available. An earlier version of the model has been subject to several rounds of calibration and validation on the basis of this data over the last ten years (Engelen *et al.* 2003; Hagen-Zanker *et al.* 2005). Preliminary runs indicated that the CORINE data offered little support for the calibration because of the short time period it covers and the small number of changes that occurred. As a resolution, the pre-existing parameter set of the Netherlands case study has become the starting point for all three case studies.

The calibration of the Netherlands case study (Hagen-Zanker *et al.* 2005) combines several optimisation strategies. It is a semi-automatic procedure combining the optimisation algorithm of Straatman *et al.* (2004) with the expert approach of White and Engelen (2003). The expert scrutinizes parameters for the cause-effect relations that they represent and only upholds those parameter values that are consistent with geographical theory or common knowledge. The evaluation of model performance is based on visual interpretation, but also a relatively new map comparison technique (Hagen 2003) that has a tolerance for small spatial errors.

The robustness of the model is evaluated on the basis of long-term simulations (2000–2035). For this period and for each application, two scenarios are created that are highly diverse in area per land-use category. One scenario represents rapid urbanization and decline of agriculture, the other urban decline and agricultural growth. The runs are evaluated on the basis of stylized facts such as cluster size distribution as well as a visual impression to judge many hard to quantify aspects of spatial structure.

A special element is that several land-use classes did not exist during the calibration, or were not recognized as separate classes, such as 'parkvilles'. These classes have been exclusively calibrated on their long-term behaviour, in particular their locations relative to other land use classes and their cluster sizes.

For each case study region, the calibration results in a model and its parameter set, representing the autonomous dynamics of the region to a sufficient degree. This model and parameter set act as the starting point for the specification of the five European scenarios.

17.3.3 Step 3: Regional Interpretation of European Scenarios

The stakeholders developed five scenarios for the future of Europe. These scenarios can be found in a dedicated web-tool on the EEA website (http://www.eea.europa.eu/multimedia/interactive/prelude-scenarios/). A summary of the scenarios is provided below:

Great Escape – Europe of Contrast

In Europe market economics is a key driver. This comes with negative consequences such as intensification and pollution. Caused by the strong financial competition, the rich get richer while the poor get poorer. This social inequity is represented by gated communities that appear on the outskirts of the cities in the early stages of this scenario. In the further future, there is also an emergence of 'parkvilles' for the extremely wealthy: spacious homes in beautifully landscaped environments, surrounded by housing for workers who provide services to the people living within the parkvilles.

Evolved Society – Europe of Harmony

After heavy floods and a large increase in energy prices, society has a change of mindset in favour of having a more rural society in which community action and social equity play a crucial role. Many people believe that lifestyles and government policies should change to better protect the environment. Total agricultural land remains constant but the sector experiences a switch from intensive to more extensive practices.

Clustered Networks – Europe of Structure

The demands of an ageing society lead to the development of coherent spatial planning policies. New 'thematic cities' with a service economy are founded in peripheral regions. Urbanisation is concentrated and rural development focuses on 'green belts' around urban centres. Internationalisation policies allow marginalisation of the agricultural sector and natural habitats develop in the wider countryside.

Lettuce Surprise U – Europe of Innovation

A major food security crisis hits Europe. As crisis management fails, people lose trust in central government and decentralisation becomes the new way of decision making. Innovative technologies provide an increased productivity which reduces the land required for food production. Environmental awareness grows, leading to strong demands for environmentally friendly produced foods.

Big Crisis – Europe of Cohesion

A series of environmental disasters highlight Europe's vulnerability and inability to adapt. As a consequence of strong support for centralised government and concerns for solidarity and equality, policies for sustainable and regionally balanced development are consolidated at the European level. Public transport is promoted as environmental awareness grows. Agricultural intensification is reversed after 2015 and over time initial environmental pressures are relieved.

Parts of the scenarios will be used throughout this chapter to clarify the applied methodology. Based on the different European scenarios and storylines, a concise regional interpretation for the case study areas is carried out. Regional characteristics (economic, political, cultural and geographical) and information from the storylines are used to interpret how scenarios at the European level will have similar or different implications at the regional level.

An example of the regionalisation of scenarios is the impact of climate change. In the Evolved Society scenario, flooding and heat waves cause people to move from northern and central Europe to eastern Europe. At the regional level, this means that the Netherlands and the Po Valley in northern Italy are both severely affected by the floods. In the Netherlands, the ring of dikes protecting the economic heart of the country is given up because of an increasing flood risk from the rising sea level. In northern Italy, the region around Venice is affected by the floods which occur further inland over time, resulting in an increase of wetlands in the Po Valley. Northern Italy also suffers from heat waves and drought, which are more severe in the hilly southern part and in the lowlands. The impact of climate change in both countries results in an emigration to the east. Estonia is largely unaffected by the floods, except for some locations along the coast that become unsuitable for economic and residential functions. Over time, Estonia becomes a safe haven and migration is stimulated by the EU.

17.3.4 Step 4: Quantification of Narrative Storylines

Storylines are narratives, with rich verbal expressions. Quantitative models on the other hand are built on state variables and parameters, which essentially are cold numbers. The process of quantification is therefore not only a process of specification, but also of reduction. It is an iterative process including the following sub-steps:

- identify driving forces and parameters in the model;
- identify developments in the storylines;
- define the activities and land uses;
- quantify growth and decline;
  - adjust the attractiveness of regions;
  - adjust the relationships between activities and land uses;
  - adjust the interaction rules describing the human behaviour;
  - adjust policy, suitability and/or infrastructure maps and parameters; and
- identify elements for which no one-to-one translation is possible.

Identify Driving Forces and Parameters in the Model

Based on the configuration of the model, a list is made of all the driving forces and the associated parameters in the model.

Identify Developments in the Storylines

An important step from narrative storylines to quantitative models is the identification of 'clues'. Clues are meaningful text fragments in the storylines that contain a statement about a state, a change in a state, a quality, a quantity, a trend, a location, an action, an interaction, a stock, a flow, a migration, a process, ..., about the geographical system, a sub-system of the latter, the region, a part of the region, the world outside the region, an agent, a group of agents, a population, groups within the population, ..., that is (are) the subject of the scenario exercise. The term meaningful has to be interpreted here both in terms of the accurate description of the system as it is available from the storyline, and the purpose of the scenario exercise.

Although clues are sought after with a view to represent a scenario by means of a model, they should be selected irrespective of the possibility of representing them effectively by the latter. Depending on the model used, it will be possible to represent more or less of the clues. The last step in this paragraph deals with clues that cannot be represented in the model. All clues are ordered in a table and linked to particular model drivers, variables and parameters. Where possible clues have a temporal and a spatial specification so that they can be ordered in time, and become effective in the model at the right moment and geographical location. An example of the links between clues and model parameters is provided in Table 17.1. Parameter values are based on past trends and EU level prognoses.

Define the Activities and Land Uses

The *Metronamica* applications are set up to represent the main demographic groups, economic sectors and land-use classes of the modelled region. Where storylines introduce as of yet unknown economic activities, new residential classes, and/or land-use classes, these are also included in the model. Three such newly introduced residential categories are gated communities, parkvilles and thematic cities. The gated communities and parkvilles play an important role in visualizing the segregation in the Great Escape scenario – with parkvilles being the even more luxurious and segregated evolution of the gated communities – while the thematic cities show how large, but pleasant cities can develop in Clustered Networks.

Table 17.1 Quantification of scenarios: linking clues from the scenarios and storylines to the parameters and variables of the model

Indication (clue)	Consequences for model input			
	Process	Northern Italy		
		Impact	Parameter	Value
By 2030, Karl and Peter had developed 15 parkvilles across Europe. Karl lived inside the one in Southern Poland, partly built by Maria's father	2020–2035: Development of 15 parkvilles across Europe	Development of a parkville in the mountains. People (wealthy elite) migrate from cities and gated communities to parkvilles	Suitability map	Suitable locations are locations away from flooding, urban centres and the south of Europe, with sufficient water supplies and high landscape values
			Rule parkville to parkville	Attraction
			Number of people living in parkvilles	Increase
			Number of people living in continuous urban fabric	Decrease
			Number of people living in gated communities	Slight decrease
Maria now lived outside the parkville walls, in a small prefab bungalow with her 19-year old son, Jan	2020–2035: Social underclass lives just outside the parkville walls	Discontinuous urban fabric near parkvilles	Rule discontinuous urban fabric to parkvilles	Attraction
Their livelihood is dependent on providing basic services for the affluent people in the parkville. She got a job as a gardener for the parkville residents, while Jan earned money cleaning the helicopters of the rich residents	2020–2035: employment in parkvilles for social underclass	Migration of people (working class) from urban centres to surroundings of parkvilles, and development of services in surroundings of parkvilles	Number of people living in dis continuous urban fabric	Increase
			Number of people living in continuous urban fabric	Decrease
			Jobs in services	Increase

Table 17.1 (continued)

Indication (clue)	Consequences for model input			
	Process	Northern Italy		
		Impact	Parameter	Value
Maria's elderly parents had both died as a result of the summer heat wave of 2029. Such heat waves had become more frequent and acute during the 2020s	2020–2035: impact of climate change: heat waves	Decrease in suitability for agriculture and living	Suitability maps for agricultural and residential functions	Decline with a gradient from north to south and high to low altitude. South and low altitude have greatest decline
Maria now stands on the ladder, against the high garden wall, pruning the roses. She is lost in thoughts. She ponders how much has changed in rural Poland during her lifetime. The fantastically beautiful landscape inside the wall reminds her of her youth. Outside, beyond the parkville's paradise gardens, she sees a stark, flat plain with wheat	2020–2035: Annual crops in large parts of (mainly eastern) Europe	Continuation of trend	Area intensive agriculture	Decline, but to a lesser extent than extensive agriculture
		The agriculture that is left in northern Italy will also become more intensive over time	Area extensive agriculture	Decline

Adjust the Attractiveness of Regions

Clues often relate to the attractiveness of regions, which is modelled at the macro level in the *Metronamica* model. Therefore, parameters determining the attractiveness of a region are set. An example is the Great Escape scenario where the market is the driving factor and policy regulations are of less importance. The attractiveness of a region for an economic sector like industry is mainly based on the job potential and the accessibility, while elements like the amount of available space as presented in zoning maps play only a limited role.

Adjust the Relationships Between Activities and Land Uses

Clues often relate to the spatial and functional relationships between economic and demographic sectors. An example is the migration of people to the countryside and the return to local food production in the Evolved Society scenario, which results in a reduction of activity in the food processing industry. Another example would be a densification of people living in rural areas, which could, for example, occur in the case of disaster: people flee from their housing which is ruined by the floods and move in with their families that live in the countryside. This would result in an increase of people living in the countryside, while the residential area would remain the same.

Adjust the Interaction Rules Reflecting Human Behaviour

Clues also relate to the locational preferences of actors. Examples of such preferences are the desire to live near the cropland, pastures and natural areas in Evolved Society, and segregation between rich people living in parkvilles and the working class living outside the gates in the Great Escape. The changes in interaction rules have played a major role in the regional modelling, especially to reflect clues relating to cultural changes.

Adjust Policy, Suitability, and/or Infrastructure Maps and Parameters

Clues in the narratives relative to the changes in the physical environment, the institutional characteristics, and the infrastructure will respectively affect the suitability, policy and infrastructure maps of the *Metronamica* model at the local level. Each change in one of the maps is entered with its associated date, thus reflecting the moment in time at which the change becomes operational.

Because of the strong focus on economic development in the Great Escape scenario, we have assumed a large increase in road construction and a development of helicopter airports in the parkvilles to enable the wealthy people to fly in after a day of work in the city. In the Evolved Society scenario, there is no increase in the road network foreseen, since cars are too expensive because of high fuel prices. For the transport of food that cannot be grown in their own region, a hover rail is constructed in northern Italy and Estonia that connects the main cities and towns. In the Netherlands, it is assumed that the current rail network is sufficient.

In four of the storylines, climate change is incorporated as an important driving force for land-use changes. Two impacts of climate change are mentioned: (i) flooding of river basins and coastal zones and (ii) increases in the temperature and heat waves in southern Europe. Both flooding and heat waves have a negative effect on the suitability for economic and residential functions. All three countries suffer from the effects of flooding. Only northern Italy is assumed to experience the negative impacts of temperature increase.

The suitability for residential and agricultural functions in northern Italy is affected by droughts. The droughts become more severe over time and the impacts are different in space. There is a gradient from south to north (the suitability in the south is more affected than in the north). And there is a gradient from low-lying to high-lying areas. In the low-lying areas, the suitability is more affected than higher in the mountains.

Based on this material and the information on heat waves and drought taken from the scenarios, a composite map for each scenario is produced which takes into consideration corridors of potentially floodable land alongside the rivers, the number of times that floods and landslides have taken place in particular areas in the past, the elevation and the slopes of the land, and the north-south gradient. During the simulation, these suitability maps gradually change over time until they reach their final condition in 2035. An example of the initial suitability map for extensive agriculture for all scenarios as well as the final suitability map

Fig. 17.1 Suitability for extensive agriculture, 2000 (See also Plate 45 in the Colour Plate Section)

Fig. 17.2 Suitability for extensive agriculture, Big Crisis, 2035 (See also Plate 46 in the Colour Plate Section)

for the Big Crisis scenario can be found in Figs. 17.1 and 17.2. Light (b/w) or green (c) colours indicate a high suitability, dark (b/w) or red (c) colours a low suitability.

Identify Elements for Which No One-To-One Translation is Possible

Narrative storylines contain clues on a wide range of subjects. It is not possible to interpret all clues as model inputs straightforwardly. One typical interpretation step is from the personal to the general. An example of such a personal clue is: *Maria now lived outside the parkville walls, in a small prefab bungalow with her 19-year old son, Jan* (Great Escape). The interpretation we have extracted from this clue is: *social underclass lives just outside the parkville walls.*

Difficulties will also occur when there is insufficient detail in the description of the storyline or when clues are relevant but cannot be related to any particular input of the model. For instance, there is no dedicated parameter to represent 'a shift in political power from the national to the local level' (LettuceSurprise U). These clues are included by their indirect consequences. In the given example, this leads to derived clues like 'lifting restrictive policy (adaptation of a policy map)' or 'ceasing protection of the agricultural sector, causing it to disappear in part (negative growth of agricultural land use)'. The risk of this approach is that modellers introduce developments and side-effects into the storyline that are not intended by the stakeholders. It is therefore crucial to discuss the interpretation and possibly adjust the model to the intention of the stakeholders.

A last type of difficulty arises because of limitations of the modelling technique and limitations of the models in use. In the project, elements that could not be modelled are, for example, social unrest and air quality. One strategy is to look at a scenario as a combination of a narrative storyline and model results. Another possibility is to extend the PSS with process relations or indicators.

17.4 Step 5: Model Runs and Analysis of Results

For each scenario, a series of model runs is carried out based on the values for the driving forces and parameters obtained in step four. The model simulates spatial developments over the period 2000–2035. Besides series of land-use maps, the PSS generates time series of indicators at the local, sub-regional and regional level. At NUTS-3 level, the model calculates the number of people in main demographic groups as well as the number of jobs in main economic sectors. At local level, it calculates the following indicators at the resolution of the land-use map: landscape identity, habitat fragmentation, quality of the landscape, disturbance of high nature value (HNV) farmland, quality of the residential environment and proximity of residential areas to green areas and open space.

One of the major challenges of the project is to see if the *Metronamica* model is able to capture the extreme drivers and shocks from the storylines as well as the introduction of new land uses and processes. Regarding the latter, we present here an example from the Great Escape (Figs. 17.3 and 17.4). In this scenario, cities become very unpleasant places to live in, causing a feeling of unsafety for the upper class who move to the gated communities (dark grey (b/w) or dark orange (c)) in the suburbs or just outside the cities. Over time, the rich become richer and a small proportion moves away from the gated communities to the even more luxurious

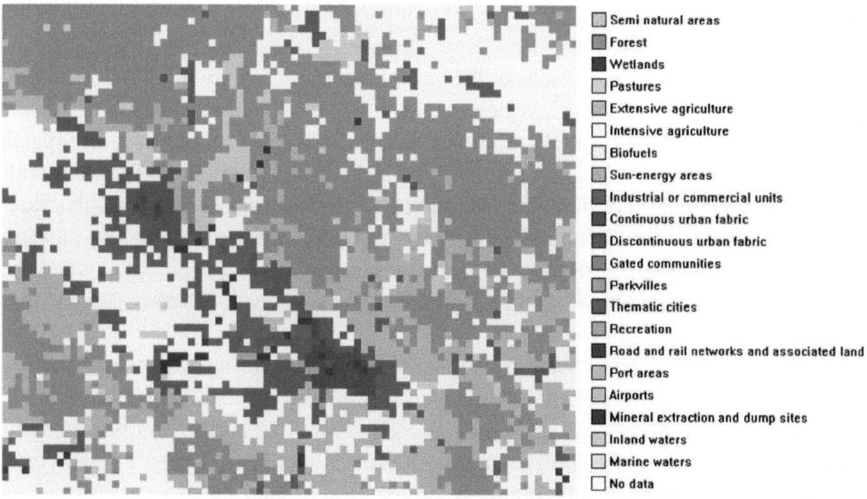

Fig. 17.3 Northern Italy, 2000, detail of the Florence region (See also Plate 47 in the Colour Plate Section)

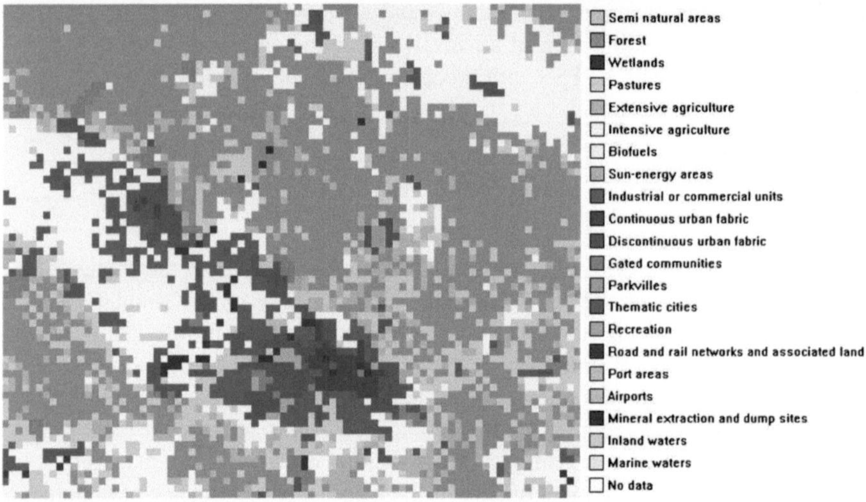

Fig. 17.4 Northern Italy, 2035, detail of the Florence region (See also Plate 48 in the Colour Plate Section)

parkvilles (black (b/w) or light orange (c)), located in the countryside, surrounded by recreation (grey (b/w) or pink (c)) and housing (grey (b/w) or red (c)) for the workers that provide services to them.

A formal data-driven validation of the models is not possible due to data limitations. Moreover the nature of the exercise would render a data-driven validation less than ideal. The purpose of the model is to give spatially and temporally explicit explications of storylines and scenarios under the influence of trends and processes that are purposefully distinct from past trends. The validation of the models is, therefore, based on the spatially explicit comparison of land-use patterns following from the different scenarios.

For comparing the result maps of the various scenarios, a moving windows based structure comparison is used, following Hagen-Zanker (2006). This comparison method applies well known and relatively simple metrics of spatial structure (patch size and Shannon diversity) on the basis of a distance weighted moving window. The patch size metric is calculated for individual land-use classes, it has high values when the class is strongly clustered in space and low values when it is fragmented. The Shannon diversity considers all land-use classes; it has a high value when the moving window contains a mixture of classes, and a low value when a single class or only a few classes are found.

The results of the cross-scenario comparisons are according to expectations and thereby increase confidence in the model's validity. The structure comparison of the scenarios Lettuce Surprise U and Great Escape (Figs. 17.5 and 17.6) illustrates this analysis. Light (b/w) or red (c) colours represent a decline in diversity; dark (b/w) or blue (c) colours represent a higher fragmentation of land.

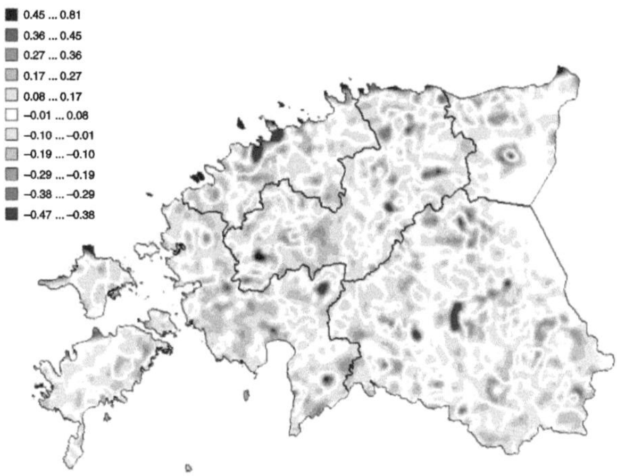

Fig. 17.5 Diversity, The Great Escape, 2035 (See also Plate 49 in the Colour Plate Section)

17.5 Step 6: Interaction with Stakeholder Group

The changes in land use under different scenarios have been presented to the stakeholders. Besides land-use maps for 2035, stakeholders also saw animations of the land-use developments over the period 2000–2035 in yearly time steps. Furthermore, we presented them with indicators representing important elements in the scenarios.

The model results confronted stakeholders with the reality of the storylines and gave cause to discussions on the impact of their scenarios on land use. The outcome of the discussions was threefold. In some cases – by seeing the impacts on the land use, stakeholders realised that their storylines perhaps exaggerated impacts (flooding of the economic heart of the Netherlands in Evolved Society) or were inconsistent, which then gave cause to adapt them. In the Lettuce Surprise U scenario for example, the initial storyline mentioned that people move to the countryside to live with several generations under one roof, while in the final storyline families live

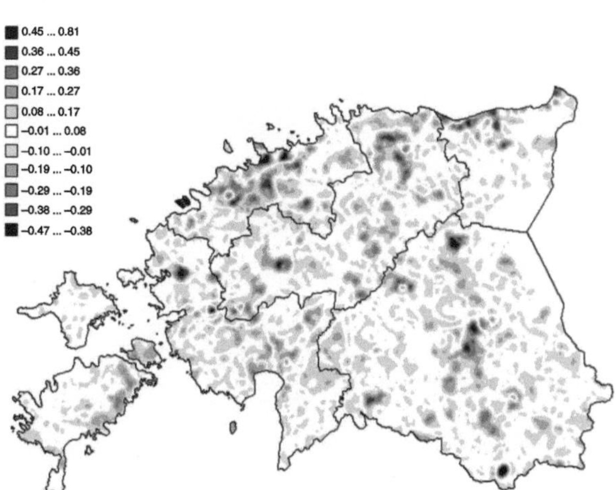

Fig. 17.6 Diversity, Lettuce Surprise U, 2035 (See also Plate 50 in the Colour Plate Section)

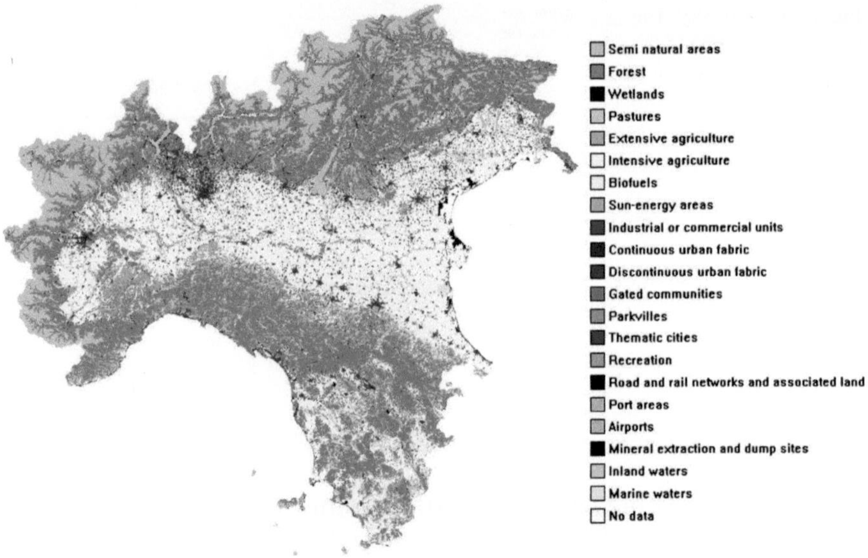

Fig. 17.7 Northern Italy, 2000 (See also Plate 51 in the Colour Plate Section)

together, but in more spacious houses. This change in the storyline is consequently reflected in the model, in particular the parameter governing population density.

In some of the scenarios, insufficient detail was provided to set some of the parameter values in the model. What does, for example, high economic growth mean? Is this 3 per cent, 8 per cent or 30 per cent? In those situations, the modellers

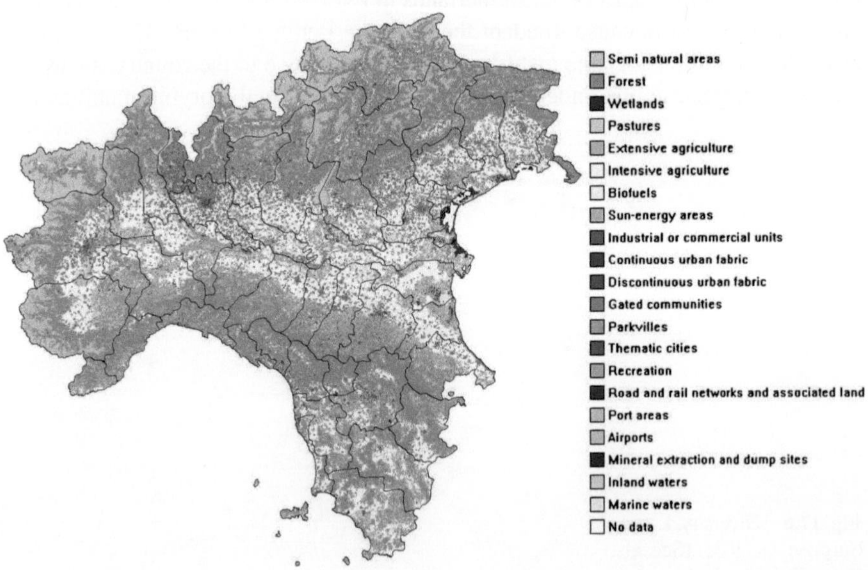

Fig. 17.8 Northern Italy, Evolved Society, 2035 (See also Plate 52 in the Colour Plate Section)

made initial assumptions and discussed those with the stakeholders in the next workshop who could then provide suggestions for model improvement.

The modelling also started some discussions. In the Evolved Society scenario, intensively managed land was to convert to small-scale, extensively managed land and a migration of people from the city to the countryside was foreseen to manage those lands and become self-sufficient (Figs. 17.7 and 17.8). When examining the model results, two things became apparent and were discussed: the scenario was not as environmentally friendly as assumed in the previous workshop because of the fragmentation of the habitat, and the question was raised as to how people in this scenario can be self-sufficient if all agriculture becomes extensive and the agricultural area cannot expand because of the geographical limitations (the Po Valley is surrounded by mountains).

Besides a discussion on the contents of the scenario, the presentation of the modelling results also started a discussion on the scale issue. An important remark in this discussion was that trends with a positive or negative impact on the environment at EU level may have the adverse impact in particular regions. A closer investigation on the scale issue would be relevant.

17.6 Reflection on the Method Used

The method described in this chapter has proven to provide a useful framework for linking qualitative scenarios and quantitative modelling. Through incremental steps (clues, ideas, impacts, parameters, values) the qualitative storylines are transformed into input required for the modelling work and afterwards model results are transformed into indicators (i.e. habitat fragmentation, quality of the residential environment) that provide information about relevant elements of the storylines. Since each step is very transparent, the translation can be discussed with stakeholders and modellers at any point in time.

In applying the methodology it is of crucial importance to allow for feedback loops between the different steps. By going through an iterative process, communication and social learning will be enhanced and will lead to a better understanding of the overall scenario that includes the narrative storylines as well as quantitative modelling.

In the present example, we made use of an existing model and further model development was not included as an option. In the development of larger integrated models it would, however, make sense to go through a similar exercise with a simple model before deciding on the exact configuration of the more complex integrated model, since the exercise provides insight into the necessity of either including more components or modelling specific components in more detail. When following this approach, the first step of the methodology can include the configuration of the different components of the integrated model as well as its application.

In the current approach, quantitative information is derived from the qualitative information based on actual data, projected values for certain future variables at higher spatial scales and a group discussion with the stakeholders. There are many other possibilities that could be explored to obtain a quantification of the qualitative information that also depend on the type of information that is provided by the qualitative storylines. If, for example, in the storyline there is a concern about air quality, criteria or standards to evaluate health risks would provide a good basis to link the results of an air quality model to the narratives. Another type of information would be the quantification of 'high economic growth'. Here it is very likely that there is a range of estimates amongst the stakeholders as to what exactly 'high' means. Is this two per cent, six, or maybe even twenty? Relevant questions are how the different opinions of the stakeholders are being taken into account in defining the figure that is being used. Is it something that we derive from a democratic process in which figures provided by stakeholders are averaged or do we use a process in which stakeholders are first tested on their economic knowledge and only the figures of those that have provided sensible answers are taken into account?

17.7 Conclusions, Recommendations and Future Directions

Carried out in an iterative process, linking qualitative storylines to quantitative modelling can improve the quality of the total scenario to a very large extent. Stakeholder discussions and narrative scenario and storyline development have the advantage that they present a very creative and flexible process. New elements and developments can be incorporated very easily. Moreover, the process of narrative storyline development is easily understandable and does not require stakeholders to have any modelling or computer knowledge. They can develop qualitative scenarios by using their own knowledge and imagination. Its drawbacks, however, are that it sometimes is difficult to oversee the complexity and to ensure consistency between all actors, factors and processes involved.

Models have the strength that they can visualise their output and thus make scenarios come alive. Furthermore, computers are very well suited to integrate different processes, because they can handle complexity and carry out objective calculations. An added benefit of spatially explicit land-use models is that they can visualise future developments including the impacts of actions taken by planners, policy makers and other stakeholders.

By combining both approaches, win-win situations can be created. Creativity and flexibility are essential elements for developing future scenarios. Stakeholder workshops in which scenarios and storylines are developed are a very good technique to incorporate both elements. Inconsistency in the storylines can easily be detected by the modelling approach. Moreover, because modelling forces us to make certain elements explicit, these elements can be incorporated in the scenario discussion. The combination of both approaches can also help in bridging the gap between scientific knowledge, captured in models, and the actual planning and decision-making process.

A strong element of the PRELUDE process has been to start with the development of the narrative scenarios while gradually incorporating the modelling elements. This has facilitated the understanding of the models by the stakeholders, because they could take a first step in developing scenarios and, subsequently, take the next step by discussing how their scenarios work out on the land use.

Another factor contributing to the success of this project was the attendance of the same stakeholders to the different workshop sessions. By attending the different workshops they undergo a process in which knowledge is shared and ownership is created. It also allows for a gradual acquaintance with the models in which knowledge is built up.

For the (modelling) support group it is a delicate process of listening and giving information. By providing too much information and listening too little, the support group loses its 'support' function and will steer the process and the scenario development; by providing too little information, stakeholders cannot make use of the (scientific) knowledge and models to the maximum extent.

Qualitative storyline development and quantitative modelling are tasks frequently carried out by different groups with different backgrounds. For future participatory processes that combine both approaches, it is recommended that the link between both is discussed from the beginning of the process. This helps the modellers to understand what they can expect of the stakeholders and their storylines and helps the facilitators to understand what information is necessary to populate the models. Decisions on the selection of indicators could be discussed and important elements for the modelling could be asked at an early stage. The latter does not mean that stakeholders should be asked to discuss parameter values at the beginning of the process, rather, that more abstract questions are dealt with in the development of the narrative scenarios and storylines. Important questions in the case of PRELUDE would, for example, have been:

- How does the scenario impact the different land use functions?
  In the PRELUDE project, some scenarios focused very much on the urban or the rural landscape, while for this exercise an incorporation of both is essential.
- What is the magnitude of change when you compare the different scenarios?
  In the PRELUDE project there was some inconsistency in the magnitude of change between the EU modelling work and the stakeholders (as well as between the stakeholders), which became apparent while showing the regional modelling. By incorporating this question into the discussion a first impression can be provided by the stakeholders. In a second phase modellers can than argue if the magnitudes are correct.

After a first stakeholder workshop, stakeholders can – in a second workshop – be asked to discuss growth and decline figures as well as the interpretations made by the modellers regarding the parameter values. Here, there is an important task for the modellers to discuss only what is most relevant and to explain the link between the scenarios and their work. It is important to understand that stakeholders and modellers normally work in two different worlds and that communication should be a special point of attention. Here is a role for both the facilitators of the workshop and the modellers.

Although the integration of participatory scenario development and modelling seems a very promising direction, some work is needed to link both approaches. In particular, the relation between the qualitative and quantitative information requires further investigation. How can qualitative information like 'moderate growth' be translated into actual figures? By whom should this be done – experts, stakeholders? And in what kind of process?

This chapter has shown that a relatively simple but flexible model is able to capture most of the elements from qualitative and creative storylines. However, there are always elements that cannot be translated in the model. How all-encompassing should a model be to provide support to the development of creative scenarios? And what level of detail (spatial, temporal, process) is required to model these types of scenario?

Very often policy makers seem to have difficulties in comprehending integrated models and PSS because of their quantitative nature and wealth of equations. It could very well be that for this reason they are not used to the extent possible in the decision-making process. On the other hand, most policy makers do seem to understand the processes that are being modelled in the systems and have a mental model that is not too different from the actual models. Would the development of scenarios by policy makers (and possibly other stakeholders) help to improve the understanding and use of integrated models and PSS? What other ways are there to bring the systems closer to the perceptions and daily practice of the policy maker?

The last suggested research direction is related to the actual use of creative scenarios in policy making. This chapter has tried to answer the question how a PSS can contribute to the development of (qualitative) storylines and scenarios in a participatory process, but has not emphasised the impact of the exercise on the decision-making process. Even though the broader perspective and the awareness of an uncertain future are often mentioned as main benefits of using creative scenarios in policy development, more emphasis could be put on the actual support of policy making when using this technique. Research questions that relate to this issue are: How to support today's policy making based on what is learned from the scenario exercises? Do very broad and creative scenarios help policy makers in making better decisions for the future? What is a 'better' decision? How can we measure if an added value is provided to the policy-making practice?

Acknowledgements The work described in this chapter was carried out in 2004 and 2005 under contract with the European Environment Agency (EEA). The results are reproduced here with permission of the EEA.

However, the EEA cannot be held responsible or liable with regard to information or opinions expressed in this chapter, which do not necessarily reflect the official opinion of the EEA and cannot be interpreted as predictions or forecasts.

The authors thank Teresa Ribeiro, Ybele Hoogeveen and Thomas Henrichs from the Scenarios & Forward Studies group of the EEA for their contributions. Their innovative ideas and enthusiasm have been crucial for the success of this project. Moreover, the authors very much enjoyed the open discussions, the critical feedback and the insightful suggestions throughout the course of the project.

References

EEA (2001) Scenarios as tools for international assessments, *Prospects and Scenarios No. 5,* European Environment Agency, Copenhagen.

EEA (2007) Land-use scenarios for Europe: Qualitative and quantitative analysis on a European scale (PRELUDE), *Technical report No 9/2007,* European Environment Agency, Copenhagen.

Engelen, G., White, R. and De Nijs, T. (2003) Environment explorer: spatial support system for the integrated assessment of socio-economic and environmental policies in the Netherlands, *Integrated Assessment*, 4(2): 97–105.

Gallopín, G., Hammond, A., Raskin, P. and Swart, R. (1997) *Branch Points: Global Scenarios and Human Choice, Pole Star Series Report No. 7*, Stockholm Environment Institute, Boston.

Hagen, A. (2003) Fuzzy set approach to assessing similarity of categorical maps, *International Journal of Geographical Information Science*, 17(3): 235–249.

Hagen-Zanker, A. (2006) Map comparison methods that simultaneously address overlap and structure, *Journal of Geographical Systems*, 8(2): 165–185.

Hagen-Zanker, A., van Loon, J., Maas, A., Straatman, B., de Nijs, T. and Engelen, G. (2005) Measuring performance of land use models: An evaluation framework for the calibration and validation of integrated land use models featuring cellular automata, Paper presented at the 14th European Colloquium on Theoretical and Quantitative Geography, Tomar, Portugal.

Haines-Young, R. and Weber, J.-L. (2006) *Land Accounts for Europe 1990–2000. Towards Integrated Land and Ecosystem Accounting, EEA Report No. 11/2006,* European Environmental Agency, Copenhagen. www.eea.europa.eu/prelude

Kok, K., Verburg, P.H. and Veldkamp, A. (2007) Integrated assessment of the land system: the future of land use, Editorial, *Land Use Policy*, 24: 517–520.

Kok, K. and Van Delden, H. (2007) Combining two approaches of integrated scenario development to combat desertification in the Guadalentín watershed, Spain, *Environment and Planning B*, forthcoming.

Straatman, B., White, R. and Engelen, G. (2004) Towards an automatic calibration procedure for constrained cellular automata, *Computers, Environment and Urban Systems*, 28(1–2): 149–170.

Van Delden, H., Luja, P. and Engelen, G. (2007) Integration of multi-scale dynamic spatial models of socio-economic and physical processes for river basin management, *Environmental Modelling and Software*, 22(2): 223–238.

White, R. and Engelen, G. (1997) Cellular automata as the basis of integrated dynamic regional modelling, *Environment and Planning B*, 24: 235–246.

White, R. and Engelen, G. (2003) A calibration procedure for constrained large neighbourhood cellular automata based land use models, Paper presented at the 13th European Colloquium on Theoretical and Quantitative Geography, Lucca, Italy.

White, R., Straatman, B. and Engelen, G. (2004) Planning scenario visualization and assessment: a cellular automata based integrated spatial decision support system, In Goodchild, M.F. and Janelle, D. (eds.) *Spatially Integrated Social Science*, Oxford University Press, New York, pp. 420–442.

Additional Reading

Shearer, A.W. (2005) Approaching scenario-based studies: three perceptions about the future and considerations for landscape planning, *Environment and Planning B*, 32: 67–87.

Steelman, T.A. and Ascher, W. (1997) Public involvement methods in natural resource policy making: advantages, disadvantages and trade-offs, *Policy Sciences*, 30: 71–90.

Vennix, J.A.M. (1999) Group-model building: tackling messy problems, *System Dynamics Review*, 15: 379–401.

Westhoek, H.J., Van Den Berg, M. and Bakkes, J.A. (2006) Scenario development to explore the future of Europe's rural areas, *Agriculture, Ecosystems and Environment*, 114: 7–20.

Xiang, W.-N. and Clarke, K.C. (2003) The use of scenarios in land-use planning, *Environment and Planning B*, 30: 885–909.

Chapter 18
Land Management with the *SMURF* Planning Support System

Marc Soutter and Alexandre Repetti

18.1 Introduction

Population growth, rapid urbanisation trends, climate change, ageing infrastructure, globalisation, emerging technologies, consumer involvement, energy use and costs, geo-political changes, changing patterns of land use, changes in the hydrological cycle, *et cetera,* are all important symptoms of global change. Core problems and negative impacts of global change have become increasingly visible and impact both on people and the environment, in developed and developing countries. In many respects, land management and planning need to be much more adaptive than they have been in the past. Cities and regions all over the world are confronted with an accelerated pace of changes that affects almost all aspects of land management. The need for instruments or systems to support planning at a strategic level is considerable. The approach presented in this chapter has potential applications that go far beyond its original context.

18.2 Background

West African mid-sized cities, as in many other parts of the under-developed world, are characterized by a rapid population growth and a lack of financial and human resources, which badly limits the working capacity of public authorities (Sawadogo 2002). This results in often anarchic urbanization and extensive development problems: economic weakness and increasing informal economies, infrastructure in limited quantity and often dilapidated, insufficient access to basic services, high poverty rate and growth of squatter settlements (slums),

A. Repetti (✉)
Cooperation@epfl
Ecole Polytechnique Fédérale de Lausanne (EPFL),
1015 Lausanne, Switzerland
E-mail: alexandre.repetti@epfl.ch

S. Geertman, J. Stillwell (eds.), *Planning Support Systems Best Practice and New Methods*,
DOI 10.1007/978-1-4020-8952-7_18,

social disparities and numerous conflicts of governance. Development policies are continuously evolving to provide solutions to the weakness of resources and the increasing difficulties of management. In 40 years, strategies have moved from a centralizing state towards decentralized models, passing through structural adjustments (World Bank 1994), through good governance policies (World Bank 1992) and through the privatization of public services (le Galès 1998). Furthermore, technical instruments for management and coordination (needs appraisal, master planning, sector-based management) also cause problems in the context of underdevelopment, due to their rigidity and to the common failures in the reality of urban planning and management (Bolay *et al.* 2000; de Graaf and Dewulf 2002; Frérot 1999).

New urban planning and management techniques and technologies are proposed. Urban strategic planning (Borja and Castells 1997; Ingallina 2001) offers promising solutions in underdeveloped contexts (Halla 2005; Steinberg 2005). It can be reinforced with planning support systems (PSS) (Allen 2001; Geertman and Stillwell 2003; Harris and Batty 2001; Harts *et al.* 2003; Klosterman 2001) or web PSS (Kingston *et al.* 2003; Mikkonen *et al.* 2003). These techniques often enable different forms of participation: participative appraisal, participative information collection, participative mapping, development forums, participative scenarios evaluation or public observatories. The lack of adequate access to information is an important aspect in the underdevelopment syndrome. One generally agrees that the improvement of information has a positive impact on management, on economic development and on local governance (Brown 2001; UNDP 2001). At a more detailed level, decision makers make their management decisions on the basis of the information they have about the land realities and about the numerous stakeholders' initiatives. In most developing cities, this information is poor and results in management and communication failures (Repetti and Prélaz-Droux 2003).

Information and communication technologies (ICT) are powerful tools that can potentially improve the sharing of information among decision makers and stakeholders. ICT can support the appropriation of information about land use and related conflicts. Thus, it can contribute to improving local knowledge. But ICT can contribute to strengthening management processes only when they are adapted to the context (James 2002). For land planning and management, geographic information systems (GIS) and the web can be relevant support tools for information and the diagnosis of territorial realities. They can support decision making, coordination and social functionalities (Dransch 1999). Like other media, these technologies are also ambivalent (Fayman and Santana 2001), with unexpected and not always positive social repercussions of their implementation: inequalities in accessibility, inequalities in information control, deviation of information for other purposes or frustrations due to unfulfilled expectations.

The existing GIS and PSS solutions are not adapted to the context and needs of land managers: they require strong skills in computing and significant financial investment for the implementation and maintenance; they are difficult to handle for the decision makers, planners and stakeholders (often without any computer skill) or are limited to data viewing functionalities. Following Klosterman (2001) and Geertman and Stillwell (2003), PSS for urban planning and management are generally not fulfilling the expectations of planners and managers, and are still underused.

18.3 The *System for Monitoring Urban Functionalities*

The *System for Monitoring URban Functionalities* (*SMURF*) is basically a modular GIS-based observatory. Although initially developed for the support of urban management in Africa, it has proved to be very appealing for the support of multi-thematic, multi-stakeholder resource management approaches in general. It is intended for the various stakeholders involved in planning and management (elected representatives and administrative services as well as business, association and population representatives) by providing a data exchange platform. As a communication tool, designed for end-users with often limited computer skills, the *SMURF* interface has to remain very simple and accessible.

The most essential feature of *SMURF* is to provide easy access to information: (i) for the user himself to get a intimate knowledge of the spatial dimension of his city or region and all its thematic features, in a self-appropriation process; (ii) among users to share concerns, transmit ideas, being able to better explain one's own point of view to others and vice versa; and (iii) for a public body, as a support to communicate and explain decisions to a larger audience. Indicators complement the data and offer monitoring, controlling and benchmarking functionalities as well as a support for the evaluation of strategies for the future. Thus, the second essential feature of *SMURF* can be seen in its ability: (i) to raise awareness about future trends and possible options in terms of planning and management; (ii) to support the definition, testing, comparison and evaluation of strategies; and (iii) to provide assistance in judgement and decision making at a strategic level.

18.3.1 Data and Indicators

SMURF is at once an information exchange platform and a decision-support tool. First, it consists of a database that allows the storage of information. Second, it includes spatial and statistical analysis components that process spatial and statistical indicators from the database. The design of the data to be stored is thus a central element of *SMURF*. Too much data leads to information overload and reduces the readability of the system. But, the lack of vital data limits the potential for decision making and for processing indicators. Thus, the data structure must be well balanced for land management and/or urban planning. Keeping the database both small and relevant is also an important aspect regarding: (i) data availability and data collection; and (ii) the maintenance and update of the database, especially in areas with limited resources. *SMURF* is therefore designed to tackle missing data and allow for data collection directly by the end-users.

The design of an appropriate data indicator structure should consider that the relevance of a given indicator is both due to its intrinsic qualities and to its relations to the whole indicator set (Repetti and Desthieux 2006). Furthermore, the following elements should be taken into consideration:

- the objective of the information system, which is monitoring local development at a strategic level;
- the priority themes of management that depend of each city or region (e.g. land management, mobility, water management, economy and production, social development, environment and governance);
- measuring elements that help users locate themselves in the territory (e.g. aerial pictures, maps, roads, squares, waterways);
- the strength and weaknesses of the particular city or region, as well as the difficulties and conflicts that the stakeholders face;
- the ongoing projects for local development, as determined by the authorities, non-governmental organisations (NGOs), associations and business representatives;
- the indicators selected to monitor the development and to evaluate planning strategies, for which statistical and spatial processing is run from basic data that must be collected and stored in the database; and
- the accessibility and availability of information. The structure of the dataset must be adapted to the existing and accessible data, must strictly limit the demand for supplementary data.

Since it is an open and dynamic process, the management of a city is influenced by numerous internal and external constraints and by uncertainty, which make its evolution hardly predictable. To be used as a decision support system, *SMURF* proposes a set of indicators relevant to the decisions to be made. The set of indicators aims at analysing and monitoring urban development and its evolution according to planning projects and decisions to be made. It allows control over the improvement of urban development toward the strategic objectives in time and comparison of the city with other cities.

The complexity of urban phenomena makes it difficult to establish an efficient set of indicators to cover all the aspects of urban development. Some stakeholders are responsible for strategic decisions (e.g. reduce CO_2 pollution level), others for management (e.g. development of public transportation) and others for operations (e.g. where to build new bus lines). The indicators must also correspond to the different scales of management: some indicators are requested to compare districts, neighbourhoods or areas inside the city (e.g. the distance to public facilities); others must give a more general view of city evolution through statistical values (e.g. evolution of the investment in public facilities). The set of indicators must be adapted to each specific situation, in order to support the comparison of the different priority themes of urban management (e.g. health, education, economic development). Finally, some indicators must be very aggregated to give a general idea of the state and evolution of the urban development (e.g. an index of education) and others must be less aggregated to allow a deeper understanding of the factors that contribute to the state and evolution (e.g. literacy rate, boys/girls education rate).

18.3.2 The SMURF Interface

The *SMURF* concept starts from information sharing between local development stakeholders to improve their global knowledge about the city. The *SMURF* software is an interactive GIS mapper, adapted to the needs of its users, to their computer skills, to the locally available infrastructures and to the data quality. User-friendly and easy to use, *SMURF* allows interaction with the database and with the indicators through basic viewing and editing (modification) functionalities. The SMURF interface offers four main visual elements to its users (Fig. 18.1):

- a graphic window for displaying maps, aerial pictures and spatial data;
- a menu bar offering some basic general functionalities;
- a toolbar with a minimum choice of tools for handling the graphic window and displaying the information; and
- a modular multi-panel pane, giving access to the different interaction modules.

The core of the system, and its main originality, consists of four complementary modules (Fig. 18.2). The first module (Mode Data viewing) is an interactive viewer

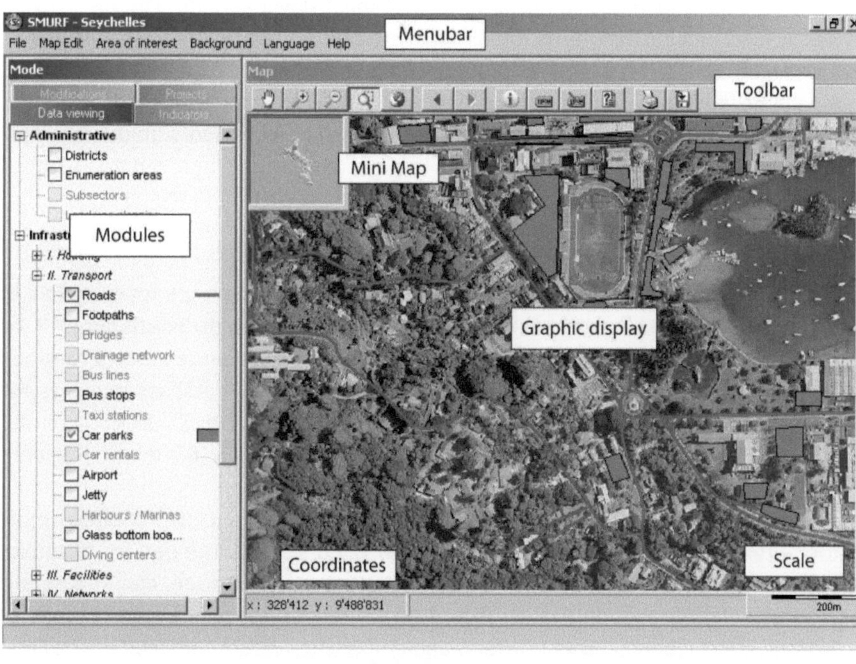

Fig. 18.1 Main elements of the *SMURF* interface

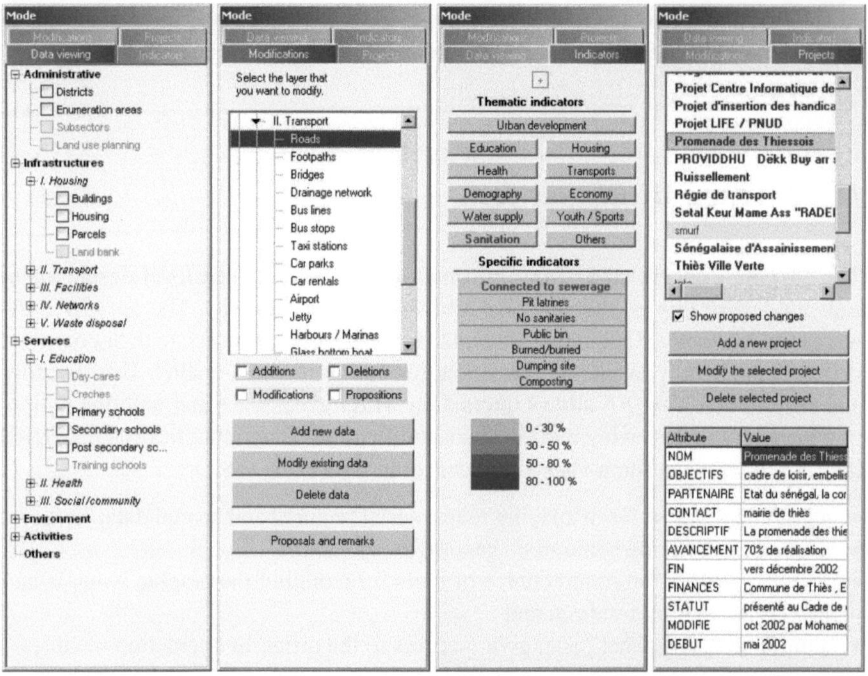

Fig. 18.2 *SMURF* interface: the four basic modules and the possible interaction for the user

for spatial and statistical data relating to the local development. The data layers are presented in a customizable hierarchical structure. Data layers identified as relevant to address the local issues, but with no available data, are present but disabled (greyed out layers). Selecting a data layer adds it to the map shown in the graphic display. The attributes of the layer's objects can then be accessed with the information button in the toolbar (Fig. 18.3). The background layers (digitized maps, aerial pictures, digital elevation model) can be switched on and off with the corresponding menu item ('Background' menu).

The second module (Modification Mode) is a data editor for both the geometry and the attributes of spatially referenced objects. The data editor allows users to add new data, update existing data, delete data and formulate open proposals (Fig. 18.4). No data layers are now accessible since they are disabled in viewer mode. Proposals for change are registered for each user and sent to the GIS database manager for validation and subsequent database update (the validation process depends on the application case). Checkboxes allow the proposals for change for the visible data layers to be shown or hidden as required.

The third module (Indicators Mode) gives access to a set of predefined indicators, organized in thematic groups (an aggregated sanitation index for example, along with its related specific indicators). More experienced users can also set up and test new tailor-made indicators (Fig. 18.5). These indicators include spatial distributions, distance indicators (buffers) and composite feature indicators (symbols' size and colour depending on combinations of attributes). Indicators' drawing

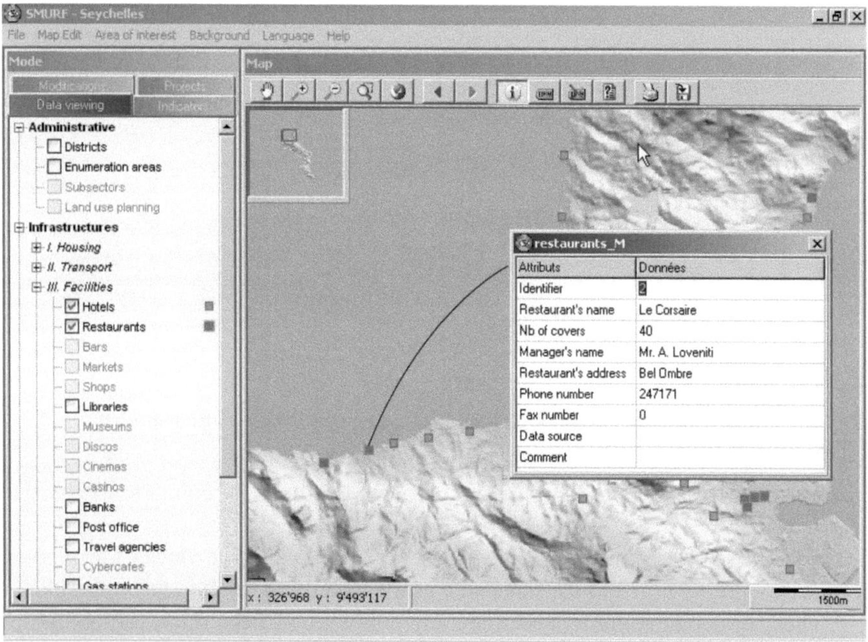

Fig. 18.3 Mode Data viewing, with the menu of the graphic display and selection of data to access the data from the graphic display

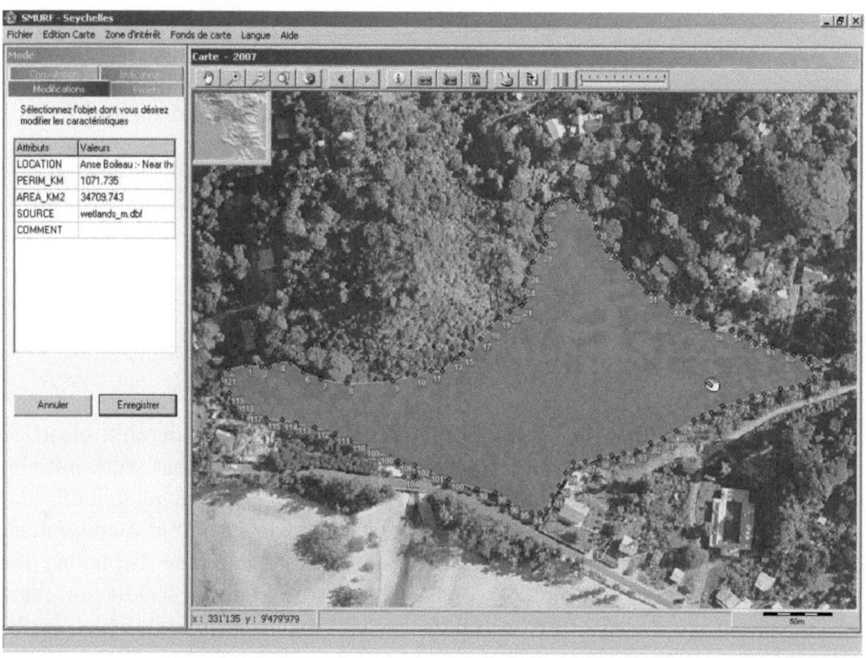

Fig. 18.4 Mode Modification, with the modification of geometry and text data for a Wetland stored in the SMURF database

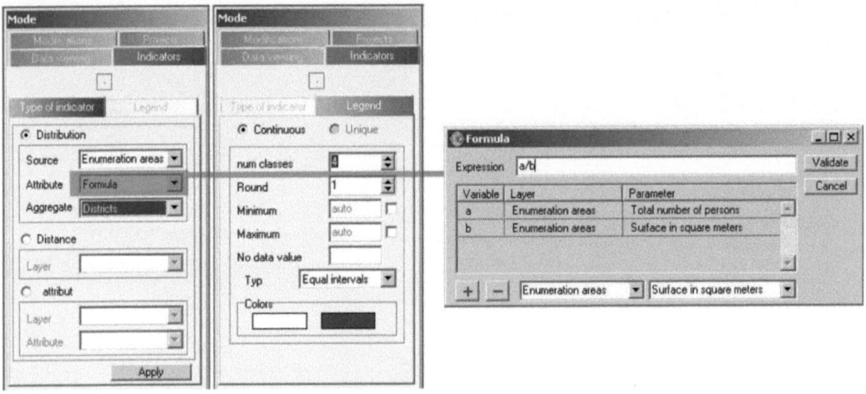

Fig. 18.5 Mode Indicators, with the possibility to create new indicators from the data

parameters such as classes' definition (number, type, limits) or colour scale can also easily be adapted to one's needs,

Finally, the fourth module (Projects Mode) is an interactive viewer for ongoing projects, working much in the same way as the data editor of the second module.

These four interaction modules are key points of *SMURF* relevance and originality. Data and indicators are complementary for the development of expertise concerning land-use. A database for ongoing projects has been setup as a response to the lack of coordination among local and international stakeholders (state agencies, multilateral and bilateral cooperation, NGO's, etc.) that results to some extent from the very limited ability of the local authorities, due to scarce human and financial resources, to manage urban development. Data editing leads to participative updating of the database. Figure 18.6 shows the main fields considered and the detailed data structure: entities, relations, attributes, spatial types (point •, line \, or area □). The general model results of several experiences and tests. It can be easily adapted through the *SMURF* setup, depending on the specific context of each city or region.

18.3.3 Setup and Management of SMURF

Starting to apply *SMURF* in different places required considerable effort to generalize and parameterize several aspects of the software that were initially very site-specific. Parameterization covers four aspects: (i) the general configuration of the software (available modules, background layers, mini-map content, language options); (ii) the organisation of the hierarchical tree displaying the available data layers; (iii) the definition of the set of predefined indicators; and (iv) the localization of the software on the basis of localization files, which facilitates the translation of the interface.

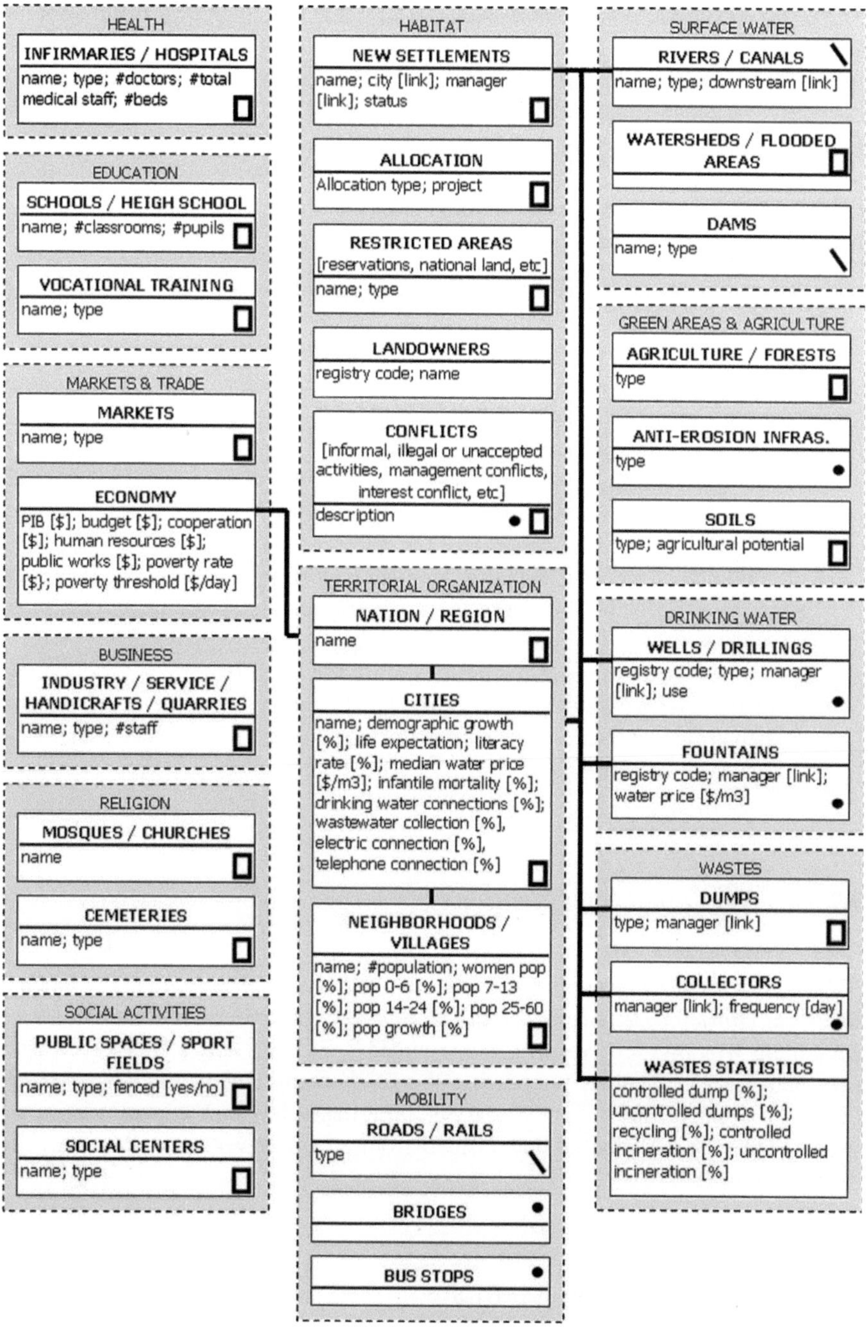

Fig. 18.6 Summary of the structure of the *SMURF* land management data

The establishment and the maintenance of *SMURF* in a city or region require some special skills, in order to ensure the operability and durability of the system. A first level of management relates to the design and coordination of the participatory process (organization, institutionalization, data validation procedures, *et cetera*). This first level will be set differently for each application case, depending on the existing structures and dynamics. At a second more technical level, *SMURF* requires some computer and GIS database management skills. A clear hierarchic assessment of the tasks is therefore essential to ease the implementation of the instrument as follows.

- A few management tasks require specific technical skills in GIS databases and computing: basically the establishment of *SMURF* for a new city or region and training of local managers. Although a *SMURF* adapted to local data and specificities can be prepared quite easily thanks to parameterization, the structures of those parameter files, even when thoroughly documented, are not that easy to understand for non-specialists.
- Ordinary data management is the main maintenance task at the city level and requires basic GIS skills. Besides the training of users, it necessitates following and centralizing the data updates that the different users propose and preparing their validation. Currently, data management is done with commercial GIS software. A specifically dedicated *SMURF* module is under development to facilitate these tasks.
- Direct contact with the *SMURF* users is a third level. It requires a good knowledge of the *SMURF* software and of the local context. It can be accomplished with the various involved organizations through a group of reference users, who get more complete training.

18.3.4 Technical Specifications

SMURF is a stand-alone application. The current *SMURF* prototype is programmed in Pascal (*Borland Delphi*) and uses the *TatukGIS* library. It runs with *Windows* (minimum *Windows 95*). Vector data are either in ESRI 'shape' format or in *MapInfo* 'tab' format, whereas images use the ECW compression standard. *SMURF* generally works offline with a CD dataset; now, depending on the local conditions, vector data files can also be downloaded from an ftp server accessible from the *SMURF* interface.

18.4 Implementation Examples

SMURF was originally developed for the city of Thies in Senegal. Some test applications were carried out in similar contexts (i.e. middle-sized cities in western Africa, such as Koudougou in Bukina Faso, Bignona and Saint-Louis in Senegal and Nakuru in Kenya) and in one very different situation, the Seychelles islands. The *SMURF* platform is now also used as a support to Learning Alliances (Moriarty *et al.* 2005) to promote integrated urban water management, within the EU SWITCH project, with recent applications in Accra, Belo Horizonte and Zaragoza. Hereafter, we describe shortly two of these case studies, the initial implementation in Thies (described with more details in Repetti *et al.* 2006) and the more recent one in the Seychelles.

18.4.1 Thies, Senegal

Thies is a regional administrative centre, with the third largest population in Senegal (300,000 inhabitants). The economy of the city is mainly based on industries and trade. This mid-sized city faces classical problems of underdevelopment: strong demographic growth (200,000 inhabitants in 1990), a weak economy, lack of infrastructure, informal and unhealthy settlements, poverty and environmental degradation, among other things. However, the city plays a leading role in the regional exchanges and supports an important series of public and private infrastructures and services.

Since the introduction of decentralization policies in Senegal (1996), urban management is a new competence of the local government. In various sectors of local development, the state, the regional government, the local associations and NGOs play a particularly important role, some with the support of multinational backers. Thus, land-use planning and management involves a wide variety of stakeholders. The reality of the urban management of Thies bears evidence of the limits of planners to manage a fast growing city (Repetti and Prélaz-Droux 2003). Lack of information and technical skills to prepare the decisions, as well as of the means to implement them, lack of knowledge and coordination between the actors, administrative services (technical) squeezed between the central State and the decentralized authorities and mostly confining themselves to the management of their current business, are some of the reasons that can explain this situation.

In the reality, the official instruments of planning are hardly ever used (a set of classic instruments of planning exists, some inherited from the pre-decentralization period), most do not meet the needs of the various actors and have not been thoroughly negotiated. In 1999, a participatory forum arose from a demand by the municipal authorities for an experimental project utilizing new urban management instruments. In order to facilitate the information exchanges and storage, *SMURF* has been made available to the main stakeholders in urban management (elected representatives, technical services, administration representatives and services, one association and one NGO) since summer 2000. The participatory forum gathers elected representatives of the City of Thies and of the Community of Fandene, the state representatives, administrative services concerned with the local development and a couple of associations and NGOs. The general objective of the forum is to improve the development in the urban area through information sharing, consultation and co-decision. Starting from this objective, *SMURF* was integrated in the process to support appropriation of the land-use, participatory planning and management. More specifically, *SMURF* aims at improving information on land use, monitoring urban development and exchanging information between the various involved stakeholders.

The design of the data indicator system (as well as of the interface itself) was an iterative process. The topics covered include territorial organization, markets and trade, business, health, education, drinking water, waste, habitat, social activities, public safety, religion, surface water, land registering and land-use, land conflicts, green and agricultural areas, mobility and governance. Figure 18.6 shows the various objects and attributes included in the Thies application. A set of 36 indicators (Table 18.1) completes the data. It presents the diverse themes, with various spatial scales, strategic and aggregation levels.

In Thies, *SMURF* is integrated into the activities of the development forum, coordinated by the City. The director of the technical services is in charge of the GIS database management. He collects the data updates from the *SMURF* users, controls the errors and presents the data modification information in the forum once or twice a year. Today, *SMURF* is installed on about 30 computers in the urban area: city halls (City of Thies and Community of Fandene), regional hall, local and regional administrative services (technical), associations and NGOs and a cyber café offering public access to *SMURF*. *SMURF* is also used to support participatory planning activities in the local forum.

18.4.2 The Seychelles

The Seychelles are a set of 115 Islands (32 granitic and 83 coralline), spread over almost 400,000 sq km in the Indian Ocean. The granitic inner islands group, with the three main islands Mahe, Praslin and LaDigue (192 sq km), hosts some 95 per cent of the overall population (~82,000 inhabitants in 2007). The largest city is the capital Victoria (Mahe) with a little more than 20,000 inhabitants. Population is 60 per cent urban, with a growth rate of 0.43 per cent. The main resources are fisheries and tourism (30 per cent of employment and 70 per cent of GDP), with an income per capita at $8,682 (2006), by far the highest in Africa. The country became independent in 1976. Socialist rule was brought to a close with a new constitution and free elections in 1993. The political evolution along with the resources gained from tourism, allowed for the development of, among others, relatively efficient health and education systems and much awareness regarding the preservation of environment.

Although there is no demographic urban explosion as in other parts of the world, land allocation is a very central issue which leads to conflicts. On the rocky tropical islands, with a narrow coastal strip and steep slopes almost everywhere else, there are great pressures on useable land: development and diversification of the economy, changes in lifestyle (more space for housing and roads needed) are in the balance with the necessity to preserve the environment (biodiversity, but mainly landscape to maintain tourist attraction). The need for flat land is so great that several development

Table 18.1 Set of indicators designed for Thies

Indicators	Type
Population density	Spatial [grid 1 × 1km] [hab/sqkm]
Age distribution	Statistical [0–6; 7–13; 14–25; 25–60; >60]
Rate of poverty	Statistical [population under poverty threshold]
Demographic growth	Statistical [% per year]
Total population	Statistical [graph]
School population	Spatial [per infrastructure] [pupils/classrooms]
School repartition	Spatial [distance to closest school]
School age population	Spatial [grid 1 × 1 km] [child 7–13/sqkm]
Education rate	Statistical [primary school pupils/child 7–13]
Illiteracy	Statistical [% of total population]
Education index*	Statistical [average (literacy, primary, secondary and university education)]
Health staff	Spatial [grid 2 × 2 km] [total staff/sqkm]
Clinics repartition	Spatial [distance to closest hospital/infirmary]
Hospital bed availability	Statistical [population/bed]
Doctor density	Statistical [population/doctor]
Medical staff density	Statistical [population/staff]
Infantile mortality	Statistical [rate of death in child 0–1]
Health index*	Statistical [0.833 (life expectation – 25) + 1.566 (32 – infantile mortality)]
Drinking water distribution	Spatial [neighbourhood] [connection rate]
Fountains repartition	Spatial [distance to closest fountain]
Water consumption	Statistical [m3/population]
Water price	Statistical [median [$/m3]]
Waste management	Spatial [solid waste collection typology]
Waste elimination	Statistical [solid waste elimination types] [%]
Waste index*	Statistical [50 wastewater treatment rate + 50 solid wastes treatment rate]
Habitat	Spatial [habitat typology]
Mobility	Spatial [access time to city centre]
Infrastructure index*	Statistical [average connection rate on the mains (water, sewage, electricity and telephone)]
Markets repartition	Spatial [distance to closest market]
Product index*	Statistical [0.833(log(PIB)–4.61)]
Decentralization	Qualitative
Public participation	Qualitative
International cooperation	Qualitative
Public spending on staff	Qualitative
Public spending	Statistical [$/population]
Urban development index*	Statistical [average (education, health, waste, infrastructure and product indexes)]

* Indicators proposed by UN-Habitat (2001).

projects on reclamation areas (gained from the sea) are underway (including a new city of 10,000 inhabitants). The mayor of Victoria, along with the ministry of local government initiated a project to adapt the *SMURF* platform to the local context. Initially the local *SMURF* was meant to support several experiences of participative planning that were going on in some of the 25 districts of the inner islands.

Implementing *SMURF* in the Seychelles was challenging, since the geographical, social and institutional background is drastically different from the one in middle-sized cities in western Africa. A first major difference related to scale, since *SMURF* had to be prepared not for a single city and its surroundings, but simultaneously for 25 districts on three islands, with quite heterogeneous spatial data coverage (among islands). The solution of one common platform was quite obvious regarding the needs and interests of the central administration, but new functionalities allowing one to work specifically on a given 'isolated' district (especially to produce district specific maps) were added.

Thanks to an efficient and open-minded administrative structure, access to the quite large set of existing data (in great part centralized at the GIS office of the Ministry of Land Use and Housing) was not a problem. This has to be emphasized since one of the major difficulties often faced in setting up a common GIS platform consists of convincing stakeholders that sharing data will be a win-win process. The data structure (layers and attributes) was developed by a group of stakeholders from district administrations, from the Ministry of Local Government that supports district administration and from several other interested ministries (Land Use and Housing, Transport, Environment, Agriculture, Fisheries, *et cetera*) during a three day workshop.

The final data structure, grouping of the various spatially referenced objects and their attributes (Fig. 18.7), is quite different than that of the Thies case, which was quite logical since major problems to be addressed and available data were not the same in this emerging country. The Seychelles data structure also includes layers identified as necessary, but without any data available so far (greyed out). Indicators were not yet matter of discussion, so that the indicator set used in Thies was kept almost unchanged: the thematic fields remained the same, but, taking advantage of the huge amount of information coming along with the detailed census data, the specific indicators available were enriched, especially regarding water use or economic activity. Furthermore, in response to the interest showed by many stakeholders that were not initially meant as end users of *SMURF,* the indicator module was extended to support the exploration of tailor-made indicators. The creation of new layers has also been added as a new functionality to allow district administrators to address local needs or interests, such as, for instance, the follow-up of Chikungunya disease cases.

In the Seychelles, the management of the system is coordinated by the Ministry of Local Government. The detailed workflow of the data update and validation process has still to be flattened down, since the responsibility of the various data layers is shared among many different public bodies. Today, *SMURF* is installed in all district administrations, as well as in many offices of the various ministries and other public services.

18.5 Experiences

In Thies, a survey was run to evaluate the interest in *SMURF*, as well as its possible influence on the local urban governance (Repetti *et al.* 2006). In the Seychelles, confirmation of interest was seen in the healthy attendance (and passionate involvement) at training workshops and in the role that will in future be given to *SMURF* in two major projects of the government, one to address natural hazards and the other on setting up an e-government structure.

In general, data viewing is the most frequently used mode. Information serves as support for knowledge, for stakeholder discussions and arguments, and is consulted for the design of on-ground projects. In Thies, the most consulted information layers are the land ownership and administrative limits, the data on surface water and on solid wastes (which correspond to the local priorities and conflicts). Information was used for designing several projects (new neighbourhoods planning, flood control reservoirs realization, new public health infrastructures creation, freeway scenarios analysis, etc.). Indicators, either in the form of aggregated statistical values or in the form of thematic maps, remain difficult to understand for the majority of the users. However, they provide an interesting added value for some technicians with a good understanding of the situation, good analytic skills and a deep knowledge of urban planning and management.

Unlike in Thies, where the interactive on-ground project viewer module is regularly consulted (mainly by local authorities), offering valuable information to all stakeholders and impacting positively on the coordination, this element only received very little echo in the Seychelles. Part of the explanation for this could be the fact that the central and local administration are efficient enough to control and coordinate on-ground projects.

The use of the data editing mode varies a lot, depending on stakeholders. Some are regularly augmenting the database as they get an interest in sustaining the instrument, but clearly without institutionalized validation process, enthusiasm might decrease rapidly, since the end-user would like to see the updates he provides becoming part of the 'real' database as quickly as possible. In the Seychelles version, the possibility to create new data layers was added in response to a demand from some district administrators willing to use the *SMURF* tool to monitor different specific aspects at the local scale, such as for instance the spreading of Chikungunya cases. In Thies, the survey also shows that database enrichment is sometimes impaired by the lack of willingness of officials to share the information they detain, this for various reasons such as hesitation to open official data to the public, fear of losing political leverage, or its link to informal or illegal activities. At a more global level, the association of *SMURF* with a forum has a positive impact on urban management in Thies: knowledge of the land realities and of the views and actions of the other stakeholders has increased. The suitability is thus reinforced between the development projects and the needs of the land users. Furthermore, the *SMURF* users are spontaneously attempting to define a common strategic plan as a result of their interactions. Their objective is to ensure a better integration of their individual initiatives at the city level. In the Seychelles, the process is too recent for such an observation.

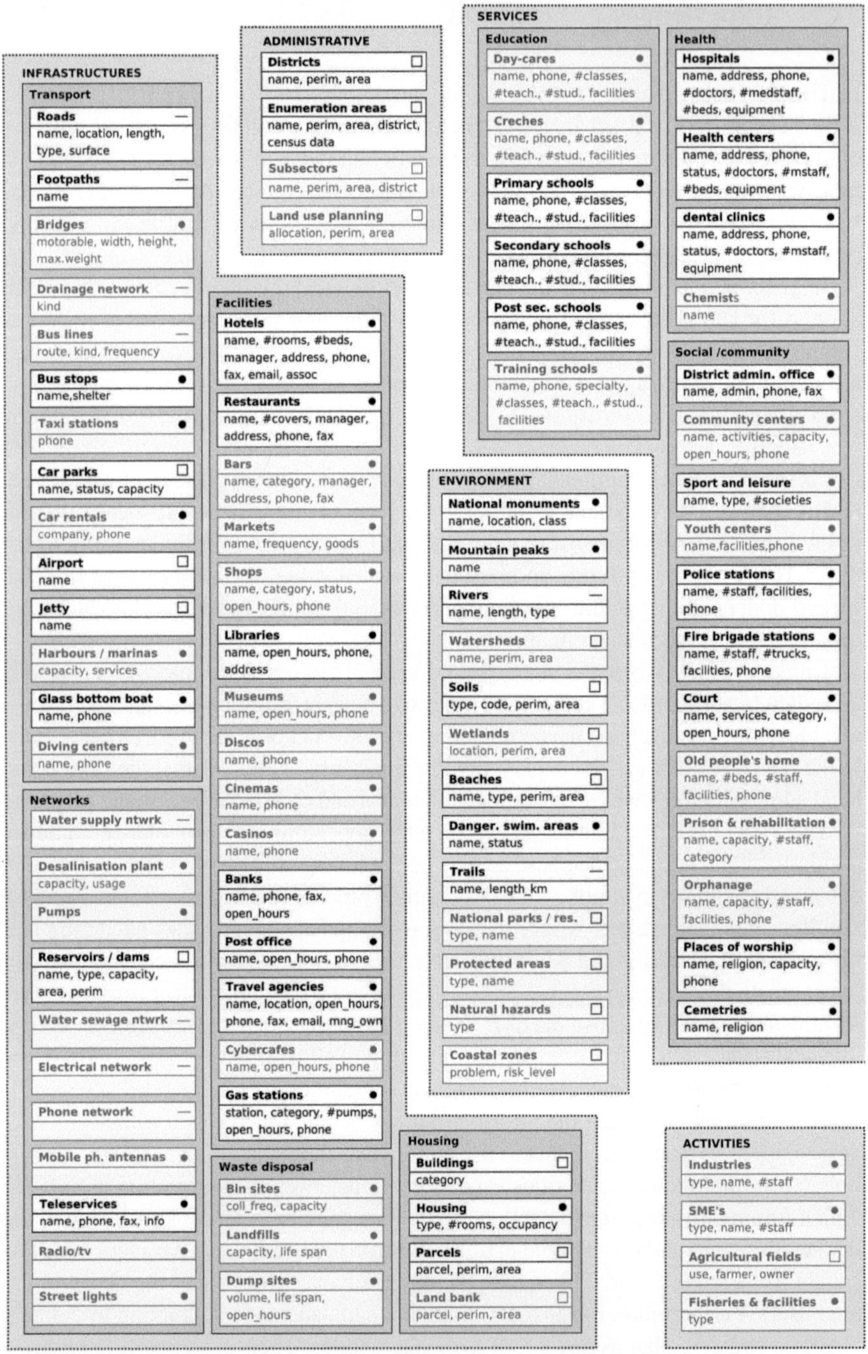

Fig. 18.7 Data structure of the Seychelles' implementation of *SMURF*

The absence of a clear and precise role assignment for the involved participants can raise high expectations that may not be fulfilled and result in some frustration. This is, however, a general issue in public participation processes: public authorities seek visibility for their action and are therefore reluctant to move towards strong consultation and co-decision. For the same reason, dealing with data about informal or illegal activities might be problematic: public authorities receive such information from other stakeholders and are not inclined to publish it under their responsibilities, as it points out complex problems and conflicts that they are not able to solve. Intended to support planning and governance, *SMURF* will not alone resolve the complex equations of underdeveloped cities dealing with legislative vagueness, with informality and with scarce means. Its relevance will depend on the political willingness to implement and use such collaborative management processes, as well as on the allocated resources.

The needs for GIS based communication tools as a support to multi-stakeholder management issues will continue to grow, since such approaches are intrinsically linked to a sound implementation of governance and sustainability principles. In the developing world, however, setting up a forum and implementing *SMURF* might often take a back seat to more immediate local priorities. The required resources, even if very modest (a half time position and a small budget line to organise forums), are still too important with respect to other more immediate tasks. Thus, the sustainability of *SMURF*, depending much on political willingness, remains an open question. In the developed world, the need for simplified GIS-based communication tools, complementing 'technical' GIS systems, is more oriented towards the support of the decision-making process. There is, therefore, a clear trend to a diversification to integrate more specialized modules into the SMURF platform (data validation, time management, generation of alternative strategies, evaluation and decision processes, simple modelling to derive indicators, links to high end models, *et cetera*).

The main lesson gained from the experiences in Thies and in the Seychelles, and the recommendations for further implementations of such tools, is the outstanding importance of the maintenance and update process. This has several implications in terms of simplicity and design of the developed tool, or in terms of availability of local resources to anchor the whole process in the local agenda. It also underlines that the tools we might develop and provide are, in themselves, only of a limited importance. The real cornerstone is in fact the institutional and social setup of the participatory approach.

18.6 Conclusions and Future Directions

The system for the monitoring of urban functionalities (*SMURF*) aims at reinforcing the decision-making processes. It has developed from the needs of the managers of mid-sized African cities and follows a complete methodological approach (diagnosis, instrument development, instrument evaluation and impacts assessment). *SMURF* does not try to automatically identify the best management scenario, but aims to reinforce the stakeholders and their interrelationships through various supplementary sub-objectives: improve knowledge of land and development projects, serve as a data exchange platform, instrumentalize participatory and collaborative structures through cartographic support, monitor the evolution of the land with indicators and inform and consult the public. *SMURF* does not try to revolutionize the stakeholder system by imposing dogmatic participation. It aims at getting public authorities, land technicians, business representatives and citizens' rights defenders to coordinate as much as possible through a formalized exchange space.

In comparison with existing works in the field, the proposed approach is focusing on building consensual information and strategic objectives, rather than on scenario evaluation which is a second step. In our experimentation, participation is used to constitute a knowledge base that supports decision making and consensus building. The process results in better knowledge for all participants and in a strong consensus about the diagnosis of the actual situation and about the strategic objectives for the local development. The role of the forum is central, as some information cannot be shared through a database. Yet, the field studies show that decision makers and stakeholders are using a SMURF-based diagnosis and the discussed strategies into account during their planning action, even if they do not model scenarios.

In addition to the implementation of a new technology, the research and user experience with *SMURF* is an attempt to integrate the social and technical sciences. *SMURF* can thus be seen as an observatory of urban development available for local stakeholders. It is complementary to the more global programs of international urban observatories. Finally, *SMURF* does not aim to reduce the complexity of urban management, but to offer an interactive tool, easy to use, which provides assessment monitoring, comparisons, communication and finally knowledge to the decision-makers.

As stated previously, *SMURF* offers a more or less formalized exchange space. Future research should aim at providing support for other aspects involved in the decision making process at a strategic level. These elements include:

- the generation of alternative scenarios, according to global trends (demography, economy, climate, etc.) and to various options, covering the political, economical and technical fields;
- the extension of the thematic fields covered by indicators and links to modelling tools to evaluate those indicators; and
- the linking to decision support schemes such as multi-criteria techniques and/or life-cycle assessment techniques.

References

Abbot, J., Chambers, R., Dunn, C., Harris, T., de Merode, E. and Poter, G. (1998) Participatory GIS: opportunity or oxymoron?, *IIED PLA Notes*, 33: 27–33.

Allen, E. (2001) INDEX: software for community indicators, In Brail, R.K. and Klosterman, R.E. (eds.) *Planning Support Systems*, ESRI Press, Redlands, pp. 229–261.

Bolay, J.-C. and Rabinovich, A. (2003) Villes intermédiaires en Amérique Latine: risques et potentiel pour un développement urbain cohérent, In Charbonneau, F. Lewis, P. and Manzagol, C. (eds.) *Villes Moyennes et Mondialisation. Renouvellement de l'Analyse des Stratégies*, Trames, Montréal.

Bolay, J.C., Pedrazzini, Y. and Rabinovich, A. (2000) Quel sens au "développement durable" dans l'urbanisation du tiers monde, *Les Annales de la Recherche Urbaine*, 86: 77–84.

Borja, J. and Castells, M. (1997) *Local and Global Management of Cities in the Information Age*, Earthscan, London.

Brown, M.M. (2001) Les TIC peuvent-elles aider les pauvres? *Choix*, Juin.

Chambers, R. (1994) Participatory rural appraisal (PRA): challenges, potentials and paradigm, *World Development*, 22(10): 1437–1454.

Decoutère, S. (1996) Finalités et modalités du MT, In Decoutère, S., Ruegg, J. and Joye, D. (eds.) *Le Management Territorial*, PPUR, Lausanne, pp. 25–38.

De Graaf, R. and Dewulf, G. (2002) Interactive urban planning, hype or reality? In Khosrowshahi, F. (ed.) *3rd International Conference on Decision Making in Urban and Civil Engineering*, London.

Dransch, D. (1999) Theoretical issues in multimedia, In Cartwright, W., Peterson, M.P. and Gartner, G. (eds.), *Multimedia Cartography*, Springer, Berlin, pp. 41–50.

Fayman, S. and Santana, L. (2001) Rapport introductif sur la ville inclusive, *Quatrième Forum International sur la Pauvreté Urbaine*, Marrakech, Morocco.

Frérot, A.-M. (1999) *Les Grandes Villes d'Afrique*. Ellipses, Paris.

Geertman, S. and Stillwell, J. (2003) Planning support systems: an introduction, In Geertman, S. and Stillwell, J. (eds.) *Planning Support Systems in Practice,* Springer, New York, pp. 3–22.

Halla, F. (2005) Critical elements in sustainable participatory planning: Bagamoyo strategic urban development planning framework in Tanzania, *Habitat International*, 29: 137–161.

Harris, B. and Batty, M. (2001) Locational models, geographic information and planning support systems, In Brail, R.K. and Klosterman, R.E. (eds.) *Planning Support Systems,* ESRI Press, Redlands, pp. 25–57.

Harts, J. J., Maat, K. and Ottens, H. (2003) A urbanisation monitoring system for strategic planning, In Geertman, S. and Stillwell, J. (eds.) *Planning Support Systems in Practice,* Springer, New York, pp. 315–329.

Hofstede, G.J. (1995), Open problems, formal problems, *Journal of Decision Systems*, 4(2): 155–165.

ICLEI (ed.) (1996) *The Local Agenda 21 Planning Guide*, ICLEI, Toronto.

Ingallina, P. (2001) *Le Projet Urbain,* Presses Universitaires de France, Paris.

James, J. (2002) Low-cost information technology in developing countries: current opportunities and emerging possibilities, *Habitat International*, 26: 21–31.

Kingston, R., Evans, A. and Carver, S. (2003) Public participation via on-line democracy, In Geertman, S. and Stillwell, J. (eds.) *Planning Support Systems in Practice,* Springer, New York, pp. 45–64.

Klosterman, R.E. (2001) Planning support systems: a new perspective on computer-aided planning, In Brail, R.K. and Klosterman, R.E. (eds.) *Planning Support Systems*, ESRI Press, Redlands, pp. 1–23.

Le Galès, P. (1998) Regulation, territory and governance, *International Journal of Urban and Regional Research*, 22(3): 482–506.

Ling, O.G. (1999) Civil society and the urban environment, In Inoguchi, T., Newman, E. and Paoletto, G. (eds.) *Cities and the Environment*, United Nations University Press, Tokyo, pp. 105–126.

Mikkonen, J., Ristimaki, M., Oinonen, K. and Hausen, H.S. (2003) A web-based support system for sustainable development, In Geertman, S. and Stillwell, J. (eds.) *Planning Support Systems in Practice,* Springer, New York, pp. 123–137.

Moriarty, P., Fonseca, C., Smits, S. and Schouten, T. (2005) *Learning Alliances for Scaling up Innovative Approaches in the Water and Sanitation Sector*, International Water and Sanitation Centre (IRC), Delft.

Noisette, P. (1996) Le marketing urbain: outils du MT, In Decoutère, S., Ruegg, J. and Joye, D. (eds.) *Le Management Territorial*, PPUR, Lausanne, pp. 166–192.

Repetti, A. and Desthieux, G. (2006) A relational indicator set model for urban land-use planning and management: methodological approach and application in two case studies, *Landscape and Urban Planning*, 77(1–2): 196–215.

Repetti, A., Soutter, M. and Musy, A. (2006) Introducing SMURF, a software system for monitoring urban functionalities, *Computers, Environment and Urban Systems,* 30: 686–707.

Repetti, A. and Prélaz-Droux, R. (2002) Planning infrastructures and resources: A DSS for sustainable development, *Journal of Decision Systems*, 11(3–4): 297–315.

Repetti, A. and Prélaz-Droux, R. (2003) An urban monitor as support for a participative management of developing cities, *Habitat International*, 27: 653–667.

Sawadogo, R.A. (2002) La refondation de l'état africain, la décentralisation et la société civile, in *Quelle Coopération pour Quel Développement*, SDC, Bern, pp. 18–28.

Srinivas, H. (1999) Urban environmental management: a partnership continuum, In Inoguchi, T., Newman, E. and Paoletto G. (eds.) *Cities and the Environment,* United Nations University Press, Tokyo, pp. 30–46,

Steinberg, F. (2005) Strategic urban planning in Latin America: experiences of building and managing the future, *Habitat International*, 29: 69–93.

UNDP (2001) *Human Development Report 2001: Making New Technologies Work for Human Development*, UNDP.

UN-Habitat (2001) *The State of the World's Cities*, Oxford University Press, New York.

World Bank (1992) *Governance and Development*, World Bank Publications, Washington.

World Bank (1994) *Adjustment in Africa,* Oxford University Press, New York.

Additional Reading

Borja, J. and Castells, M. (1997) *Local and Global Management of Cities in the Information Age*, Earthscan, London.

Brail, R.K. and Klosterman, R.E. (eds.) *Planning Support Systems,* ESRI Press, Redlands.

Frérot, A.-M. (1999) *Les Grandes Villes d'Afrique*, Ellipses, Paris.

Geertman, S. and Stillwell, J. (eds.) *Planning Support Systems in Practice*, Springer, New York.

IIED (2006) *Mapping for Change: Practices, Technologies and Communication*, PLA (54), London.

Repetti, A. (2004) *Un Concept de Monitoring Participatif au Service des Villes en Développement: Démarche Méthodologique et Réalisation d'un Observatoire Urbain,* Ecole polytechnique Fédérale de Lausanne.

Chapter 19
SoftGIS as a Bridge-Builder in Collaborative Urban Planning

Maarit Kahila and Marketta Kyttä

19.1 Introduction

Current trends in urban planning and recent changes in planning culture (Puustinen 2006; Staffans 2004) support the development of a new methodology for collaborative planning. Collaboration in Finland is supported by European Union (EU) directives and the Finnish Building and Land Use Act (1999). The latter aims to ensure wide participation and to support open and high-quality planning decisions and processes. In practice, such wide participation is difficult to realize as resources are inadequate. The participatory approach needs also to be evident in all phases of the planning process, which is rarely the case in reality.

The views of participants can potentially smooth the planning and design processes and reduce the number of conflicts (Taylor 1998), promote adaptation to continuously changing societal conditions (Innes and Booher 1999), strengthen recognition of the differing values of various actors (Healey 1997), support learning and knowledge-building (Friedmann 1998) and, finally, help to achieve tangible regeneration outcomes (Beresford and Hoban 2005). To implement these objectives, planners have to acquire not only new skills and professional roles (Forester 1989; Puustinen 2006), but also develop more usable and effective participation methods, as well as a deeper understanding of the knowledge hidden in the experiences of the inhabitants.

In the first sections of this chapter, a new soft geographic information system methodology, softGIS, is introduced that has the potential to enhance participation by allowing the residents the possibility of sharing their knowledge of their living environment with urban planners and researchers. SoftGIS methodology refers not only to a whole set of individual softGIS methods but also to the special collection

M. Kahila (✉)
Helsinki University of Technology
Centre for Urban and Regional Studies
PO Box 9300, 02015 TKK, Finland
E-mail: mkahila@cc.hut.fi

S. Geertman, J. Stillwell (eds.), *Planning Support Systems Best Practice and New Methods*,
DOI 10.1007/978-1-4020-8952-7_19, © Springer Science+Business Media B.V. 2009

of theories, concepts and ideas behind the development of these methods. SoftGIS methods will therefore be evaluated in the light of the comments O'Sullivan (2006) makes in his critical GIS discourse. That includes the observation that GIS is not merely a technical tool for handling geographical information, but also takes into account its societal impact. The aim of this chapter is to analyse to what degree and in which ways softGIS methods can build a bridge between residents and urban planners. Because softGIS methodology is in the early stages of development, we will focus on the main ideas behind the innovation, the few existing experiences of the methodology's use and the future development prospects concerning the use of softGIS methodology in collaborative planning practices.

19.2 SoftGIS Methods

SoftGIS methods allow residents to produce localized experiential knowledge. SoftGIS leads us to ask how the everyday lives of the residents are organized, what kind of place-based positive and negative experiences (Manzo 2003) residents have and how they behave in their physical environment. This knowledge is collected through user-friendly internet-based applications (Kyttä and Kahila 2006; Rantanen and Kahila 2008). Finally, the localized experiential knowledge comprises a special layer in the GIS used by experts such as urban planners and researchers. SoftGIS methods are built on the following principles:

- the operationalization of perceived knowledge is grounded in the theories of humanistic geography and environmental psychology;
- the perceived knowledge is gathered through scientifically valid, reliable and ethical methods;
- softGIS methods are developed in cooperation with urban planners, who can use this novel knowledge in their planning practices;
- the database makes systematic GIS and statistical analyses possible; and
- the methods provide a user-friendly internet platform for residents to evaluate their everyday living environment.

Our aim is to build a bridge between residents, researchers and urban planners by promoting the participation and collaboration of citizens with the help of softGIS methods. These methods promote the sharing of the residents' experiences and behaviour concerning their living environment. We believe that new technology, and especially web-based GIS applications, can become an important way of communication between inhabitants and planners (Fig. 19.1).

The technical development of these web applications has been realized by varied groups of media technology students in the Institute of Technology of the Espoo-Vantaa University of Applied Sciences. The current technical requirements for softGIS methods are presented in Table 19.1.

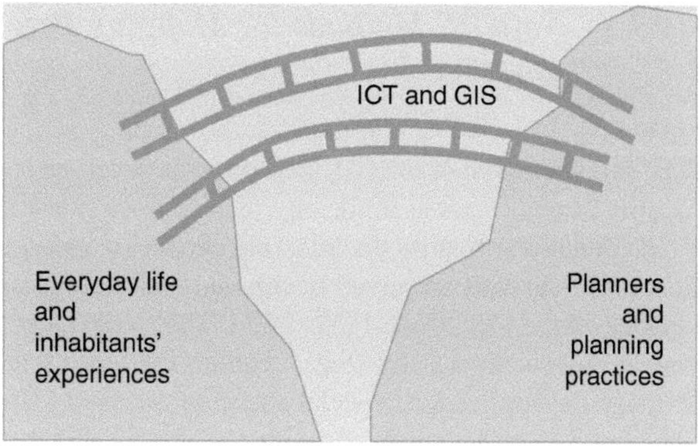

Fig. 19.1 SoftGIS method as a bridge-builder between planning practices and inhabitants' experiences of everyday life

The first softGIS prototype was launched in 2004 in Jarvenpaa, Finland. By fall 2007 five different softGIS methods had commenced in six different municipalities. Up to today these web questionnaires have reached about 3,000 respondents. In the following sections we introduce five softGIS methods that we have categorized as: (i) softGIS methods capturing the general perceived quality of the environment, (ii) softGIS methods concentrating on a specific theme, and (iii) softGIS methods for a special user group.

Table 19.1 The technical requirements for softGIS methods

Implementation techniques	Requirements
Programming of the application • PHPS (http://www.php.net) **Needed extensions to PHP** • MapServer PHP Map Script (http://www.maptools.org/php_manuscript) **Database Program** • MySQL 5.0.18 or later version (http://www.mysql.com) **Implementation of the user interface Either** • HTML (http://www.w3.org/TR/REC-html40/) • CSS (http://www.w3.org/Style/CSS/) • JavaScript (http://www.ecma-international.org/publication/standards/Ecma-262.htm) **Or** • Flash 7.0 (http://www.adobe.com/products/flashplayer/) • XML (http://w.w3org/TR/2006/REC-xml-20060816/).	**Server** • Apache HTTP server (http://httpd.apache.org/) • MapServer 4.8.4 (http://pserver.gis.umn.edu/) **Web browser** • Netscape 6.0 or other combatible browsers • JavaScript-support on

19.2.1 SoftGIS Methods for Mapping Perceived Environmental Quality

The first prototype of softGIS methods was developed to study broadly the perceived quality of environment in the city of Jarvenpaa (*www.softgis.fi*). To study the localized, perceived quality factors (affordances), individual eco-social niches, used local services and perceived health and well-being, an internet-based GIS method was developed in 2004 (Table 19.2). This web questionnaire had a tube structure and it proceeded step by step, meaning that each visitor followed the same route through the application. Special attention was paid to the quality of the maps used in the application. Aerial photos with a scale of 1:4,000 were used and the orientation was aided by highlighted roads and landmarks. Owing to technical restrictions, it was only possible to mark point information on the map, not routes or areas.

Later, in 2006, the Jarvenpaa pilot study was replicated in three additional municipalities, Mäntsälä, Kerava and Nurmijarvi (Fig. 19.2). All of these applications were designed in close collaboration with urban planners. The results in each case have been published in traditional research reports and in web summaries but part of the data has also been delivered as a CD-rom to urban planners to be utilized in their planning projects.

Four hundred and twenty-seven inhabitants answered the original Jarvenpaa survey and about 1,300 inhabitants participated in the following study in three additional Finnish towns. In each case, the web questionnaire was available on the front page of the websites of the cities for three months and all inhabitants were eligible to answer the questionnaire. The web surveys were advertised in local newspapers and libraries. According to our studies, middle-income households and middle-aged women with children living in single family houses were over-represented in these web questionnaires.

The evolution of a softGIS method to study perceived environmental quality has moved from more open questions to more structured ones and from longer questionnaires to shorter ones. To make the analysis easier, originally freely expressed, personal affordances of the original questionnaire were later pre-classified, but we still allowed the respondents to name their own affordances (Kyttä and Kahila 2006). To combine the qualitative and quantitative approaches in each softGIS questionnaire the inhabitants could, for example, write down stories or memories about personally meaningful places.

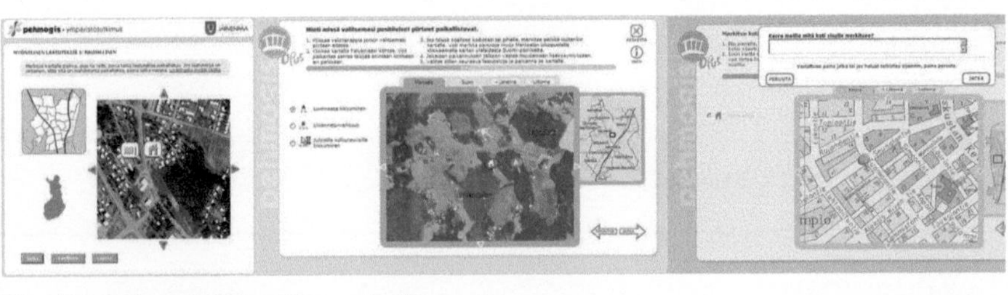

Fig. 19.2 The softGIS quality methods

Table 19.2 The contents of the softGIS survey on the perceived quality of the environment

1.	Introduction	General information about the study
2.	Background information	Age*, Gender*, Family type*, Occupation, Car ownership, Income, Housing type*, House type*, Size of the dwelling, Childhood environment, Situation when filling the questionnaire, Participation earlier
3.	Positive quality factors*	Introduction, choose 1–5 quality from the list, a possibility to name your own quality factor
4.	Actualization of the quality factors*	Actualization of the chosen positive quality factors
5.	Negative quality factors*	Information, choose 1–5 quality factors from the list, a possibility to name your own quality factor
6.	Actualization of the quality factors*	Actualization of the chosen negative quality factors
7.	Map application: Location of home*	Mark the location of home to the map and a possibility to express the personal meaning of home
8.	Map application: Location of positive quality factors	• mark the location of those quality factors that where chosen eariler • every factor can be located max. in three places • quality factors can be located to the map of municipality or a map of Finland • maps can be zoomed
9.	Additional query	After every location a small query appears • accessibility of the place • means of transport • possibility to describe the place
10.	Map application: Location of neagative quality factors	• mark the location of those qaulity factors that where chosen earlier • every factor can be located max. in three places • quality factors can be located to the map of municipality or a map of Finland • maps can be zoomed
11.	Additional query	After every location a small query appears • possibility to avoid the place • disturbance at some time of day • possibility to describe the place
12.	Map application: The location of basic services	• mark the location of workplace, day care, study place, daily grocery store, schools and other activities • after every location, a possibility to describe the place
13.	Map application The questiuons concerning the community	• mark the location of those places or builiding that should be conserved • place that could be attractive to move to • after every location, a possibility to describe the place
14.	Additional questions	Floor height for new buildings and the possibility to write down characters that make the centre attractive
15.	The perceived well-being*	Three question from the Health 2000 survey
16.	The perceived health*	Euroqol-scale (EQ-5D)
17.	The ending	Feedback about the softGIS questionnaire

19.2.2 Thematic SoftGIS Methods

In addition to methods that collect the experiential data broadly, various thematic softGIS methods can be developed to study, for example, perceived safety, urban mobility or experiences concerning the green environment. In each case, relevant research literature and expert researchers should be consulted to produce valid, reliable, theoretically and practically well-operationalized methods. These applications need to be compact and tested modules that can also be combined with each other. The themes should support current planning challenges and represent themes which planners find interesting and relevant. Planners should also always have the possibility of adding to questionnaires some questions of their own.

The *softGISsafety* method was developed to study the perceived safety of a neighbourhood (*www.pehmogis.fi/muotiala*). This web survey was used in Muotiala, a neighbourhood in the city of Tampere in Finland. The planning of Muotiala has been inspired by the CPTED-model (Crime Prevention Through Environmental Design). Through the *softGISsafety* method we gathered profound feedback from the residents about actual perceived safety, sense of community and everyday life infrastructure in the area. The residents localized places of perceived social and traffic dangers, pleasant and unpleasant routes and signs of disorder and care. The results give detailed information to planners how some realizations of CPTED principles, for example, lighting, public space design or light traffic routing are reflected in the residents' localized experiences and use of their environment (Fig. 19.3). The localizations of perceived danger can also be compared with the police maps of the actual scenes of crime. The data can then also be utilized by police in their preventative work in neighbourhoods.

The *softGISsafety* questionnaire was not publicly accessible as were the earlier applications. Instead, we gathered the addresses from the census register and sent a letter to every household asking all inhabitants over 15 years old to answer the questionnaire on the web. In the second round, a pen and pencil questionnaire was also offered. One hundred and eighty-two (39 per cent) inhabitants returned the questionnaire either on the web or in the mail. About 30 per cent of the subjects chose the traditional questionnaire; the majority of them were over 60 years old. The next new themes for thematic softGIS applications could be the mobility patterns and perceived health of urban inhabitants. Ewing *et al.* (2003) have highlighted the need for this kind of micro-level research to study the health-promoting qualities of urban settings as well as the use of GIS-based tools.

19.2.3 SoftGIS as a Method for a Special Group

The softGIS methods have to be sensitive to the varying skills, interests and orientations of different user groups of the environment. Either special tools for different users or differing interfaces for the same applications must be available

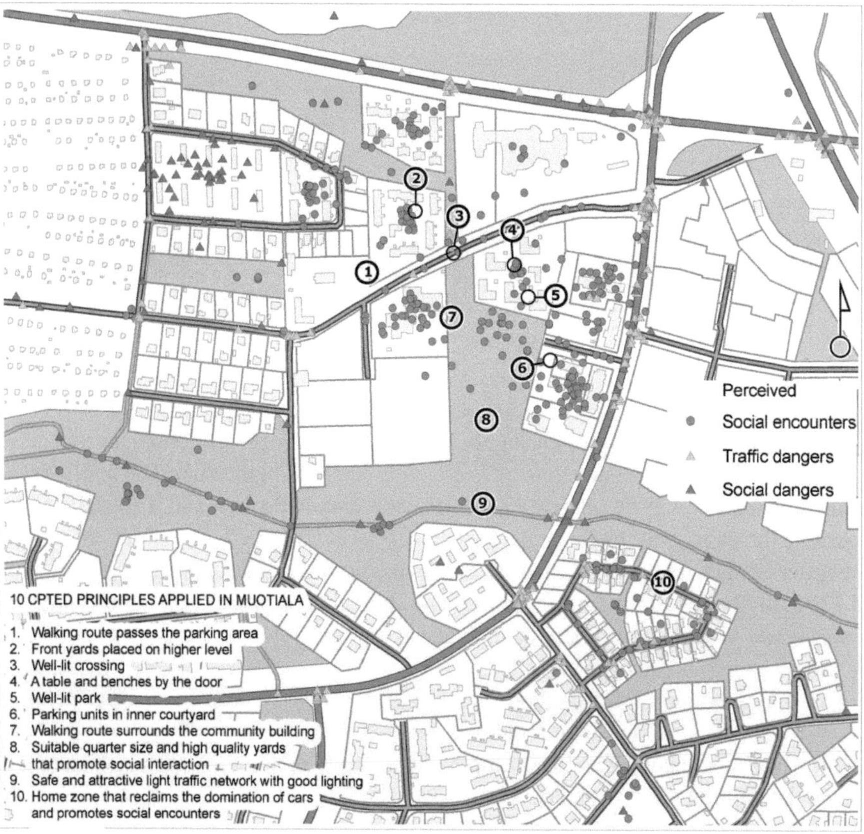

Fig. 19.3 Ten CPTED principles applied in Muotiala and inhabitants' perceptions of danger and social interaction (See also Plate 53 in the Colour Plate Section)

to guarantee as wide participation as possible. The interfaces are designed to be easy-to-use because they are built for people who are not familiar with web-based GIS services.

The *softGISchildren* method in Fig. 19.4 is specially designed for the use of children and young people. An internet-based design game *www.kaupunginosat.net/seikkailu/* that mediatedchildren's environmental visions preceded this method (Kyttä *et al.* 2003). Theoretically, *softGISchildren* is based on the definition of environmental childfriendliness by Kyttä (2003), where the diversity of environmental resources and access to play and exploration were chosen as the two central criteria of a child-friendly environment. In the *softGISchildren* method, this two-level model was enlarged with a third level, namely the well-being and perceived health of the children. In the *softGISchildren* method, 9–15 year-old children can, for example, locate places (affordances) that are functionally, emotionally or socially meaningful, draw a route from home to school and speak about their perceived health and well-being. This method grasps the essential qualities of child-environment relationships that are theoretically-based and empirically-tested in earlier studies.

Fig. 19.4 The *softGISchildren* method

Currently, we are collecting data with the *softGISchildren* method in the city of Turku with the help of local comprehensive schools. The local actors, city authorities and schools have been active during the method development and are eager to receive the results. These different actors have differing interests in the data. The *softGISchildren* application will be open after the research data collection and a special tool to explore the experiential knowledge in web has been developed. An online, interactive map-based data analysis tool, *softGISview*, is currently under construction that can be used by both planning professionals and the public.Privacy needs to be carefully considered in the development of this online tool to explore the experiential data of children and young people in web.

A future example of a method targeting a specialgroup can be a *softGISelderly* tool. Although the elderly are a challenging group to approach through the web, an interesting theme could be, for example, the mobility constraints of the environment from the point of view of ageing women.

19.3 The Critical Evaluation of the SoftGIS Methods

The interface between geographic information science and geographical social theory has created an interesting critical debate concerning the different uses and theorizing around GIS and especially about the possibility to 'straddle the fence' between human geography and GIS (O'Sullivan 2006). In the core of this critical GIS discourse, three viewpoints are especially relevant for the softGIS methods. We will evaluate the softGIS critically following these three themes.

The first critique of the traditional, technical use of GIS is grounded on the work that supports laypersons' and communities' capabilities to utilize GIS in their everyday life. The development of public participation GIS (PPGIS) or participatory GIS (PGIS), that allows lay persons to take part in GIS, has been significant in recent years. On the other hand, the planning support systems (PSS) aim mainly to support experts' work. We will first compare the softGIS method with some examples of PPGIS and PSS and argue how softGIS can build a bridge between these two systems.

The second step in the critical GIS literature concerns the need to adapt more qualitative information to GIS. As Kwan (2000, 2004) and Pavlovskaya (2005), who represent feminist GIS research, have pointed out, GIS data should be completed and enriched with the details of everyday life. Talen (2000), who has developed a special bottom-up GIS (BUGIS) tool, says that: *"The goal of a BUGIS should not be to capture all meaning but to strengthen the quality and depth of communication about residents' views and preferences"* (Talen 2000, p. 281). Because softGIS is about attaching experiential knowledge to GIS, we will discuss the nature of the knowledge collected with softGIS methods theoretically and consider the possibilities for handling this information with expert systems.

The critical issue for softGIS methodology is its ability to support the structures of democratic society. In collaborative planning, the flow of information between different actors is critical (Healey 1997). The competence of softGIS methodology is challenged to a degree by how experiential knowledge can be transferred to various actors in the planning process. In the critical 'GIS and society' discussion, issues like privacy, access and ethics are essential. In softGIS methods, where the mapping of individual data is central, a lot of concern arises about data collection, transfer, analysis and storage (O'Sullivan 2006).

19.4 The Enlarged Use of GIS

GIS/PSS and PGIS/PPGIS are developed to ease and support experts' work with geographical information or to enhance laypersons' access to geographical information and communication between different stakeholders. In the next section we will study the similarities and differences between softGIS and the existing methods. SoftGIS aims to form a bridge between existing traditions in the fields of participative GIS and PSS (Fig. 19.5).

19.4.1 SoftGIS and PSS

Our target is the rich use of GIS incollecting, storing, analysing and visualizing place-based experiential knowledge. GIS should not only be seen as a container of maps in digital form, but also as a supporter of spatial decision-making (Nedovic-Budic 2000), a facilitator of spatial thinking throughout the process (Pavlovskaya 2005) and, above all, a tool for revealing what is otherwise invisible in geographic information (Kwan 2000, 2004; Pavlovskaya 2005), such as social spatial patterns or in our case the soft spatial knowledge (Rantanen and Kahila 2008).

Owing to its technical development work and commercial applications (for example, *MapInfo* and *ArcGIS*), GIS is known as a support system both among researchers and in the professional field like urban planning. The very place-based nature of urban planning encourages the adaptation of GIS techniques also to communicate with the public and decision-makers (Nedovic-Budic 2000). PSS traditionally consists of three important components: data, models and geo-visualization

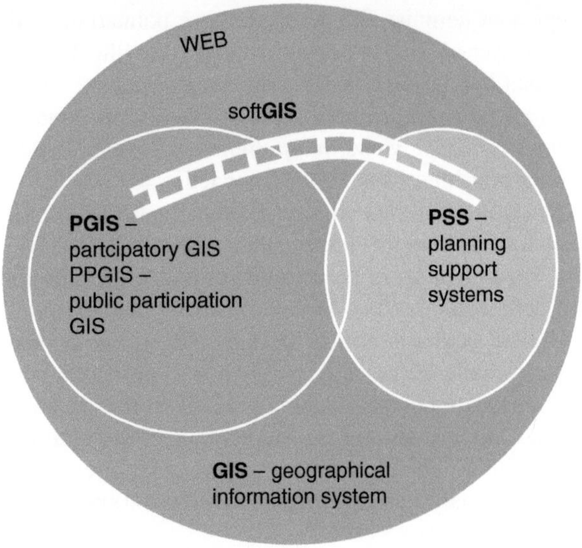

Fig. 19.5 SoftGIS: building the bridge between participative GIS and PSS

(Klosterman 1999). Most of the systems are tools that support specific tasks within planning processes (Geertman and Stillwell 2004). There are several examples of web-based PSS, such as area-specific Planning System for Sustainable Development (PSSD), and more task-specific systems such as SPARTACUS. These systems are toolboxes that share information among experts like urban planners (Geertman and Stillwell 2004). Some of the systems also enhance public participation. For example, games that give laypersons a possibility to familiarize themselves with planning problems in certain areas, like the *Baas op Zuid* game (*/www.baasopzuid.nl/*) in Rotterdam, can be seen as an example of PSS.

SoftGIS aims to support urban planning by gathering experiential knowledge systematically and allowing the urban planners to take part in the development of softGIS applications. Still, we have to ask: in what way or to what degree can softGIS be PPS? Can softGIS really assist planners in their work? One of our next steps is to study how softGIS as a method and the produced soft information can be integrated in the urban planning process? According to our experience, urban planners seem to be keen to have place-based experiential knowledge integrated into their systems.

19.4.2 SoftGIS, PGIS and PPGIS

Quite often residents are neglected in PSS developments. This has not been the case in PGIS or PPGIS discourse, which aims to give lay persons the possibility to be integrated in GIS and to give privilege and legitimacy to local and indigenous knowledge of certain areas. In these socially-oriented GIS applications, the aim has been more to enrich the content than to develop the technique. The target has been society's greater participation in the production and use of geographical information (Dunn 2007) although some PPGIS developers still rely only on publicly-accessible government data (Tulloch 2007). On the web there are a lot of examples where the aim is to offer laypersons a possibility to explore maps and different information. *Van Map* (*/www.city.vancouver.bc.ca/vanmap/*) represents one example of map services offered by cities. *Virtual Slaithwaite* (*/www.geog.leeds.ac.uk/papers/99-8/*) is an online PPGIS application that allows a two-way flow of information (Carver *et al.* 1999). This system allows residents to express their opinions, which builds up a community database. Many PGIS examples are not web-based like BUGIS (Talen 2000) but allow a small group of residents to describe their living environment and, with the help of facilitators, this information is transferred to a GIS.

PPGIS applications often rely technically on expert systems and only a few methods are designed for lay persons. These methods have been mainly application-driven and the theoretical basis is often inadequate (Balram and Dragićević 2006). Sieber (2003) has commented that if conventional GIS technology remains a core component of PPGIS it may play a restrained role in enabling democratic participation. We nevertheless think that new, easy-to-use internet-based systems can be useful for both residents and for some urban planners who find conventional GIS applications hard to use.

According to O'Sullivan (2006), the breadth and depth of the work done in the PGIS realm cannot be exaggerated, as is the case in the PSS realm. In Fig. 19.6, we have classified the various, participative GIS applications according to the level of information, the application uses and the level of openness or accessibility of the application. In this context, data, information and knowledge represent the levels of information that can be either official, register-based geographical information or knowledge produced by lay persons. Another level of the differentiation depends on whether the systems are permanently web-based, only part-time in the web or only PC-based.

The softGIS methods represented so far are questionnaires available only temporarily on the web which gather knowledge produced by residents. More development work is needed for softGIS to become a proper two-way channel of continuous communication between residents and planners. To enhance the role of PSS and PGIS in the future, the common development interests should be recognized (Jiang *et al.* 2003; Klosterman 2005).

19.4.3 Formulating SoftGIS Knowledge for Planners

One of the most crucial questions is how to analyse and present the experiential knowledge in planning practice in sufficiently digestible ways. If this task does not succeed, the experiential dimensions are unlikely to become embedded in design and planning practice (Thwaites and Simkins 2007). When transferring new information to urban planning practice, we should be aware that the planners

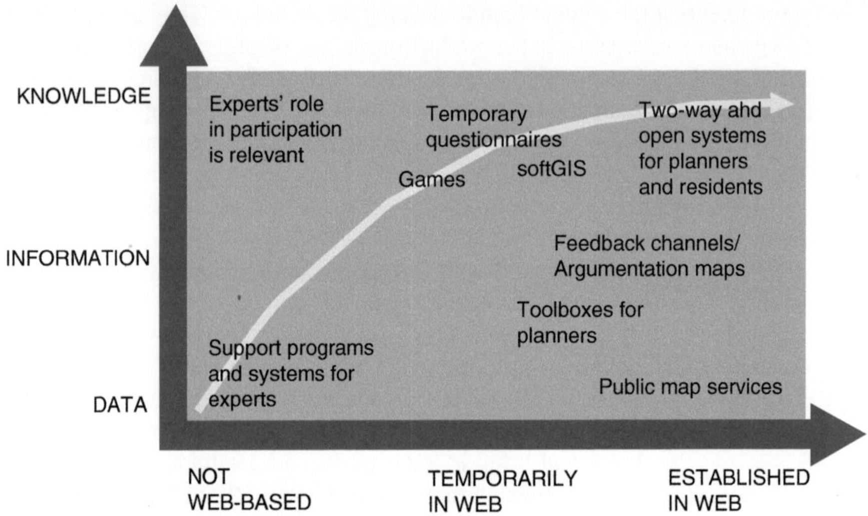

Fig. 19.6 The Existing Participative GIS Applications

are already struggling with a huge information overload. Visualization plays a key role in this challenging task, not least because urban planning is graphical by its very nature. Owing to more effective GIS capabilities, the role of mapping has changed from the end-product to a tool that is actively used in all the phases of the process. In this sense, visualization and mapmaking can be seen as a bridge between research and design. In addition to the traditional focus on available statistical or register-based information, information produced by different actors is also needed (Laurini 2001). Van Herzele (2004) calls for new 'interpretative frameworks' that should be available for planners to help them better understand the different forms of non-expert knowledge and views. We agree with McCall and Minang (2005) that this kind of knowledge must be seen to be as legitimate as experts' knowledge. In participation planning GIS can be an important enabling tool (Horelli 2002).

We created a small web questionnaire for politicians and urban planners and asked about current participation possibilities and the perceived value of the data gathered with softGIS methods. Respondents thought that the current participation procedure is adequate but the experiential, localized data could offer a useful addition to the knowledge available. The contexts where softGIS data could be utilized varied from town and general planning projects to development projects in build-up areas and new residential areas. Traffic planning was also perceived as a potential user of softGIS knowledge, for example, concerning the location of basic services.

Visualizations of softGIS data aimed at helping planners can be, for example, experiential knowledge of the green environment presented on top of the map of the green structure of the town. Similarly, traffic-related positive and negative comments can be shown on the traffic network map (Fig. 19.7). Naturally, the planners of our partner municipalities have received research reports of the soft-GIS projects and also some located data in a CD-rom to be utilized in their GIS programs. At the moment, we are building a web-based online tool, for the interactive analysis of the data from the children's softGIS project. With the help of this *softGISview* tool, urban planners and all other interested parties have the chance to explore the data.

The success and impact of the softGIS approach in the planning processes will eventually depend on the willingness of the planners and decision-makers to use the produced experiential knowledge and the new methods in their work. In the near future we will study in what form and in what part of the process different actors are ready to use the experiential knowledge produced by the softGIS methods.

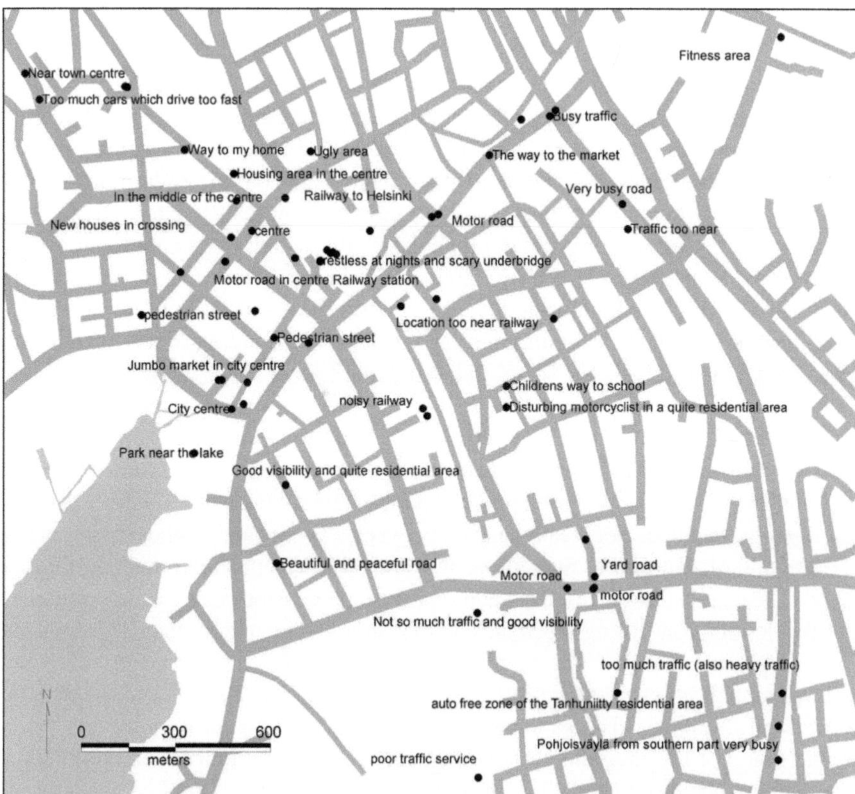

Fig. 19.7 Positive and Negative comments shown on the traffic network map from the city centre of Järvenpää

19.5 The Theoretical and Analytical Challenges of GIS-Based Individual Data

19.5.1 The Theoretical Basis of Experiential Knowledge Gathered with SoftGIS

The theoretical background of the softGIS methods lies in the humanistic geography, environmental psychology and urban planning theory that emphasize the experiential approach to planning (Thwaites and Simkins 2007). In these traditions we have looked for transactional approaches where the person-environment relationship is seen as a dynamic, interactive system. The transactional approach stresses the active role of both parties in this interactive relationship. People are active agents in their environments and can influence their environments and change them. In the same way, the material, social and cultural environment actively influences human beings by providing prerequisites for certain functions or by facilitating social encounters (Altman and Rogoff 1987).

There is a long 'mentalist' tradition in environmental psychology and social sciences of ignoring the physical environment itself, even in people-environment studies. Consequently there is a lot of theoretical and empirical research within humanistic geography and environmental psychology concerning, for example, cognitive representations, cognitive maps or mental maps of the environment without reference to the physical features of the environment that they try to represent. Existing studies on perceived environmental quality share the sameessential flaw (Bonnes and Bonaiuto 1995; Skjaeveland and Gärling 1997). In social sciences, too, the material world has long been disregarded. The 'new wave' of humanistic geography and social scientific studies aims at a better understanding of the relationship between the material and the social (Gieryn 2000; Latour 2002; Thrift 1996).

As a whole, the dissolution of the dualism between the mental, experiential world and the physical world has presented a challenge. The few non-dualistic approaches that aim to bring the material environment 'back' to the man-environment research offer a relevant basis for our work. An example of a truly transactional, non-dualistic theoretical concept, that does not create dualism between a person and his/her environment, is the notion of affordances that is used in ecological perceptual psychology (Gibson 1979). The term 'affordances' has traditionally referred to the perceived opportunities and restrictions concerning a person's actions in a given environment. This concept can be expanded to include the emotional, social and socio-cultural opportunities and restrictions that an environment offers (Gibson 1979; Heft 2001; Kyttä 2003). The concept of affordances breaks the subject-object dichotomy; an affordance is not a characteristic of the environment, nor a characteristic of the individual, but rather it is something between them. When the concept of affordances is applied, the transactional relationship between the person and the environment can be operationalized. That is exactly what we have attempted to do in softGIS methodology.

SoftGIS attempts to capture both the experiential knowledge of the inhabitants and the information about behaviour in the environment as well as the use of

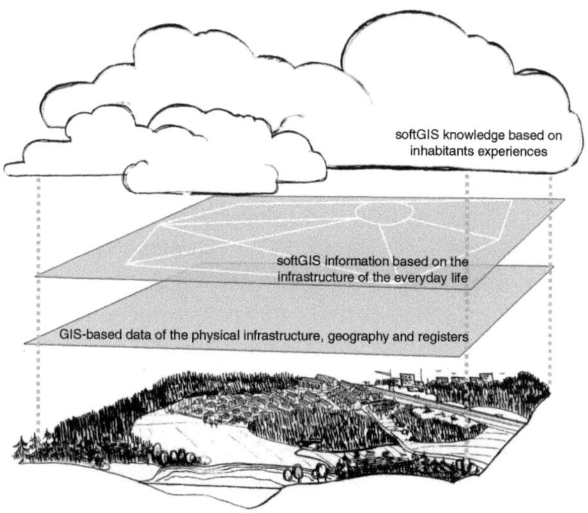

Fig. 19.8 The various layers representing the experienced, lived and physical environment

the infrastructure of everyday life (Horelli 2002). On the other hand, GIS offers a unique possibility to operationalize the depiction of the physical environment, although GIS is not an objective representation of the environment. Still, it is so far the most useful description of the physical qualities of the settings that we trace from the point of view of everyday life and individual experiences (Fig. 19.8). We strive to link the human experience with spatial expression in ways that are relevant to design and planning decision-making. This is one of the crucial tasks, if we aim to build a bridge between the planning practice and experiential knowledge.

19.5.2 Processing the SoftGIS Data

The personally meaningful affordances of different user groups (adults, children) in various types of urban and rural settings are distinguished and located in softGIS. The actualization and the accessibility of these affordances are studied, together with the characteristics of the physical environments. The construction of the softGIS methods and the analysis of the data produced by the residents are based on scientific principles concerning the validity and reliability of the data. With the help of GIS techniques, the perceptions of the residents are combined and analysed along with the information concerning the physical structure of the city, for example, the density, the type of land use, the amount and quality of the green areas, the connectivity of urban structures etc. Experiential knowledge is also thoroughly analysed both with qualitative and statistical quantitative methods.

The softGIS methods allow the construction of a database that contains a rich set of background information and both located and non-located experiential knowledge. The database can be transferred to various statistical and GIS programs.

In GIS analysis, the geographically imprecise and fuzzy softGIS data are challenging, which has to be taken into account when the results are interpreted. Often the GIS data are assumed to be 'perfect' and only technical problems receive major attention. Nevertheless, GIS data, including softGIS data, are not free from social and human influences (Taylor and Johnston 1995).

The map material that is used in the softGIS applications affects the residents' opportunities to read and orientate on the maps. It also affects the accuracy of the gathered information. In our applications we have used aerial photographs, address maps and geocoded axonometric projections. So far we know that the softGIS respondents find address maps generally easier to orientate and use than aerial photographs. On that basis the use of aerial photographs are problematic, although we have manipulated the photographs by highlighting roads and adding street names. Originally we thought that the aerial photographs could work well in the mapping of experiential knowledge, because it is possible to zoom even to the yards, locate trees, etc. There is certainly a lot of room for further research in this area. The map tool is especially difficult for the oldest and the youngest respondents. In the *softGISchildren* method, a technical mapping aid is used to help children to orientate themselves (Fig. 19.9). In the future, we will study systematically the usability of different types of maps and the map-reading skills of different user groups.

In the earliest softGIS applications it was only possible to locate point information. Respondents, nevertheless, would have preferred to locate their experiences areally. Technically, and in the analysis phase, this type of localized information is challenging. In the *softGISchildren* and *softGISsafety* applications also it has been possible to locate route information (Fig. 19.10). Some inhabitants found the route-drawing tool quite difficult to use. In the *softGISchildren* application, where a Flash

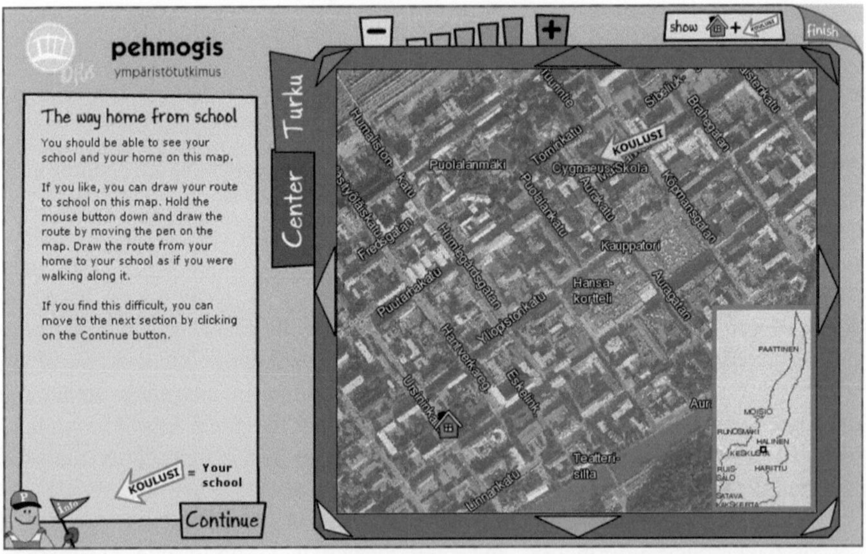

Fig. 19.9 The *softGISchildren* application actively aids a child to orientate in the map (See also Plate 54 in the Colour Plate Section)

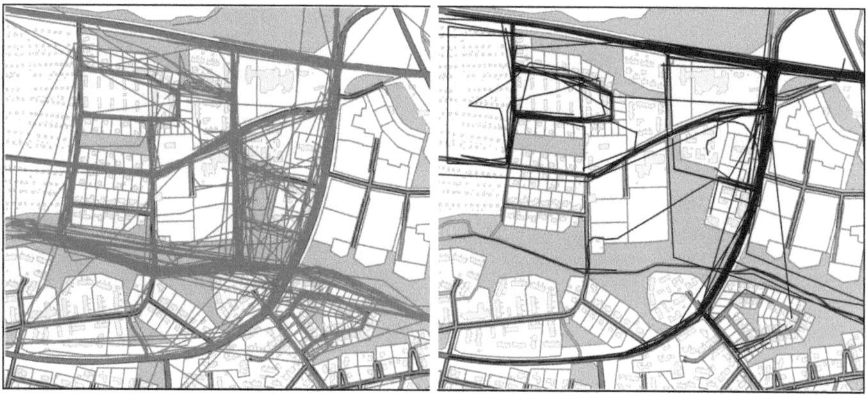

Fig. 19.10 The pleasant (left) and unpleasant (right) routes can be located in the *softGISsafety* method

Player is applied in the user interface, the route-drawing tool is more user-friendly. Maybe the most relevant way to locate the soft fuzzy knowledge would be the spray-can tool used, for example, to map the high crime areas in the city of "Ligh Crime" areas in the city o feeds (Waters and Evans 2003).

Compared with the traditional 'hard' GIS data, all our softGIS data are qualitative by nature because they are based on the residents' experiences and behaviour (cf. Kwan 2000, 2004; Pavlovskaya 2005). Nevertheless, from the point of view of social sciences, part of our 'qualitative' softGIS data can be analysed quantitatively (the classified, closed questions) and another part qualitatively (stories that the residents attach to their meaningful places). Although it may be tempting only to collect qualitative information with softGIS methods, the analysis of these argumentation maps (Rinner 2001) is quite laborious. We seek to combine the 'soft' and 'hard' data in our softGIS data analysis, so the quantitative analysis of soft data is, in general, more useful for that purpose.

The rich, multidimensional and imprecise softGIS data challenge the spatial GIS analysis. The visual exploration of softGIS data on the various types of maps is the basic step in the analysis. Maps that show how the affordances cluster and what kinds of hotspots can be traced allow visual data-mining and further considerations of the following analysis (Fig. 19.11). These maps tell us 'stories' in visual format about the deeper understandings of the living environment. In this sense, maps can be used in the analysis process, and not only as end-products, and visualization should be used especially because it has been proven to be an effective way to deliver knowledge (Wood 1992).

The buffering technique was utilized to analyse the physical qualities of the respondents' immediate surroundings and the degree of urban density: the amount of green space was calculated within the buffers surrounding respondents' homes and a sufficiently sensitive measure for the physical qualities of individual surroundings was found. Our results showed a significant negative correlation between urban density and perceived overall quality of the living environment (Kyttä and Kahila 2006). The distances between localized affordances

Fig. 19.11 Visualisation of positive quality factors (left) located on a map of Järvenpää and (right) on an aerial photo clipped from the city centre of Järvenpää (See also Plate 55 in the Colour Plate Section)

and respondents' homes were also calculated and the individual quality networks compared in different neighbourhoods. Spatial interpolation and thematic mapping were used to study, for example, the strength of the clustering of the positive and negative affordances.

19.6 SoftGIS and Society/Political GIS

In a democracy, the possibilities for public participation form an important cornerstone. In urban planning cultures, the communicative or collaborative turn denotes the types of planning practice whose emphasis is on interaction and communication between various stakeholders. According to Horelli (2002), enabling tools refer to any techniques that enhance the transaction and knowledge-creation of the stakeholders during the different phases of planning. Horelli distinguishes a number of different enabling tools that can be classified as diagnostic, expressive, organizing and political. Though researchers have developed and studied a lot of different participation tools, they are rarely implemented in urban planning practices. In Finland, public hearings are the most common participation methods. The collection of softGIS methods can be seen as a methodological package that is best suited to the evaluation phase of the planning process where diagnostic tools and traditional research methods should dominate. In the future, softGIS can also become a continuous method for monitoring during the whole planning process from initiation to planning, implementation, evaluation and maintenance.

As current participation methods inadequately support democratic participation possibilities (Healey 1997; Kingston 2007), web-based methods can make participation more democratic, because they free participation from the limits of time and place and they can potentially reach large numbers of inhabitants anonymously. Face-to-face meetings and conversations are certainly needed, but internet-based methods have an increasingly important role in the communicative planning processes (e.g. Kangas and Store 2003; Yeh and Webster 2004).

In collaborative planning process many different actors work together. Planners and designers are essential – but not only – potential utilizers of softGIS knowledge. Politicians and local city authorities strive to find ways to influence neighbourhood development projects. SoftGIS data can potentially help find empirical evidence of themes that interest these actors. In the city of Turku, where we have realized the *softGISchildren* study, different authorities have expressed varying interests in the softGIS data (Table 19.3).

Table 19.3 The varying interests of different actors in *softGISchildren* project

The softGIS project partners in the city of Turku	The various interests towards the softGIS data
Environment and city planning department	• utilization of the data in various planning projects e.g in land use plan for green environment • information on the places where young people spend their free time
Real estate department	• thematic and regional reports concerning the experiential quality of environmet • willingness to transfer the original data into their own geographical information system
Centre of sports services	• what children and young people do in Turku, how they spend their free time and what kind of social network they have • softGIS knowledge could be utilized e.g. in departmental statements of the city plans • the exploration of the data in web, online queries of the experiential knowledge according to background information
Social centre	• the preventative use softGIS knowledge • the well-being of the young people • the need to transfer the opinions of the young people directly to decision-makers
Health office	• mapping the regional differences of the perceived well-being and symptoms of the young people
School centre	• possible use of the method to support environmental education
Cultural centre	• the location of the most library services • the collection of localized memories of the young people
Youth work centre	• the places where young gather together • things they like to do in Turku and mapping the differences between region's abilities to offer these activities • The perceived safety of the young people; regional differences in well-being

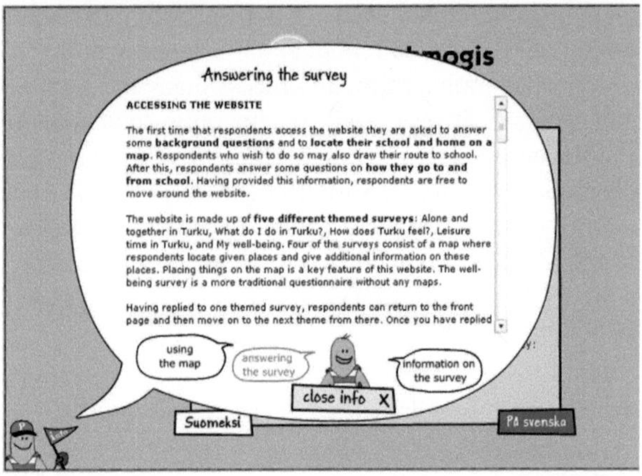

Fig. 19.12 Picture of the Info-pages in the *softGIS-children* application

It is obvious that the digital divide can weaken equality between different users. In Finland, the readiness to adopt web-based participation methods is quite pleasing, because the majority of households have access to the internet and every school and public library offers access to the web. On the whole, reviews and comparisons between web and traditional surveys show significant differences in neither the characteristics of the respondents (Gosling *et al.* 2004) nor in the results of the surveys (Rittel *et al.* 2004). According to our experience, softGIS is able to reach those inhabitants that are underrepresented in traditional participation gatherings. While older people dominate in public hearings, middle-aged, busy inhabitants are the majority of softGIS respondents. We have found little evidence of repeated answering or result manipulation in softGIS studies. There are also effective ways to reduce that threat (Gosling *et al.* 2004).

In the collection of large databases an important ethical issue is the protection of privacy. The confidentiality of GIS-based individual data collected on the internet is especially vulnerable. The privacy questions do not apply only to the data collection phase of the softGIS surveys, but also to the analysis, storage and delivery of the data. Data are usually always aggregated, so that individual respondents cannot be identified, the captured data are stored carefully and for urban planners only a part of the raw data can be transferred. Visualization requires special care in order not to stigmatize certain areas. Because residents need to be convinced of privacy protection, information concerning ethical issues can always be found in the softGIS web pages (Fig. 19.12).

19.7 Conclusion and Future Visions

Web-based GIS services have a lot of potential to become established information frameworks for city authorities, urban planners and lay persons in the future. At the moment, many good practices in the field remain diverse and dispersed. While the utilization of the web more effectively has huge capacity, the dangers and challenges of web-based GIS services should be recognized. One challenge relates to the information overload that is a result of the improving possibilities of knowledge production (Curry 1998). Although many useful services are already developed, most of them are not established practices. Further ICT and GIS technology innovations are needed but they are not enough: social and societal innovations are also required. Development work must therefore be transdisciplinary, and the practical knowledge of different actors of urban planning must be applied therein.

The softGIS methods, among other participatory GIS methods, have the potential for extensive applications in the future in environmental management and in traffic and community planning. In the future, the collection of various softGIS methods could form a methodology that provides a new platform for collaborative planning. To guarantee the usability of the methods and the quality of the information they produce, serious development of the methods and their theoretical basis is needed. Currently, we are studying further, how softGIS can form a bridge between participative GIS methods and planning support systems, between research and practice, and above all between planning practices and inhabitants' experiences of everyday life. To advance the methodology, research and development is needed on three different levels. First, the methodological development work aims to achieve a stabile technical platform based on a sound theoretical framework. Second, we are continuously evolving new, relevant ways to analyse and reveal the located experiential knowledge. Finally, we are keen to study how the knowledge of every day life can be assimilated in planning practices and decision making.

Acknowledgements The development work of the softGIS methodology has been realized through various research projects financed by the Finnish Ministry of the Environment, the National Technology Agency of Finland, the Finnish Cultural Foundation, the graduate school KIRSU and the cities of Järvenpää, Mäntsälä, Kerava, Nurmijärvi, Tampere and Turku. The authors are grateful to all the financial backers and the researchers involved in these projects. Special thanks belong to Liisa Horelli, Ari Jolma, Timo Heikkinen, Jenni Kuoppa and Anna Broberg for valuable critical comments.

References

Altman, I. and Rogoff, B. (1987) World views in psychology: trait, transactional, organismic and transactional perspectives, In Stokols, D. and Altman, I. (eds.) *Handbook of Environmental Psychology,* Vol 1, Wiley, New York, pp. 7–40.

Balram, S. and Dragićević, S. (2006) Collaborative geographic information systems: origins, boundaries, and structures, In Balram, S. and Dragićević, S. (eds.) *Collaborative Geographic Information Systems*, Idea Group Publishing, UK, pp. 1–22.

Beresford, P. and Hoban, M. (2005) *Participation in Anti-poverty and Regeneration Work and Research. Overcoming Barriers and Creating Opportunities*, Joseph Rowntree Foundation, New York. Retrieved October 9, 2007, from http://www.jrf.org.uk/bookshop/eBooks/1859353738.pdf.

Bonnes, M. and Bonaiuto, M. (1995) Expert and layperson evaluation of urban environmental quality: the 'natural' versus the 'built' environment, In Guerriei, Y., Alexander, N., Chase, J. and O'Brien, M. (eds.) *Values and the Environment. A Social Science Perspective*, Wiley, Chichester.

Building and Land Use Act (Maankäyttö- ja rakennuslaki) (1999) Finlex, valtion säädöstietopankki. Retrieved October 12, 2007, from http://www.fi nlex.fi /fi /laki/ajantasa/1999/19990132.

Carver, S., Evans, A., Kingston, R. and Turton, I. (1999) *Virtual Slaithwaite: A Web Based Public Participation 'Planning for Real System'*, Virtual Slaithwaite Case Study Report. Retrieved October 18, 2007, from http://www.geog.leeds.ac.uk/papers/99–8/.

Curry, M.R. (1998) *Digital Places. Living with Geographic Information Technologies*, Routledge, London.

Dunn, C.E. (2007) Participatory GIS – a people's GIS, *Progress in Human Geography*, 31(5): 616–637.

Ewing, R., Schmid, T., Killingsworth, R., Zlot, A. and Raudenbush, S. (2003) Relationship between urban sprawl and physical activity, obesity and morbidity, *American Journal of Health Promotion*, 18(1): 47–57.

Forester, J. (1989) *Planning in the Face of Power*, University of California Press, Berkeley and Los Angeles.

Friedmann, J. (1998) The new political economy of planning: the rise of the civil society, In Douglass, M. and Friedmann, J. (eds.) *Cities for Citizens*, Wiley, Chichester.

Geertman, S. and Stillwell, J. (2004) Planning support systems: an inventory of current practice, *Computers, Environment and Urban Systems*, 28: 291–310.

Gibson, J.J. (1979) *The Ecological Approach to Visual Perception*, Houghton Mifflin, London.

Gieryn, T.F. (2000) A space for place in sociology, *Annual Review of Sociology*, 26: 463–495.

Gosling, S.D., Vazire, S., Srivastava, S. and John, O.P. (2004) Should we trust web-based studies? A comparative analysis of six preconceptions about internet questionnaires, *American Psychologist*, 59(2): 93–104.

Healey, P. (1997) *Collaborative Planning. Shaping Places in Fragmented Societies*, McMillan Press Limited, Hampshire and London.

Heft, H. (2001), *Ecological Psychology in Context: James Gibson, Roger Barker, and the Legacy of William James's Radical Empiricism*, Lawrence Erlbaum Associates, Publishers Mahwah, New Jersey, London.

Horelli, L. (2002) A methodology of participatory planning, In Bechtel, R. and Churchman, A. (eds.) *Handbook of Environmental Psychology*, Wiley, New York, pp. 607–628.

Innes, J.E. and Booher, D. E. (1999) Consensus building and complex adaptive systems. A framework for evaluating collaborative planning, *APA Journal*, 65(4): 412–423.

Jiang, B., Huang, B. and Vasek, V. (2003) Geovisualisation for planning support systems, In Geertman, S. and Stillwell, J. (eds.) *Planning Support Systems in Practice*, Springer, Berlin, pp. 177–191.

Kangas, J. and Store, R. (2003) Internet and teledemocracy in participatory planning of natural resources management, *Landscape and Urban Planning*, 62: 89–101.

Kingston, R. (2007) Public participation in local policy decision-making: the role of web-based mapping, *The Cartographic Journal*, ICA Special Issue, 44(2): 138–144.

Klosterman, R.E. (1999) The What if? collaborative planning support system, *Environment and Planning B*, 26: 317–320.

Klosterman, R.E. (2005) Guest Editorial, *Environment and Planning B*, 32: 477–484.

Kwan, M.P. (2000) Analysis of human spatial behavior in a GIS environment: recent developments and future prospects, *Journal of Geographic Systems*, 2: 85–90.

Kwan, M.P. (2004) Beyond difference: from canonical geography to hybrid geographies, *Annals of Association of American Geographers*, 94: 756–763.

Kyttä, M. (2003) Children in outdoor contexts. Affordances and independent mobility in the assessment of environmental child friendliness, *Yhdyskuntasuunnittelun tutkimus- ja koulutuskeskuksen julkaisuja A 28*. Otamedia, Espoo. Retrieved September 18, 2007, from http://lib.tkk.fi/Diss/2003/isbn9512268736/isbn9512268736.pdf.

Kyttä, M., Kaaja, M. and Horelli, L. (2003) An internet-based design game as a mediator of children's environmental visions, *Environment & Behavior*, 1: 127–151.

Kyttä, M. and Kahila, M. (2006) PehmoGIS – elinympäristön koetun laadun kartoittajana, *Yhdyskunta-suunnittelun tutkimus- ja koulutuskeskuksen julkaisuja B 90*, Otamedia, Espoo. English summary: http://opus.tkk.fi/softgis/documents/IAPS%202006.pdf.

Latour, B. (2002) Morality and technology. The end of the means, *Theory, Culture & Society*, 19(5–6): 247–260.

Laurini, R. (2001) *Information Systems for Urban Planning. A Hypermedia Co-operative Approach*, Taylor & Francis, London and New York.

Manzo, C.L. (2003) Beyond house and haven: toward a revisioning of emotional relationships with places, *Journal of Environmental Psychology*, 23: 47–61.

McCall, M.K. and Minang, P. (2005) Assessing participatory GIS for community-based natural resource management: claiming community forests in Cameroon, *Geographical Journal*, 171: 340–356.

Nedovic-Budic, Z. (2000) Geographic information science implications for urban and regional planning, *URISA Journal*, 12(2): 81–93.

O'Sullivan, D. (2006) Geographical information science: critical GIS, *Progress in Human Geography*, 30(6): 783–791.

Pavlovskaya, M. (2005) Theorizing with GIS: a tool for critical geographies? *Environment and Planning A*, 38(11): 2003–2020.

Puustinen, S. (2006) Suomalainen kaavoittajaprofessio ja suunnittelun kommunikatiivisen käänteen haaste, *Yhdyskuntasuunnittelun tutkimus- ja koulutuskeskuksen julkaisuja A 34*, Frenckellin kirjapaino Oy, Espoo.

Rantanen, H. and Kahila, M. (2008) The SoftGIS approach to local knowledge, accepted for publication in *Journal of Environmental Management,* Special Issue on Collaborative GIS: Perspectives on decision support and visualization.

Rinner, C. (2001) Argumentation maps: GIS-based discussion support for on-line planning, *Environment and Planning B*, 28(6): 847–863.

Rittel, P., Lorig, K., Laurent, D. and Matthews, K. (2004) Internet versus mailed questionnaires: a randomized comparison, *Journal of Medical Internet Research*, 6(3). Retrieved June 9, 2006, from http://www.jmir.org/2004/3/e29/.

Sieber, R.E. (2003) Public Participation geographic information systems across borders, *The Canadian Geographer*, 47: 50–61.

Skjaeveland, O. and Gärling, T. (1997) Effects in interactional space on neighbouring, *Journal of Environmental Psychology*, 17: 181–198.

Staffans, A. (2004) Vaikuttavat asukkaat, *Yhdyskuntasuunnittelun tutkimus- ja koulutuskeskuksen julkaisuja A 29,* Yliopistopaino, Helsinki.

Talen, E. (2000) Bottom-up GIS. A new tool for individual and group expression in participatory planning, *Journal of the American Planning Association*, 66: 279–294.

Taylor, N. (1998) *Urban Planning Theory Since 1945*, Sage, London.

Taylor, P.J. and Johnston, R.J. (1995) GIS and geography, In Pickles, J. (ed.) *Ground Truth. The Social Implications of Geographic Information Systems*, The Guilford Press, New York, pp. 51–67.

Thrift, N. (1996) *Spatial Formations,* Sage, London.

Thwaites, K. and Simkins, I. (2007) *Experiential Landscape. An Approach to People, Place and Space*, Routledge, UK.

Tulloch, D.L. (2007) Many, many maps: empowerment and online participatory mapping, *First Monday*, 12(2). Retrieved February 16, 2007, from http://www.firstmonday.org/issues/issue12_2/tulloch/.

Van Herzele, A. (2004) Local knowledge in action. Valuing non-professional reasoning in the planning process, *Journal of Planning Education and Research*, 24: 197–212.

Waters, T. and Evans, A.J. (2003) Tools for web-based GIS mapping of a "fuzzy" vernacular geography. Retrieved September 10, 2007, from http://www.geocomputation.org/2003/Papers/Waters_Paper.pdf.

Wood, D. (1992) *The Power of Maps*, Guilford Press, New York.

Yeh A.G. and Webster C. (2004) Planning, government, information, and the Internet, (editorial), *Environment and Planning B*, 31(2): 163–165.

Chapter 20
The *Mainport Planning Suite*: Planning Support Software for Studio-Based Planning

R. Chin

20.1 Introduction

The main ports in the Netherlands, such as the Port of Rotterdam (PoR) and Amsterdam Airport Schiphol (AAS), are important for the national economy. Sustainable growth and the commercial success of 'Mainport Holland', located in Europe's most densely populated area, is threatened by a lack of available land, a congested infrastructure, and an increasingly complex social, economic and political reality. To deal with these threats, the main ports are reengineering their planning processes. Instead of making plans based on an extrapolation of current trends, the aim is now to find answers to what-if questions which are applied to concurrent scenarios. 'Mainport planning' is like solving a large jigsaw puzzle, but unlike a conventional jigsaw, the pieces used to solve the puzzle are not available beforehand, and there is no single optimum solution. Solving the mainport planning puzzle is a difficult, lengthy, knowledge and information intensive, multi-actor process.

In this research we have developed the *Mainport Planning Suite* (*MPS*), an innovative planning support system (PSS) which supports user engagement and stakeholder participation in a mainport planning project. The underlying paradigm is what we call studio-based planning. Core to this paradigm is the notion of a studio: a facilitative environment which supports the blend between people, processes and technology. Hence we do not consider planning support as a solely technical challenge but instead we consider it from a process-oriented, multi-actor perspective which can be supported by information and communications technology. The challenge is to bring process and technology together to effectively support mainport planning projects.

R. Chin (✉)
Delft University of Technology
Faculty of Technology, Policy and Management
Jaffalaan 5, 2628BX Delft
The Netherlands
E-mail: info@roychin.nl

S. Geertman, J. Stillwell (eds.), *Planning Support System Best Practice and New Methods*, DOI 10.1007/978-1-4020-8952-7_20, © Springer Science+Business Media B.V. 2009

The research method is based on case studies as the main research instrument and design science as the research strategy. Whereas natural scientists and behavioural scientists try to understand reality, design science attempts to create things that serve human purposes. Natural science often consists of discovery and justification. Design science revolves around building and evaluating objects and systems. Hevner *et al.* (2004) stress the complimentary nature of both paradigms, as design relies on using existing theories. They state that design science is proactive with respect to technology, whereas natural sciences, or behavioural sciences, take technology as given. They strongly advocate a research cycle where design science is used to create artefacts aimed at solving specific information problems, based on relevant theory from the field of natural and behavioural sciences.

The relevance of this research became clear in a case study at the PoR (Smits *et al.* 2005). In 2004, the PoR and the Delft University of Technology jointly defined a case study which was focused on improving the effectiveness of port planning by means of planning support technology. We will use this case study to illustrate the ideas presented in this chapter. The PoR is responsible for developing and exploiting about 10,000 acres of the Rotterdam port and industrial complex. Customers of the PoR are, for example, container companies, petrochemical companies, bulk materials companies and distribution centres.

To be able to continuously control the development of the port, planning processes are of increasing importance. In the search for growth opportunities, the PoR continuously faces the scarcity of resources. For instance, the amount of available land within the port industrial complex has substantially decreased over the past few years. Nowadays there are hardly any large and unused areas left. Available land is mainly found in small sizes and widespread and is not always accessible from the waterside. Another issue is the increasingly dynamic, unpredictable and complex environment in which the PoR has to operate. The world is changing fast. Asia is booming, events like 9/11 lead to new requirements regarding safety and security, and new EU regulations increase the complexity of daily operations. Considering the scarcity of resources and the increasingly dynamic, unpredictable and complex environment, the PoR is actively developing more effective port planning processes. The PoR defines port planning as: *"the process of balancing the mutual relations between spatial planning, business development and environmental constraints and opportunities"* (Smits *et al.* 2005). In order to plan and design the port industrial complex taking into account the scarcity of resources, professional port planning is required. During our case study, we focused on a type of planning called area planning which deals with the development of a specified port area for a five to ten year period into the future.

This chapter is structured as follows. In Section 20.2, the concept of studio-based planning is introduced. In Section 20.3, the details of the research approach are elaborated. In Section 20.4, the design of our *MPS* is outlined, followed by the research results and an identification of future trends and future research directions in Section 20.5.

20.2 Studio-Based Planning

This research is based on a longstanding research tradition in Collaborative Business Engineering (CBE) at the Delft University of Technology. Business engineering deals with "*organizational transformation focusing on integral design of both information technology and organizational processes and structures*" (Hammer 1990; Maghnouji *et al.* 2001). In CBE, a combination of simulation models and group support systems (GSS) are used in a dynamic modelling approach (Babeliowsky 1997; Maghnouji *et al.* 2001). Simulation models are used to develop dynamic models of organizational processes and information systems. Simulation modelling tools provide an environment in which one can understand, analyse and improve the business process. Simulation outcomes are usually visualized using animations and graphs which allow the stakeholders to understand the model and communicate about it. GSS are used to assist a group in structuring activities, generating ideas and improving group communications (Hengst and Vreede 2004).

The research efforts in CBE lead to a richer image of a support environment for planning and decision making as introduced by Keen and Sol (2007) known as decision enhancement studios, or studios in short. Studios are facilitative simulation environments designed to enable executives to rehearse the future. As Keen and Sol explain, rehearsing the future rests on vision, envisioning shared images, collaboration and communication among people scattered across the organization. Studios are not simulation environments in the traditional sense of the word, instead in studios *"tools of technology and the analytic sciences are embedded in the processes and interactions"* (Keen and Sol 2007, p. 12). The concept of a studio can be compared with a television studio: a facilitative environment which merges people, processes and technology. Studio-based planning provides both horizontal and vertical support, i.e. it supports a wide range of stakeholders in the planning process, and it supports the leveraging of information through the different organizational layers. Studio-based planning is also closely related to ubiquitous computing. For example, Russell *et al.* (2005) explain how new ways of interacting with computers can lead to interactive workspaces and smart environments where people can work together in technology-rich spaces. Hence, we argue that studio-based planning focuses on two sides: the organizational, i.e. processes, roles, decision context, *et cetera,* and the technological, i.e. computer visualizations, simulation models, group support, etc.

Figure 20.1 shows the concept of studio-based planning. The studio is not at one location, but is dispersed over meeting rooms and offices in the organization. In that sense, a planning studio differs from a television or recording studio; the meeting rooms are not standard and equipped with tools that support planning. Instead, when a planning team occupies a meeting room they transform the room into a studio. A suite of software services changes a meeting room into a planning studio in a digital way; it offers services, tools, documents, maps, electronic agendas and so forth, through an enterprise information portal which hooks up to the corporate IT infrastructure. It sets up a dynamic network of IT services through which the

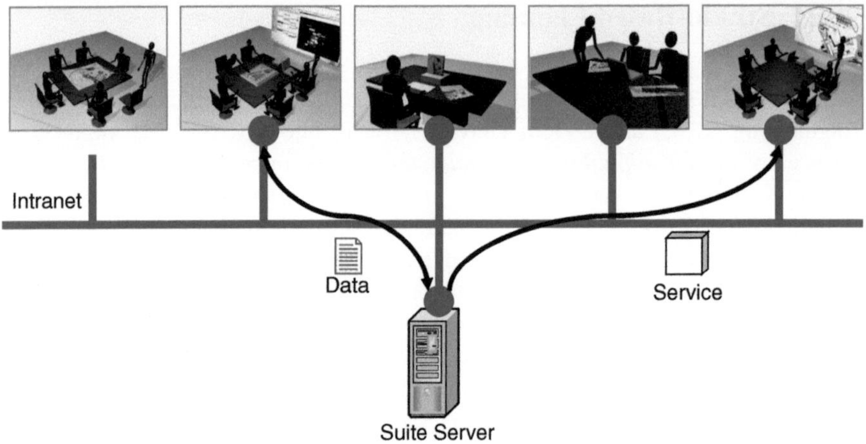

Fig. 20.1 A planning studio schematic

involved actors can access, edit and share information. It offers services for analysis, designing creative solutions, evaluating and choosing preferred solutions.

Studio-based decision support and planning encompasses:

- people: the actors representing a whole scope of disciplines and interdisciplinary dimensions in the context of the problem to be solved;
- process: a process founded in the method of simulation, guiding search towards a solution; and
- technology: suites of interrelated information and communication services, simulation instruments, analytic methods and visualization interfaces.

As Keen and Sol (2007) explain, a suite of software services is the foundation for meshing technology and process. It consists of domain-specific information and communication services, which form building blocks that support recipes for repeatable processes. In a way, a suite is a toolbox for studio-based decision making. A surgeon uses a specific toolbox to operate on a specific patient. Like the surgeon, an executive, planner or expert often faces different problems which require using a specific set of instruments. The choice of instruments will depend on the context, the roles of the involved actors, and possibly, the personal preferences of the managers involved. This is why we focussed this research on the development of a suite to support mainport planning. In the next section the research approach is introduced to explain how the *MPS* was developed.

20.3 Research Approach

This research was conducted in close cooperation with the PoR, enabling the design of the *MPS* in a realistic case study which lasted for about three years. Our focus was to develop a suite of software services to support area planning projects. In area planning, an inter-disciplinary project team of about six to eight experts develops a plan for a specific port area for a target period of five to ten years. This plan provides input to the various departments of the PoR.

The objective of this case study was to explore the needs of mainport planners in practice, which would help to sharpen our insights into the requirements of a suite to support mainport planning. Initially a project team was formed which consisted of four researchers from the University and three managers from the PoR, referred to respectively the TU Delft team and the PoR team. The author participated as one of the four researchers from the TU Delft team. The PoR team members were the first point of contact for gaining more insights into area planning. Regular meetings were held every two or three weeks to present design ideas to the PoR team. Furthermore the PoR Team was responsible for making this project known throughout the PoR's departments involved in area planning. They also provided names of experts who were available for interviews and discussions, and helped in organizing presentations of design ideas to the PoR's departments. In addition, at a later stage, a graduate student from the University was added to the PoR team and given the unique opportunity to participate, mainly as an observer, in a real planning project. The role of the student was that of a business analyst, i.e. to investigate and report on how the current planning process was conducted and to propose ideas for improving the planning process (Schalkwijk 2005).

The introduction of new technologies in area planning played an important role in this research. To provide a realistic impression of what studio-based planning means, extensive use was made of prototyping. Working software prototypes were used alongside slideshow presentations, which enabled ideas to be effectively transferred to the experts in the PoR. Throughout this research, prototypes were used in meetings with the PoR team, in interviews and discussions with individual experts, and in presentations for the different PoR departments. The software prototypes went through a number of iterations, partly based on a trial and error approach. Eventually, after about a year, the prototype converged to a more or less stable solution that could be used in an evaluation session.

Together with the PoR team, several ideas for a prototype evaluation were discussed. Ideally, the prototype would be used and evaluated directly in an ongoing area planning project. However, this would impose considerable risks to the continuity of such a project. Therefore it was decided to first evaluate the prototype in a controlled environment. A one-day evaluation session was prepared: a fictitious but realistic area planning meeting in which the different studio-based technologies would be introduced. The time to prepare this evaluation session was considerable, nearly half a year, for a number of reasons. First, a team of area planners was needed to participate in the session. Typically an area planning team consists of people from different departments. The PoR team needed to contact the different departments to find participants for the evaluation session. Second, a realistic storyboard needed to be developed in which the different technologies would be introduced. Third, after a team of participants was set up, they were actively involved in creating realistic content for the artificial area planning meeting. Fourth, based on the inputs of the PoR team, the prototype was fine-tuned for the evaluation session and the prepared content needed to be entered in an electronic database.

In the week before the evaluation session, the participants were asked to complete an online survey about their current way of working. On the day of the evaluation session, the area planners were guided through the storyboard supported by the prototype. Participant feedback was gathered through an electronic survey system, a group discussion, and individual interviews afterwards. Furthermore, the whole session was recorded on video.

After the evaluation of our prototype, it was decided to improve the prototype based on the feedback gathered. The improved prototype was presented to and discussed with the participants during several meetings about a year after the evaluation session, after which this research was concluded. In the next section, the details of the *MPS* design and prototype are outlined.

20.4 Design of the Mainport Planning Suite

We had the unique opportunity to develop a prototype *Mainport Planning Suite* in close cooperation with its intended end users. We extensively researched how area planning in the PoR is conducted in practice, by having interviews and discussions with experienced area planners and by observing area planning projects in progress.

20.4.1 The Area Planning Process

Area planning is a process in which the actors, the area planners, play different roles depending on their field of expertise and their contribution to the planning process. For example, roles that can typically be identified are project leader, domain expert, analyst, and facilitator. Area planning involves expertise in domains such as infrastructure engineering, geographical information systems, economics, strategic planning, environmental issues. Furthermore, area planning projects can last for about ten months, during which the project goes through several stages. First a project team is formed; then the area is analyzed and opportunities and constraints are inventoried. A number of workshops are scheduled to discuss alternative planning solutions for the area, and finally a final report, the area plan, is written.

Area planning is an iterative process in which the people work both individually in their offices, and together in multi-disciplinary team meetings. During this process a wide variety of information is used and produced. A common denominator throughout an area planning project is the use of geographic maps. A map of the area provides a common frame of reference to all involved area planners. Furthermore, paper maps are used as writing pads to sketch and communicate design ideas.

20.4.2 Challenges on Area Planning

During our investigation, a number of challenges in area planning came to light, both with respect to the planning process and the planning content. With respect to the planning process it was observed that, due to the duration of an area planning project, it is difficult to manage and keep track of the abundance of information and resources involved. The information is of many kinds and maintained by different individuals and departments. For example, geographic maps are prepared by the drawing department. During a meeting, area planners make sketches and annotations on the map, and after the meeting, the map is stored in the office of one of the area planners. When, after ten months, the area plan is written, it turns out to be difficult to find out when the map was made, who annotated it, and what was discussed in that particular meeting.

With respect to the content, it was clear that information of many different kinds needed to be integrated, compared and evaluated. Often changes in the area over time and the transition to the future situation are considered to be important, but it is difficult to incorporate dynamic effects and alternative scenarios in the planning process. Furthermore, creating an integral view on all the relevant aspects leads to an information overload.

20.4.3 Overview of the Design

The *MPS* was designed as a web-based solution, because of the fact that area planners work both asynchronously at their offices and synchronously in meetings (Chin 2007). The web-based solution requires the area planners to login to the project web-portal using a web browser. After logging in they are assigned a role based on their function in the area planning project. A distinction in roles makes it possible to assign functionalities to area planners which match their information needs.

Figure 20.2 shows a selection of the main software services provided by the *MPS*. Each of these services supports a different activity in the problem solving cycle (Sol 1982). We focused on the activities which fall inside the shaded box, because these are the activities which are conducted by multi-disciplinary planning teams, while the other activities are more mono-disciplinary in nature. The supported activities are analysis, finding solutions, *ex ante* evaluation and choice. A *Map* service was designed to support analysis because, in mainport planning, geographic maps are commonly used to analyze mainport areas. The *Sketchbook* service represents a digital sketchbook, or scrapbook to keep track of, compare and view alternative planning solutions. *Matchbox* is a visualization tool which helps multi-disciplinary teams in evaluating alternative solutions. Finally, *Aspect Explorer* is a tool which helps multi-disciplinary teams in choosing a preferred alternative solution. In the following sections each of the suite's services is explained in more detail.

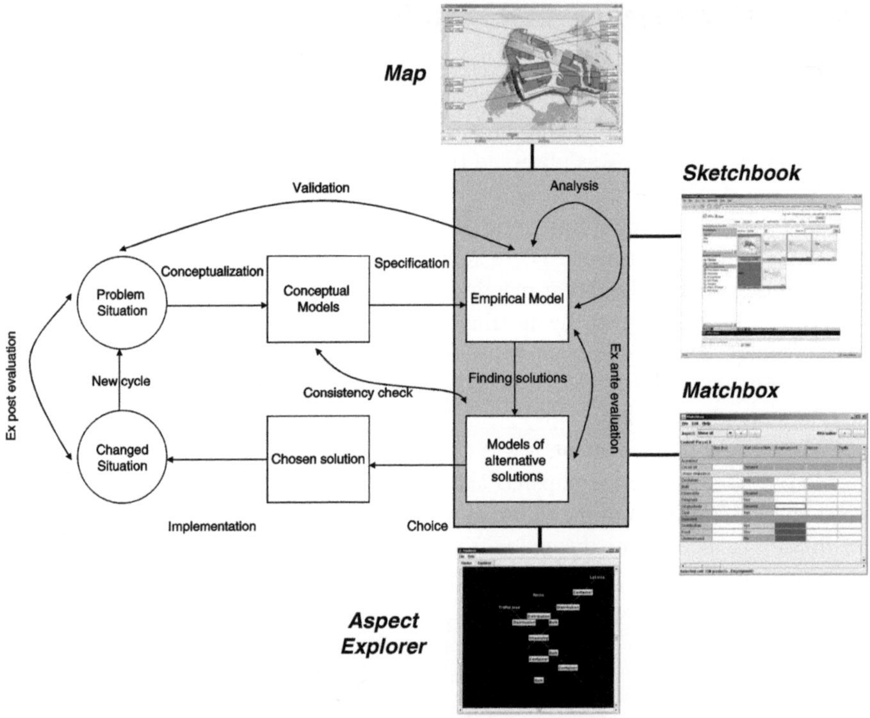

Fig. 20.2 *MPS* services and their relation to the stages and activities in problem solving

20.4.3.1 *Map* Service

During our case study at the PoR, it was observed that geographic maps are commonly used in area planning projects. During project meetings the team members use paper maps to analyse the port area and sketch design ideas. The disadvantage of using paper maps is that these are not properly documented and archived. The paper maps are produced by the PoR's drawing department on request of the area planning team. The drawing department is responsible for managing all geographic material that is used in the PoR. However, the paper maps used in area planning meetings tend to end up in the offices of the individual team members, because these are not 'official' maps. Yet, these maps contain important sketches and annotations. The *Map* service was designed specifically to address the needs of mainport planners. In close cooperation with experts in the PoR, the concept of an interactive map viewer was iteratively rethought and redesigned. The *Map* service is a web-based interactive viewer for both static and animated geographic maps as is shown in Fig. 20.2.

The Map service is meant to be a lightweight creativity tool instead of a full GIS system. The problem with the advanced GIS systems that we observed in the PoR, was that mainport planners were not trained to use these advanced systems. Hence, GIS analysts had to produce ready-to-use paper maps for planning meetings.

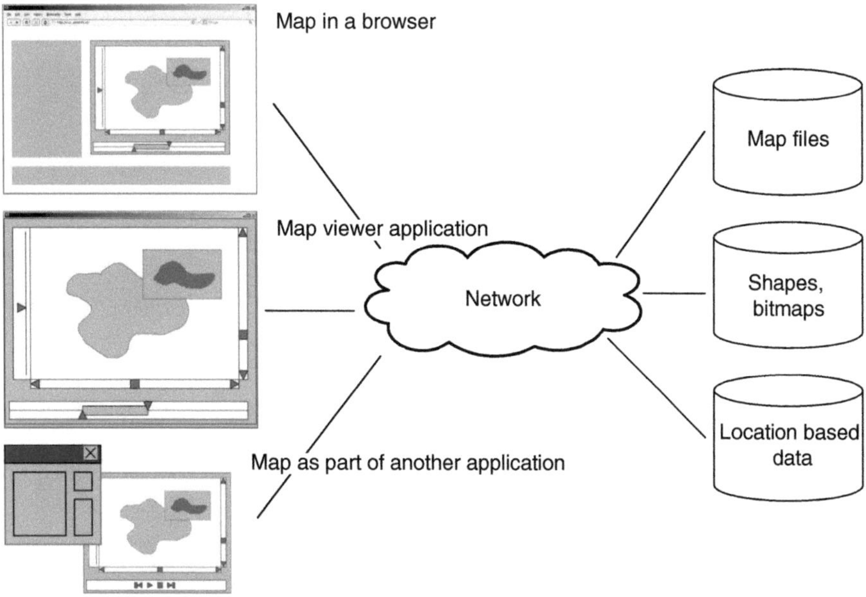

Fig. 20.3 Using *Map* in different ways to visualize and interact with distributed information

The mainport planners had no easy way of passing their annotated paper maps and design sketches back into the system. The Map service was designed to support the distinction between providers of maps and users of these maps, and to provide a relatively easy way of processing and displaying 'unofficial' maps. The distinction between providers and users of maps makes it possible to formalize the use of unofficial maps that are created by the mainport planners. Sketches and annotated maps are passed to the GIS analysts who can further process and archive the maps. Hence, instead of ending up in desk drawers, the annotated maps and sketches are made an integral part of the total collection of maps in a planning process, and they are managed by the GIS analysts. Furthermore, the Map service is a web-based tool. It makes geographic information available to mainport planners who work dispersed over the mainport's organization and who require web-based access to, visualization of and interaction with completely different types of information. Finally, the Map service is designed with the animation of temporal information in mind. While cartographic material usually consists of static images, data sets with a time-axis are not uncommon in mainport planning. The Map service provides functionality for binding layers of geographic information to time intervals. Hence the Map service can be used to display how a mainport changes over time geographically, and to display the animation of simulation outcomes directly on a map.

Figure 20.4 shows one specific example of how the *Map* service can be used in area planning, and especially how the Map service can be used to display temporal information. An area planning team may be interested to know which other projects overlap in terms of time and location within the area they are planning for. In this example, the *Map* service displays project boundaries for a certain time interval,

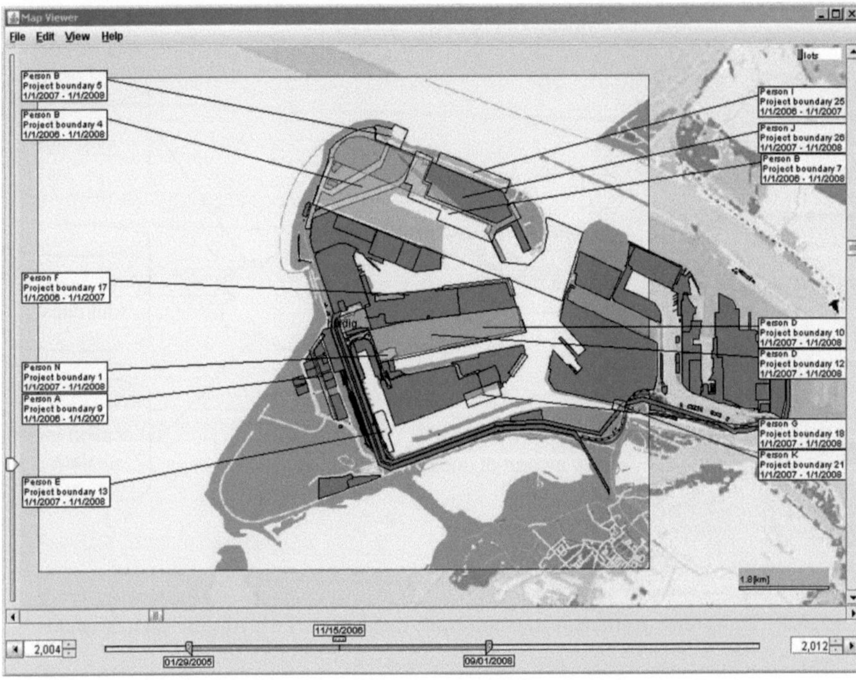

Fig. 20.4 Example of the *Map* service with contents of labels purposefully blurred

within a specific port area. The labels indicate the name of these projects, contact information and the time-span. The slider below the map represents a timeline that is used to manually select a time interval. When a time interval is selected, the map is automatically updated by displaying the corresponding project boundaries and labels. The selected time interval can also be dragged along the timeline to animate changes on the map over time.

On request of the PoR the contents of the labels contains only fictitious information. Furthermore this is just an example of one of the uses of the *Map* service. The same *Map* service has the flexibility to display totally different types of information when a planning team requires this for their specific planning project.

20.4.3.2 Sketch Book

The *Sketchbook* is meant to support indexing, filtering, viewing and interacting with information used and produced during a mainport planning process. In mainport planning, an abundance of information is used and produced. A substantial amount of information exists in the form of images such as geographic maps, diagrams, and graphs. Other forms of multi-media may also be used such as photographs, movies and animations. Typically, the available information is stored in distributed databases which fall under the responsibility of different departments and individuals.

It is not easy to manage and work with the abundance of information used and produced in mainport planning.

Sketchbook uses the metaphor of a paper sketchbook or scrapbook. Sketching, drawing up ideas and annotating images are common tasks for designers and mainport planners. In a paper sketchbook, all pages are connected, i.e. one needs to flip the pages to go from one sketch to another. Yet, this is not how designers like to browse through their information (Keller *et al.* 2004, 2006; Lugt 2001; Steenbergen *et al.* 1999). Designers like to create many different perspectives on the same information and they like to combine and compare very different types of information. They put images on the table, stick them to the wall, show them on an overhead projector, use a computer to generate 3D images or do all this simultaneously. They create storylines by putting images behind each other and annotating them. They make choices by comparing images spread out on the table and marking them with sticky notes. They create structure by creating stacks of images in different categories. Designers rip the pages from their sketchbook, but this makes it difficult to keep track of and organize the information in long term mainport planning processes. Furthermore, *Sketchbook* is meant to offer mainport planners the freedom they need to combine and compare information of many types creatively, without the danger of loosing track of information during the mainport planning process. Sketchbook does not have pages in a fixed order, but provides many different viewpoints on information.

The design of *Sketchbook* is shown in Fig. 20.5. Three main sections can be distinguished in the user interface: viewpoints, virtual folders and a content area at the right side. The viewpoints section is used to select a specific view on the data that is shown in the content area. The virtual folders section is used to filter and structure the data, such that a user can select different subsets of data. Although *Sketchbook* looks like a file manager such as those found in the major operating systems, it differs from a file manager in the sense that the data itself can be aggregated from various dispersed file systems. Virtual folders are not directories in a files system but represent queries over possibly a number of directories. The content area displays the selected data using a specified viewpoint. In Fig. 20.5, the data are displayed as little thumbnail images, similar to a file manager. However, *Sketchbook* can, for example, also display the data based on geographical location using a map. The data represents relevant content which is used and produced during a planning project, e.g. maps, photos, simulation models, and images.

For our test session we implemented a web-based version of *Sketchbook* that supported a limited set of viewpoints. It was capable of indexing distributed data such as images, geographic maps, and models. The user could define virtual folders and organize the data in these virtual folders. Furthermore the data could be displayed in different viewpoint implementations. We could display the data as little thumbnails and on a geographic map when it was bound to a geographic location. Consequently we consider our Sketchbook prototype as a working proof of concept. Later on other viewpoints should be implemented to better support the needs that were described above.

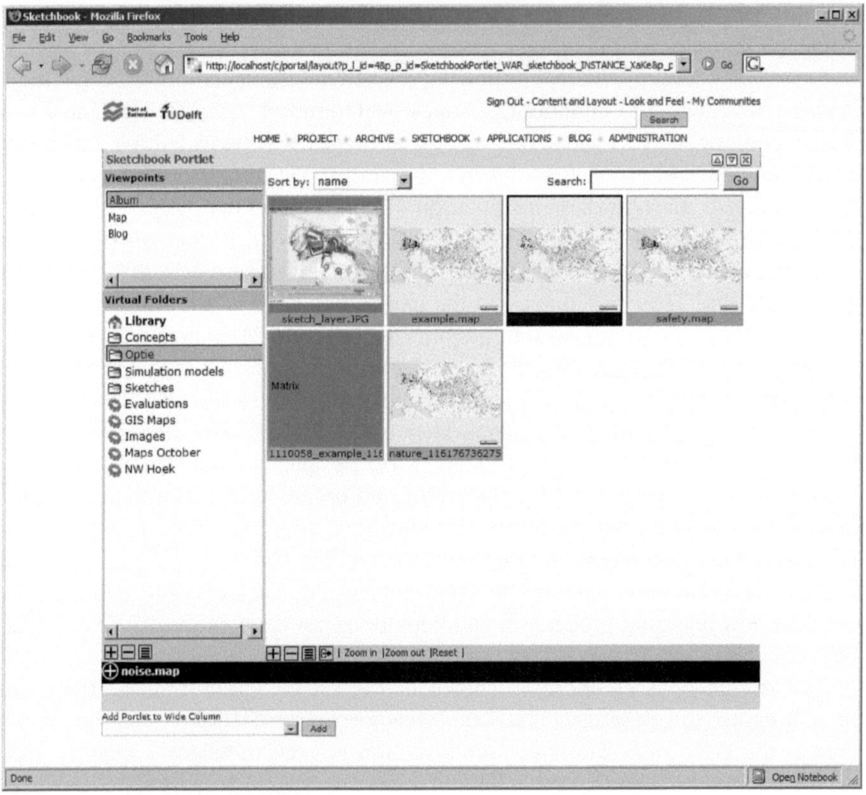

Fig. 20.5 The *Sketchbook* interface

20.4.3.3 Matchbox

Matchbox aims to support the evaluation of alternative solutions in mainport planning. It has been designed in close cooperation with the mainport planners in the PoR and visualizes the status of a multi-disciplinary evaluation process in which numerous aspects need to be compared. *Matchbox* supports different domain experts in the evaluation of a large variety of aspects and alternatives. *Matchbox* supports both quantitative and qualitative considerations in the evaluation of alternative-aspect combinations. The knowledge and information to make an evaluation is related to both possibilities and preferences. In some cases there are hard quantitative constraints which may not be violated by an alternative solution. Sometimes, counter measures can be taken to overcome hard constraints. Yet in other cases, the constraints are more related to what is desirable instead of what is possible (or impossible), i.e. the constraints are more qualitative in nature. *Matchbox* supports a multi-disciplinary team of mainport planners in making a distinction between accepted and rejected alternatives. *Matchbox* visualizes how far an evaluation has progressed and where there are still open issues that need to be addressed.

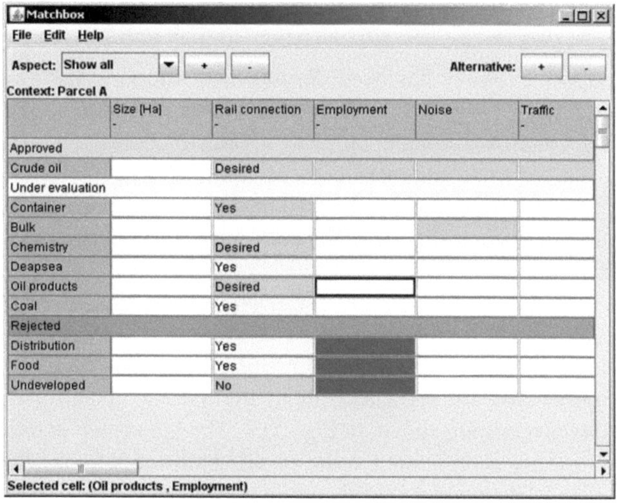

Fig. 20.6 Example screen in *Matchbox*

The design of *Matchbox* will be explained based on Fig. 20.6. Suppose an area planning team made an extensive analysis of a port area, and they came to the conclusion that a land parcel 'A' will be rented to a new customer. Now the team wants to evaluate possible industry types for parcel A. In *Matchbox,* they define the industry types as rows, and they define the different aspects under consideration as the columns. As such they specified a matrix of alternative industry types versus relevant aspects. Next, the individual experts can enter key values in the cells of the matrix. These key values indicate the expected requirements of each alternative-aspect combination. The values can be numeric values, ranges, or textual indicators. Next they compare the key values with the characteristics of land parcel A, and decide if there is a match between the required value and the respective characteristic value. A colour coding is used to visualize the result: green: possible or desirable; yellow: unclear/needs further research; red: impossible/undesirable; and white: not evaluated. An expert can also type 'behind' each cell comments as to why he or she applied a certain colour coding.

After the individual cells are evaluated the planning team can approve or reject alternative solutions. The evaluation of all aspects by a multi-disciplinary team is supported by three distinct categories: approved (alternative solutions that fit the context); under evaluation (alternative solutions that should still be evaluated); and rejected (alternative solutions that are rejected for the specified context). These categories are used to group the available alternatives. In the beginning of the evaluation, all alternative solutions are in the category 'under evaluation'. As the evaluation progresses and more information becomes available, the alternatives are either moved to the 'approved' category or moved to the 'rejected' category. This is done manually by a team member based on the outcome of a mainport planning team negotiation. Finally, no alternative solutions should remain under evaluation. Note that Matchbox only visualizes the status of an evaluation process. No weight factors are used, and no answer is provided by the tool itself.

20.4.3.4 Aspect Explorer

Aspect Explorer has been developed to support making a choice between viable alternative solutions. Whereas the outcome of *Matchbox* is a list of viable alternative solutions, a planning team still has to choose a preferred solution. When there is only a small number of viable solutions and relevant alternatives, the team can probably easily decide which solution they prefer. As the number of alternatives and aspects becomes larger, it becomes more difficult to make a choice among alternative solutions.

The design of *Aspect Explorer* is shown in Fig. 20.7. In essence *Aspect Explorer* is an exact implementation of Spence's neighbourhood explorer. Spence (2001) demonstrated how this visualization concept can be used to make a choice among houses that are for sale without the need to quantify the available information. Two screens are shown in Fig. 20.7. The left screen is used by an individual expert to specify a preference order of alternative solutions when considering a specific aspect. For example, as shown in the figure, a preference can be specified for 'lot size'. In this case the expert has a preference for 'distribution', followed by 'chemistry', 'container' and finally 'bulk' as his/her least preferred alternative. Next, the right screen shows how the alternatives compare based on the different preferences that were specified. The alternative solutions are presented on each axis. The alternative solutions are sorted top to bottom for each aspect with the most preferred alternative on top and the least preferred alternative at the bottom. When sliding one of the alternatives to the centre of the diagram, one can see how it relates to the other alternatives for each aspect.

Spence's visualization concept was chosen to support making choices in mainport planning because it provides a shared visual representation that allows for a purely qualitative comparison of alternative solutions. *Aspect Explorer* can be used to show the preferences of domain experts regarding viable alternative solutions in a single

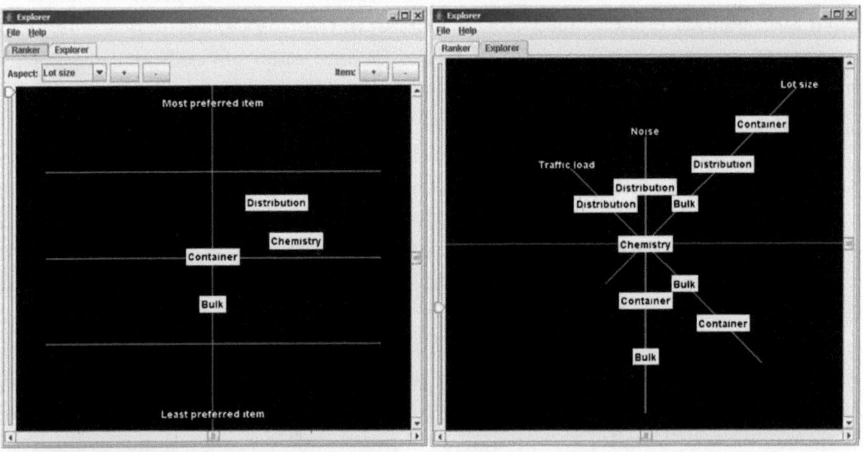

Fig. 20.7 Example interactive screens of *Aspect Explorer*

(interactive) image. The result is that it is possible to show specifically how the mainport planning team as a whole sees the solution space.

20.5 Conclusions and Future Directions

Based on the feedback, the *MPS* was perceived as a useful and innovative contribution to mainport planning (Chin *et al.* 2005, 2006). We conclude, based on the outcomes of the questionnaires, discussions and interviews that the *MPS* is expected to positively contribute to the awareness of the information used and produced in area planning. An *MPS* is expected to change the way of thinking about area planning in the port: it will lead to thinner, more to the point plans which can be updated in a 'rolling manner', i.e. at regular intervals. While planning outcomes are traditionally documented in a static planning report, it is expected that in the future advanced IT systems can be used to dynamically query specific and up-to-date planning data. Experts in the PoR clarified that, in their opinion, an *MPS* will become a 'memory of the organization'. Furthermore the participants of the evaluation sessions said that the *MPS* was also useful in planning projects other than area planning.

Designing an easy to use *MPS* is a balancing act between introducing structure and maintaining creative freedom. On the one hand, a *MPS* should support the structuring and archiving of the information which is used and produced during the planning process, but on the other hand, an *MPS* should also stimulate the creativity needed to find alternative solutions. The participants of the evaluation session said that they felt that the bureaucracy introduced by an *MPS* could potentially harm their creativity. However, the same participants also recognized the need for a more structured and rational approach towards area planning. Thus the support offered by an *MPS* should be carefully tuned to the roles of individual mainport planners, i.e. a secretary should be provided with different services than, for example, a GIS analyst.

Another issue with respect to the ease of use of the *MPS* solution deals with the available hardware. During the evaluation session, laptops and projectors were used to support the area planners. Laptops and PCs are typically designed to support the interaction between a single human and the machine, and not a group of people. To overcome this limitation, projectors were used to present information to the whole team of area planners. However, the participants indicated that they felt that they lost touch with each other, i.e. body language and face-to-face conversations, when watching a projection on the wall. A future trend which may help to overcome these limitations is the recent interest in multi-touch displays. For example, *Microsoft*® recently introduced *Surface*™ computing, which uses a computer in the shape of a coffee table where the table top is a large multi-touch display. Hence, the users do not sit in front of a monitor, but they sit around a display that is integrated in a table. Furthermore the display is touch-sensitive, meaning that users can use their fingers to click on, or drag displayed items. Such new types of human-computer interface will be more suitable for supporting teams of people during complex planning processes.

The iterative design approach, in close cooperation with mainport planners and experts in the PoR, positively contributed to their attitude towards using of the *MPS*. The response team at the PoR could provide quick insights into best practices, open doors to departments that would otherwise remain closed, interest people throughout the organization in studio-based mainport planning, and provide feedback on design ideas and prototype implementations.

Based on this research, a number of future research directions can be identified. First, mainport planning depends, for a substantial part, on data which is produced by computer models, e.g. simulation or analytical models. These models are managed and executed by individual experts inside departments, while during an interdisciplinary team meeting only the outcomes of the models are made available. Modelling in real-time, in a planning meeting, still remains a challenge for future research. Second, it was mentioned earlier that an *MPS* could become a system that, instead of a static planning report, can be queried by different actors for specific and up-to-date information. This raises the question if it is possible to automatically generate useful reports and presentations of the latest planning data available. In other words, it now takes month of work to create a planning report as the outcome of a planning project. Would it be possible to dynamically generate useful reports while planning is being conducted? Third, a structured evaluation session was used to evaluate the *MPS* prototypes, but we did not achieve an institutionalization of the *MPS* in the PoR during this research. Although a substantial amount of new insights were gained from this research, further research is still required to study how an *MPS* can be used on a daily basis, and to evaluate its long-term contribution to mainport planning projects. Finally, several participants at the PoR mentioned that an *MPS* could prove to be useful in projects other than area planning; hence we conclude that further research is needed to investigate the generalizability of the *MPS* solution.

Acknowledgement We want to thank the Port of Rotterdam for contributing to this research project. This research was not possible without their support and cooperation.

References

Babeliowsky, M.N.F. (1997) Designing interorganizational logistic networks: a simulation based interdisciplinary approach, Doctoral Thesis, Delft University of Technology, Delft.

Chin, R.T.H. (2007) Mainport Planning Suite: software services to support mainport planning, Doctoral Thesis, Delft University of Technology, Delft.

Chin, R.T.H., Smits, J.J., Verbraeck, A. and Weststrate, J.W. (2006) Area Planning Studio, a decision enhancement studio to support area planning in the Port of Rotterdam, Port Research Centre Rotterdam-Delft, Delft.

Chin, R.T.H., Houten, S.P.A. van, Verbraeck, A., Smits, J., Veenstra, P., Weststrate, J.W. and Schalkwijk, E. (2005) A decision enhancement studio for area planning in the Port of Rotterdam, Port Research Centre Rotterdam-Delft, Delft.

Hammer, M. (1990) Reengineering work: don't automate, obliterate, *Harvard Business Review*, July-August: 104–112.

Hengst, M. den and Vreede, G.J. de (2004) Collaborative business engineering: a decade of lessons from the field, *Journal of Management Information Systems*, 20(4): 85–114.

Hevner, A.R., March, S.T., Park, J. and Ram, S. (2004) Design science in Information Systems research, *MIS Quarterly*, 28(1): 75–105.

Keen, P.G.W. and Sol, H.G. (2007) *Decision Enhancement Services*, Forthcoming.

Keller, A.I., Hoeben, A. and Helm, A. Van Der (2006) Cabinet: merging designers' digital and physical collections of visual materials, *Personal and Ubiquitous Computing*, 10(2-3): 183–186.

Keller, A.I., Stappers, P.J. and Vroegindeweij, S. (2004) Supporting informal collections of digital images: organizing, browsing and sharing, In *Proceedings of the csonference on Dutch directions in HCI*, Vol. 65, Amsterdam.

Lugt, R. Van Der (2001) Sketching in design idea generation meetings, Doctoral Thesis, Delft University of Technology, Delft.

Maghnouji, R. De Vreede, G.J. Verbraeck, A. and Sol, H.G. (2001) Collaborative simulation modeling: experiences and lessons learned, In *Proceedings of the 34th Annual Hawaii International Conference on System Sciences*.

Russel, D.M., Streitz, N.A. and Winograd, T. (2005) Building disappearing computers, *Communications of the ACM*, 48(3): 42–48.

Schalkwijk, E. (2005) *A process design and tool-suite to enhance matching in the Port of Rotterdam*, Master thesis, Port Research Centre, Rotterdam-Delft, Delft.

Smits, J., Veenstra, P., Weststrate, J.W., Chin, R.T.H., Van Houten, S.P.A., Verbraeck, A. and Schalkwijk, E. (2005) *A challenge in port planning and design, Towards an Area Planning Studio for the Port of Rotterdam*, International Conference on Port-Maritime Development and Innovation, Rotterdam.

Sol, H.G. (1982) *Simulation in information systems development*, Doctoral Thesis, University of Groningen, Groningen.

Spence, R. (2001) *Information Visualization*, ACM Press, Essex.

Steenbergen, C., Mihl, H., Reh, W. and Aerts, F. (1999) *Architectural Design and Composition*, TU Delft, Faculty of Architecture, THOTH Publishers, Bussum.

Additional Reading

Carlsson, C. and Turban, E. (2002) DSS: directions for the next decade, *Decision Support Systems*, 33(2): 105–110.

Coutaz J., Crowley, J.L., Dobson, S. and Garlan, D. (2005) Context is key, *Communications of the ACM*, 48(3): 49–53.

Davis, F.D. (1989) Perceived usefulness, perceived ease of use, and user acceptance of information technology, *MIS Quarterly*, 13(3): 319–340.

Geertman, S. (2002) Inventory of Planning Support Systems in planning practice: conclusions and reflections, In *Proceedings of the 5th AGILE Conference on Geographic Information Science*, Palma, pp. 1–5.

Oosterhuis, K. (2003) *Hyperbodies: Towards An E-motive Architecture,* Birkhäuser – Publishers for Architecture, Basel.

Chapter 21
Developing Computer-Based Participatory Approaches to Mapping Landscape Values for Landscape and Resource Management

Steve Carver, Alan Watson, Tim Waters, Roian Matt, Kari Gunderson and Brett Davis

21.1 Introduction

The last 50 years or so have seen a steady increase in the rate of destructive wildfires across the world, partly as a result of climate change and partly as a result of encroachment of human settlement on fire-based ecosystems (Russell *et al.* 2004; Westerling *et al.* 2006). Years of active fire suppression in such areas has inevitably led to the build-up of hazardous fuel loads, creating ideal conditions for destructive wildfires (Johnson *et al.* 2001). Recently, serious wildfires have occurred in Australia, southeast Asia and the Mediterranean, as well as those occurring in the USA in California, Montana, Idaho and Alaska. Current thinking on fire management is very much focused on re-instating natural fire regimes and allowing fire, as nearly as possible, to function in its natural ecological role (Miller 2006), thereby reducing the occurrence of destructive fires. Mechanical fuel treatments (e.g. thinning) and prescribed burning are being used to reduce fuel loads to near natural conditions, after which natural fire regimes can be allowed to operate. There are two main types of thinning that either remove selected trees to create a more widely spaced forest consisting of trees of different sizes/ages or remove all smaller trees and brush within the understory to leave a more uniform forest of more widely spaced older trees. Prescribed burning uses small managed fires, rather than mechanical means, to achieve the latter. This is a long and involved process and often has the potential to create conflict between the different management regimes associated with adjacent lands and between the different inhabitants and stakeholders affected in the short to medium term. This requires a high degree of collaboration and participatory planning if acceptable fuel reduction strategies and management plans are to be developed.

S. Carver (✉)
School of Geography
University of Leeds
Leeds LS2 9JT, UK
E-mail: s.j.carver@leeds.ac.uk

S. Geertman, J. Stillwell (eds.), *Planning Support Systems Best Practice and New Methods*,
DOI 10.1007/978-1-4020-8952-7_21, © Springer Science+Business Media B.V. 2009

Throughout the western USA and in Alaska, rural communities and federal, state and local agencies are working cooperatively to initiate plans for reducing risks associated with the build-up of hazardous fuel on forest and grass lands. Most of the emphasis of this collaborative planning is on reducing these risks in the wildland/urban interface, where public lands and private, developed lands meet (Gunderson 2006). Another interface that is critically important in accomplishing fuel treatment and fire management objectives is the interface between wilderness and non-wilderness lands. In designated wilderness, the use of wildland fire, including prescribed burns, can help restore the natural role of fire as well as create a buffer along the wilderness boundary to protect non-wilderness values associated with adjacent lands.

This chapter describes a participatory process and methodology that has been developed to help capture personal and community meanings (or values at risk) associated with different landscapes and uses that are otherwise difficult to document. The case study is described in detail together with initial results from a participatory planning process. This is a project conducted collaboratively between the Confederated Salish and Kootenai Tribes (CSKT), the Aldo Leopold Wilderness Research Institute (ALWRI), the University of Montana (UM) and the University of Leeds (UoL). It represents Phase 2 of a three part study to incorporate personal and community meanings associated with a buffer zone for the planning and potential application of fuel treatment and fire management.

The aims of the project are to:

- contrast the meanings that tribal and non-tribal residents place on the protected landscapes of the Tribal Wilderness and the Tribal Buffer Zone within the Flathead Indian Reservation with the ultimate goal of using this local knowledge to better inform forest fuel reduction strategies implemented by the Tribal Forestry Department; and
- map these meanings for the Tribal Buffer Zone to understand the intensity and spatial distribution of individuals and communities associated with this controversial landscape and to describe how the potential application of fuel treatments in this wilderness/non-wilderness interface may affect them.

The objectives of this chapter are to review the development and use of GIS-based participatory mapping tools within this case study and use the results to better understand these values and meanings associated with Tribal lands within the Mission Mountains Tribal Buffer Zone. From the Phase 1 qualitative research, five layers of meanings were mapped: wildlife and water values; access and functional benefits; wilderness protection; personal and cultural values; and recreational and scenic values. This chapter discusses the efficacy and utility of such an approach in obtaining a range of views from diverse cultural backgrounds and in capturing local knowledge with a view to evaluating its wider applicability to similar problems in other locations.

21.2 Background

On the Flathead Indian Reservation in Montana, an interface area exists between the Mission Mountains Tribal Wilderness and the non-wilderness valley floor. The area, formally protected as a 'buffer zone' (referred to as the Mission Mountains Tribal Buffer Zone), was established in 1987 in the foothills of the Mission Mountains with the intention of protecting the wilderness from outside influences; primarily human civilization and development on the valley floor. The interface area provides opportunities for multiple uses corresponding to both wilderness and non-wilderness, including cultural uses, home sites, recreation, and other consumptive and non-consumptive resource uses. Figure 21.1 shows a map of the Mission Mountains Tribal Wilderness and the Buffer Zone. Figure 21.2 provides a glimpse of this Rocky Mountain landscape.

One of the primary concerns within the study area is the effect that fire suppression has had on the structure and general health of the forest within significant portions of the buffer zone. The widespread policy of fire suppression throughout much of the twentieth century in the U.S., as popularized by the U.S. Forest Service mascot 'Smokey Bear', has led to a forest ecosystem in both the Tribal Wilderness and the Tribal Buffer Zone that exhibit heavy and hazardous accumulations

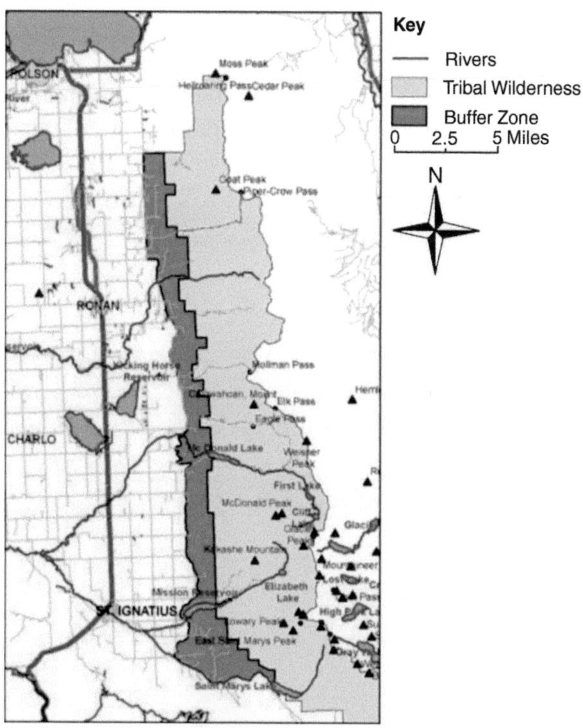

Fig. 21.1 Map of the Mission Mountains Tribal Wilderness and the Buffer Zone

Fig. 21.2 View of the Mission Mountains

of dead wood and down timber on the forest floor, a dense understory of brush and young trees, and closely spaced trees and closed forest canopy (Fig. 21.3a) (CSKT 2005c).

Such forests are increasingly regarded as 'unhealthy' and highly susceptible to destructive wildfires, as well as diseases such as mountain pine beetle *Dendroctonus ponderosa*. In contrast, the traditional culture of local tribes included the use of fire as a tool to nurture and sustain the natural resources of the land. Tribal people used fire on a vast scale, applying it to the land at different times of the year for various reasons. This included the thinning of dense brush in forested areas to keep them 'clean' for wildlife and hunting, clearing of trails for travel-

Fig. 21.3 (a) 'Unhealthy forest' and (b) 'clean forest' maintained by mechanical thinning

ling, and nurturing of food and medicinal plants for subsistence use (Fig. 21.3b). In the nineteenth century, native use of fire began to be altered as European settlement expanded into the West, and by the early twentieth century, following some devastating fires in 1910, the traditional use of fire by tribal people was largely eliminated. This greatly impacted tribal knowledge of traditional use of fire with the result that tribal resource managers now employ modern fire management, though under the guidance and direction of the cultural values ascribed

Fig. 21.4 Smokey Bear with tribal elder (U.S. Forest Service)

to the land that have been passed down through the generations (CSKT 2005a) (Fig 21.4).

The current focus of resource managers is on how to reduce potentially hazardous fuel loads in the Tribal Buffer Zone in order to address "*emerging issues of public and firefighter safety, reduce the risk of large high intensity fires, and restore the forests to a more biologically balanced condition*" (CSKT 2005b, p. 1). However, past proposals to reduce hazardous fuel loads in the Buffer Zone have failed due to lack of support among tribal members, especially for commercial thinning operations. The current process, therefore, uses participatory methods adapted from a previous project in the adjacent Bitterroot National Forest (Gunderson *et al.* 2004) and expanded to use web-based mapping methods developed in the UK (Carver *et al.* 2001; Evans and Waters 2007) to better capture how tribal and non-tribal Reservation residents view the Tribal Buffer Zone and how they might respond to new fuel treatment proposals in this area.

21.3 Participatory Approaches to Mapping Landscape Values

One of the key problems in developing a better understanding of different responses to resource management proposals, such as the fuel treatments dealt with here, is being able to faithfully record and accurately spatially delineate the meanings that stakeholders ascribe to the landscape in question. This needs to be done in a consistent and repeatable manner if cross-comparisons are to be made between different stakeholder groups and/or different areas.

Previous work on mapping landscape values and meanings relative to landscape-level fuels treatment has been carried out in cooperation with the Bitterroot National Forest. The Bitterroot Ecosystem Management Research Project (BEMRP) conducted a baseline assessment of individual and community meanings attached to the Bitterroot landscape using a rapid appraisal methodology (Gunderson 2006; Gunderson and Watson 2007). Semi-structured interviews, key informant interviews and focus group interviews were conducted to encourage people to talk about their local landscape, to differentiate between the meanings associated with places they commonly went to and those they have seldom or never visited, and to discuss how these meanings interact with alternative fuel treatments. Critically, as part of these interviews, respondents were asked to spatially define these areas and the meanings attached to them by drawing on paper base maps. These were then digitised and imported into GIS for analysis and integration in decision making along with existing ecological modelling efforts used to evaluate alternative fuel treatments on the Bitterroot landscape. Example output maps from the Bitterroot project are shown in Fig 21.5. This has allowed managers to better understand the juxtaposition of both social and ecological/resource data in the development of better and more acceptable fuel treatments.

Being able to actually map and compare the different meanings people place on the landscape in this fashion has a number of advantages over more basic place-based techniques of rapid appraisal. These include the ability to link meanings to specific locations or landscape units, and perform advanced analyses on responses by looking at spatial relationships based on proximity, adjacency, containment, connectivity and visibility. An area index describing the overall importance of a particular landscape unit is defined by the BEMRP methodology as a summative combination of scores based on specificity of area (broad, medium or specific), site replication (the number of times different respondents select the same site), frequency of visit, and level of importance as specified by the respondents themselves (Fig. 21.6). This has been used to identify areas with high levels of significant meanings attached by local community members.

Despite the success of the BEMRP work, a number of shortcomings were identified in the mapped-based aspects of the methodology adopted. These include the varying degrees of detail used by respondents when delineating areas on the paper map and the labour intensive nature of the digitising process (Carver *et al.* 2005). While the latter is just a matter of access to sufficient time and resources (i.e. labour), the former issue is more fundamental in nature and concerns how people actually conceptualise their environment. An individual's relationship with their local landscape is essentially fuzzy and cannot be easily captured using traditional map-based features or entities such as points, lines and polygons. Rather, people usually talk about places they know, use or have visited either by name (i.e. place-based nomenclature) in more vague terms such as 'the head of valley beyond the lake' or 'the woods out the back of my acreage' (Waters and Evans 2007). Where these places begin and end is hard to define in definitive terms and therefore is not easily captured and incorporated within the strictures of Cartesian-based GIS data models. Inevitably, when given a map and a pencil and asked to define areas which have value or meaning, respondents will adopt a variety of approaches ranging in scale and detail from broad sweeping circles that indicate an approximate area to small, crisply outlined regions that attempt to follow the landscape as closely as possible.

In order to address these issues, the current project adopts more fuzzy methods of capturing the landscape areas that people value or for which they hold a particular meaning. This is based around the application of a Java-based mapping Applet called 'Tagger' that uses a spray can tool, similar to that found in most desktop image processing/manipulation packages, to allow users to define areas over a base map in a manner that allows them to easily vary the density, extent and shape of the sprayed area (Evans and Waters 2007). This is used to capture information about fuzzy spatial concepts such as vagueness and approximation in defining spatial pattern and extent, as well as (un)certainty and importance in the relative values and meanings attached to these. The system can be used both online over the internet and offline on a standalone laptop facilitated by a member of the research team. Individual spray patterns are logged by the system together with their associated text describing relative values and meanings and converted to raster GIS maps for subsequent combination and analysis.

21.4 Methodology

An expanded rapid appraisal technique (Beebe 1995) is used to direct three phases of data collection. Each phase is guided by a specific objective as follows: In Phase 1 key knowledgeable informant interviews were used to develop an understanding of

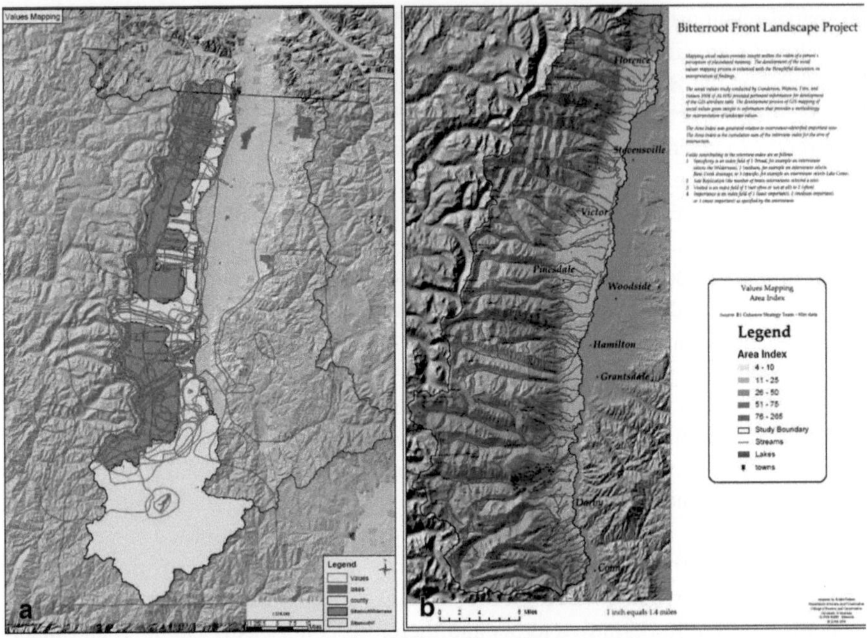

Fig. 21.5 Example of (a) a digitised polygon net and (b) a composite map from the BEMRP work (See also Plate 56 in the Colour Plate Section)

the range and types of values attached to the Mission Mountains landscape (focussing both on the Tribal Wilderness and the Tribal Buffer Zone for contrast) by tribal members and non-tribal residents on the Flathead Indian Reservation (Watson *et al.* in press). In Phase 2, participatory GIS methods were used to visually illustrate the range, intensity and spatial distribution of the meanings differentiated in Phase 1 within only the Buffer Zone. In Phase 3, focus group interviews will be used to

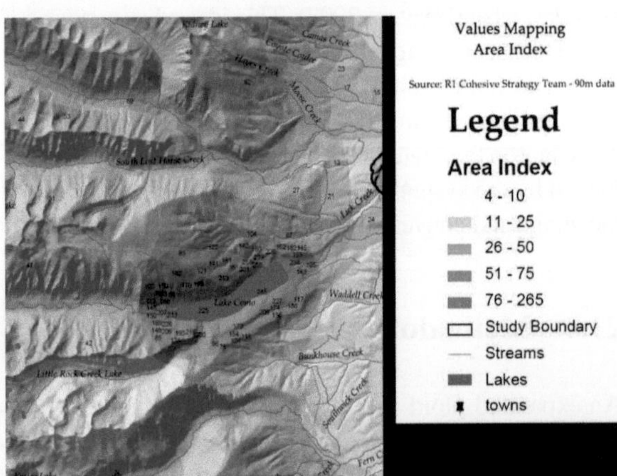

Fig. 21.6 Area index for Lake Como, BEMRP project (See also Plate 57 in the Colour Plate Section)

better understand the interaction between mapped landscape meanings from Phase 2 and potential fuel treatment options at the wilderness/non-wilderness interface.

21.4.1 Phase 1: Key Informant Interviews

Semi-structured interviews were conducted with tribal members and non-tribal Reservation residents to solicit information on the range and types of meanings associated with this landscape with varying levels of protection (Wilderness and Buffer Zone). Following the work of Lewis and Sheppard (2005), key informants were selected to meet the following criteria:

- knowledge, understanding and appreciation of traditional tribal and non-tribal meanings associated with the landscape;
- roles in the community that require a wide exposure to a range of tribal and non-tribal perspectives and perceptions of the landscape; and
- ability to communicate with outsiders and ability to discuss relevant research issues in detail.

Twenty-two interviewees were asked to provide information about the range and type of meanings attached to the landscape in interviews scheduled and conducted at the convenience of the interviewee. These analyses and results are described in detail by Watson *et al.* (2007; in press).

21.4.2 Phase 2: Landscape Mapping

In this phase, the objective is to understand where local meanings associated with the landscape are actually located or the regions/places they are attached to, together with the intensity of attachments. This phase is guided by the findings from the Phase 1 interviews, with the emphasis on capturing the spatial patterns in the meanings using fuzzy rather than traditional discrete mapping methods. The actual methods used here to capture spatially fuzzy regions and their ascribed attributes draw strongly on previous work on mapping place meanings (Gunderson and Watson 2007) and on participatory GIS (Carver *et al.* 2001; Waters and Evans 2007). These methods are brought together in developing a fuzzy GIS-based tool for collecting qualitative, but spatially referenced, local knowledge and meanings from a range of key informants and local people representing both tribal and non-tribal residents and stakeholder groups. These are analysed by creating composite maps of the fuzzy attribute-tagged maps generated by survey respondents and linking these to more in-depth interview transcripts from the Phase 1 key informant interviews. The result of this phase of the project is a GIS dataset (and hard copy maps) that provides a visual representation of the range, types, intensity and spatial distribution of the meanings associated with the Buffer Zone. Where possible, the same interviewees from Phase 1 of the project are used, with additional input from other residents within the region via an online version of the system. Whereas it is believed possible to reach saturation in regard to the range and types of landscape meanings through 15–20 key informant interviews, research by Gunderson (2006) would suggest that greater numbers of Phase 2 respondents are required to capture the richness and variation in the spatial distribution of these meanings across the whole of the study area. Initially, 34 respondents were solicited to test the system for possible errors and software problems, with some of them using the online method and some completing the exercise offline on a stand-alone laptop computer. In both cases, everything appeared exactly the same to the subject, rather it was just the method of recording and storing their results that differed. All offline exercises were assisted by one of the research team, sometimes to make it easier for high priority participants to engage and sometimes due to deficiency in computer skills or internet access.

Data were collected in a way that generated five map layers of themed meanings. These were driven by the Phase 1 findings, and collected to represent the meanings of the Buffer Zone for themed topics covering 'wilderness protection', 'wildlife and water values', 'recreation', 'access', and 'personal and cultural meanings'. For demonstrative purposes, this paper focuses on 154 images developed by these 34 respondents across the five themed layers. For the overall map and categorization process, these 154 images are averaged and images produced using classes based on natural groupings inherent in the data with break points identified by picking the class breaks that group similar responses and maximize the differences between classes using the ArcGIS software (Jenks 1967). For further insight, the five individual meaning layer images are also presented, based on the same classification scheme used to develop the 'overall mean' image. Whilst conflicts between respondents are handled at this stage by simple averaging, future research will look at areas where conflicting opinions exist with particular reference to any potential tribal and non-tribal differences.

21.4.3 Phase 3: Focus Group Interviews

The objectives of the third phase of the project are to present the GIS representations outlined above and collaboratively evaluate these with tribal members and non-tribal residents in regard to the potential effects of different fuel treatment methods in the Tribal Buffer Zone. At the time of writing, this work is still ongoing. At least two focus group interviews will be conducted facilitated by the research team members and the Tribal liaison. The information gathered from these focus groups will be used to understand how the meanings associated with the landscape will likely interact with potential application of fuel treatments in the Buffer Zone. Phase 3 also serves the very positive purpose of informing the public that their views are valued and are being incorporated into the planning process thereby helping to break down one of the main barriers in accomplishing resource management objectives.

21.4.4 Development of the Landscape Mapping Tool and User Interface

The main user interface is based around the 'Tagger' fuzzy area definition software developed by Evans and Waters (2007) as a Java applet running within a standard web browser environment. This uses a spray can tool that will be familiar to many users from popular desktop image editing packages, and with which a user can define fuzzy areas of varying density on a map. Variations in intensity or importance can be easily made by spraying more or less over an area, making it correspondingly darker or lighter. Attribute information can be attached to the fuzzy area through the use of free-format text input boxes. The basic interface is shown in Fig. 21.7 and consists broadly of:

- a spray can tool that enables the user to spray fuzzy areas on a map;
- Radio buttons to select small or large spray sizes;
- an Erase button to remove any spraying from the current map;
- a New Area button to save the current sprayings and allow the user to spray another area;
- a Paint All button that opens up a dialog window asking for a number from 0-100 per cent to uniformly spray over the whole area;
- a Send Everything button that sends all the saved sprayings and attached attributes to the server; and
- text areas where the user is asked to type about 'Why the areas are important', and 'What are the threats to these areas?'

Users progress through the application in a linear fashion as follows:

1. Welcome and outlining of the tasks and any technical requirements needed.
2. Background information about the project, fire and fuel treatments and the area of the Buffer Zone.
3. Introduction to the mapping system, showing the graphical user interface, explaining concepts such as 'The longer you hold down the mouse button the darker it gets. The darker it gets means it has more importance' showing examples, and how users can use the mapping tool to indicate intensity (Fig. 21.8).
4. Practice page, which helps people become familiar with the spray can mapping tool using a map of the USA and asking participants to 'Spray the areas that are the warmest'—and in the text areas type 'Where is the hottest area you've been to?' and 'How hot was it?'
5. Questionnaire, with questions about age group, gender, residency of the Flathead Indian Reservation, tribal membership, zip code, and a question asking if they were doing the exercise individually, or in a group.
6. After the questionnaire, a short explanation of the task is given i.e. 'Please show on the next map those areas that are most important to you for either personal or cultural reasons, for example, childhood memories, family outings, memorable experiences, ancestral ties, and/or spiritual and symbolic activities'. An option is given to skip a map, if desired.

7. Maps were given for each category:
   - Wildlife habitat and water quality
   - Access and functional benefits
   - Protection of the wilderness
   - Personal and cultural
   - Recreation and scenic values
8. Each user is given these maps in random order, to help reduce any bias.
9. Participants are asked to spray on each map showing where the most important areas to them were and write in the text areas 'Why these areas are important?' and 'What are the threats to these areas?'

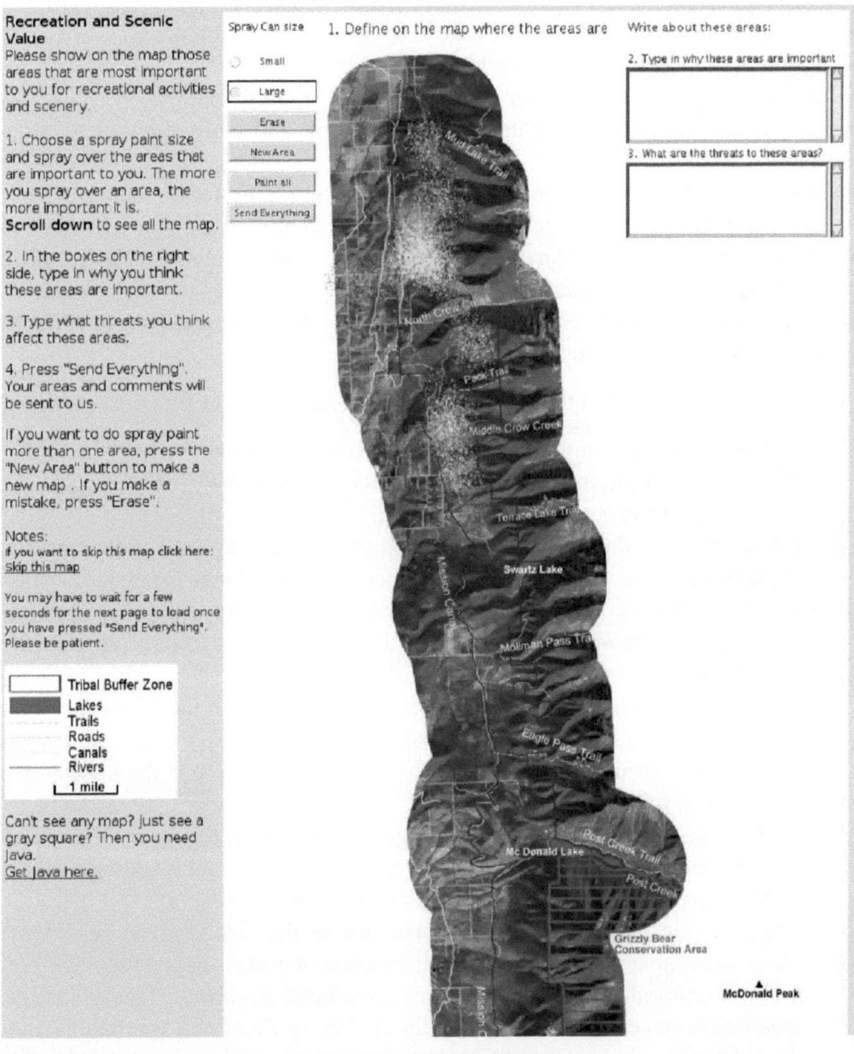

Fig. 21.7 Landscape mapping tool interface (See also Plate 58 in Colour Plate Section)

10. After completing all the maps, a thank you page is shown together with further details about the project.

Each sprayed area entered by the user is processed by the Tagger software into a standard image format (gif and GeoTIFF) and compressed. The image and associated attribute information are stored, and can be viewed either as an individual entity or combined into an aggregate 'average' map based on all the users' responses. The Tagger system also includes a query functionality, where researchers can click on the composite image of everyone's responses and query this area for the text associated with that area, weighted by intensity. Additional administrative tools include the ability to examine, delete or edit individual maps and textual responses. The Tagger code is open source and may be freely used by academic and non-profit, non-governmental organisations, as well as private individuals (see http://www.ccg.leeds.ac.uk/software/tagger/).

21.5 Results

The results from the Phase 2 mapping exercise are shown in Fig. 21.9 grouped into the five themed topics and including the composite of all meanings for all users. These data are preliminary, representing the initial effort at obtaining public participation in this process. A mixture of online (web-based) and offline (laptop-based) participants were included and results monitored to assure quality of presentation, data capture and to plan more complex analyses that make the distinction between tribal and non-tribal groups as well as standard geo-demographic differences of age and gender. After this preliminary data collection with the 34 participants, the wider public was invited to participate in this process via the local media including announcements in the local weekly newspaper and on the tribal government intranet web site appearing on every employee's start-up login page.

It can be seen from the averaged maps in Fig. 21.9 that there are marked differences in the spatial pattern and intensity of meanings associated with the five themes. It should be noted that in order to aid comparison, the same classification and colour scheme is used for all five theme maps and the composite map. It should also be noted that whilst the focus of the mapping exercise is the Buffer Zone, users were not limited to this area by the Tagger software and can spray extensively outside of the Buffer Zone boundary. Marked edge effects can

Fig. 21.8 Spray pattern examples

therefore be seen in some maps such as for the wildlife and water theme shown in Fig. 21.9b. Although the system records all spray patterns, those falling outside the Buffer Zone are largely ignored in the following analysis as they are outside the area of interest.

Of the five themes, wilderness protection (Fig. 21.9a) shows the greatest uniformity, perhaps as a result of the general view among respondents of the Buffer Zone as a policy zone with equal protection along its length. A significant proportion of users used the 'Paint All' button to apply a uniform spray density over the whole area for this theme to indicate value and meaning that is both broad in scale (covering the whole Buffer Zone) and not specific to any one area. The access theme (Fig. 21.9d) exhibits the greatest localized differences which is most likely due to the limited number of maintained trails within the Buffer Zone, with most people associating higher levels of importance to those areas along and adjacent to these trails. Hot spots within the recreation (Fig. 21.9c) and personal/cultural (Fig. 21.9e) theme maps reflect the nature of the access corridors shown in Fig. 21.9d and are likely to be driven by frequency of visits to popular sites such as McDonald Lake, Mission Reservoir, Saint Mary's Lake, Swartz Lake and North Crow Creek trail (Fig. 21.1). This said, it is clear that other more remote and inaccessible areas are also highly valued and have meaning for both recreational and personal/cultural use. The map for wildlife and water (Fig. 21.9b) shows perhaps the greatest spatial variation of all the five themes with users identifying areas not on their ease of access and frequency of use, but on their environmental credentials for wildlife and water protection. Many of the hotspots on this map relate to watersheds, rivers and lakes.

21.6 Conclusions and Further Research

This chapter describes work in progress on a new planning support system aimed at collecting spatial and contextual information about public perceptions of landscapes with an emphasis on developing better understandings of place-based values and associated meanings. The three phase approach based around key informant interviews, landscape mapping and focus group interviews demonstrates a rigorous approach to participation and data collection in developing publicly acceptable management strategies and actions in the face of multiple, spatially variable and sometimes conflicting viewpoints. The process described involves a mix of both face-to-face and automated methods of data collection using key informant interviews (Phase 1) to inform the development and application of automated methods of spatial and attribute data capture using map-based computer interfaces (Phase 2). Both initial phases are now complete and the project is currently in the final stage of focus group interviews to evaluate the resulting maps and suggest appropriate fuel treatments for the Buffer Zone area.

Experience from Phase 2 and the results gained suggest that this combined approach has been successful in capturing the rich picture of meanings attached

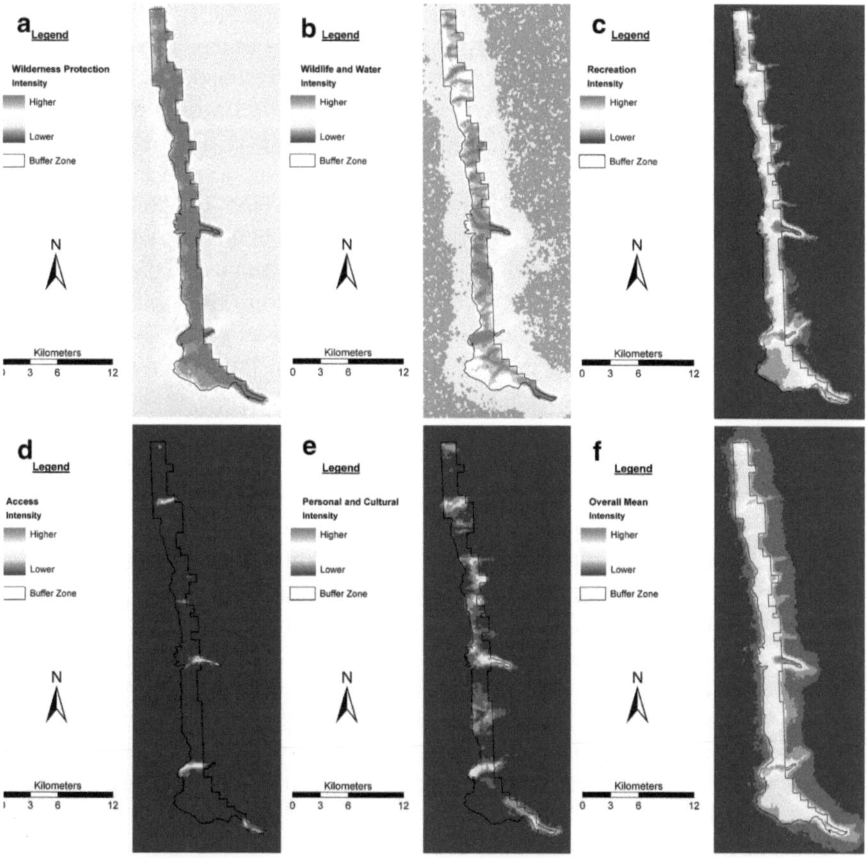

Fig. 21.9 Results for (a) wilderness protection theme, (b) wildlife and water theme, (c) recreation theme, (d) access theme, (e) personal and cultural theme, and (f) composite map for overall averaged results

to the complex landscape mosaic represented in the Buffer Zone. In particular, the combination of user defined fuzzy areas and qualitative data collection methods has enabled a wide range of personal and community meanings to be spatially delimited across a broad area in a manner and detail that would be impossible by verbal and/or manual means. Automated data collection and 24/7 availability of the online version of the system also means that the work can span a far greater audience than would be possible via traditional face-to-face or paper-based social survey methods, making detailed and disaggregated analyses (e.g. comparisons of tribal and non-tribal responses) possible with the higher numbers of respondents realised, even though it is understood that data collected online might not be as reliable as that collected face-to-face (Carver *et al.* 2001). In addition, the linkage of themed spray patterns and tagged attributes with user profiles makes detailed analysis of differences between different demographic and/or stakeholder groups possible. For example, in the final analysis, focus will likely be on tribal member responses, though understanding of non-tribal responses is also of interest to land managers and the Tribal Council. The resulting dataset and possibilities for further analysis represents an extremely rich resource for policy development in regard to landscape resource management that takes both public perceptions and the realities of the physical world into account.

The approach adopted is not without its problems however. The initial setting up of the mapping software requires customization for each application and a relatively high level of competence with web/Java programming. From the point of view of usability, experience suggests that there is a wide range in the levels of familiarity with IT among members of the public. Although independent use of the system only requires knowledge of how to use a web browser (i.e. simple mouse and keyboard skills), there is still a significant proportion of the population who have had little or no previous exposure to computers and the internet. Previous research in the field of web-based participatory approaches suggests that lower levels of computer literacy and internet access/usage are strongly associated with rural and/or older demographics (Carver *et al.* 2001). Work by Hoffman *et al.* (2001) suggests that noticeable differences exist between racial groups in urban areas that may extend to rural communities, but note that this is mainly a function of access to a computer and the internet within the home environment. While some barriers to internet access appear to persist in rural areas and on tribal lands, particularly in respect to culturally sensitive issues, Tribal Councils and Reservations are actively embracing digital technologies with relative enthusiasm (Gordon *et al.* 2003; Monroe 2002) and use of IT is increasing. Nonetheless, it is clear from the current research that many older members of the community, including tribal elders who are the holders of much important traditional knowledge, benefit from the use of facilitated offline systems where problems arising from unfamiliarity with IT or reluctance to use it can be avoided.

The case study described here has demonstrated that the approach is well suited to developing a better understanding of indigenous peoples' relationships with the land, and is particularly useful for contrasting meanings attached to different classifications of land, articulating what is worth protection and what it should be protected from (i.e., threats from development and alternative uses or values) and how it is viewed by people from different cultures and/or stakeholder groupings. In dealing with complex human/landscape relationships, we need in the twenty-first century to fully understand and appreciate the different orientations toward landscapes and wilderness in particular as the original or primary landscape type. One area of significant future research will be in how global climate change will impact on both forest ecosystems and the communities that rely on them. Methods such as those described here could usefully be adapted and expanded on to provide inputs to further our understanding of transition periods and allow society to proactively increase the resilience of society to wider changes in global climate and shifting ecosystems.

As a general conclusion, the positive experiences and results from the work described here would indicate that the approach adopted will find wider application across a range of spatial problems requiring an element of public participation involving vague or spatially fuzzy concepts. Examples might include studies of

NIMBYism, landscape, urban and regional planning and mental mapping, in fact, almost anything where there is a need to better understand the spatial and contextual elements of human relationships with the social, cultural, built and natural environment that do not fit neatly into the sharp, crisp world of traditional mapping methods. The Tagger software has already been successfully tested in projects aimed at mapping perceptions of crime and wilderness qualities (Evans and Waters 2007; Carver *et al.* 2005) and some aspects of the work described here are already being considered for use by other U.S. Forest Service projects and similar organisations elsewhere in the world. On a technical level, the Tagger software could be developed to work with commonly used web mapping interfaces such as *Google Earth* or NASA's WorldWind. This would make the tool more readily available to a larger audience and make it easier to incorporate into a wider range of applications.

Acknowledgements The collaboration between the Confederated Salish and Kootenai Tribes, the Aldo Leopold Wilderness Research Institute, the University of Leeds and the University of Montana was primarily supported by these institutions with funding from the U.S. National Fire Plan and the Bitterroot Ecosystem Management Research Project.

References

Beebe, J. (1995) Basic concepts and techniques of rapid appraisal, *Human Organization*, 54(1): 42–51.

Carver, S., Evans, A. and Watson, A. (2005) Mapping sense of place and associated wild land values using online GIS and fuzzy methods: some examples from Europe and the US, Paper presented at the 8th World Wilderness Congress, Anchorage, Alaska, October.

Carver, S., Evans, A., Kingston, R. and Turton, I. (2001) Public participation, GIS and cyberdemocracy: evaluating on-line spatial decision support systems, *Environment and Planning B,* 28(6): 907– 921.

Confederated Salish and Kootenai Tribes (CSKT) (2005a) *Confederated Salish and Kootenai Tribes Forestry Department Wilderness Buffer Zone Reclassification Environmental Assessment,* Report prepared by the Buffer Zone Interdisciplinary Team, Confederated Salish and Kootenai Tribes of the Flathead Indian Reservation, Pablo, MO.

Confederated Salish and Kootenai Tribes (CSKT) (2005b) *Fire on the Land: An Interactive Educational DVD,* Confederated Salish and Kootenai Tribes of the Flathead Indian Reservation, Pablo, MO.

Confederated Salish and Kootenai Tribes (CSKT) (2005c) *Ecosystem Management on the Flathead Indian Reservation* Confederated Salish and Kootenai Tribes Forestry Department, Pablo, MO.

Evans, A.J. and Waters, T. (2007) Mapping vernacular geography: web-based GIS tools for capturing "fuzzy" or "vague" entities, *International Journal of Technology, Policy and Management,* 7(2): 134–150.

Gordon, A.C., Gordon, M. and Dorr, J. (2003) Native American technology access: the Gates Foundation in Four Corners, *The Electronic Library,* 21(5): 428–434.

Gunderson, K. (2006) Understanding place meanings for wilderness: personal and community values at risk, *International Journal of Wilderness* 12(1): 27–31.

Gunderson, K. and Watson, A. (2007) Understanding place meanings on the Bitterroot National Forest, Montana, *Society and Natural Resources,* 20: 705–721.

Hoffman, D.L., Novak, T.P. and Schlosser, A.E. (2001) The evolution of the digital divide, In Compaine, B.M. (ed.) *The Digital Divide,* MIT Press, Cambridge, MA, pp. 47–97.

Jenks, G.F. (1967) The data model concept in statistical mapping, *International Yearbook of Cartography,* 7: 186–190.
Johnson, E.A., Miyanishi, K. and Bridge, S.R.J. (2001) Wildfire regime in the Boreal Forest and the idea of suppression and fuel buildup, *Conservation Biology* 15(6): 1554–1557.
Kimmerer, R.W. and Lake, F.K. (2003) The role of indigenous burning in land management, *Journal of Forestry,* 99(11): 36–41.
Lewis, J.L. and Sheppard, S.R.J. (2005) Ancient values, new challenges: Indigenous spiritual perceptions of landscapes and forest management, *Society and Natural Resources* 18: 907–920.
Miller, C. (2006) Wilderness fire management in a changing world, *International Journal of Wilderness,* 12(1): 18–21.
Monroe, B. (2002) The internet in Indian country, *Computers and Composition*, 19: 285–296.
Russell, G.T., McCaffrey, S.J. and Theresa, B. (2004) Science basis for changing forest structure to modify wildfire behavior and severity, *General Technical Report RMRS-GTR-120*, U.S. Department of Agriculture, Forest Service, Rocky Mountain Research Station, Fort Collins, CO.
Watson, A., Knotek, K., Matt, R. and Yung, L. (2007) *Understanding landscape meanings, attributes, and threats for the planning and application of fuel treatment and fire management on the Flathead Indian Reservation, Montana,* Report to the Confederated Salish & Kootenia Tribes of Montana, Aldo Leopold Wilderness Research Institute, Missoula, MO.
Watson, A., Matt, R., Knotek, K., Williams, D. and Yung, L. (In press) Traditional wisdom—protecting wilderness as a cultural landscape, *Proceedings Sharing Indigenous Wisdom International Conference*, Green Bay, Wisconsin, June 2007.
Westerling, A.L., Hidalgo, H.G., Cayan, D.R. and Swetnam, T.W. (2006) Forest wildfire activity warming and earlier spring increase Western US, *Science*, 313(5789): 940–943.
Wray, J. and Anderson, M.K. (2003) Restoring Indian-Set Fires to Prairie Ecosystems on the Olympic Peninsula, *Ecological Restoration* 21(4): 296–301.

Additional Reading

Confederated Salish and Kootenai Tribes (CSKT) (2007) *Fire on the Land*, Retrieved October 2, 2007, from www.cskt.org/fire_history.swf.
Confederated Salish and Kootenai Tribes (CSKT) (2005) *Mission Mountains Tribal Wilderness. A Case Study,* Confederated Salish and Kootenai Tribes of the Flathead Indian Reservation, Pablo, MO.
Craig, W.J., Harris, T.M. and Weiner, D. (2002) *Community Participation and Geographic Information Systems,* CRC Press, Boca Raton, FL.
Fritz, S., Carver, S. and See, L. (2000) New approaches to wild land mapping in Europe, *Proceedings of the Wilderness Science in a Time of Change Conference,* USDA Forest Service Proceedings RMRS-P-15-VOL-2 (2000) Missoula, MO, pp. 120–127.

Index

A
2D, 212, 262
3D, 13, 14, 25, 262, 319, 423
Accessibility, 6, 8, 31, 41, 54, 55, 57, 59, 123, 134, 199, 201, 204, 206, 207, 222, 227, 246, 277, 356, 370, 372, 399, 403
Aerial photograph, 257
Agent Based Modelling/ABM, 20, 178, 183, 188, 211–213, 219–226
Agroforestry, 318, 323, 343
Algorithm, 25, 65, 72, 100, 143, 166, 198, 199, 231, 240, 241, 350
Allocation, 32, 54–57, 63, 65, 73–75, 78, 86, 142, 166, 173, 179, 200, 234, 381
Analysis, 3, 12, 13, 16–22, 33, 63, 69, 94, 97, 116, 119–122, 138, 141, 150, 162, 167, 170, 172, 176–178, 186, 192, 201, 229, 230, 233, 237, 251, 298, 299, 308, 317, 321, 333, 347, 359, 371, 383, 396, 405
Animation, 10, 76, 421
Anticipation, 71, 135
Application, 1, 12, 14–22, 24, 27, 33, 54, 70, 119, 123, 128, 134, 146, 203, 211–213, 219, 220, 222, 223, 236, 249, 262, 271, 289, 298, 342, 349, 377–378, 399, 404, 432, 441, 444, 446
Application Programming Interface/API, 14, 73, 227, 236–239, 242
Appraisal, 162, 370, 386, 436, 437, 447
Appraisal methodology, 436
Assessment, 16–20, 22, 27, 30, 42, 127, 137, 160, 162, 168, 179, 191, 205, 222, 243, 250, 251, 262, 266, 287, 308, 317–323, 339, 385, 436
Atlas, 214, 228, 229, 237, 238, 302, 303
Awareness, 6–8, 10, 20, 167, 199, 207, 219, 220, 223, 224, 233, 313, 323, 332, 335, 341, 351, 352, 366, 371, 381, 427

B
Behaviour, 18, 20, 31, 67, 84, 102–104, 107, 109, 116, 121, 166, 181, 186, 191, 192, 195, 198, 201, 203, 205, 207, 227, 343, 348, 350, 352, 356, 390, 402, 405, 410, 448
Benefit, 39, 95, 101, 102, 107, 191, 198, 204, 205, 211, 302, 312, 364, 446
Best practice, 1, 7, 8, 10, 16, 315, 317, 431
Biodiversity, 315, 319, 335, 381
Bottleneck, 100, 126
Bottom up, 6, 33, 149, 225, 226, 397, 411
Built up, 73, 88, 176, 180, 182, 183, 185, 186, 272, 283, 365

C
Calibration, 17, 55, 56, 437
Case Based Reasoning/CBR, 31, 50
Case study, 23, 24, 70, 71, 73, 74, 131, 135, 189, 192, 203, 204, 225, 263, 315, 333, 340, 342–344, 348, 350, 352, 378, 387, 410, 414, 416, 417, 420, 432, 446, 448
Catchment area, 61, 239, 241
Cell, 31, 32, 54, 143, 145, 146, 166, 194, 196–201, 256, 304–306, 425
Cellular Automata/CA, 16, 17, 24, 31, 51, 55, 66, 67, 72, 84, 88, 116, 173, 178, 179, 211, 214, 263, 264, 290, 291, 349, 367, 399
Climate, 19, 50, 51, 72, 87, 89, 113, 114, 117, 126, 130, 133, 135, 156, 159–161, 164, 169, 170–174, 211, 213, 250, 276, 352, 357, 369, 386, 431, 446

Clustering, 100, 176, 231, 406
Collaboration, 9, 21, 23, 87, 117, 129, 135, 220, 293, 295–297, 317–319, 389, 390, 392, 415, 431, 447
Collaborative decision making, 315
Collaborative planning, 10, 26, 50, 88, 89, 179, 213, 296, 298, 314, 389, 390, 397, 407, 409, 410, 432
Collaborative Planning System, 10
Communication, 9, 11, 22, 71, 87, 119, 126, 127, 134, 195, 212, 214, 239, 271, 298, 343, 349, 363, 365, 370, 371, 384, 386, 388, 390, 397, 399, 406, 415, 416
Communicative planning, 296, 314, 406
Community, 5, 8, 9, 12–14, 21, 25, 70, 212, 213, 227, 232, 295–297, 318, 394, 432, 436
Competition, 219, 351
Complexity, 3, 4, 29, 50, 68, 69, 118–120, 123, 124, 126, 127, 156, 162, 178, 187, 189, 199, 213, 214, 223, 224, 226, 342, 364, 372, 386, 414
Comprehensive plan, 295, 299, 300, 302, 309, 310–314
Computational grids, 13
Computer Aided Design/CAD, 25
Concentration, 54, 58, 59, 61–64, 99, 108, 176, 269
Concept, 2, 3, 15, 21, 23, 76, 135, 136, 168, 189, 196, 197, 199, 207, 230, 243, 249, 250, 272, 273, 277, 281, 283, 284, 290, 295, 296, 300, 303, 309, 311, 344, 373, 388, 402, 414, 415, 420, 423, 426, 448
Conceptual framework, 5, 195, 205, 289
Conceptual model, 266, 267, 271, 273, 288, 289
Configuration, 44, 46, 114, 144, 155, 196, 278, 287, 353, 363, 376
Consensus, 18, 22, 33, 39, 49, 84, 87, 175, 220, 296, 297, 314, 385, 410
Constraint, 20, 31, 56
Continuity, 176, 417
Cost, 39, 43, 86, 94, 95, 101–104, 122, 125, 128, 135, 147, 154, 160, 203, 204, 236, 239, 304, 312, 387
Cost Benefit Analysis/CBA, 312
Coupling, 33, 40, 42, 49, 115, 121, 125, 126, 224
Crash, 18, 76, 93–102, 104–111
Crash Information System, 108
CrimeStat, 18, 98–100, 106, 110
Critical GIS, 10, 23, 390, 396, 411
Cultural heritage, 319

D

Dasymetric mapping, 143, 144, 150, 156
Data, 2, 3, 6, 8–15, 17, 18, 21, 30, 35, 65, 70, 73, 83, 93, 96–98, 105–108, 114, 121–126, 131–135, 141, 153, 192–198, 203–205, 213–215, 229–240, 244–248, 250, 265–273, 276–278, 281–284, 288–290, 303, 307, 318, 323, 339, 350, 360, 371–379, 381–385, 399–405, 408, 423, 428, 443–445
Data availability, 11, 219, 371
Database, 15, 19, 22, 96, 98, 137, 144, 163, 193, 195, 197, 205, 214, 234, 236, 243, 245, 246, 250, 251, 253, 262, 263, 266, 269, 271, 272, 277, 283, 371–377, 381, 384, 385, 390, 399, 403, 418
DataBase Management System/DBMS, 193, 195
Data integration, 10, 11, 323
Data model, 307, 448
Decision, 1–3, 9, 13, 18–20, 22, 24, 47–49, 73, 76, 82, 94, 114–118, 123, 124, 133, 162, 179, 180, 188, 191, 195, 203–206, 211–213, 216, 250, 262, 276, 296, 312, 338–342, 347, 366, 370–372, 379, 397, 415, 416
Decision making, 3, 9, 18–20, 22, 24, 71, 94, 114–116, 162, 177, 192, 195, 211, 213, 244, 297, 313–315, 317, 318, 347, 370, 371, 415, 416, 436
Decision support, 1, 2, 13, 20, 24, 25, 30, 76, 79, 114, 116, 135, 136, 157, 162, 179, 189, 191, 205–208, 250, 262, 276, 289, 312–315, 341, 344, 367, 372, 386, 411, 416, 429, 447
Decision Support System/DSS, 3, 19, 24, 113–116, 118–136, 157, 191, 193, 205–208, 276, 284, 285, 312, 367, 372, 387, 429
Demographic, 11, 13, 16, 32, 114, 179, 197, 198, 227–233, 238, 239, 271, 302, 318, 342, 353, 356, 359, 379, 381, 386, 443, 445
Density, 43, 57, 143, 156, 157, 176, 181, 186, 241, 274, 308, 339, 362, 403, 405, 437, 441, 444
Deprivation, 13, 234, 238
Design, 3, 17–20, 24, 32, 70, 102, 116–123, 127, 129, 134, 135, 137, 154, 155, 162–164, 169, 180, 183, 192, 193, 196, 197, 202, 203, 205, 206, 236, 252, 262, 265, 271, 275, 298, 318, 323, 340, 341, 385, 389, 395, 400, 403, 414–421, 423, 425, 426, 428

Developer, 5, 75–78, 86, 87, 128, 131, 214, 215, 221, 222
Diffusion, 5–8, 25, 26, 72, 235, 236, 239

E

Ease of use, 312, 427, 429
Ecological, 21, 22, 50, 58, 113, 116, 175, 226, 264, 317–323, 339, 343, 402, 410, 431, 436, 448
Economic, 8, 11–13, 19, 20, 22, 41, 44, 47–49, 84, 133, 138, 165–170, 175, 176, 179, 204, 211–213, 217, 219, 222, 285, 317, 319, 330–335, 352, 353, 356, 357, 359, 362, 364, 369, 370, 382, 413
Ecosystem, 277–279, 289, 321, 343, 367, 433, 436, 447
Embedded modelling, 38, 47
Enhanced Compressed Wavelet/ECW, 378
Environment, 5–9, 13, 17, 21–23, 53–57, 59, 60, 63–65, 74, 76, 109, 120, 124, 125, 132, 138, 166, 176, 179, 183, 188, 191–193, 195, 202, 205, 212–214, 219–222, 238, 245, 265–268, 271, 275–277, 279, 284, 286, 288, 289, 321, 333, 337, 347, 351, 356, 359, 363, 366, 369, 381, 389–392, 394–396, 402, 403, 405, 413–415, 437
Environmental planning, 20, 30, 33, 51, 116, 132, 190, 243, 249, 262, 263, 271, 291, 345
Error, 55, 255, 417
E-society, 233, 235, 238, 239, 241
ESRI, 11, 24, 50, 51, 73, 88, 137, 141, 156, 195, 240, 243, 244, 263, 264, 303, 315, 378, 386–388
Ethnicity, 229, 241
European Environment Agency/EEA, 156, 347, 350, 366, 367
European Union/EU, 11, 21, 113, 124, 135-137, 154, 157, 265, 271, 280, 319, 331, 343, 348–350, 352, 353, 363, 365, 367, 378, 389, 414
Evaluation, 2, 10, 11, 16–18, 20, 21, 24, 53, 55, 63, 64, 76, 80, 84, 87, 88, 124, 131–133, 176–178, 180, 184–188, 203–205, 211, 222, 235, 281–283, 286, 287, 289, 303, 309, 311, 335, 350, 370, 371, 385, 389, 396, 406, 417–419, 424, 425, 427, 428
Evidence base, 10
Evolution, 12, 17, 30, 31, 50, 66, 72, 149, 153, 175, 180, 181, 188, 317, 331, 340, 353, 372, 373, 381, 385, 392, 447
Experience, 2, 6, 7, 10, 20, 23, 30, 38, 49, 50, 73, 93, 95, 107, 108, 133, 134, 162, 219–221, 223, 230, 236, 243, 267, 288, 291, 307, 320, 335, 337, 338, 349, 357, 385, 398, 403, 408, 444, 446
Experiential knowledge, 390, 396–404, 409
Experiment, 9, 207, 217, 223, 264, 288, 289
Expert, 18, 31, 50, 59, 74, 86, 93, 98, 123, 166, 169, 172, 185, 205, 220, 253, 262, 290, 312, 350, 394, 397, 399, 400, 409, 416, 418, 425, 426
Extensible Markup Language/XML, 12, 123

F

Feasibility, 17, 260, 319, 64, 114, 127
Feedback, 33, 48, 49, 86, 87, 117, 130, 132, 133, 160, 170, 171, 182, 215, 220, 223, 295, 302, 309, 321, 338, 348, 349, 363, 366, 394, 418, 427, 428
Fitness for use, 5
Flood risk, 19, 159–164, 168–173, 352
Forecast, 17, 29, 166, 185, 281
Framework, 3, 5, 14, 18, 20, 22, 23, 32–37, 53, 76, 103, 106, 107, 113, 120, 124, 128, 129, 135, 155, 161, 175, 177–180, 188, 192, 193, 195, 199, 205, 227, 245, 253, 266, 278, 280, 289, 301, 302, 308, 312, 317, 319, 321–323, 331, 339, 363, 409

G

Gaming, 21, 311
Geocoding, 110
Geodatabase, 20, 73, 84, 246
Geographic, 1, 2, 9–14, 20–24, 31, 71, 76, 83, 93, 138, 140, 143, 149, 160, 177–179, 188, 227, 229–231, 235, 236, 238, 243, 245, 246, 265–268, 270–274, 276, 278, 287–290, 295, 298–300, 302–304, 317–319, 342, 350, 352, 353, 363, 370, 381, 387, 389, 390, 396, 397, 399, 418–423
Geographic Information System/GIS, 1, 2, 9, 10, 13, 14, 19, 20, 23, 71–73, 76, 86, 93, 96–98, 105, 118, 134, 137, 138, 141, 142, 146, 150, 154, 155, 165, 177, 179, 188, 192, 193, 195, 205, 219–224, 235–240, 243–249, 262, 265, 266, 271, 280, 281, 288–290, 298, 339, 370, 371, 373, 374, 377, 378, 381, 382, 384, 389, 390, 392, 394–411, 420, 421, 427, 432, 436–440

Geographic Virtual Urban Environments/ GeoVUE, 13, 237
Geography, 11, 12, 24, 25, 56, 66, 128, 157, 211, 229, 235, 241, 342, 367, 390, 396, 402, 410, 411, 447, 431
Geography Markup Language/GML, 12
Geoinformation, 3, 4, 6–8, 16, 21, 26, 114, 178, 190, 207, 264, 265, 289–291, 298
Geoprocessing, 73, 243–245, 262–264
Geosimulation, 179, 189, 224
Geotechnology, 1, 2, 12
Geovisualisation, 71, 173, 235, 397, 410
Google, 12, 14, 18, 20, 70, 71, 76, 86–88, 145, 227, 234–240, 242, 447
Google Earth, 12, 14, 18, 70, 71, 76, 86–88, 145, 234, 236, 239, 447
Google Maps, 12, 14, 20, 227, 234–240, 242
Governance, 160, 175–177, 188, 332, 370, 372, 379, 382, 384, 387
Government, 2, 7, 24, 53, 79, 98, 101, 105, 106, 109, 112, 132, 133, 141, 143, 161, 171, 175, 177, 185, 187, 188, 203, 215, 218, 222, 229, 234, 238, 243, 249, 262, 279, 289, 297, 299, 300, 302, 312–314, 347, 351, 372, 379, 381, 382, 399, 411, 443
Graph, 14, 35, 111, 216
Graphical interface, 243, 160
Graphical User Interface/GUI, 3, 74, 114, 128, 129, 193, 244, 441
Grid, 12, 13, 25, 54, 72, 100, 141, 143, 159, 166, 187, 188, 245, 247, 291, 304, 307
Growth efficiency modelling, 303, 304, 307, 312

H

Harmonisation, 11, 12
High Performance Computing/HPC, 13
HyperText Markup Language/HTML, 12, 89, 109–111, 136, 157, 206, 224, 225, 237, 242, 314, 344

I

Impact analysis, 39, 172, 251, 298, 313
Impact assessment, 17, 30, 32, 50, 168, 308
Index, 11, 136, 157, 234, 314, 343, 372, 374, 386, 437, 438
Indicator, 117, 150–152, 199, 201, 202, 207, 273, 274, 309, 348, 349, 372, 379, 382, 387
Information, 1–16, 18, 19, 21–26, 33, 44, 65, 70, 71, 93, 105–108, 127, 137, 156–159, 227, 233, 243, 265, 266, 295, 298, 338–341, 347, 365, 370, 389–390, 399, 400, 413–414, 419, 421–428
Information and Communication Technology/ ICT, 12, 271, 370, 409
Information provision, 10, 25
Information technology, 3, 8, 76, 178, 297, 314, 387, 415, 429
Infrastructure for Spatial Information in Europe/INSPIRE, 11, 262
Infrastructure planning, 219
Innovation, 6, 8, 25, 171, 240, 244, 343, 351, 390, 429
Institutional, 132, 135, 235, 276, 280, 299, 314, 318, 342, 344, 356, 381, 385
Instrument, 5, 7, 21, 114, 122, 128, 137, 150, 153, 154, 160, 169, 172, 195, 262, 266, 377, 384, 385, 414
Integrated Spatial Decision Support System/ ISDSS, 19, 24, 50, 51, 66, 69, 88, 89, 107, 113, 130, 133, 135, 154, 157, 159, 162, 164, 170, 172, 175, 189, 191, 205, 208, 211, 262, 276, 295, 312, 315, 367, 369, 372, 387, 397, 410, 413, 444
Interactive, 10, 12, 14, 18, 19, 24, 26, 114, 120, 122, 123, 126, 128, 137, 179, 196, 291, 296, 314, 340, 349, 350, 373, 376, 383, 386, 396, 401, 402, 415, 420, 426, 427, 447
Interface, 3, 14, 73, 74, 114, 120, 123–125, 128, 129, 132, 133, 139, 160, 172, 184, 185, 192–194, 196–200, 203–206, 214, 215, 219–222, 227, 236–239, 243, 244, 246, 247, 312, 371, 373, 374, 377–379, 396, 405, 423, 424, 427, 432, 433, 439, 441, 442
Internet, 9–12, 15, 21, 50, 70, 188, 235, 236, 238–240, 243, 262, 390, 392, 395, 399, 406–408, 410, 411, 437, 440, 446, 448
Interoperable, 11, 188
Interpolation, 100, 156, 195, 288–290, 406
Intranet, 443
Iterative process, 17, 32, 117, 213, 352, 363, 364, 379, 418
Iterative Proportional Fitting/IPF, 198

K

Keyhole Markup Language/KML, 76, 86, 234, 235, 238, 240
Keypad polling, 302, 310, 312, 314, 315

Knowledge, 3, 7, 8, 14, 16, 18, 19, 21, 22, 30, 31, 71, 107, 108, 114, 119, 121–124, 132, 134, 135, 160, 227, 235, 265–267, 288–290, 296–299, 317, 364, 365, 370, 385, 389, 390, 396–409, 432, 446
Knowledge-based reasoning, 31
Knowledge integration, 114, 121, 124, 132, 134, 135
Knowledge management, 8, 14, 134, 344

L
Land allocation, 32, 74, 179, 381
Land cover, 44, 57, 89, 156, 188, 225, 269, 282, 283, 317, 319, 321, 323, 330, 343, 344, 350
Land degradation, 156, 317, 319, 343, 344
Land management, 333, 336, 342, 369, 371, 372, 377, 448
Land resources, 21, 264, 317, 319, 331, 342
Landscape, 18, 21–23, 53, 69, 70, 191, 212, 243, 265–267, 269, 270, 347, 432, 433, 436
Landscape change, 212, 318, 335, 342, 343
Landscape management, 318
Landscape planning, 21, 83, 264–266, 317, 318, 331, 335, 342, 344, 367
Landscape sensitivity model, 303–305, 307
Landscape visualisation, 340, 345
Land suitability, 74, 75, 86, 179
Land-use change, 2, 6, 8, 11–13, 15, 17–19, 22, 29, 31, 53, 54, 69, 113, 116, 159–161, 195, 197, 211–213, 218, 229, 243, 318, 369, 431
Land-use model, 17, 19, 22, 30, 54, 65, 159, 165, 188, 349
Land-use monitoring, 20
Land-use planning, 16, 26, 30, 50, 69, 70, 74, 88, 208, 264, 295, 314, 339, 341, 367, 370, 379, 387
Land-use simulation, 17, 54, 167, 173
Learning organization, 8, 26

M
Map, 13–15, 17, 23, 45, 54–57, 114, 193, 196–198, 231, 246, 392, 396, 432, 433, 436–438, 440–445
Mapping, 9, 10, 12, 14, 23, 71, 72, 83, 109, 141, 143, 144, 150, 156, 157, 177, 228, 232, 235–242, 257, 303, 342, 370, 388, 392, 397, 400, 404, 406, 410, 411, 431, 432, 436, 437, 439, 441–444, 446–448
Maptube, 13, 14, 240
Mashup, 14, 236–238
Meaning, 21, 102, 123, 140, 152, 272, 288, 392, 397, 427, 437, 440, 444
Metadata, 3, 12, 14, 135
Methodology, 21–23, 26, 63, 84, 93, 95, 97, 98, 101–103, 110, 114, 117, 118, 132, 138, 155, 156, 224, 229–231, 243, 262, 266, 271, 277, 283, 287, 288, 319, 333, 340, 342, 348, 349, 352, 363, 389, 390, 397, 402, 409, 410, 432, 436, 437
Metrics, 48, 74, 176, 179–181, 184, 185, 190, 360
Microsimulation, 14, 16, 19, 24, 192, 201, 205–207
Model, 6, 16, 17, 19–22, 24, 30, 31, 54–56, 70, 114, 116, 159, 192–196, 198–203, 211–213, 243–248, 266, 267, 269, 271, 347, 348, 394, 395, 415
Model Based Management System/MBMS, 193, 195, 196
Model based reasoning, 31, 50
Model integration, 114, 119–121, 123–126
Modelling, 3, 6, 13, 14, 16, 18, 20, 22, 24, 29, 192, 193, 195, 198, 211, 212, 243–245, 347, 348, 415, 436
Model parameters, 35, 114, 216, 223, 248, 353
Model variables, 221
Monitoring, 10, 13, 17, 19, 20, 22, 50, 64, 65, 108, 134, 137, 140, 149, 153, 154, 177, 178, 188, 189, 192, 193, 195, 206, 207, 227, 228, 230, 262, 303, 344, 371, 372, 379, 385–388, 406
Monte Carlo simulation, 17, 55, 64
Morphological, 21, 56, 265, 270, 278, 287–290
Multi-Agent Model, 25
Multi-Criteria Evaluation/MCE, 76, 203, 205, 298, 303–305, 312
Multilevel, 317, 320, 321
Multimedia, 422

N
Narrative storylines, 349, 352, 353, 358, 363
NASA, 76, 447
National Science Foundation/NSF, 13
Natural resources, 20, 34, 43, 51, 135, 176, 249–251, 253, 277, 309, 319, 410, 434, 447, 448
Neighbourhood, 54, 55, 60, 67, 185, 193, 207, 208, 227–230, 232, 281, 284, 367, 394, 407, 426
Nomenclature des Units Territoriales Statistiques/NUTS, 61, 359

Non-Governmental Organisation/NGO, 132, 376, 379
Not In My Back Yard/NIMBY, 23

O
Off-the-shelf, 4, 47, 73
Open Geospatial Consortium/OGC, 12, 120
OpenGIS, 12
Open Platform for Urban Simulation/OPUS, 179, 410
Open standard, 12
Optimisation, 25, 117, 133, 166, 350

P
Paradigm, 8, 24, 71, 121, 314, 386, 413
Parameter, 55, 56, 72, 73, 216, 243, 245–248, 260, 262, 350, 353, 358, 362, 365, 378
Participation, 2, 8–10, 14, 17, 21–23, 25, 124, 127, 171, 179, 188, 202, 233, 234, 293, 296, 297, 301, 302, 310, 313–315, 317–319, 331, 332, 335, 339–345, 370, 384, 385, 387, 389, 390, 395, 396, 398–400, 406–411, 413, 443, 444, 446–448
Participatory GIS/PGIS, 9, 25, 386, 396–399, 409–411, 438, 439
Participatory planning, 4, 9, 21, 22, 24, 206, 225, 296, 314, 379, 381, 387, 410, 411, 430–432
Pattern analysis, 321
Physical planning, 159, 300
Pixel, 72
Plan, 3, 13, 21, 33, 67, 71, 150, 178, 185, 187, 188, 244, 250, 265, 269, 275–280, 284, 288, 295–297, 299–303, 307–314, 333, 344, 384, 414, 417–419, 443, 447
Plan adoption, 21, 296, 299, 301, 310, 311
Plan implementation, 297, 310–313
Planner, 74, 88, 150, 222, 255, 263, 307, 311, 312, 416
Planning, 1–4, 6–11, 14, 16–25, 29–31, 53–55, 57, 69–71, 93, 113–117, 137, 138, 140, 162, 175–177, 191–193, 195, 203, 211–213, 227–230, 243, 244, 248, 265–267, 269, 270, 295–297, 317–319, 347, 348, 369–371, 389–392, 394, 396, 413–416, 431, 432
Planning practice, 1, 2, 4, 6–8, 18, 87, 175, 187, 207, 208, 220, 221, 223, 347, 348, 391, 399, 403, 406, 429
Planning process, 2–4, 9, 21, 71, 109, 111, 115, 178, 244, 265–267, 270, 271, 276, 281, 289, 290, 295–302, 309, 310, 312, 319, 332, 335, 340, 342, 389, 397, 398, 406, 407, 411, 415, 417–419, 421–423, 427, 432, 440
Planning support, 1, 3, 4, 9, 17, 19–21, 24–26, 29, 48, 50, 69, 70, 88, 89, 93, 107, 113, 115, 117, 135, 138, 153, 154, 156, 175, 178, 189, 208, 209, 211, 212, 219, 225, 227–229, 237, 239, 240, 243, 262, 263, 265, 295–299, 311–315, 317, 347, 369, 370, 386–388, 396, 409, 410, 413, 414, 429, 431, 444
Planning Support Instrument/PSI, 295, 298, 299, 301, 302, 307, 312
Planning Support System/PSS, 1–10, 12, 13, 16–21, 23, 24, 27, 29–31, 33, 47–50, 69–71, 73, 76, 79, 84, 86–89, 91, 93, 95, 97, 98, 101–109, 113–116, 122, 132, 135, 137–141, 146–150, 152–155, 175–181, 183, 185–188, 206, 208, 211–214, 219, 222–224, 262, 265, 269, 293, 295, 297–299, 301–305, 308–315, 347–349, 359, 366, 369–371, 396–399, 410, 413, 444
Policy, 1, 2, 4, 10, 11, 14, 17–20, 29, 30, 53, 93, 113, 115–117, 137, 138, 140, 160, 161, 176, 177, 211–213, 296, 319, 348, 413, 433
Policy maker, 2, 366
Portal, 111, 112, 143, 214, 344, 415, 419
Practitioner, 231, 313
Prediction, 3, 10, 16, 18, 66, 70–73, 78–84, 86–88, 189, 190
Probability, 17, 31, 55, 61–64, 160, 163, 166, 168, 169, 171, 185, 199–201
Processing model, 243, 244, 252, 253, 263
Professional, 3, 4, 89, 136, 156, 220, 298, 301, 314, 318, 340, 345, 389, 397, 411, 414
Projection, 6, 73, 74, 178, 307, 427
Prototype, 17, 19, 20, 114, 120, 131, 132, 136, 137, 161, 179, 180, 183–185, 191, 192, 205, 215, 237, 378, 391, 392, 417, 418, 423, 428
Proximity, 55, 58, 74, 176, 253, 257, 258, 303, 308, 359, 437
Public, 3, 4, 9, 10, 12–14, 18, 20–22, 53, 93, 94, 159, 176, 191, 192, 227–232, 243, 244, 265, 295–297, 317–319, 369–371, 394, 396, 432, 436

Public participation, 9, 10, 22, 25, 171, 295–297, 299–301, 309–312, 317, 319, 331, 332, 340, 342–344, 384, 387, 396, 398, 406, 410, 411, 443, 446, 447
Public Participation GIS/PPGIS, 225, 313, 342, 396–399

Q
Qualitative, 10, 21–23, 29, 127, 128, 134, 250, 271, 276, 288, 347, 348, 363–366, 392, 396, 403, 405, 424, 426, 432, 439, 445
Quality, 3, 5–8, 17, 19, 21, 114, 137, 229, 233, 295–297, 319, 389, 391, 392
Quantitative, 10, 22, 23, 32, 66, 170, 199, 203, 250, 271, 288, 303, 304, 333, 338, 347, 348, 352, 353, 363–367, 392, 403, 405, 424

R
Raster, 72, 141, 143, 246, 253, 256, 258, 259, 276, 281, 307, 437
Regional planning, 17, 25, 29, 30, 43, 47, 49, 51, 69, 70, 105, 176, 179, 207, 243, 244, 250, 281, 315, 411, 447
Remote Sensing/RS, 17, 65, 88, 134, 145, 177, 180, 188–190, 318, 339, 344
Resolution, 14, 32, 41, 50, 65, 118, 121, 143, 145, 154, 155, 166, 170, 179, 190, 194, 253, 330, 349, 350, 359
Risk, 18–20, 53, 65, 66, 97–100, 103, 107, 108, 157, 159–164, 168–173, 223, 232, 233, 238, 276, 278, 281, 286, 289, 314, 341, 352, 358, 432, 436, 447
Robustness, 348, 350

S
Satellite, 14, 180, 188, 189, 231, 234, 236, 321, 330, 344
Scalable Vector Graphics/SVG, 12
Scenario, 19, 20, 22, 32, 33, 44, 49, 51, 54, 57, 58, 60–62, 65, 73, 74, 84, 87, 89, 126, 133, 136, 155, 162, 164, 166, 167, 169, 171, 173, 178, 179, 186, 192, 195, 198, 201, 203, 205, 211, 213, 215, 219, 262, 281, 283, 287, 289, 303, 304, 315, 334, 335, 348, 350, 353, 356, 361, 363, 367, 385
Scrapbook, 419, 423
Script, 247
Semantic, 12, 14, 15, 24, 25, 176, 271, 290
Sensitivity modelling, 303, 307
Service planning, 20, 227, 228, 238–240
Simulation, 13, 14, 16, 17, 19, 24, 29, 31, 33, 35, 40, 42, 44, 51, 54, 56, 64, 65, 87, 88, 116, 124–126, 133, 136, 149, 160, 167, 173, 177, 179, 183, 185, 187, 188, 193, 206, 211, 212, 215, 222, 224, 225, 262, 264, 313, 347–350, 357, 415, 416, 421, 423, 428, 429
Site analysis, 20
Sketch, 71, 309, 418, 420, 422, 423
Sketchbook, 23, 419, 422, 424
Sketch planning, 71, 309
Social learning, 363
Society, 19, 65, 66, 135, 159, 170, 190, 224, 227, 233, 235, 238, 239, 241, 268, 315, 341, 344, 351, 352, 356, 357, 361–363, 387, 397, 399, 406, 410, 411, 446, 448
Socioeconomic, 13, 19, 22, 66, 67, 84, 135, 155, 156, 173, 207, 212, 317, 319, 332, 333, 342
SoftGIS, 22, 23, 389, 390, 392, 394, 396, 411
Software, 3, 9, 13, 14, 23, 29, 70, 193, 236, 243, 247, 319, 413, 415
Spatial analysis, 13, 16, 51, 67, 97, 99, 106, 111, 177, 189, 190, 237, 239, 241, 242, 264, 291
Spatial configuration, 44, 46, 155
Spatial Decision Support System/SDSS, 1–3, 19, 24, 116, 191–193, 205, 207, 367
Spatial interaction model, 304, 349
Spatial metrics, 176, 181
Spatial modelling, 50, 86, 136, 193, 241
Spatial planning, 4, 17, 19, 26, 53, 55, 57, 64, 65, 172, 173, 175, 178, 189, 204, 207, 298, 313, 351, 414
Spatial Planning Support System/SPSS, 19, 175, 176, 180, 181, 185, 187
Spatial temporal, 201
Sprawl, 19, 50, 54, 56, 57, 59, 60, 62, 63, 68, 167, 175, 181, 182, 185, 190, 211, 410
Spreadsheet, 203, 204, 216, 303
Stakeholder, 8, 9, 17, 22, 23, 29, 32, 44, 47, 49, 65, 79, 82, 84, 127, 188, 212, 297, 318, 319, 332, 333, 335, 339, 341, 342, 347, 349, 361, 364, 365, 371, 383, 385, 413, 436, 439, 445, 446
Standard, 12, 59, 74, 123, 163, 167, 179, 188, 193, 237, 238, 245, 262, 299, 337, 378, 415, 441, 443

Standard Generalized Markup Language/SGML, 12
Statistic, 55, 185, 204
Statistical, 31, 106, 143, 155, 157, 160, 169, 178, 206, 213, 241, 250, 288, 333, 371, 372, 374, 383, 390, 400, 403, 448
Storyline, 348, 353, 358, 359, 361, 362, 364, 365
Storytelling, 298
Strategic decision, 114, 118
Strategic planning, 3, 8, 115, 116, 135, 313, 319, 370, 387, 418
Strategy, 17, 19, 53, 55, 57, 59, 62, 64, 79, 162, 163, 165, 167–169, 177, 196, 204, 278, 321, 343, 359, 414
Studio, 23, 413, 417, 428, 429
Suitability, 54, 55, 57, 58, 73, 75, 78, 80, 81, 84, 86, 87, 166, 167, 179, 185, 253, 255, 260, 287, 352, 356, 358, 384
Supercomputer, 13
Survey, 4, 26, 51, 95, 144, 162, 167, 201, 204, 214, 301, 302, 304, 311, 314, 317, 318, 322, 323, 335, 337, 339, 382, 384, 392, 394, 418, 439, 445
Sustainability, 2, 38, 50, 74, 124, 159, 160, 191, 224, 280, 319, 331, 342, 384
Sustainable development, 173, 249, 250, 253, 255, 263, 279, 318, 342, 344, 387, 398
System architecture, 119, 128, 134, 135
System design, 123, 127, 164
System development, 30, 33, 47, 122
System Dynamics/SD, 135, 177, 178, 180, 181, 189, 219, 367

T

Tagger software, 443, 447
Technique, 31, 125, 127, 143, 201, 203, 289, 301, 302, 312, 350, 359, 364, 366, 399, 405, 437
Technology, 1, 3, 6, 8, 10, 12, 15, 25, 76, 88, 104, 129, 135, 161, 170, 178, 179, 207, 220, 222, 225, 229, 235, 236, 239–241, 243, 244, 262, 263, 271, 288, 297, 298, 302, 312, 314, 315, 320, 323, 340, 369, 385, 387, 389, 390, 399, 409, 411, 413, 416, 428, 429, 447
Territory, 139, 140, 150, 152, 155, 266, 271, 273, 275, 284, 288, 372, 387
Thematic map, 236, 237, 239, 240
Theory, 3, 22, 25, 30, 33, 69, 70, 72, 73, 154, 207, 211, 213, 219, 225, 226, 266, 267, 298, 314, 350, 390, 396, 402, 411, 414
Time, 1, 3, 13, 16, 18, 29, 31, 69, 94, 175, 176, 192, 194, 195, 198, 202, 211, 212, 244, 348
Tool, 17, 20, 23, 29, 33, 114, 137, 162, 192, 195, 197, 203, 211, 213, 227, 229, 231, 233, 244, 247, 317, 319, 371, 390, 396, 434
Toolbox, 18, 23, 246, 314, 315, 416
Toolkit, 69
Top down, 6, 86, 240
Topological, 21, 265, 270, 290
Tourism, 20, 207, 211, 215, 217, 219, 220, 222, 223, 225, 226, 279, 381
Transdisciplinary, 409
Transition rule, 59
Transparency, 86, 133, 231, 238, 240, 339
Trial and error, 417

U

Uncertainty, 17, 55, 64, 65, 67, 119, 122, 135, 162, 167, 169, 170, 172, 173, 213, 223, 264, 297, 372
United Kingdom/UK, 11–14, 111, 206, 224, 225, 227, 229–232, 234, 238, 241, 317, 344, 399, 409, 411, 431, 436, 443
United States of America/USA, 13, 21, 23, 25, 29, 93, 183, 188, 190, 224, 225, 295, 299, 431, 432, 441
Urban growth, 21, 24, 51, 56, 67, 71, 72, 76, 78, 87, 89, 175, 178, 189, 190, 243, 248, 253, 260, 264
Urbanisation, 53, 60, 65, 175, 176, 190, 213, 271, 279, 350, 351, 369, 386, 387
Urban model, 178
Urban planning, 14, 22, 50, 84, 177, 188, 189, 192, 193, 195, 206, 207, 263, 289, 291, 315, 344, 370, 371, 383, 386, 387, 389, 397, 399, 400, 402, 406, 409–411
Urban sprawl, 19, 50, 167, 175, 180, 182, 187, 190, 211, 410
Urban zoning, 194
Usefulness, 6, 138, 165, 429
User acceptance, 5, 429
User community, 5, 299, 312
User friendliness, 5, 6, 120, 122, 138, 155, 164
User interface, 3, 74, 114, 123, 128, 129, 132, 133, 172, 184, 193, 194, 196, 199, 200, 203, 206, 215, 219, 236, 244, 246, 247, 405, 423, 441
Utility scores, 82, 83

V

Validation, 17, 56, 63, 66, 119, 132, 140, 143, 146, 148, 166, 173, 187, 189, 213, 215, 241, 312, 349, 350, 360, 367, 374, 377, 378, 382, 384, 385
Values, 21, 23, 117, 197, 245, 246, 295, 296, 319, 389, 431, 432, 435
Vector, 12, 142, 143, 146, 194, 246, 253, 276, 281, 307, 378
Verification, 49, 132, 312
Virtual, 13, 18, 24, 25, 70, 76, 86, 88, 262, 319, 332, 335, 337, 338, 342, 344, 399, 410, 423
Virtual environment, 337, 338
Virtual Reality/VR, 319, 332, 335, 342
Vision, 33, 119, 310, 311, 415
Visioning, 33, 211, 213, 298
Visualisation, 3, 10, 12, 13, 17, 20, 24, 30, 50, 70, 71, 88, 89, 97, 119, 125, 173, 177, 178, 183, 184, 186, 188, 214, 221, 227, 231, 235, 236, 238, 239, 242, 262, 287, 298, 309, 315, 318, 319, 331, 333, 335, 336, 343, 345, 367, 397, 400, 405, 406, 408, 411, 416, 419, 421, 426, 429
Vulnerability assessment, 135, 162

W

Web, 4, 9, 11, 12, 14, 16, 19, 24, 26, 50, 98, 101, 109, 126, 134, 157, 188, 214, 215, 221, 229, 234, 238, 241, 242, 262, 350, 370, 387, 390–392, 394, 396, 398, 401, 406, 408, 410, 411, 419–421, 423, 436, 441, 443, 446, 447
Web based, 12, 14, 19, 24, 26, 109, 188, 214, 215, 221, 238, 242, 387, 390, 395, 398, 399, 401, 406, 408, 410, 411, 419, 421, 423, 436, 443, 446, 447
WebGIS, 235, 236, 239, 240
Web Ontology Language, 14
Web services, 12, 15, 134
Website, 14, 15, 234, 236, 237, 240, 350
Weighted Linear Combination/WLC, 203, 315
Weighting, 256, 259, 262, 305, 306
World Wide Web/WWW, 4, 11, 13, 15, 24, 26, 50, 73, 88, 89, 98, 109, 110, 112, 120, 126, 143, 156, 157, 189, 206, 214, 224, 225, 229, 234, 237, 240, 242, 314, 315, 342, 344, 350, 367, 392, 394, 395, 398, 399, 409–411, 443, 448
World Wide Web Consortium/W3C, 12, 14
Worldwind, 76, 447

Colour Plate Section

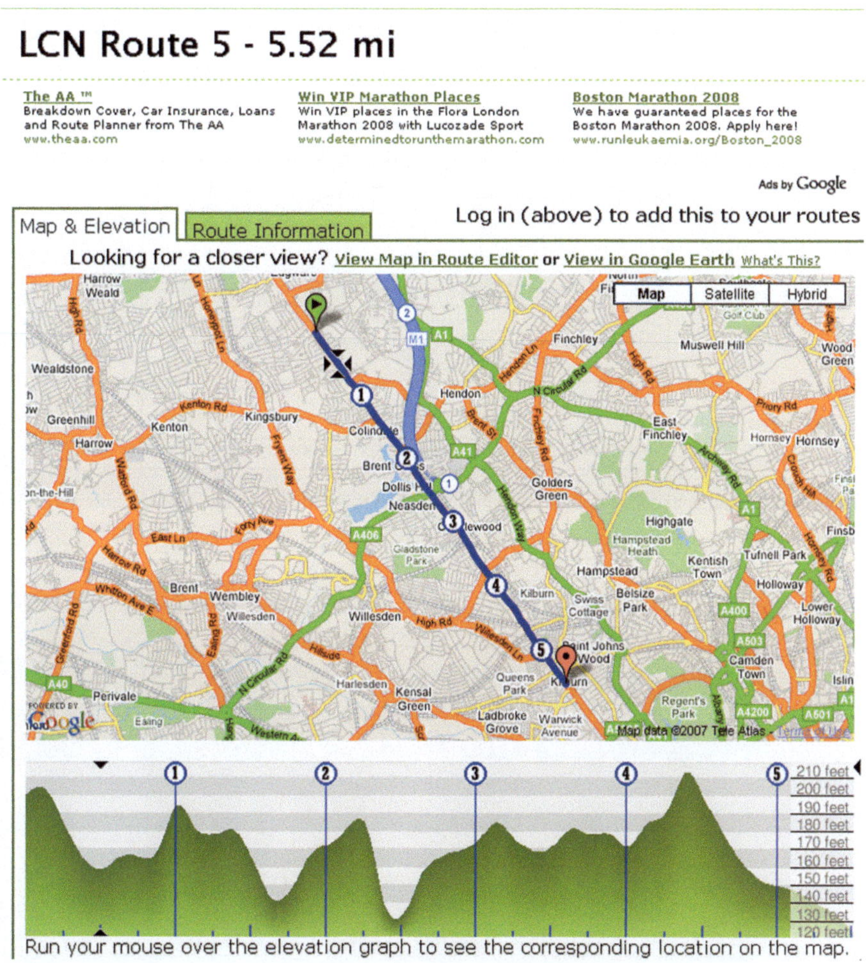

Plate 1. Map view of Brent's runstoppable website (See also Figure 1.3 on page 15)

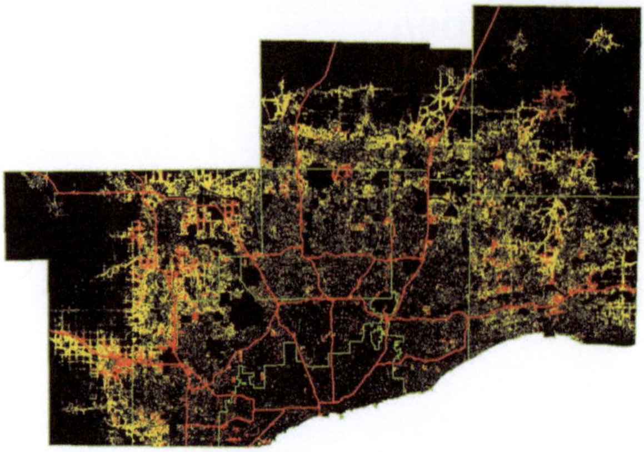

Plate 2. Land-use change in the Chicago region: *Baseline* simulation (See also Figure 2.1 on page 34)

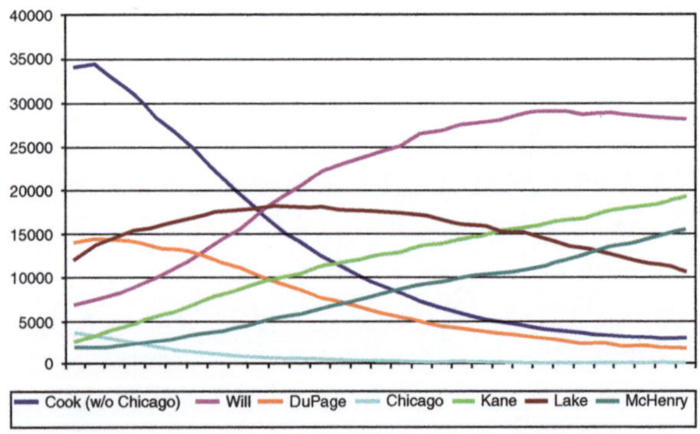

Plate 3. Annual land consumption by county: *Baseline* simulation (See also Figure 2.2 on page 35)

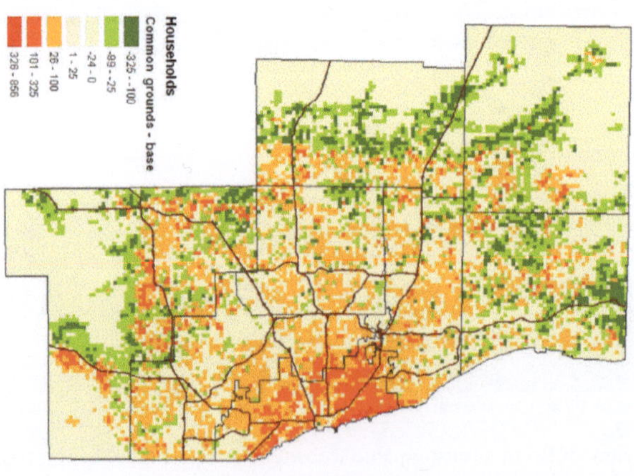

Plate 4. Difference in household location between *Baseline* and *Future Framework* simulations (See also Figure 2.3 on page 36)

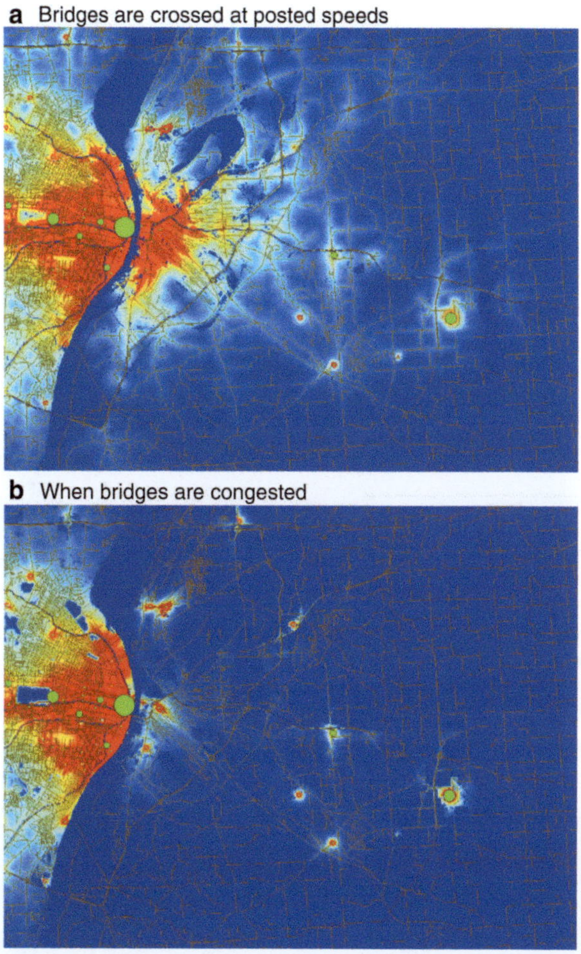

Plate 5. The change in the attractiveness (red is high – blue is low) of large scale employers in the regions (*green dots*) when (a) bridges are crossed at posted speeds and (b) when they are congested (See also Figure 2.6 on page 39)

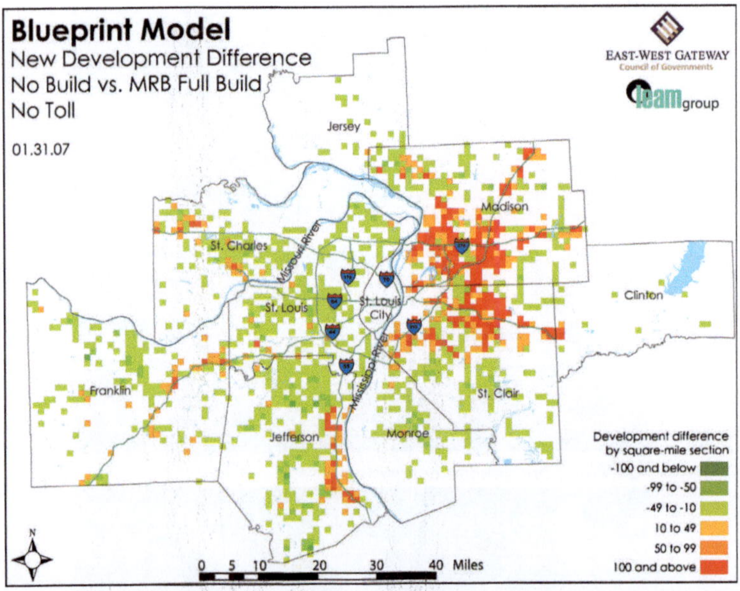

Plate 6. Differences in land-use change between *Full Build* and *No Build* simulations (See also Figure 2.8 on page 41)

Plate 7. The location of Mitchell Shire in South Eastern Australia (See also Figure 4.1 on page 71)

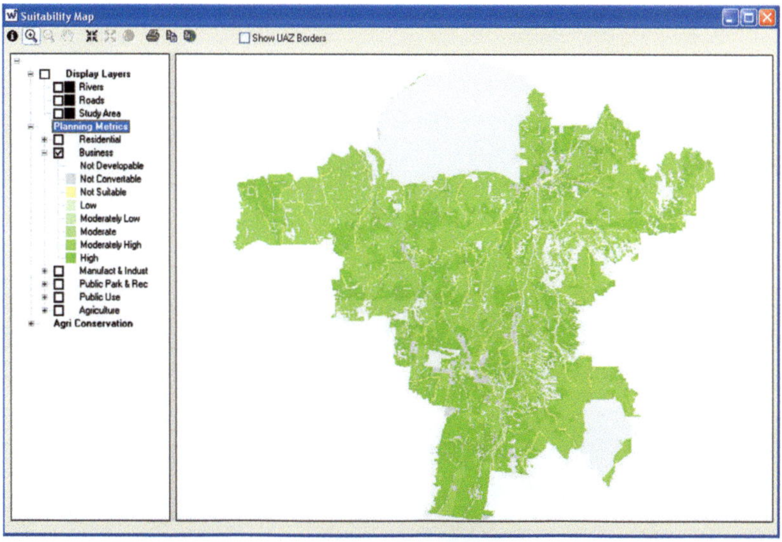

Plate 8. Land suitability for business purposes with all agricultural land use convertible: a developer-centric, suitability-based scenario (See also Figure 4.2 on page 75)

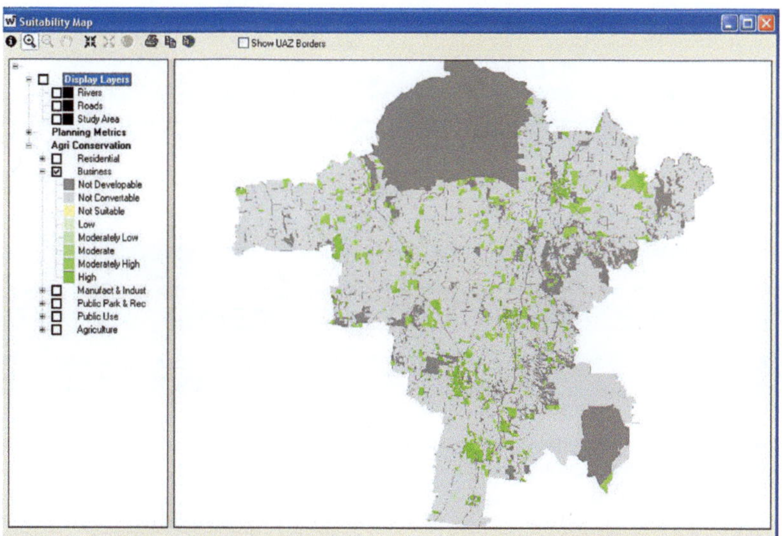

Plate 9. Land suitability for business purposes with no agricultural land use convertible: a conservation-oriented scenario (See also Figure 4.3 on page 75)

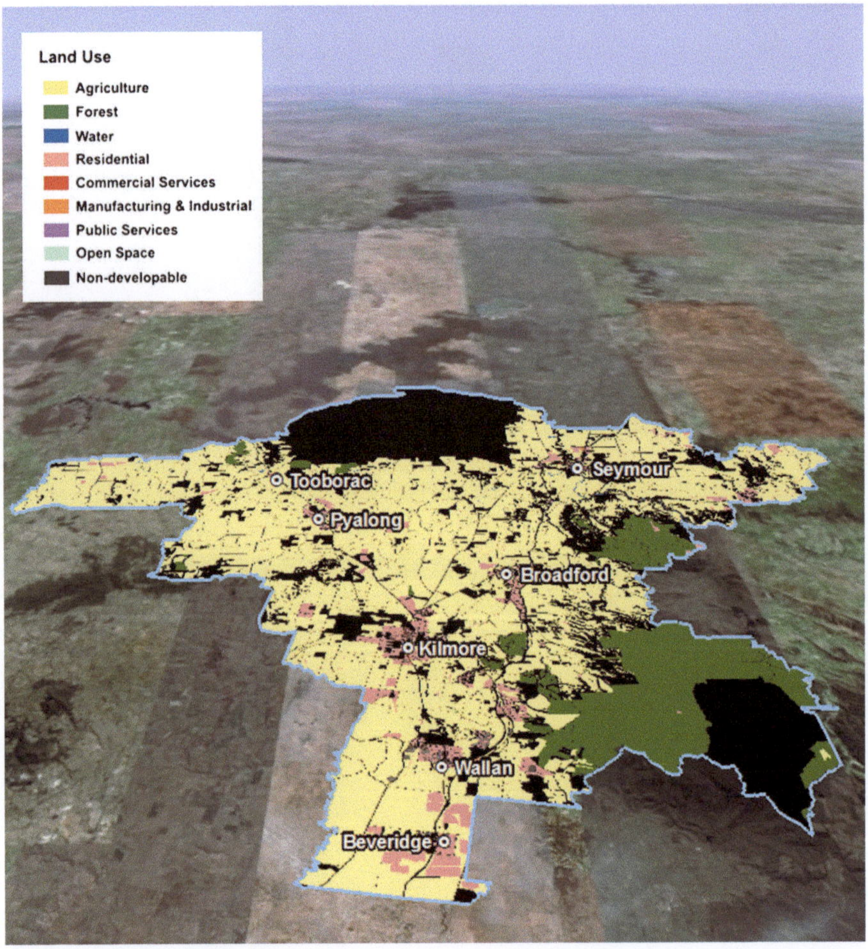

Plate 10. The *What if?*™-generated, developer-centric (Suitability) scenario for Mitchell Shire in 2030 (See also Figure 4.5 on page 78)

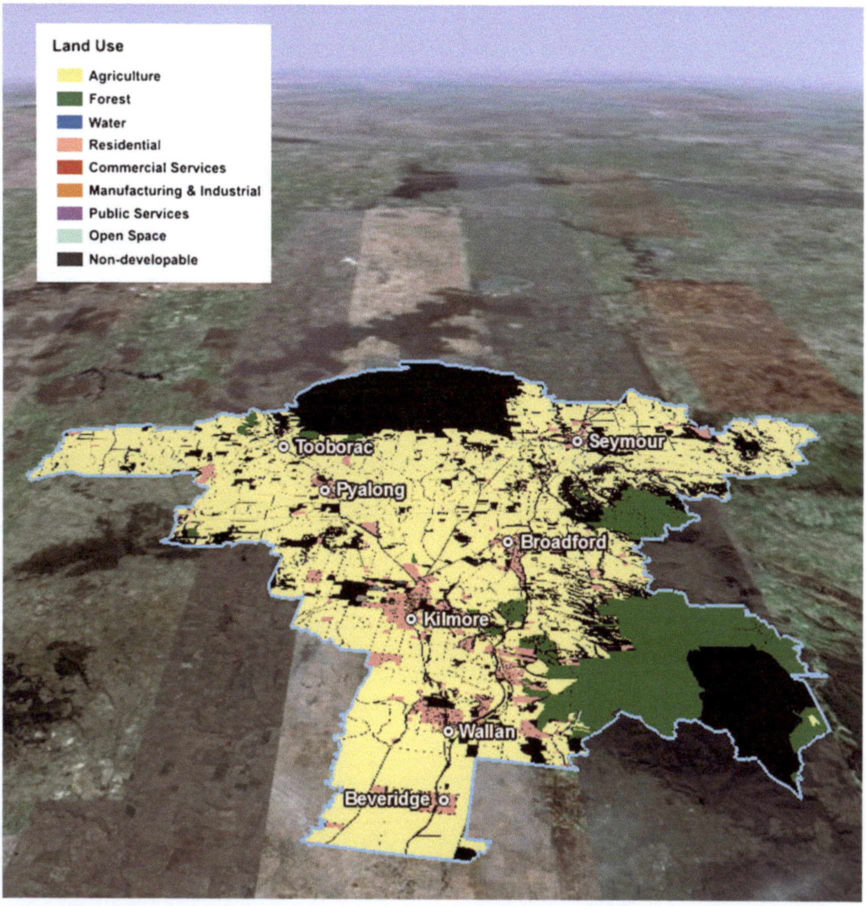

Plate 11. The *What if?*™-generated, agriculture protection (Conservation) scenario for Mitchell Shire in 2030 (See also Figure 4.6 on page 79)

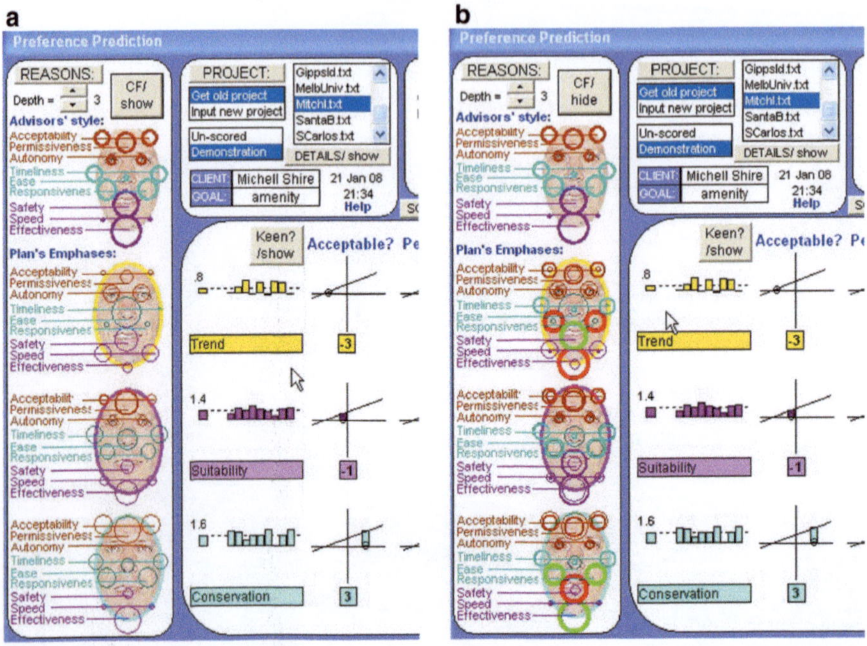

Plate 12. Scores for utility that are likely to be assigned to our three scenarios by all of the past software users taken together (See also Figure 4.8 on page 81)

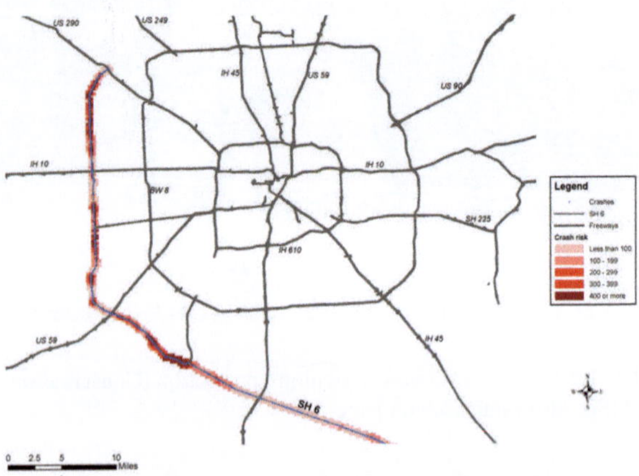

Plate 13. Crash risk on State Highway 6, 1999-2001 (See also Figure 5.4 on page 99)

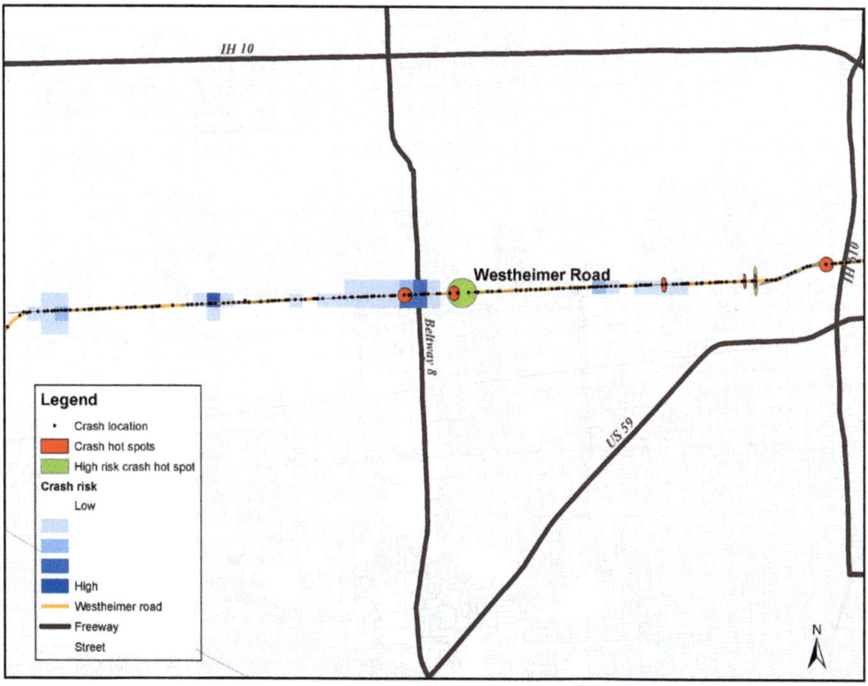

Plate 14. Crash hot spots and crash risk on Westheimer Road (See also Figure 5.5 on page 100)

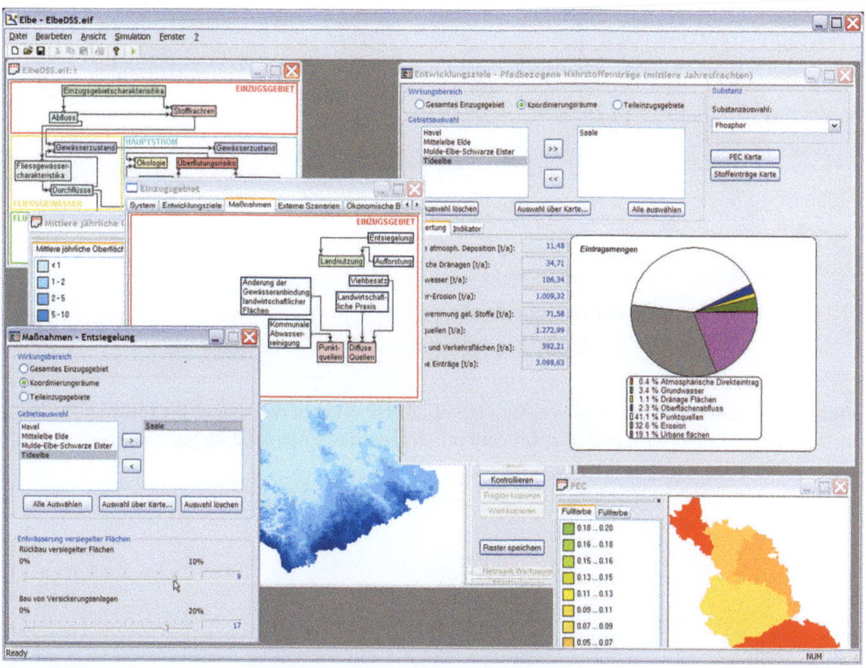

Plate 15. Graphical user interface of the *Elbe DSS* (See also Figure 6.4 on page 129)

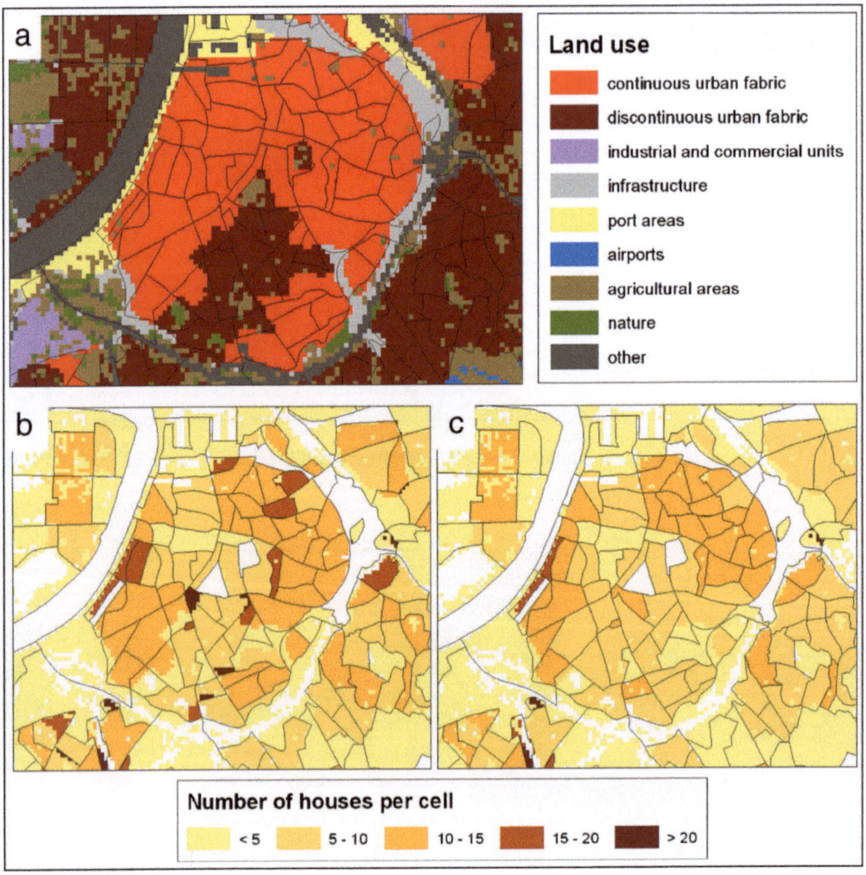

Plate 16. Land use in the centre of Antwerp (a), distribution of housing as the result of the dasymetric mapping before (b) and after (c) calibration (See also Figure 7.3 on page 144)

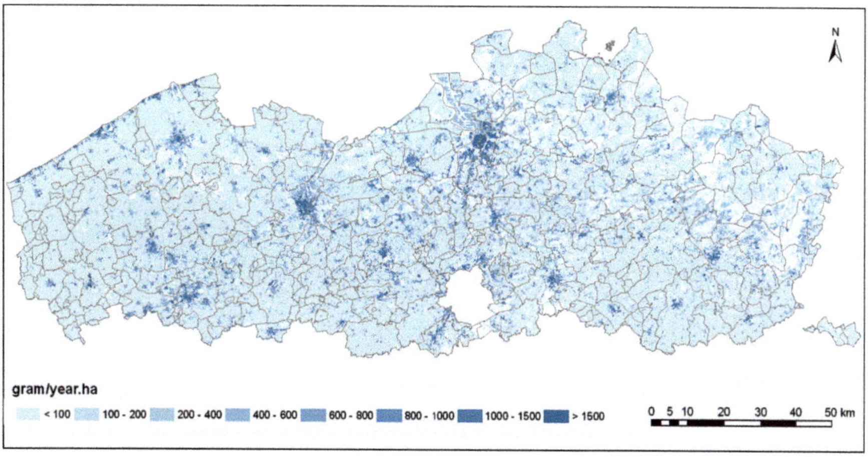

Plate 17. Zinc: gross emissions (See also Figure 7.4 on page 145)

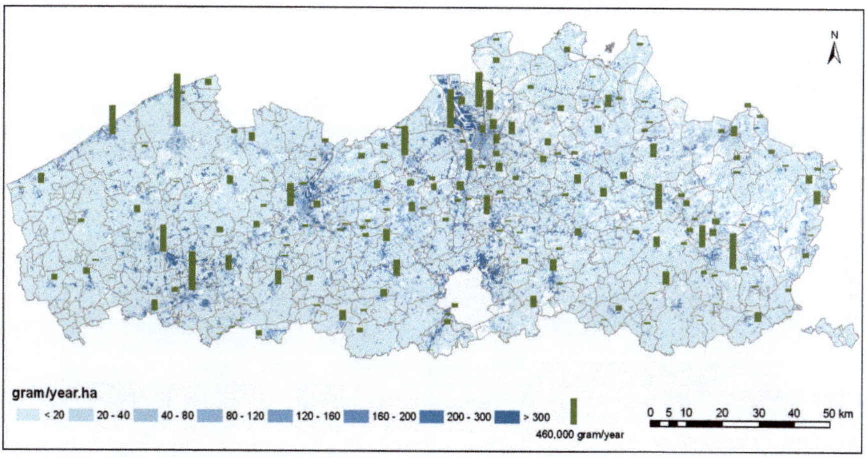

Plate 18. Zinc: net emissions (See also Figure 7.5 on page 147)

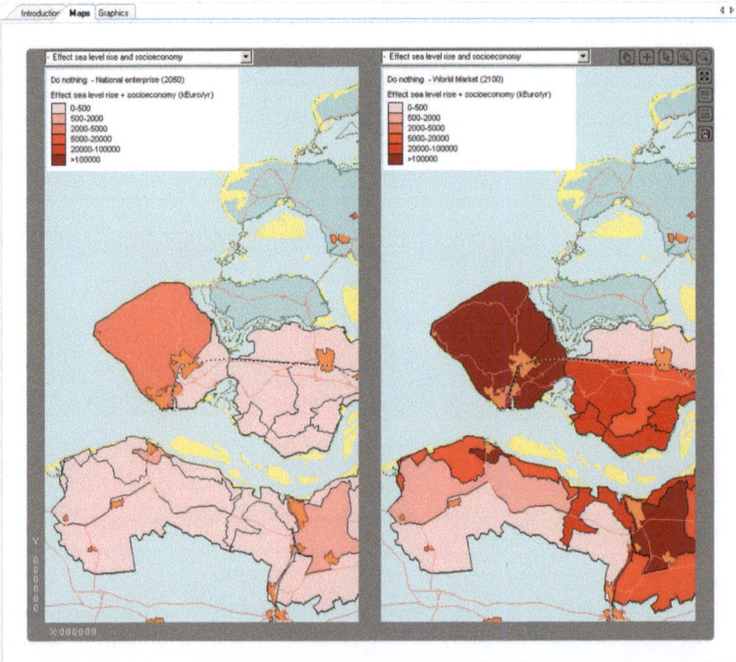

Plate 19. Comparison of economic damage in two future scenarios in the system (See also Figure 8.3 on page 165)

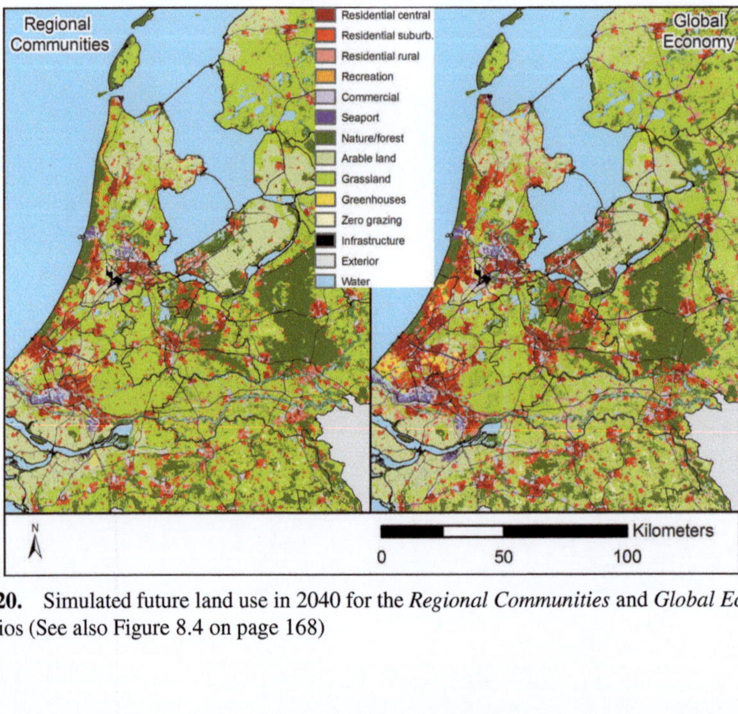

Plate 20. Simulated future land use in 2040 for the *Regional Communities* and *Global Economy* scenarios (See also Figure 8.4 on page 168)

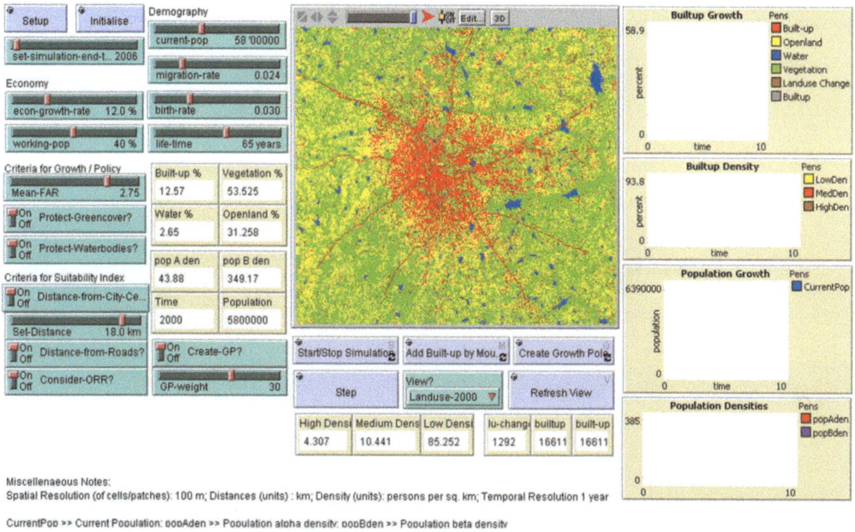

Plate 21. Prototype of the SPSS (See also Figure 9.6 on page 185)

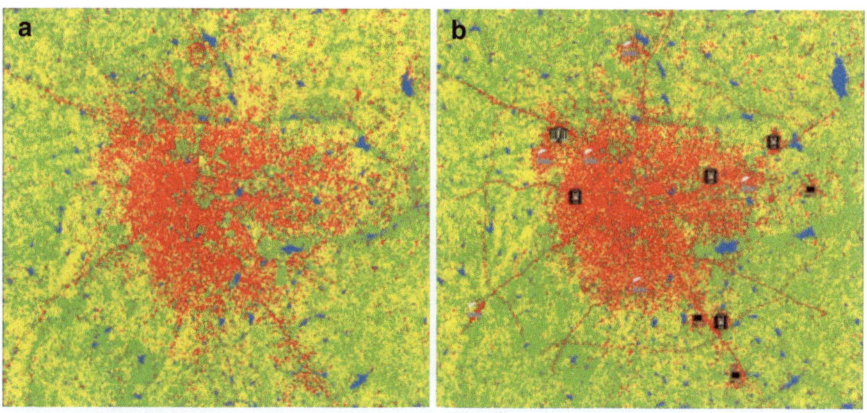

Plate 22. Classified and simulated land-use of Bangalore for 2006. (a) Classified land-use (2006), (b) Simulated land-use (2006) (See also Figure 9.7 on page 186)

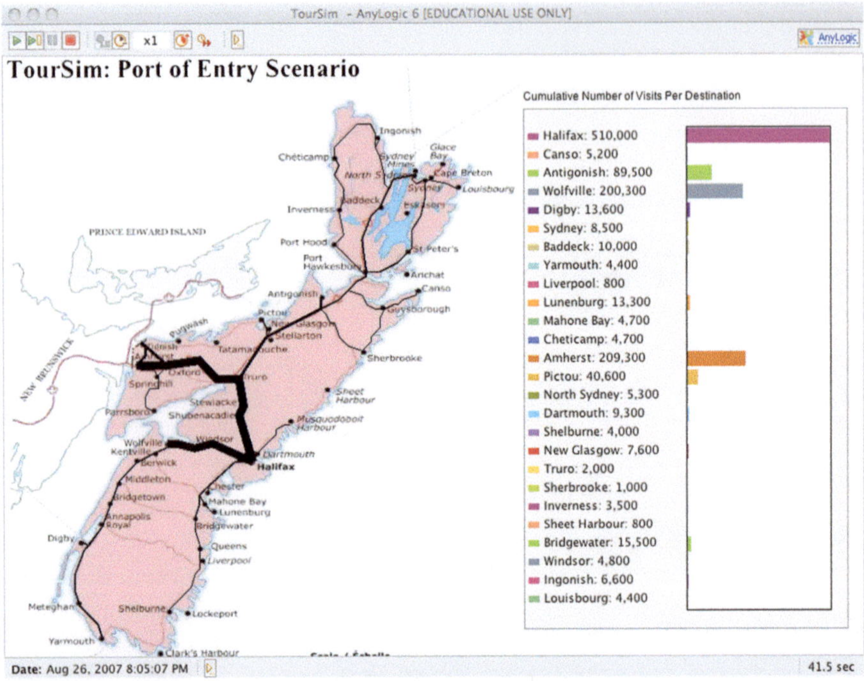

Plate 23. Main *TourSim* model page, including map of Nova Scotia; the width of each route represents the number of tourist trips with bar chart showing cumulative visitation (See also Figure 11.2 on page 217)

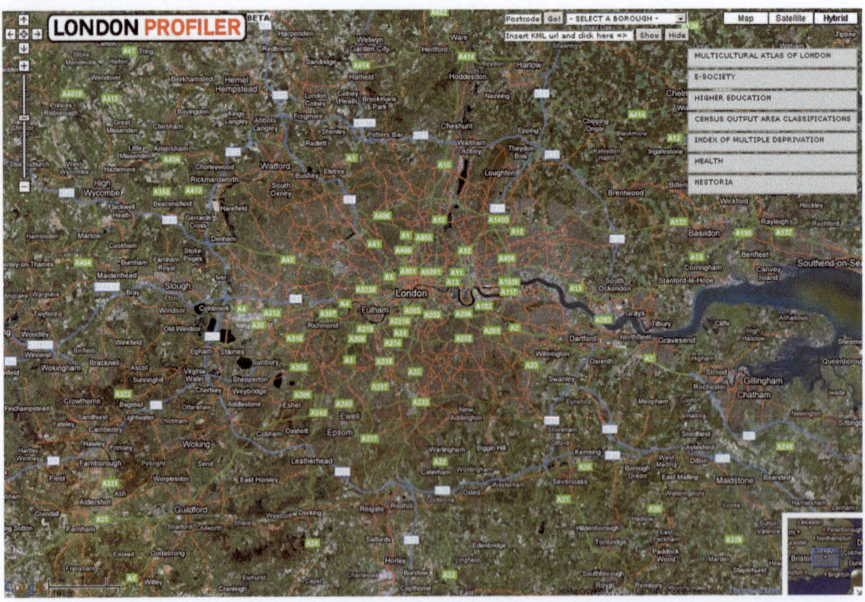

Plate 24. *London Profiler*: full London view, hybrid layer (See also Figure 12.1 on page 228)

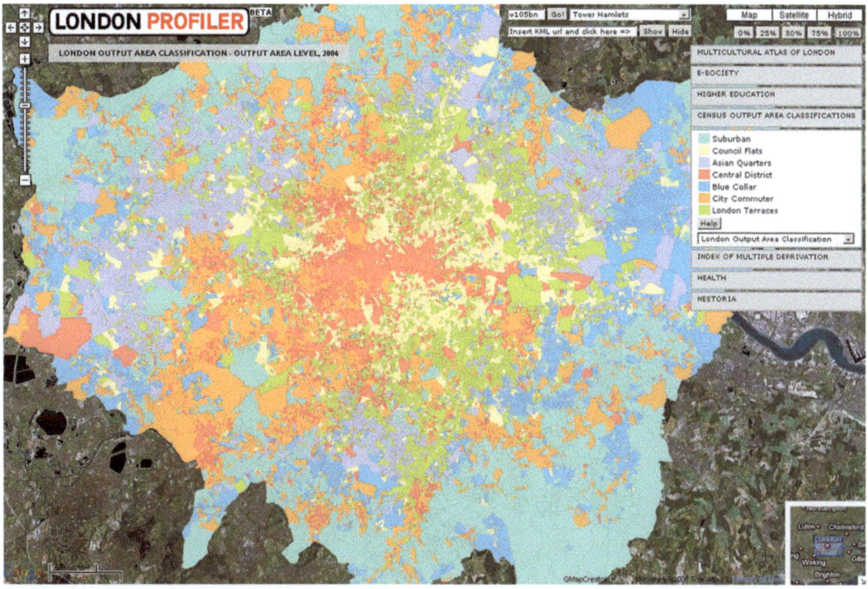

Plate 25. *London Profiler*: full London view, London Output Area Classification layer, 100 per cent transparency, satellite layer (See also Figure 12.2 on page 231)

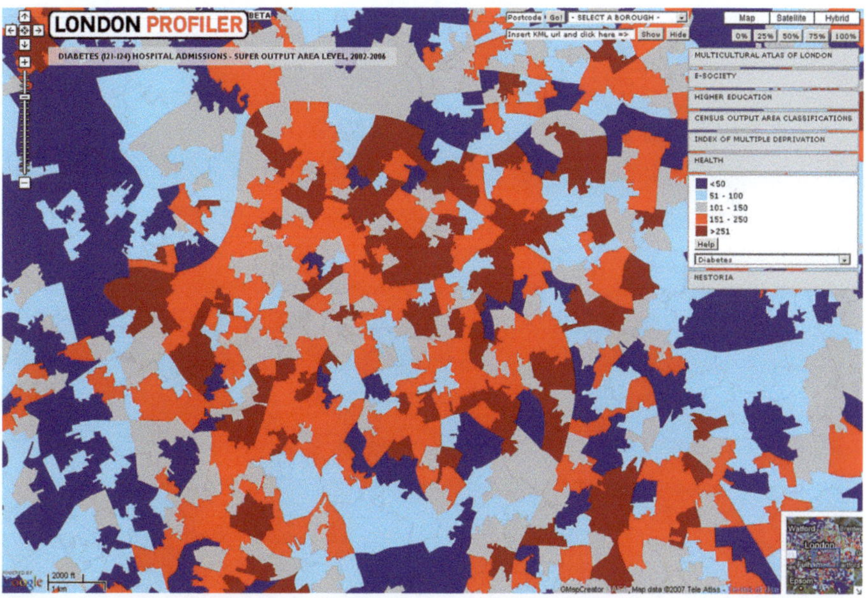

Plate 26. *London Profiler*: health tab, HALT dataset, diabetes hospital admission ratio, 100 per cent transparency (See also Figure 12.3 on page 232)

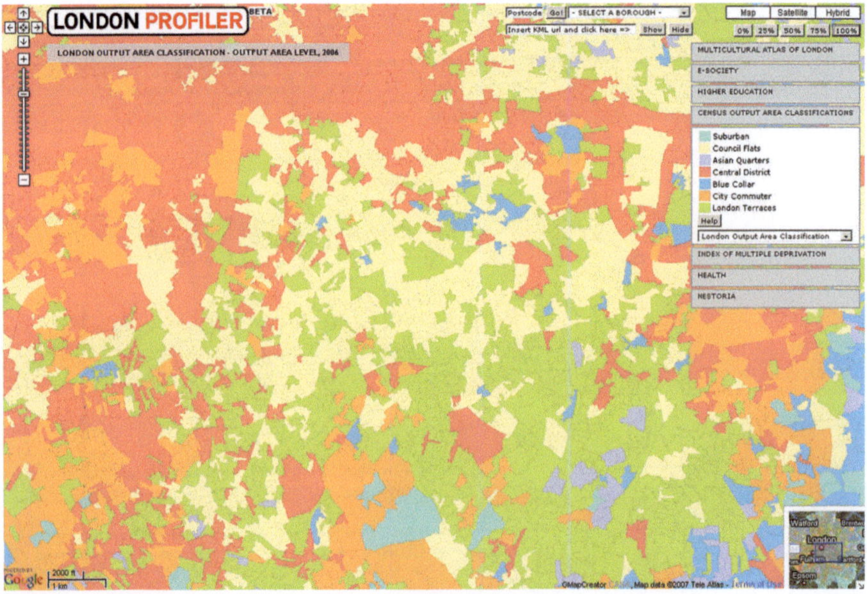

Plate 27. *London Profiler*: London Output Area Classification layer, 100 per cent transparency (See also Figure 12.4 on page 233)

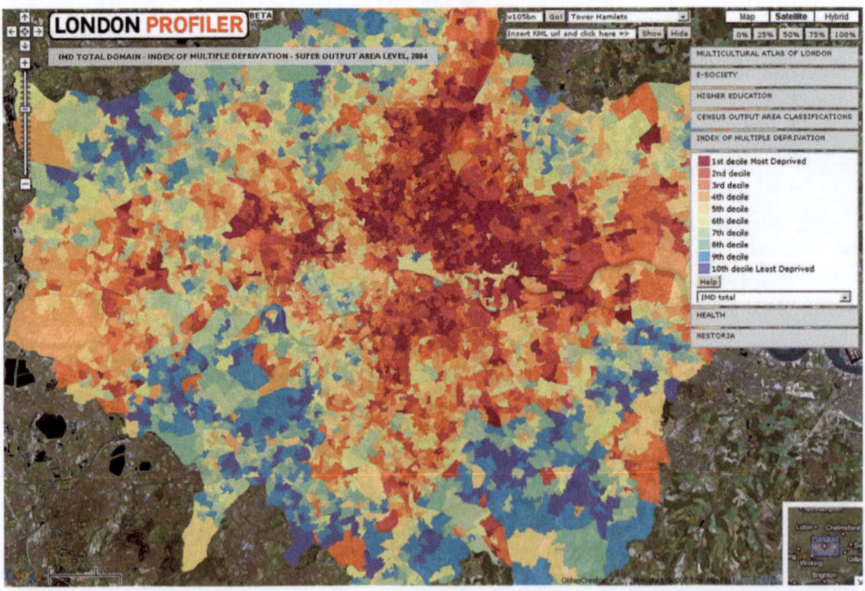

Plate 28. *London Profiler*: full London view, Index of Multiple Deprivation tab, IMD total layer, 75 per cent transparency, satellite layer (See also Figure 12.5 on page 234)

Plate 29. *London Profiler*: postcode level, e-society classification tab and layer with a KML file overlay from Nestoria data, 100 per cent transparency, hybrid layer (See also Figure 12.6 on page 235)

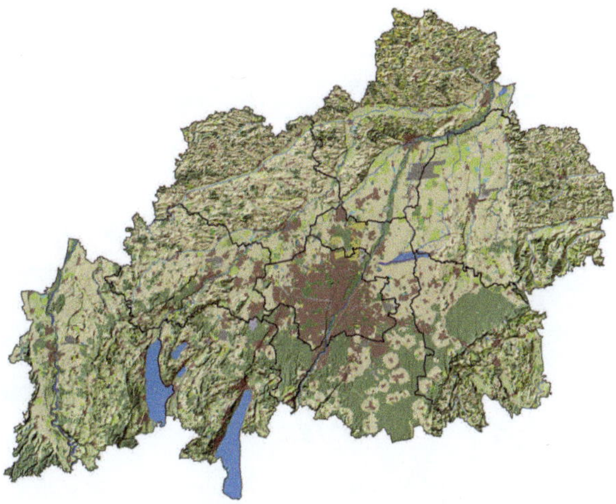

Plate 30. Land use map and digital terrain model (DTM) of the Munich planning region (Schaller and Schober 2007) (See also Figure 13.4 on page 249)

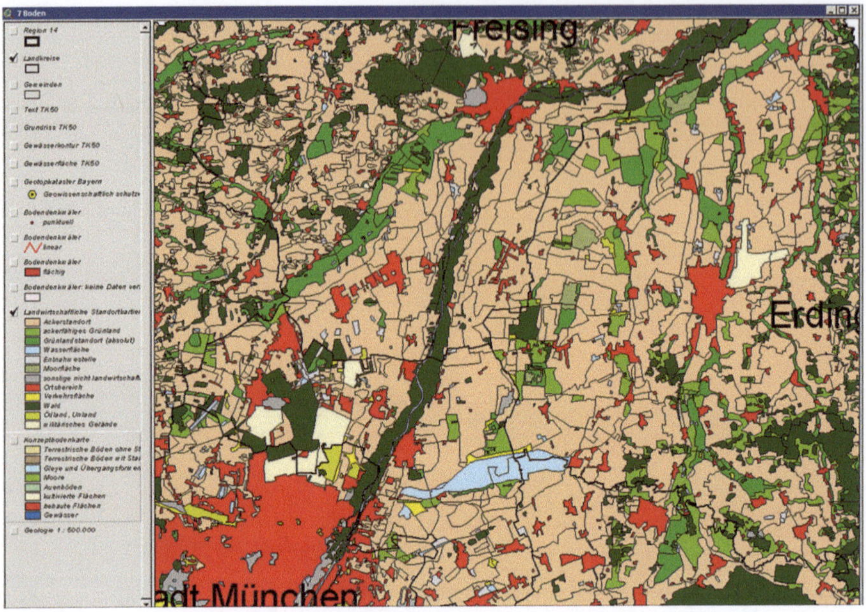

Plate 31. Soil potential and yield classes for agricultural land use (See also Figure 13.6 on page 252)

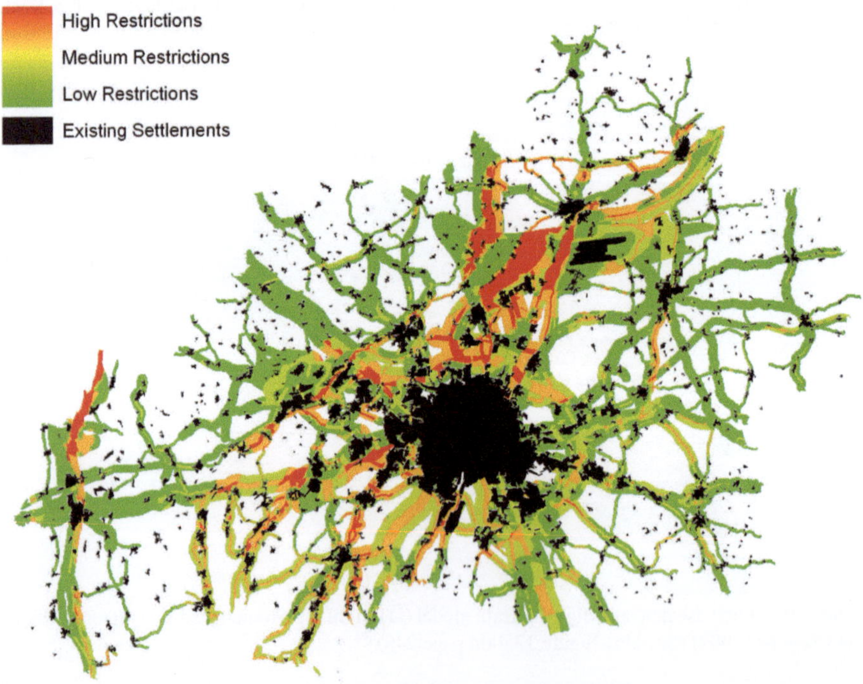

Plate 32. The final restrictions raster (white spaces = no restriction) (See also Figure 13.9 on page 256)

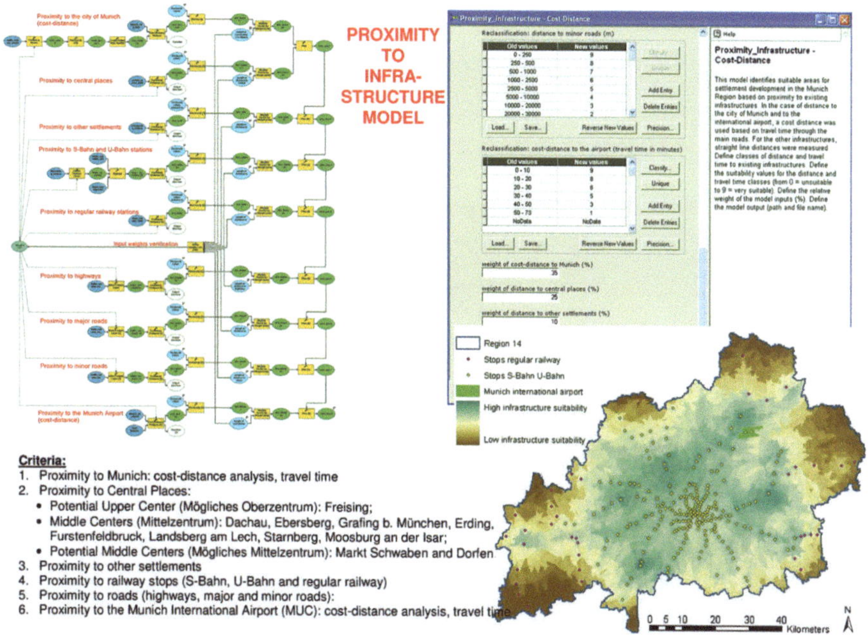

Plate 33. The final proximity model and map result (Mattos 2007) (See also Figure 13.10 on page 257)

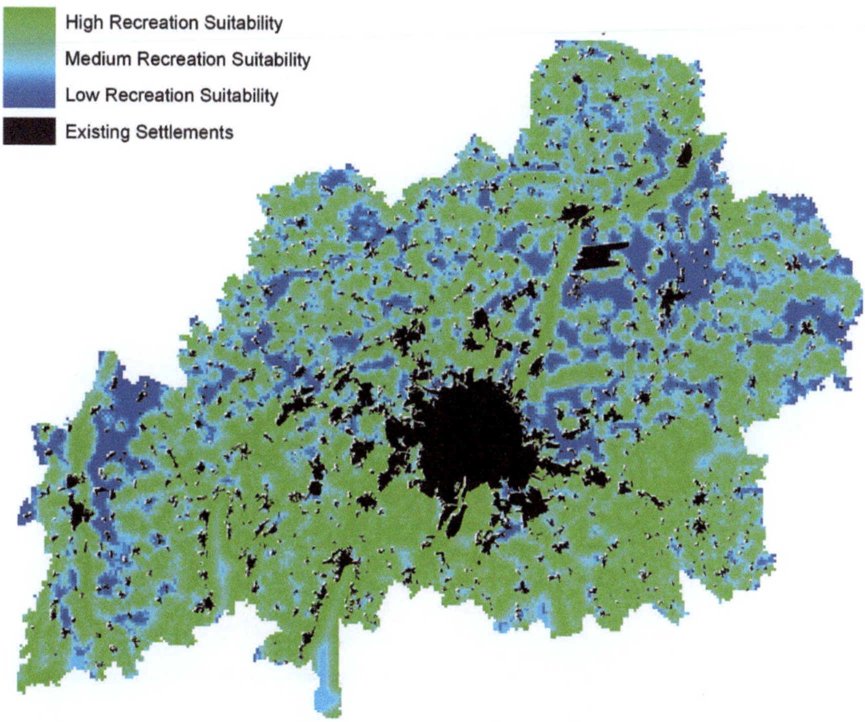

Plate 34. The final recreation raster map (See also Figure 13.12 on page 259)

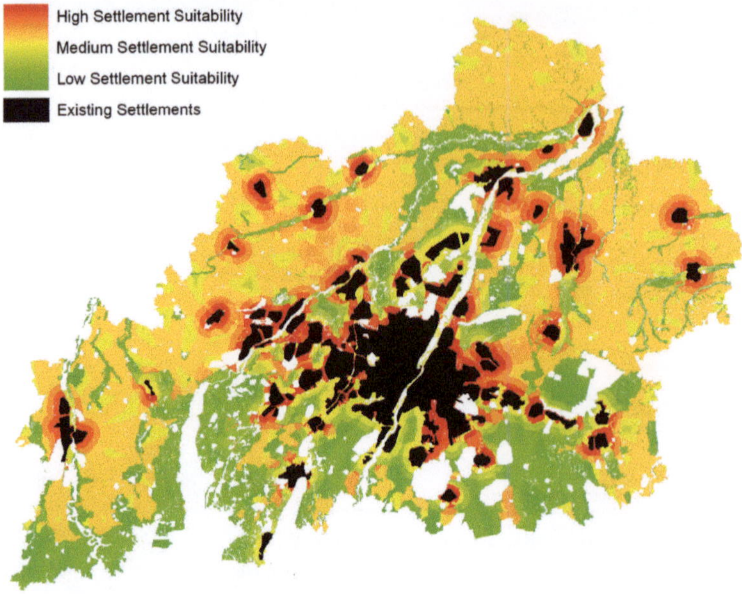

Plate 35. The resulting map for settlement suitability (white spaces = exclusions) (See also Figure 13.13 on page 260)

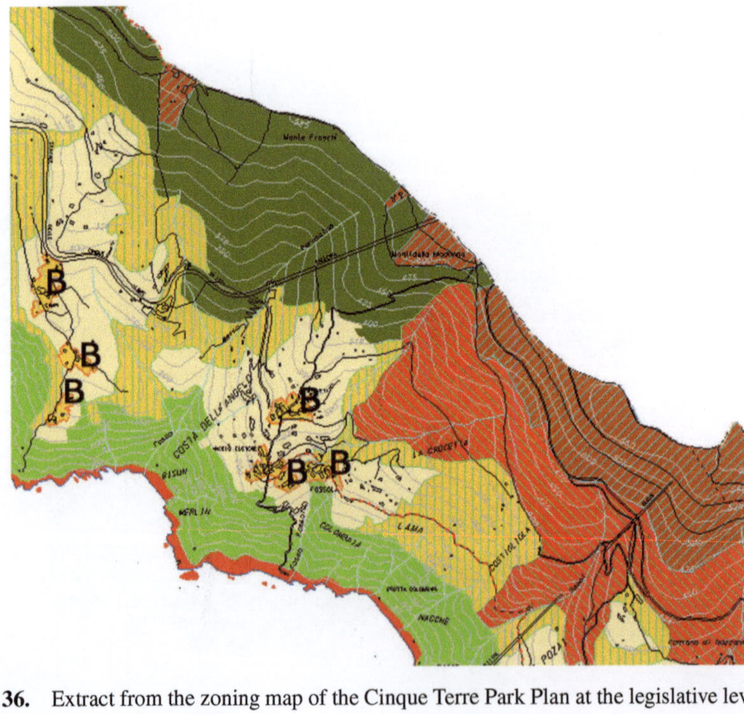

Plate 36. Extract from the zoning map of the Cinque Terre Park Plan at the legislative level (See also Figure 14.11 on page 280)

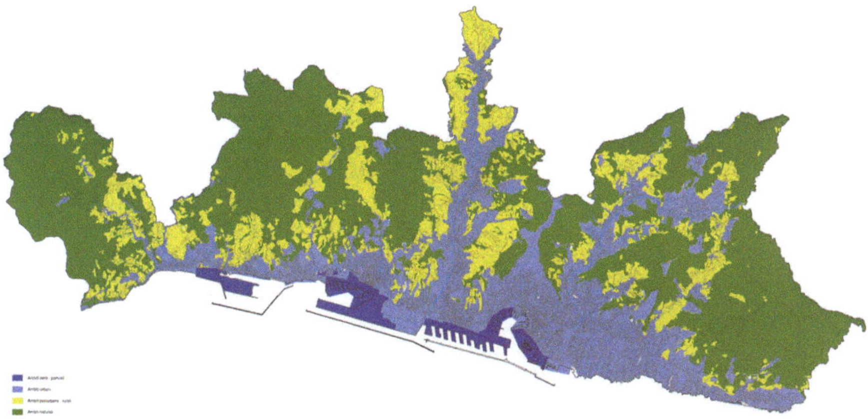

Plate 37. Urban region environmental system scenario (See also Figure 14.16 on page 285)

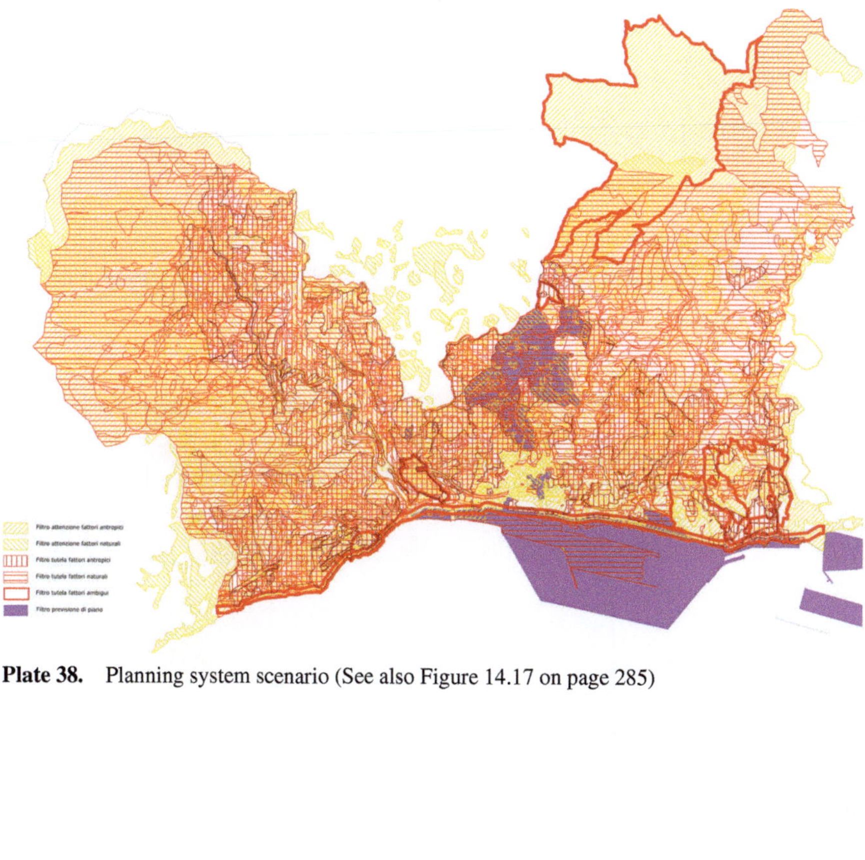

Plate 38. Planning system scenario (See also Figure 14.17 on page 285)

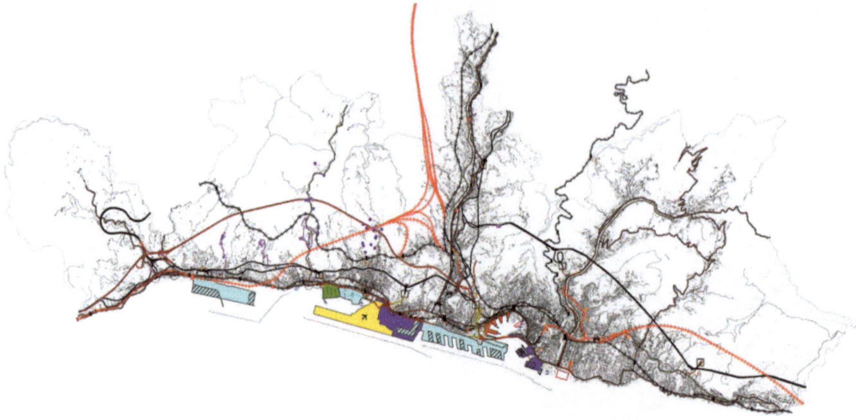

Plate 39. Network infrastructure scenario (See also Figure 14.18 on page 286)

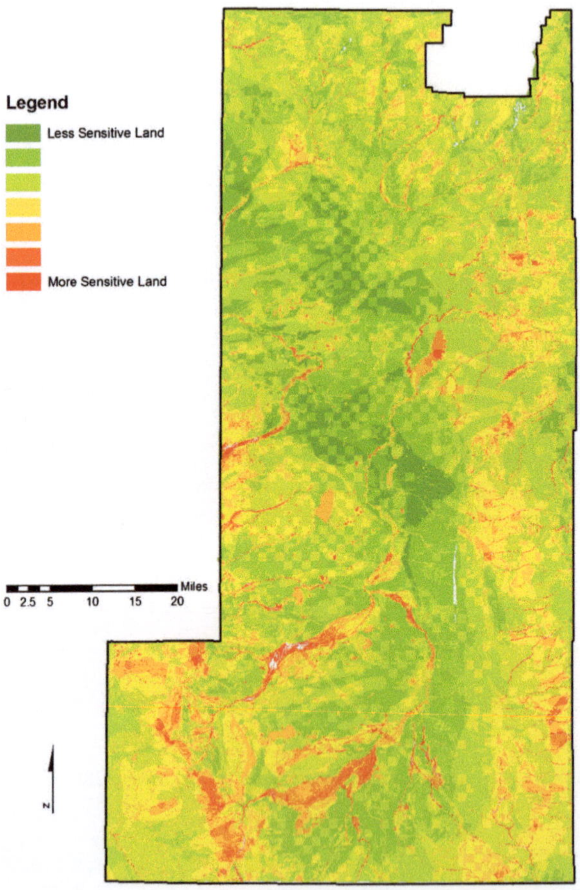

Plate 40. Value weighted landscape sensitivity (See also Figure 15.3 on page 305)

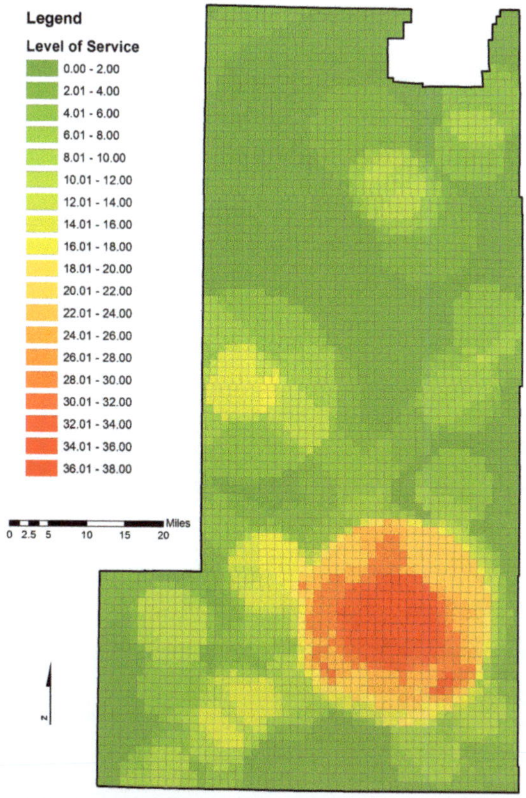

Plate 41. Value weighted growth efficiency (See also Figure 15.4 on page 306)

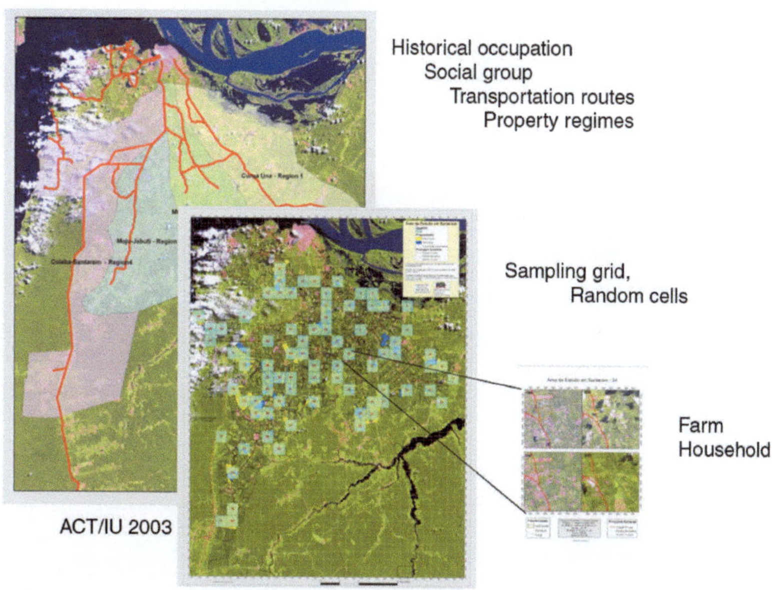

Plate 42. Stratifying levels of analysis; images of the ACT study areas, Santarem-Belterra Region in the Brazilian state of Para (See also Figure 16.2 on page 322)

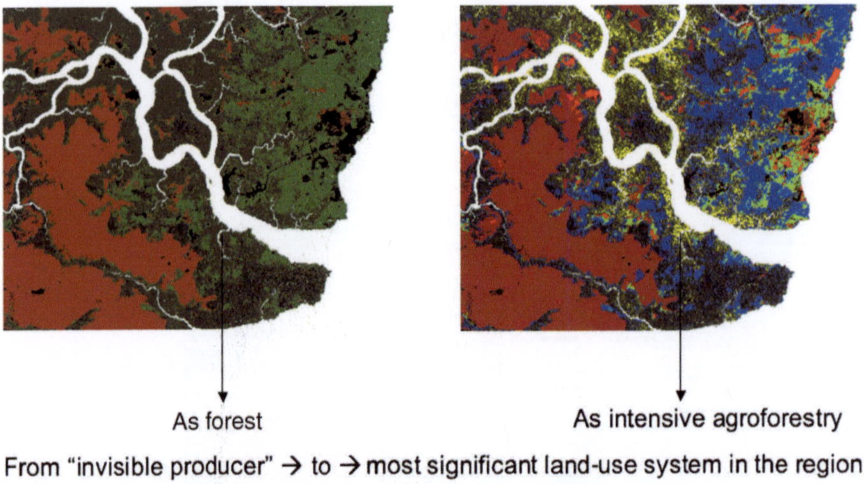

Plate 43. Outputs from an ethnographic (participatory assessment) in which local input improved the classification process with the identification of areas of intensive agro-forestry (right-hand image) (See also Figure 16.3 on page 332)

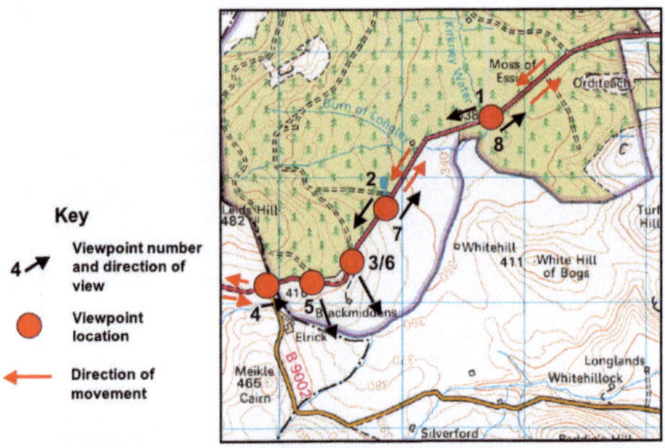

Plate 44. Route of walk-through of landscape using the Virtual Landscape Theatre (Map reproduced from Ordnance Survey, Crown Copyright, under license to Macaulay Institute GD27237ˊ2002) (See also Figure 16.8 on page 338)

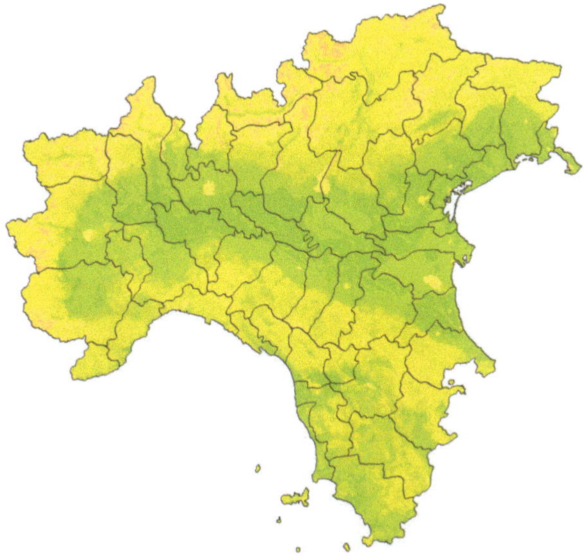

Plate 45. Suitability for extensive agriculture, 2000 (See also Figure 17.1 on page 357)

Plate 46. Suitability for extensive agriculture, Big Crisis, 2035 (See also Figure 17.2 on page 358)

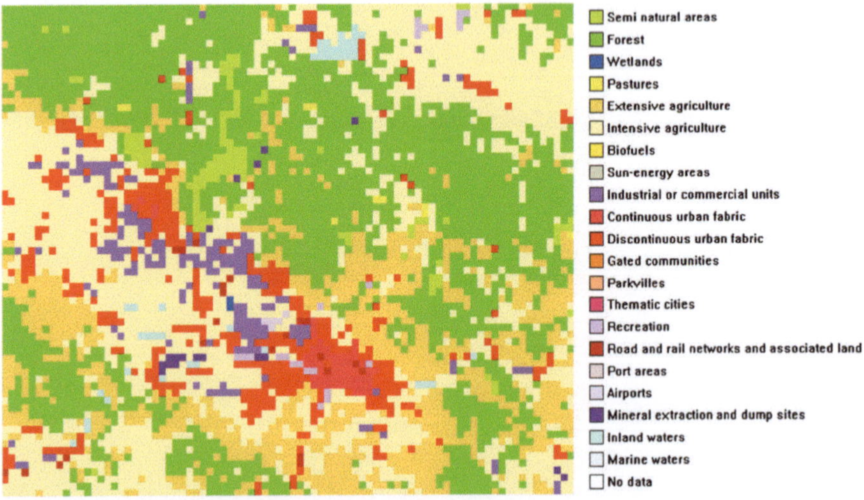

Plate 47. Northern Italy, 2000, detail of the Florence region (See also Figure 17.3 on page 359)

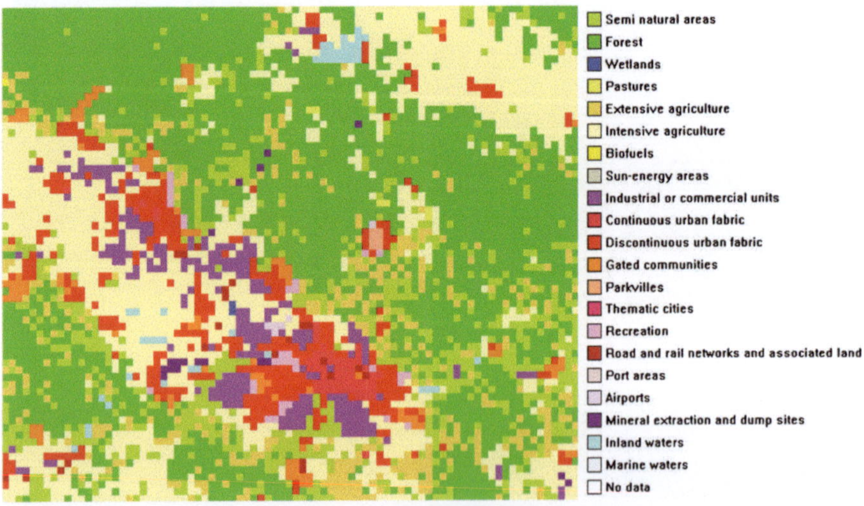

Plate 48. Northern Italy, 2035, detail of the Florence region (See also Figure 17.4 on page 360)

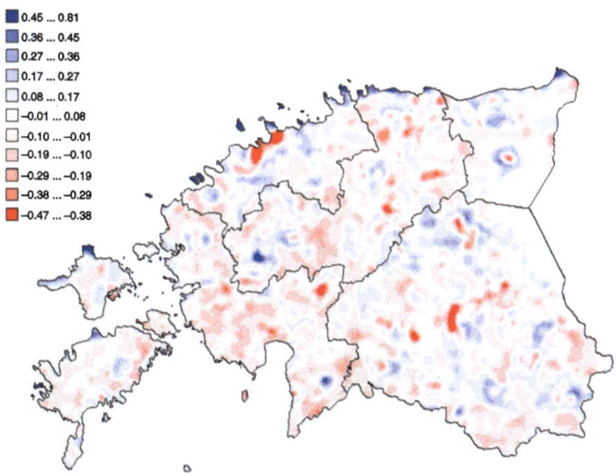

Plate 49. Diversity, The Great Escape, 2035 (See also Figure 17.5 on page 361)

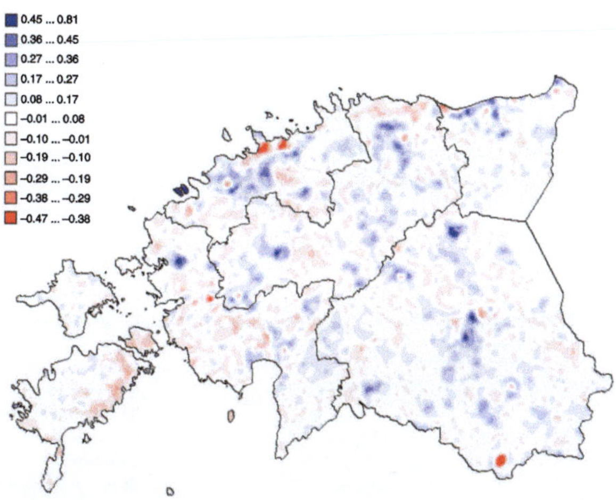

Plate 50. Diversity, Lettuce Surprise U, 2035 (See also Figure 17.6 on page 361)

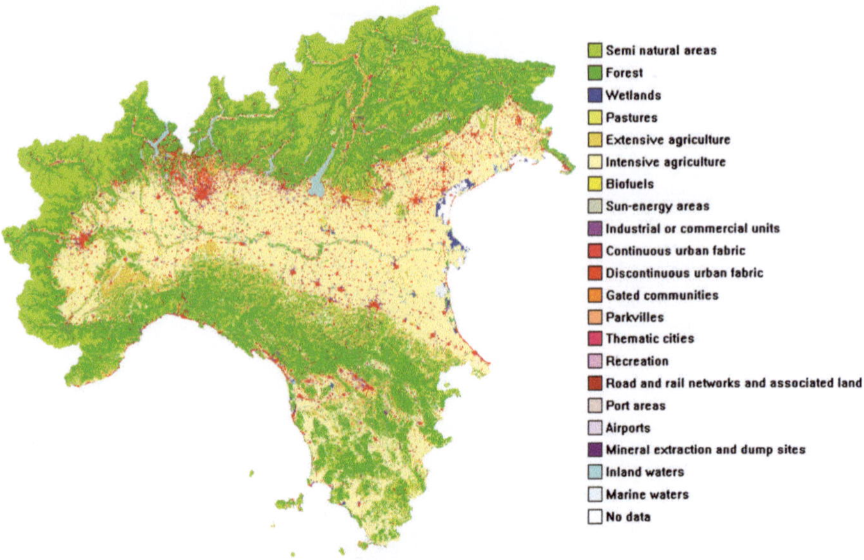

Plate 51. Northern Italy, 2000 (See also Figure 17.7 on page 362)

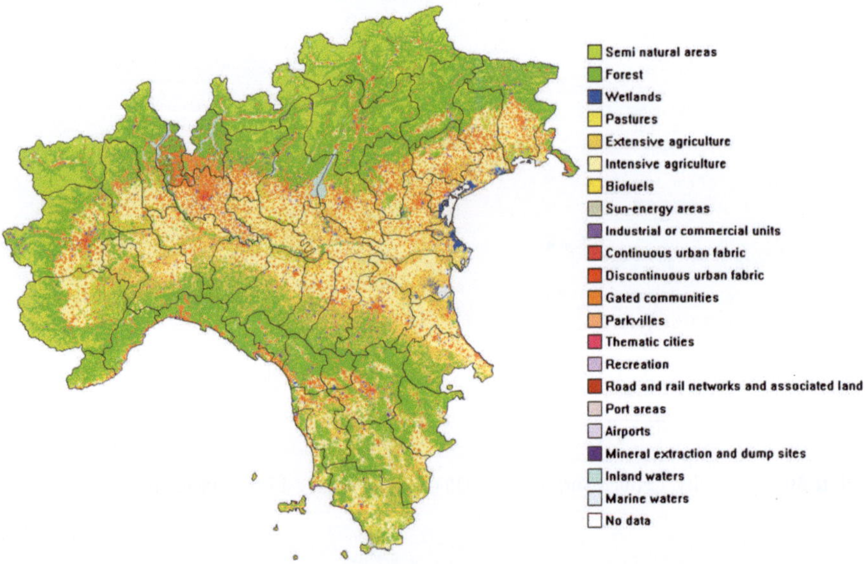

Plate 52. Northern Italy, Evolved Society, 2035 (See also Figure 17.8 on page 362)

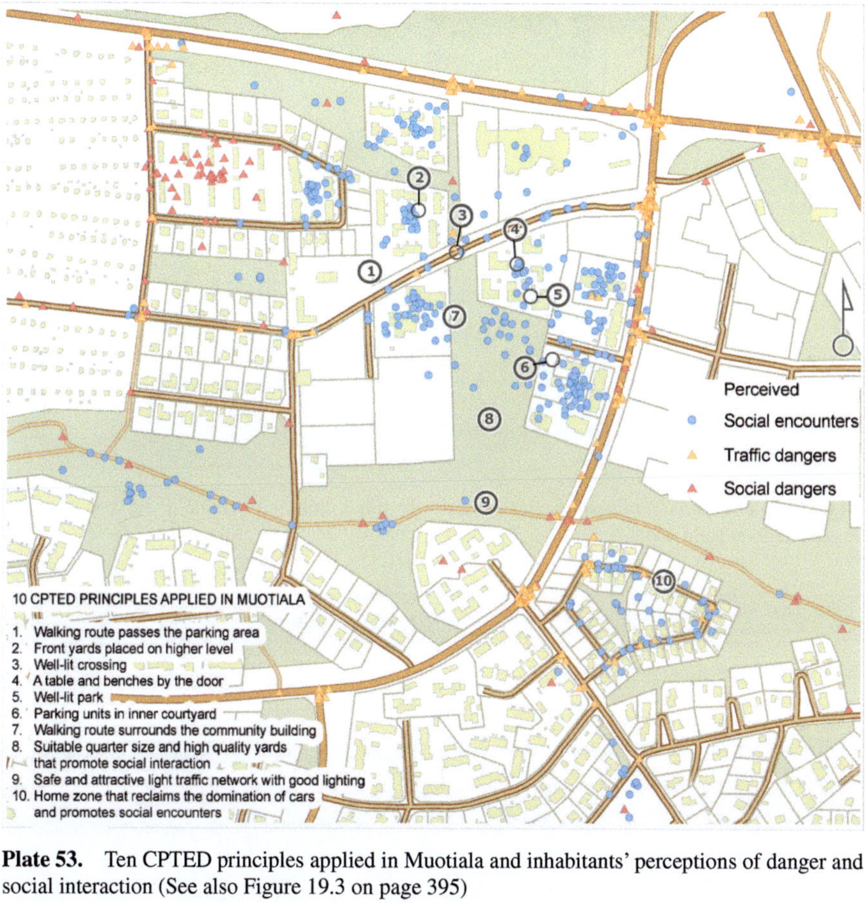

Plate 53. Ten CPTED principles applied in Muotiala and inhabitants' perceptions of danger and social interaction (See also Figure 19.3 on page 395)

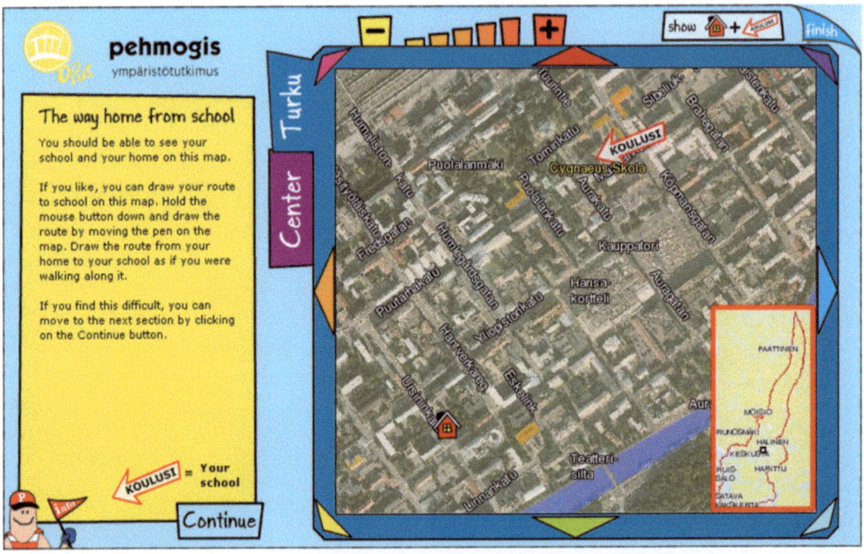

Plate 54. The *softGISchildren* application actively aids a child to orientate in the map (See also Figure 19.9 on page 404)

Plate 55. Visualisation of positive quality factors (left) located on a map of Järvenpää and (right) on an aerial photo clipped from the city centre of Järvenpää (See also Figure 19.11 on page 406)

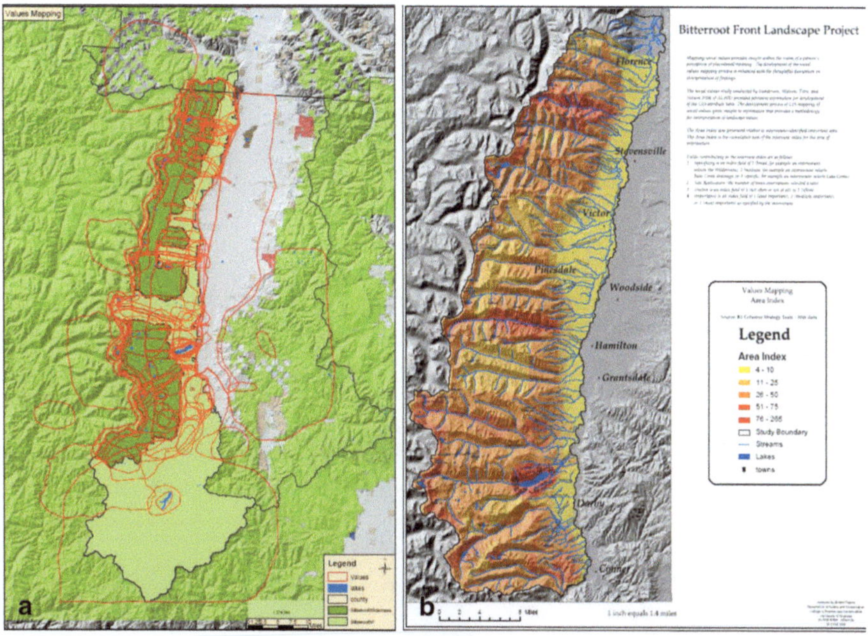

Plate 56. Example of (a) a digitised polygon net and (b) a composite map from the BEMRP work (See also Figure 21.5 on page 438)

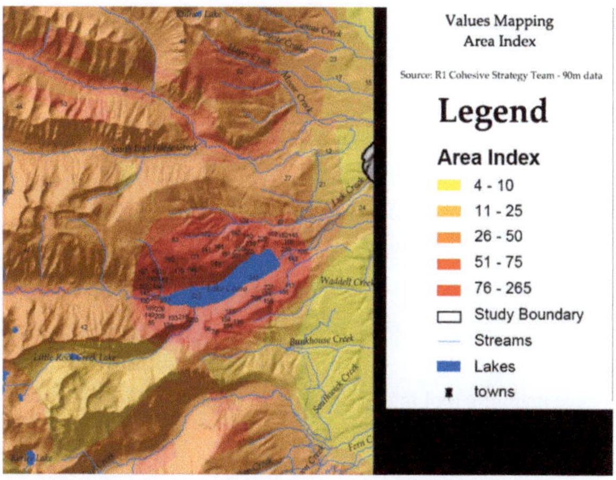

Plate 57. Area index for Lake Como, BEMRP project (See also Figure 21.6 on page 438)

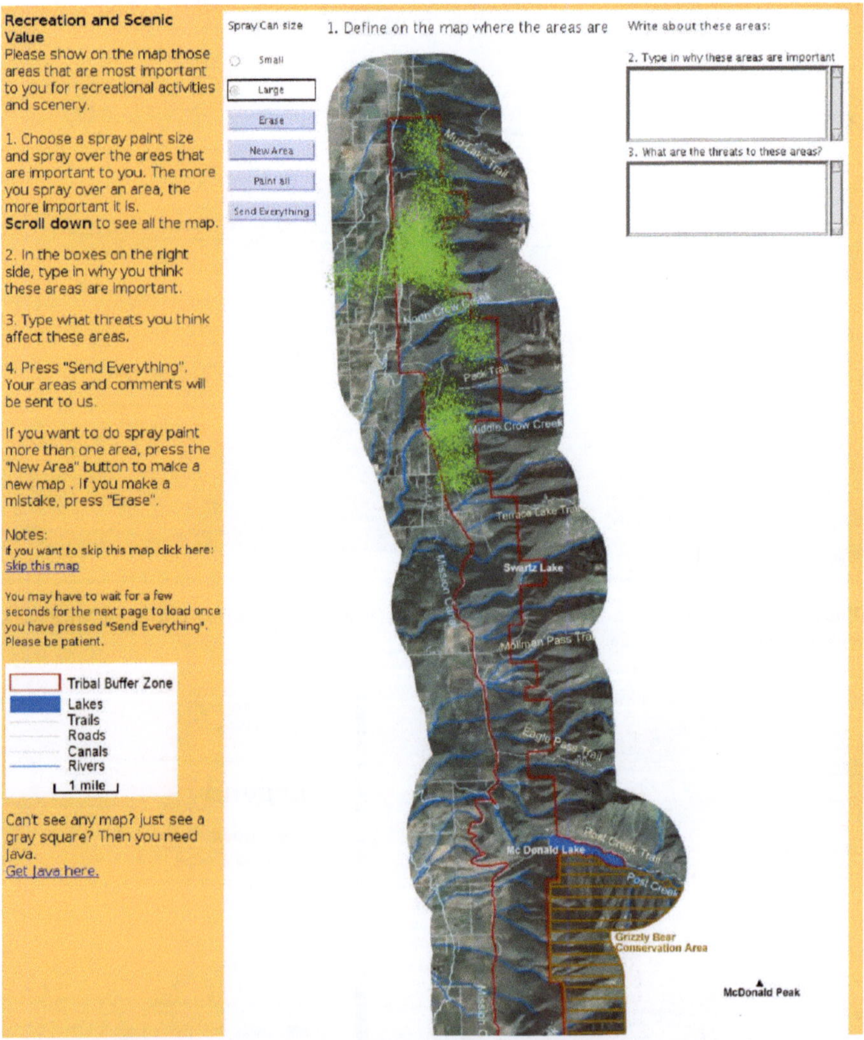

Plate 58. Landscape mapping tool interface (See also Figure 21.7 on page 442)

MIX
Papier aus verantwortungsvollen Quellen
Paper from responsible sources
FSC® C105338

If you have any concerns about our products,
you can contact us on
ProductSafety@springernature.com

In case Publisher is established outside the EU,
the EU authorized representative is:
Springer Nature Customer Service Center GmbH
Europaplatz 3, 69115 Heidelberg, Germany

Printed by Libri Plureos GmbH
in Hamburg, Germany